Edward J. Tarbuck
Frederick K. Lutgens
Illinois Central College

THE EARTH
An Introduction To
Physical Geology

Charles E. Merrill Publishing Company
A Bell & Howell Company
Columbus Toronto London Sydney

Published by Charles E. Merrill Publishing Company
A Bell & Howell Company
Columbus, Ohio 43216

This book was set in Souvenir
Production Editor: Rex E. Davidson
Text Designer: Cynthia Brunk
Art Coordination and Layout: James H. Hubbard
Cover Design: Tony Faiola
Cover Photo: David Muench Photography, Inc.
Illustrations by Dennis Tasa, Tasa Graphic Arts, Inc., Minneapolis

Library of Congress Catalog Card Number: 83–61873
International Standard Book Number: 0–675–20051–2
Printed in the United States of America
1 2 3 4 5 6 7 8 9 10—88 87 86 85 84

PREFACE

In recent years, media reports have made us increasingly aware of the geological forces at work in our physical environment. News stories graphically portray the violent force of a volcanic eruption, the devastation created by a strong earthquake, and the large numbers left homeless by mudflows and flooding. Such events, and many others as well, are destructive to life and property, and we must be better able to understand and deal with them. However, our natural environment has an even greater importance, for the earth is our home. The earth not only provides the mineral resources so basic to modern society, but it is also the source of most of the ingredients necessary to support life. Therefore, as many members of society as possible should acquire a basic understanding of how the earth works.

With this in mind, we have written a text to help people increase their understanding of our physical environment. We hope this new knowledge will encourage some to actively participate in the preservation of the environment, while others may be sufficiently stimulated to pursue a career in the earth sciences. Equally important, however, is our belief that a basic understanding of earth will greatly enhance appreciation of our planet and thereby enrich the reader's life.

The Earth: An Introduction to Physical Geology is intended for both majors and non-majors taking their first course in geology. We have attempted to write a text that is not only informative and timely, but one that is highly usable as well. The language is straightforward and written to be understood by a student with little or no college-level science experience. We have deliberately refrained from using excessive jargon and when new terms are introduced, they are placed in boldface and defined. Further, a list of key terms with page references is found at the end of each chapter, and a glossary is included at the conclusion of the text for easy reference

to important terms. Other learning aids include chapter by chapter review questions and a list of supplementary readings at the end of the text. Useful information on metric conversions, the periodic table of the elements, common minerals, and topographic maps is also provided in the appendices.

Since geology is a visual science, we gave special attention to the quality of the photographs and artwork used in the text. The many color and black and white photographs were selected to add realism to the subject as well as to heighten the interest of the reader. Moreover, the line art was carefully selected, planned, and produced to provide the most accurate representations possible. We believe that the clarity of the art will significantly aid student understanding by making difficult concepts less abstract.

As much as possible, we have attempted to provide the reader with a sense of the observational techniques and reasoning processes that constitute the discipline of geology. As with other sciences, geology is much more than a mere collection of facts. At its heart are the various methods of probing the earth aimed at uncovering its secrets. These methods involve the collection of the necessary data used to test hypotheses about the nature of the forces that shape our changing planet. In addition to gaining a better understanding of these natural processes, this activity often leads to a re-evaluation of ideas formulated at a time when less information was available. An excellent example of the way geological "truths" are uncovered and reworked is found in Chapter 16. Here we trace the historical formulation and subsequent rejection of the hypothesis that continents drift about the face of the earth and then examine the data that led to the "rebirth" of this idea as part of a more encompassing concept known as plate tectonics.

The organization of the text is intentionally traditional.

Following the overview of geology in the introductory chapter, we turn to a discussion of earth materials and the related processes of volcanism and weathering. Next, the geological work of gravity, water, wind, and ice in modifying and sculpturing landscapes is treated. After this look at external processes, we examine the earth's internal structure and the processes that deform rocks and give rise to mountains. Finally, the text concludes with chapters on geologic time and the solar system. This particular organization was selected largely to accommodate the study of minerals and rocks in the laboratory, which usually comes early in the course. Realizing that some instructors may prefer to structure their courses somewhat differently, we made each chapter self-contained so that it may be taught in a different sequence. Thus, the instructor who wishes to discuss earthquakes, plate tectonics, and mountain building prior to dealing with erosional processes may do so without difficulty. We also chose to introduce plate tectonics in the first chapter so that this revolutionary theory could be incorporated in appropriate places throughout the text. Although plate tectonics is an integral part of this book, we have not included it at the expense of other topics. While it is true that plate tectonics is fascinating and of utmost importance in understanding the dynamics of the earth, other topics are equally worthwhile and interesting to the beginning student. We have made one departure from many of the current texts. Rather than including separate chapters on environmental problems and earth resources, we have incorporated these topics into the text at appropriate places. For example, the discussion of fossil fuels is found in the chapter on sedimentary rocks, the pollution of wells is treated in the chapter on groundwater, and attempts at controlling beach erosion is examined in the chapter on shorelines.

As with any project of this scope, the contributions of others were very important and too numerous for us to give proper credit to each and every person involved.

The credit for the content must go to our teachers, colleagues, and students, who challenged us to search for a deeper understanding. We wish to express our thanks to the many individuals, institutions, and government agencies that provided information, photographs, and illustrations for use in this text. At the risk of possible omission, we would like to acknowledge Professor James E. Patterson, Illinois State University; Dr. John Shelton; Professor Garrett Deckert, University of Wisconsin, Richland Center; Professor Kenneth Hasson, East Tennessee State University; Professor Emeritus John Montagne, Montana State University; Dr. Warren Hamilton, United States Geological Survey; Mr. Dennis Tasa, Tasa Graphics; and Mr. Stephen Trimble, for diligently searching through their photo collections so that we could have just the right photograph. A special debt of gratitude also goes to those colleagues who were kind enough to review all or part of the manuscript. Their critical comments greatly improved this project. We thank Professor Larry Agenbroad, Northern Arizona University; Professor Gary Allen, University of New Orleans; Professor Rex Crick, University of Texas, Arlington; Professor Tom Freeman, University of Missouri; Professor Robert Hatcher, University of South Carolina; Professor Jerry Horne, San Bernardino Valley College; Professor David Lageson, Montana State University; Professor Michael J. Neilson, University of Alabama, Birmingham; Professor Darlene Richardson, Indiana University of Pennsylvania; and Professor Jack Schindler, St. Petersburg Junior College. Our thanks also go to Diane Weber who typed much of the manuscript, Dennis Tasa for his imaginative production of the line art, and to Rex Davidson, our production editor, who along with the many other fine people at Charles E. Merrill Publishing, skillfully transformed our manuscript into a finished product.

EJT
FKL

BRIEF CONTENTS

CONTENTS

12 DESERTS AND WINDS 291

13 SHORELINES 315

14 EARTHQUAKES 339

1

AN INTRODUCTION
TO GEOLOGY

The spectacular eruption of a volcano, the terror brought by an earthquake, the magnificent scenery of a mountain valley, and the destruction created by a landslide are all subjects for the geologist. The study of geology deals with many fascinating and practical questions about our physical environment. What forces produce mountains? Will there soon be another great earthquake in San Francisco? What was the Ice Age like? Will there be another? What created this cave and the stone icicles hanging from its ceiling? Should we look for water here? Is strip mining practical in this area? Will oil be found if a well is drilled at that location? What if the landfill is located in the old quarry?

The subject of this text is **geology,** a word that literally means "the study of the earth." To understand the earth is not an easy task because our planet is not an unchanging mass of rock but rather a dynamic body with a long and complex history.

The science of geology is traditionally divided into two broad areas—physical and historical. **Physical geology,** which is the primary focus of this book, examines the materials composing the earth and seeks to understand the many processes that operate beneath and upon its surface. The aim of **historical geology,** on the other hand, is to understand the origin of the earth and its development through time. Thus, it strives to establish an orderly chronological arrangement of the multitude of physical and biological changes that have occurred in the geologic past. The study of physical geology logically precedes the study of earth history, because we must first understand how the earth works before we attempt to unravel its past.

View of the earth from *Apollo 17.* (Courtesy of NASA)

SOME HISTORICAL NOTES ABOUT GEOLOGY

The nature of our earth—its materials and processes—has been a focus of study for centuries. Writings about such topics as fossils, gems, earthquakes, and volcanoes date back to the Greeks, more than 2300 years ago. Certainly the most influential Greek philosopher was Aristotle. Because Aristotle was a philosopher, his explanations were not always based on keen observations and experiments but often were arbitrary pronouncements. He believed that rocks were created under the "influence" of the stars, and that earthquakes occurred when air crowded into the ground, was heated by central fires, and escaped explosively. When confronted with a fossil fish, he explained that "a great many fishes live in the earth motionless and are found when excavations are made." Although Aristotle's explanations may have been adequate for his day, they unfortunately continued to be expounded for many centuries, thus thwarting the acceptance of more up-to-date accounts. Frank D. Adams states in *The Birth and Development of the Geological Sciences* (New York: Dover, 1938) that "throughout the Middle Ages Aristotle was regarded as the head and chief of all philosophers; one whose opinion on any subject was authoritative and final."

CATASTROPHISM

During the seventeenth and eighteenth centuries the doctrine of **catastrophism** strongly influenced the formulation of explanations about the dynamics of the earth. Briefly stated, catastrophists believed that the earth's landscape had been modeled primarily by great catastrophes. Features such as mountains and canyons, which today we know take great periods of time to form, were explained as having been produced by sudden and often worldwide disasters produced by unknowable causes that no longer operate. This philosophy was an attempt to fit the rate of earth processes to the then-current ideas on the age of the earth. In 1654, Archbishop James Usher, a scholar of the Bible, concluded that the earth was approximately 6000 years old, having been created in 4004 B.C. Later, another biblical scholar named Lightfoot was even more specific, declaring that the earth had been created at 9:00 A.M. on October 26, 4004 B.C.

The relationship between catastrophism and the age of the earth has been summarized nicely as follows:

3

That the earth had been through tremendous adventures and had seen mighty changes during its obscure past was plainly evident to every inquiring eye; but to concentrate these changes into a few brief millenniums required a tailor-made philosophy, a philosophy whose basis was sudden and violent change.[1]

THE BIRTH OF MODERN GEOLOGY

The late eighteenth century is generally regarded as the beginning of modern geology, for it was during this time that James Hutton, a Scottish physician and gentleman farmer, published his *Theory of the Earth* in which he put forth a principle that came to be known as the doctrine of **uniformitarianism** (Figure 1.1). Uniformitarianism is a basic concept in modern geology. It simply states that the physical, chemical, and biological laws that operate today have also operated in the geologic past. That is to say that the forces and processes that we observe presently shaping our planet have been at work for a very long time. Thus, to understand ancient rocks, we must first understand present-day processes and their results. This idea is commonly stated by saying "the present is the key to the past."

Prior to Hutton's *Theory of the Earth*, no one had effectively demonstrated that geology had to deal with extremely long periods of time. However, Hutton persuasively argued that forces which appear small could, over long spans of time, produce effects that were just as great as those resulting from sudden catastrophic events. Unlike his predecessors, Hutton carefully cited verifiable observations to support his ideas. For example, when he argued that mountains are sculptured and ultimately destroyed by weathering and the work of running water, and that their wastes are carried to the oceans by processes that can be observed, Hutton said, "We have a chain of facts which clearly demonstrates . . . that the materials of the wasted mountains have traveled through the rivers"; and further, "There is not one step in all this progress . . . that is not to be actually perceived." He then went on to summarize this thought by asking a question and immediately providing the answer: "What more can we require? Nothing but time."

Since Hutton's literary style was cumbersome and

FIGURE 1.1
A contemporary caricature of James Hutton, the 18th century Scottish geologist who is often called the "father of modern geology." The faces scowling at Hutton from the rocky cliff are believed to be profiles of Hutton's most vocal critics. (Courtesy of the Library of Congress)

difficult, his work was not widely read nor easily understood. However, that began to change in 1802, when Hutton's friend and colleague, John Playfair, published *Illustrations of the Huttonian Theory,* a volume in which he presented Hutton's ideas in a much clearer and attractive form. The following well-known passage from Playfair's work, which is a restatement of Hutton's basic principle, illustrates this style:

Amid all the revolutions of the Globe, the economy of nature has been uniform and her laws are the only things which have resisted the general movement. The rivers and the rocks, the seas and the continents have been changed in all their parts; but the laws which direct those changes, and the rules to which they are subject, have remained invariably the same.

Although Playfair's book gave impetus to Hutton's ideas and aided the cause of modern geology, it is the

---

[1]H. E. Brown, V. E. Monnett, and J. W. Stovall. *Introduction to Geology* (New York: Blaisdell, 1958).

English geologist Charles Lyell who is given the most credit for advancing the basic principles of modern geology (Figure 1.2). Between 1830 and 1872 he produced eleven editions of his great work, *Principles of Geology*. As was customary, Lyell's book had a rather lengthy subtitle that outlined the main theme of the work: *Being an Attempt to Explain the Former Changes of the Earth's Surface, by Reference to Causes now in Operation*. In the text, he painstakingly illustrated the concept of the uniformity of nature through time. He was able to show more convincingly than his predecessors that the geologic processes which are observed today can be assumed to have operated in the past. Although the doctrine of uniformitarianism did not originate with Lyell, a fact that he openly acknowledged, he is the person who was most successful in interpreting and publicizing it for society at large.

Despite its importance in modern geology, the doctrine of uniformitarianism should not be taken too literally. To say that geologic processes in the past were the same as those occurring today is not to suggest that they always operated at precisely the same rate. Although the processes have remained essentially the same, their rates have undoubtedly varied during geologic time [1]

The acceptance of the concept of uniformitarianism meant the acceptance of a very long history for the earth, for although processes vary in their intensity, they still take a very long time to create or destroy major features of the landscape.

For example, rocks containing fossils of organisms that lived in the sea more than 15 million years ago are now part of mountains that stand 3000 meters (9800 feet) above sea level. This means that the mountains were uplifted 3000 meters in about 15 million years, which works out to a rate of only 0.2 millimeter per year! Rates of erosion (the processes that wear away land) are equally slow (Figure 1.3). Estimates indicate that the North American continent is being lowered at a rate of just 3 centimeters per 1000 years. Thus, as you can see, tens of millions of years are required for nature to build mountains and wear them down again. But even these time spans

FIGURE 1.2
Charles Lyell. Lyell's book, *Principles of Geology*, did much to advance modern geology. (Courtesy of the Institute of Geological Sciences, London)

are relatively short on the time scale of earth history, for the rock record contains evidence that shows the earth has experienced many cycles of mountain building and erosion. Concerning the everchanging nature of the earth through great expanses of geologic time, Hutton stated, "We find no sign of a beginning, no prospect of an end." A quote from William L. Stokes sums up the significance of Hutton's basic concept:

> In the sense that uniformitarianism implies the operation of timeless, changeless laws or principles, we can say that nothing in our incomplete but extensive knowledge disagrees with it.[1]

In the chapters that follow, we shall be examining the materials that compose our planet and the processes that modify it. It will be important to remember that although many features of our physical landscape may seem to be unchanging in terms of the tens of years we might observe them, they are nevertheless changing, but on time scales of hundreds, thousands, or even many millions of years.

---

[1] It should be pointed out that during the earth's formative period, when our planet was very different than it is today, some processes were at work that are no longer operating.

[1] *Essentials of Earth History* (Englewood Cliffs, New Jersey: Prentice-Hall, 1966), p. 34.

A.

B.

FIGURE 1.3
Geologic processes often act so slowly that changes may not be visible during an entire lifetime. These two photographs were taken from the same vantage point nearly one hundred years apart. Photograph A was taken by J. K. Hillers in 1872 and photograph B was taken in 1968 by E. M. Shoemaker. The photos reveal practically no visible signs of erosion. (Photos courtesy of U.S. Geological Survey)

GEOLOGIC TIME AND THE GEOLOGIC CALENDAR

Although Hutton, Playfair, Lyell, and others recognized that geologic time is exceedingly long, they had no methods to accurately determine the age of the earth. However, with the discovery of radioactivity near the turn of the twentieth century and the continuing refinement of radiometric dating methods that were first attempted in 1905, geologists are now able to assign fairly accurate, specific dates to events in earth history.[1] Current estimates put the age of the earth between 4.6 and 4.8 billion years.

The magnitude of geologic time is a most difficult concept to grasp, because we must learn to think in spans of time that far exceed our common experience. Earth features, which seem to be everlasting and unchanging to us and in fact to generations of people, are indeed slowly changing. Thus over millions of years, mountains rise and are eroded to hills, and rivers excavate deep canyons. How long is 4.6 billion years? If you were to begin counting at the rate of one number per second and continued 24 hours a day, 7 days a week, and never stopped, it would take about two lifetimes (150 years) to reach 4.6 billion! Don L. Eicher gives another basis for comparison:

> **Compress for example, the entire 4.5 billion years of geologic time into a single year. On that scale,**

[1] A more complete discussion of this topic is found in Chapter 19.

the oldest rocks we know date from about mid-March. Living things first appeared in the sea in May. Land plants and animals emerged in late November and the widespread swamps that formed the Pennsylvanian coal deposits flourished for about four days in early December. Dinosaurs became dominant in mid-December, but disappeared on the 26th, at about the time the Rocky Mountains were first uplifted. Manlike creatures appeared sometime during the evening of December 31st, and the most recent continental ice sheets began to recede from the Great Lakes area and from northern Europe about 1 minute and 15 seconds before midnight on the 31st. Rome ruled the Western world for 5 seconds from 11:59:45 to 11:59:50. Columbus discovered America 3 seconds before midnight, and the science of geology was born with the writings of James Hutton just slightly more than one second before the end of our eventful year of years.[1]

During the nineteenth century, long before the advent of radiometric dating, a geologic calendar was developed using principles of relative dating. **Relative dating** means that events are placed in their proper sequence or order without knowing their absolute age in years. This is done by applying principles such as the **law of superposition,** which states that in an undeformed sequence of sedimentary rocks or lava flows, each layer is older than the one above it and younger than the one below it (Figure 1.4). Today such a proposal appears to be quite elementary, but 300 years ago, it amounted to a major breakthrough in scientific reasoning by establishing a rational basis for relative time measurements. However, since no precise rate of deposition can be determined for most rock layers, the actual length of geologic time represented by any given layer is unknown.

Fossils, the remains or traces of prehistoric life, were also essential to the development of the geologic calendar (Figure 1.5). A fundamental principle in geology, one that was laboriously worked out over many years by collecting fossils from many rock layers in many places, is known as the **principle of**

---

[1]Don L. Eicher. *Geologic Time,* 2nd ed. (Englewood Cliffs, New Jersey: Prentice-Hall, 1978), pp. 18–19. Reprinted by permission.

A.

B.

FIGURE 1.4
A. By applying the law of superposition, the relative ages of these undeformed layers are easily determined. **B.** If rock layers are highly contorted, their relative ages may be very difficult to determine. (Photos by Warren Hamilton, U.S. Geological Survey)

faunal succession. This principle states that fossil organisms succeed one another in a definite and determinable order, and therefore any time period can be recognized by its fossil content. Once established, this principle allowed geologists to identify rocks of the same age in widely separated places and to build the geologic calendar as shown in Figure 1.6 on page 9.

By examining the geologic calendar you can see that the largest subdivisions are called **eras.** The three eras currently recognized are the **Paleozoic** ("ancient

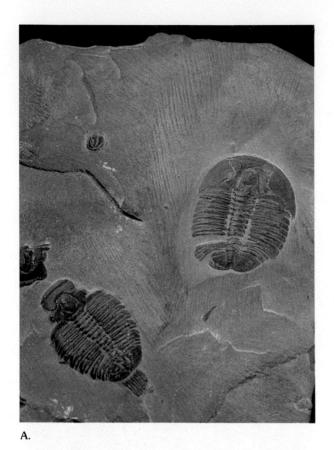

A.

B.

FIGURE 1.5
Fossils are important tools for the geologist. In addition to being very important in relative dating, fossils can be useful environmental indicators. **A.** Trilobites in shale. **B.** A coiled cephalopod of Cretaceous age. (Photos by E. J. Tarbuck)

life"), **Mesozoic** ("middle life"), and **Cenozoic** ("recent life"). As the names imply, the eras are bounded by quite profound worldwide changes in life forms. Each era is divided into time units known as **periods.** The Paleozoic has seven, the Mesozoic three, and the Cenozoic two. Since we currently live in the Cenozoic era, there will likely be more periods yet to come. Each period is characterized by a somewhat less profound change in life forms as compared with the eras. Finally, each of the twelve periods is further divided into still smaller units called **epochs.** Except for the seven epochs which have been named for the two periods of the Cenozoic era, those of other periods have not been given distinctive names. Rather, the terms *early, middle,* and *late* are used to distinguish these smaller time units.

Notice that the detail of the geologic calendar does not begin until about 600 million years ago, the date for the beginning of the first period of the Paleozoic era, the Cambrian period. The more than 4 billion years prior to the Cambrian is simply referred to as the **Precambrian.** This vast expanse of time, representing more than 85 percent of earth history, is not extensively divided into smaller units. The meager Precambrian fossil record is the primary reason for the lack of detail on this portion of the calendar. Without abundant fossils, geologists lose their primary calendar-building tool.

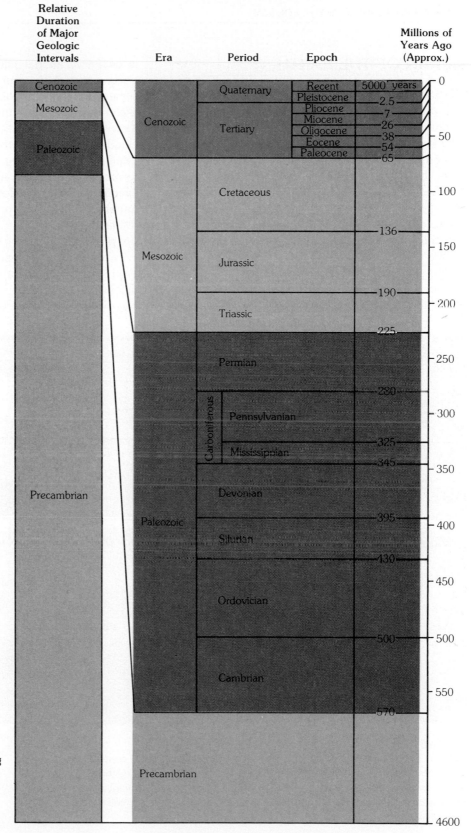

FIGURE 1.6
The geologic calendar. Absolute dates were added quite recently, long after the calendar had been established using relative dating techniques. The Precambrian accounts for more than 85 percent of geologic time.

ORIGIN OF THE EARTH

The earth is one of nine planets that along with several dozen moons and numerous smaller bodies revolve around the sun. The orderly nature of our solar system led most astronomers to conclude that its members formed at essentially the same time and from the same primordial material. This theory, known as the **nebular hypothesis,** suggests that the bodies of our solar system formed from an enormous cloud composed mostly of hydrogen and helium with only a small percentage of all the other heavier elements.

About 5 billion years ago, and for reasons that are not yet fully understood, this huge cloud of minute rocky fragments and gases began to contract under its own gravitational influence (Figure 1.7). The contracting material is assumed to have had some component of rotational motion, which, like a spinning ice skater pulling in her arms, rotated faster and faster as it contracted. This rotation in turn caused the nebular cloud to assume a flattened disklike shape. Within the rotating disk, relatively small eddylike contractions formed the nuclei from which the planets would eventually develop. However, the greatest concentration of material was pulled toward the center of this rotating mass and gravitationally heated, forming the hot *protosun*.

In a relatively short time after the formation of the protosun, the temperature within the rotating disk dropped significantly. This decrease in temperature caused substances with high melting points to con-

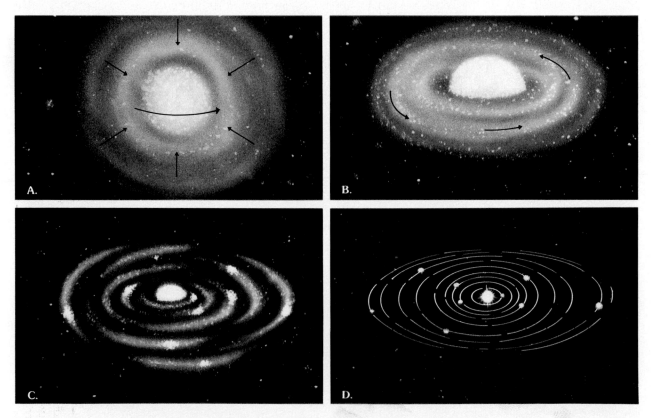

FIGURE 1.7
Nebular hypothesis. **A.** A huge cloud of dust and gases begins to contract. **B.** Because of its rotational motion it forms into a flattened disk. **C.** The planets then begin to accrete along the disk while most of the material is gravitationally swept toward the center, producing the sun. **D.** In time most of the remaining debris is collected into the nine planets and their moons.

dense into small particles, perhaps the size of sand grains. Materials such as iron and nickel solidified first. Next to condense were the elements of which rocky substances are composed. As these fragments collided, they joined into larger objects that in a few tens of millions of years accreted into the planets. In the same manner, but on a lesser scale, the processes of condensation and accretion acted to form the moons and other small bodies of the solar system.

As the *protoplanets* (planets in the making) accumulated more and more debris, the solar system began to clear. The removal of debris allowed sunlight to heat the surfaces of the newly-formed planets. The resulting high surface temperatures of the inner planets, coupled with the fact that these bodies possessed comparatively weak gravitational fields, meant that the earth and its neighbors, Mercury, Venus, and Mars, were unable to retain appreciable amounts of the lighter components of the primordial cloud. These materials, which included hydrogen, helium, ammonia, methane, and water, vaporized from their surfaces and were eventually wisked from the inner solar system by the solar winds. At distances beyond Mars temperatures are quite low. Consequently the large outer planets, Jupiter, Saturn, Uranus, and Neptune, accumulated huge amounts of hydrogen and other light materials from the primordial cloud. The accumulation of these gaseous substances is thought to account for the comparatively large sizes and low densities of the outer planets.

Shortly after the earth formed, the decay of radioactive elements, coupled with heat released by colliding particles, produced at least some melting of the interior. Melting, in turn, is thought to have allowed the heavier elements, principally iron and nickel, to sink, while the lighter rocky components floated upward. This segregation of material, which began early in the earth's history, is believed to still be occurring, but on a much smaller scale. As a result of this chemical differentiation, the earth's interior is not homogeneous. Rather, it consists of shells or spheres composed of materials having different properties. The principle divisions of the earth include: (1) the **inner core,** a solid iron-rich zone having a radius of 1216 kilometers (756 miles); (2) the **outer core,** a molten metallic layer some 2270 kilometers (1410 miles) thick; (3) the **mantle,** a solid rocky layer having a maximum thickness of 2885 kilometers (1789 miles);

and (4) the **crust,** a relatively light outer skin that ranges from 5 to 40 kilometers (3 to 25 miles) thick (Figure 1.8).

A very important zone exists within the mantle and deserves special mention. This region, called the **asthenosphere,** is located between the depths of approximately 100 and 700 kilometers. The asthenosphere is a hot, weak zone that is capable of gradual flow. Situated above the asthenosphere, geologists recognize a zone called the **lithosphere** ("sphere of rock"), which includes the crust and uppermost mantle (Figure 1.9). In contrast to the asthenosphere upon which it rests, the lithosphere can be considered to be cool and rigid.

An important consequence of the period of chemical differentiation is that gaseous materials were allowed to escape from the earth's interior, similar to what happens today during volcanic eruptions. By this process an atmosphere composed chiefly of gases expelled from within the planet gradually evolved. It is on this planet, with this atmosphere, that life as we know it came into existence.

A VIEW OF THE EARTH

Figure 1.10 is the view of earth that greeted the *Apollo 8* astronauts as their spacecraft came from behind the moon after circling it for the first time in December, 1968. A view such as this from a distance of 160,000 kilometers (100,000 miles) provided the astronauts, as well as those of us back on earth, a unique perspective of our planet. For the first time we were able to see the earth from the depths of space as a small, fragile-appearing sphere surrounded by the blackness of an infinite universe. Such views were not only spectacular and exciting, but also humbling, for they showed us as never before what a tiny part of the universe our planet occupies.

As we look more closely at our planet from space, the most conspicuous features are not the continents, but the swirling clouds suspended above the surface and the vast global ocean (see chapter-opening photo). From such a vantage point, we can appreciate why the earth's physical environment is traditionally divided into three major parts: the envelope of air called the **atmosphere;** the **hydrosphere,** or water portion of our planet; and, of course, the solid earth.

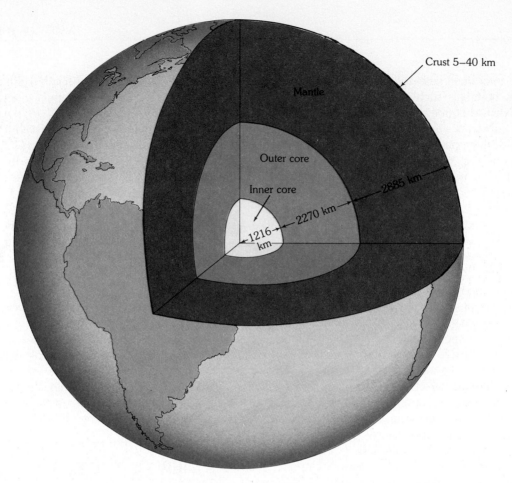

FIGURE 1.8
View of the earth's layered structure. The inner core, outer core, and mantle are drawn to scale, but the thickness of the crust is exaggerated by about five times.

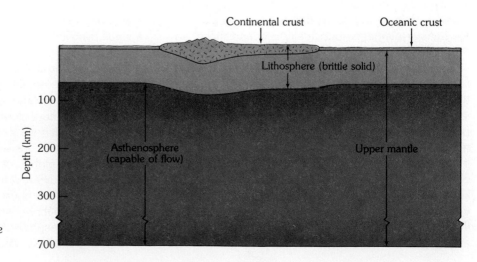

FIGURE 1.9
Schematic of the respective positions of the asthenosphere and lithosphere.

FIGURE 1.10
View of the earth that greeted the *Apollo 8* astronauts as their spacecraft came from behind the moon. (Courtesy of NASA)

Ours is a dynamic planet that is not dominated by rock, water, or air alone. Rather, it is characterized by continuous interactions as air comes in contact with rock, rock with water, and water with air.

The atmosphere, the earth's life-giving blanket of air that is hundreds of kilometers thick, is an integral part of the planet. It not only provides the air that we breathe, but also acts to protect us from the sun's intense heat and dangerous radiation. The energy exchanges that continually occur between the atmosphere and the earth's surface, and between the atmosphere and space produce the effects we call weather and climate.

The hydrosphere is a dynamic mass of liquid that is continually on the move, from the oceans to the air, to the land, and back again. The global ocean is obviously the most prominent feature of the hydrosphere, blanketing 71 percent of the earth's surface and accounting for about 97 percent of the earth's water. However, the hydrosphere also includes the fresh water found in streams, lakes, and glaciers, as well as that found in the ground. Although these latter sources constitute just a tiny fraction of the total, they are much more important than their meager percentage indicates, because they are responsible for sculpturing and creating many of our planet's varied landforms.

Lying beneath the atmosphere and the ocean is the solid earth (see Figure 1.11 on pp. 16–17). The study of the solid earth is largely confined to the more accessible surface features. Fortunately, these observable features represent the outward expressions of the dynamic behavior of the subsurface materials. By examining the most prominent surface features and their global extent, we can obtain clues to the dynamic processes which have shaped our planet.

The two principle divisions of the earth's surface are the continents and the ocean basins. Surprisingly, the present shoreline is not the boundary between these quite distinct regions. Rather, along most coasts a gently sloping platform of continental material, called the **continental shelf,** extends seaward from the shore. The extent of the continental shelf has varied greatly from one period to another. For instance, during the most recent ice age, when more of the world's water was stored on land in the form of glacial ice, the level of the sea was about 150 meters lower than it is today. Consequently, during this period, more of the earth's surface was dry land. The boundary between the continents and the deep-ocean basins is perhaps best placed about half way down the **continental slopes,** which are steep dropoffs that lead from the edge of the continental shelves to the deep-ocean basins. Using this as the dividing line, we find that about 60 percent of the earth's surface is represented by the ocean basins, while the remaining 40 percent exists as continental masses.

The most obvious difference between the continents and the ocean basins is their relative levels. The average elevation of the continents above sea level is about 840 meters, whereas the average depth of the oceans is about 3800 meters. Thus, the continents stand on the average about 4.6 kilometers above the level of the ocean floor. The elevations of these crustal layers is largely a reflection of their densities. The continental blocks are composed of material which has properties similar to those of granite, a common rock with a density about 2.7 times that of water. The crust of ocean basins, on the other hand, is thought to have a composition similar to that of basalt, a rock that is about 3 times denser than water. This difference alone cannot account for the elevated positions of the continents. However, the rocky material located below 100 kilometers is weak and capable of flow. Thus, the rigid outer layer can be thought of as floating on this weak layer, much like an ice cube floats on water. The continental blocks, which consist of thick slabs of less dense rock, float higher than the thinner, more dense oceanic materials.

Within these two diverse provinces, great variations in elevation exist. The most prominent features of the continents are linear mountain belts (Figure 1.11). Although the distribution of mountains appears to be random, this is not the case. When the youngest mountainous terrains are considered, we find they are located principally in two zones. The circum-Pacific belt includes the mountains of the western Americas and continues into the western Pacific in the form of volcanic island arcs. Island arcs are active mountainous regions composed largely of deformed volcanic rocks. Included in this group are the Aleutian Islands, Japan, the Philippines, and New Guinea. The other major mountain belt extends eastward from the Alps through Iran and the Himalayas, and then dips southward into Indonesia. Careful examination of mountainous terrains reveals that most are places where thick sequences of rocks have been squeezed and highly deformed, as if placed in a gigantic vise.

Older mountains are also found on the continents. Examples include the Appalachians in the eastern United States and the Urals in the Soviet Union. Their once lofty peaks are now worn low, the result of millions of years of erosion. Still older are the stable continental interiors. Within these stable interiors are areas known as shields, extensive and relatively flat expanses composed largely of crystalline material. Radiometric dating of the shields has revealed that they are truly ancient regions. The ages of some samples exceed 3.8 billion years. Even these oldest known rocks exhibit evidence of enormous forces that have folded and deformed them.

Not many years ago the ocean basins were thought to be rather nondescript regions with only an occasional volcanic structure emerging from the depths. This view of the ocean floor could not be more incorrect. Indeed, the ocean basins are now known to contain the most prominent mountain range on earth, the **oceanic ridge system** (Figure 1.11). This broad elevated feature forms a continuous belt that winds for nearly 65,000 kilometers (40,000 miles) around the globe in a manner similar to the seam on a baseball. Rather than consisting of highly deformed rock, such as most of the mountains found on the continents, the oceanic ridge system consists of layer upon layer of once molten rock which has been fractured and uplifted.

The ocean floor also contains extremely deep grooves that are occasionally more than 11,000 meters (36,000 feet) deep. Although these deep-ocean **trenches** are relatively narrow and represent only a small portion of the ocean floor, they are nevertheless very significant features. Trenches are located adjacent to the young mountains which flank the continents, such as the Andes of western South America, or they are found paralleling the volcanic island arcs.

What is the connection, if any, between the young, active mountain belts and the oceanic trenches? What is the significance of the enormous ridge system that extends through all the world's oceans? What forces crumple rocks to produce majestic mountain ranges? These questions must be answered as we attempt to discover the dynamic processes which shape our planet.

THE DYNAMIC EARTH

The earth is a dynamic planet. If we could go back in time a billion years or more, we would find a planet whose surface was dramatically different than it is today. Such prominent features as the Grand Canyon, the Rocky Mountains, and even the much older Appalachian Mountains did not exist. Moreover, we would find that the continents would have different shapes and be located in different positions than we find them in today. On the other hand, a billion years ago the moon's surface was almost the same as we now find it. In fact, if viewed telescopically from earth, perhaps only a few craters would be missing. Thus, when compared to the earth, the moon is a lifeless body wandering through space and time.

The processes that alter the earth's surface can be divided into two categories. Those forces which wear away the land include weathering and erosion. Unlike the moon, where weathering and erosion progress at infinitesimally slow rates, these processes are continually altering the landscape of the earth. In fact, these destructive forces would have long ago leveled the continents had it not been for opposing constructional processes. Included among the constructional processes are volcanism and mountain building, which increase the average elevation of the land in opposition to gravity. As we shall see, these forces depend upon the earth's internal heat for their source of energy.

Within the last few decades, a great deal has been learned about the workings of our dynamic planet. In fact, many have called this period a revolution in our knowledge about the earth which has been unequalled at any other time. This revolution began in the early part of the twentieth century with the radical proposal that the continents had drifted about the face of the earth. Because this idea contradicted the established view that the continents and ocean basins are permanent and stationary features on the face of the earth, it was received with great skepticism. More than 50 years passed before enough data was gathered to transform this relatively simple hypothesis into a working theory which weaved together the basic processes known to operate on the earth. The theory that finally emerged, called **plate tectonics**[1], provided geologists with a comprehensive model of the earth's internal workings.

According to the plate tectonics model, the earth's rigid outer shell, the lithosphere, is broken into several individual pieces called **plates** (Figure 1.12). It is further thought that these rigid plates are slowly, but nevertheless continually, in motion. This motion is believed to be driven by a thermal engine, the result of an unequal distribution of heat within the earth. As hot material wells up from deep within the earth and spreads laterally, the plates are set in motion (Figure 1.13 on page 20). Ultimately, this movement of the

---

[1]Tectonics is the study of large scale deformation of the earth's lithosphere that results in the formation of major structural features such as those associated with mountains.

FIGURE 1.11 →

The topography of the earth's solid surface is shown on the following two pages. (Copyright © by Marie Tharp)

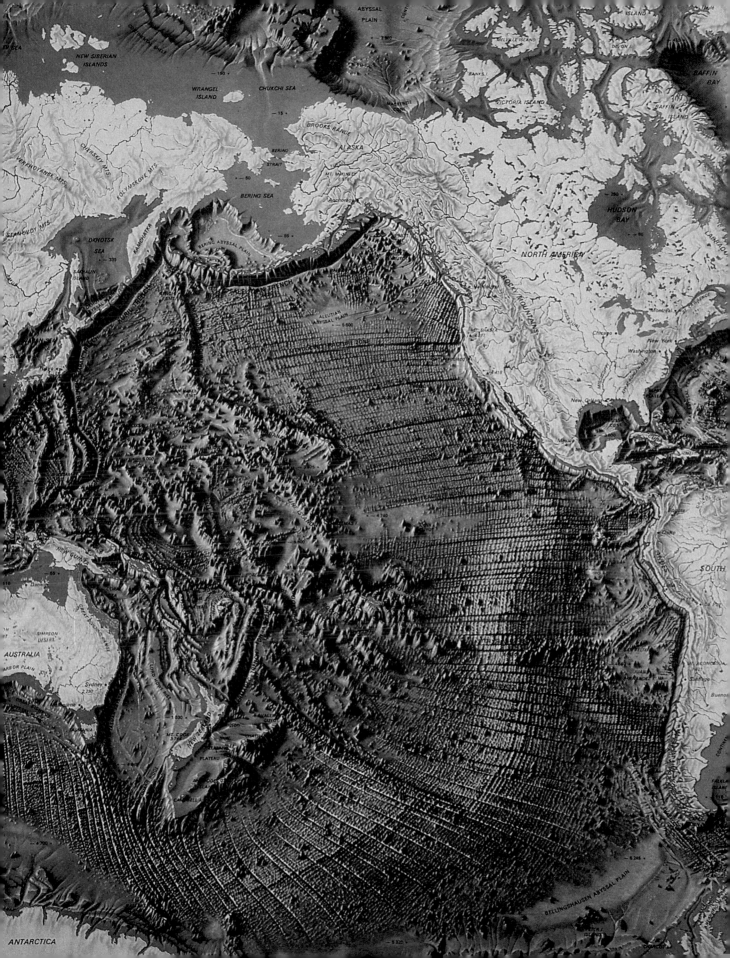

earth's lithospheric plates generates earthquakes, volcanic activity, and the deformation of large masses of rock into mountains.

Because each plate moves as a distinct unit, all interaction among individual plates occurs along their boundaries. The first approximations of plate boundaries were made on the basis of earthquake and volcanic activity. Later work indicated the existence of three distinct types of plate boundaries, which are differentiated by the movement they exhibit (Figure 1.14). These are:

1 **Divergent boundaries**—zones where plates move apart, leaving a gap between them.

2 **Convergent boundaries**—zones where plates move together, causing one to go beneath the other, as happens when oceanic crust is involved; or where plates collide, which occurs when the leading edges are made of continental crust.

3 **Transform fault boundaries**—zones where plates slide past each other, scraping and deforming as they pass.

Each plate is bounded by a combination of these zones (Figure 1.12). Movement along one boundary requires that adjustments be made at the others.

Plate spreading (divergence) is believed to occur at the oceanic ridges. As the plates separate, the gap created is immediately filled with molten rock that wells up from the hot asthenosphere (Figure 1.15 on pages 22–23). This material slowly cools to produce a new sliver of sea floor. Successive separations and fillings continue to add new oceanic lithosphere between the diverging plates. This mechanism, which has produced the floor of the Atlantic Ocean during the past 200 million years, is appropriately called **sea-floor spreading.** The typical rate of sea-floor spreading is estimated to be 5 centimeters (2 inches) per year, although it varies considerably from one location to another. This seemingly slow rate of movement is nevertheless rapid enough so that all of the existing ocean basins could have been generated within the last 5 percent of geologic time.

Although the earth's rigid outer layer is constantly being generated at the oceanic ridges, the total surface area of the earth remains constant. Therefore, lithosphere must be destroyed at the same rate that it is created. The zone of plate convergence is the site of this destruction. As two plates move together, the leading edge of one of the slabs is bent downward,

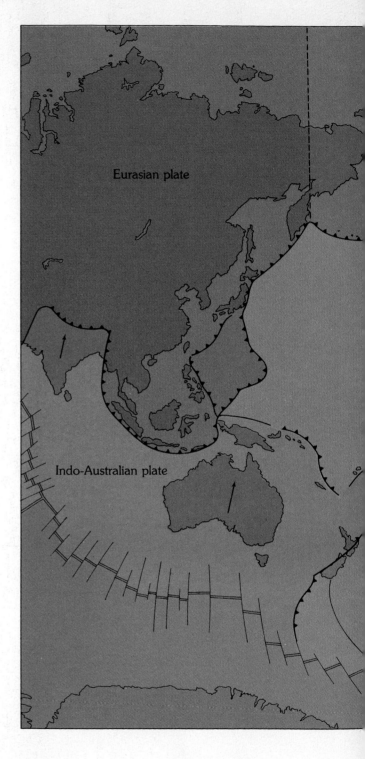

Eurasian plate

Indo-Australian plate

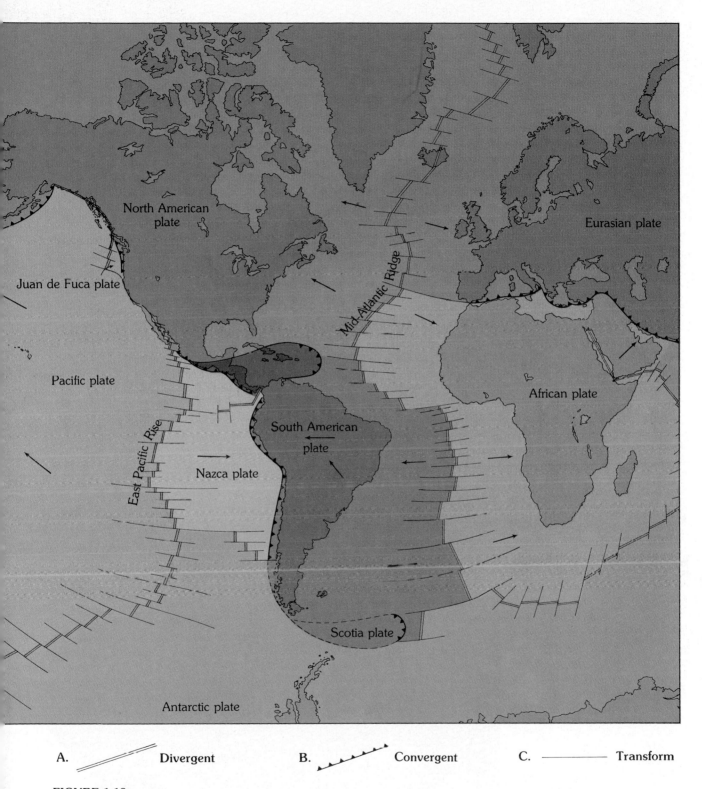

A. ———— **Divergent**   B. ����� **Convergent**   C. ———— **Transform**

FIGURE 1.12

Mosaic of rigid plates that constitute the earth's outer shell. **A.** Divergent boundary. **B.** Convergent boundary. **C.** Transform fault boundary. (After W. B. Hamilton, U.S. Geological Survey)

19

allowing it to slide beneath the other. Whenever continental and oceanic lithosphere collide, it is always the denser oceanic material that plunges into the weak asthenosphere below (Figure 1.15).

The regions where oceanic lithosphere is being consumed are called **subduction zones.** Here, as the solid plates move downward, they enter high-pressure and high-temperature environments. Some of the subducted material is thought to melt and migrate upward into the overriding plate. Occasionally this molten rock may reach the surface where it gives rise to volcanic eruptions such as those of Mount St. Helens.

Other boundaries, represented by transform faults, are located where plates slip past each other without producing or destroying crust. These faults form in the direction of plate movement and were first discovered in association with offsets in the oceanic ridges (Figure 1.15). Although most transform faults are located within the ocean basins, a few slice through the continents. The San Andreas fault of California is a famous example. Along this fault the Pacific plate is moving toward the northwest, past the North American plate. The movement along this boundary does not go unnoticed. As these plates pass, strain builds in the rocks on opposite sides of the fault and is occasionally released in the form of a great earthquake of the type that devastated San Francisco in 1906.

It has only recently been realized that the interaction of plates along their boundaries initiates most of our planet's volcanism, earthquakes, and mountain building. Further, these boundaries do not remain constant through time. For example, a divergent boundary which runs through eastern Africa appears to have developed in the relatively recent past. If spreading continues there, Africa will split into two continents separated by a new ocean basin. At other locations continents are presently moving toward each other and may eventually join into a "supercontinent." When continents collide, the thick accumulations of rocks and sediments along their margins are gradually thrust into majestic mountain ranges.

As long as the temperatures deep within the earth remain significantly higher than those near the surface, the material within the earth will continue to move. This internal flow, in turn, will keep the rigid outer shell of the earth in motion. Thus, as long as the earth's internal heat engine operates, the positions and shapes of the continents and ocean basins will

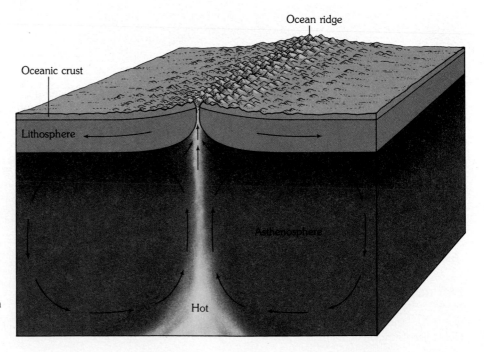

FIGURE 1.13
Unequal temperature distribution in the earth's interior is thought to produce convection currents that move the earth's rigid outer shell.

change, and the earth will remain a dynamic planet.

In the remaining chapters we will examine in more detail the workings of our dynamic planet in light of the plate tectonics model.

THE ROCK CYCLE

The **rock cycle** is one means of viewing many of the interrelationships of geology. By studying the rock cycle we may ascertain the origin of the three basic rock types and gain some insight into the role of various geologic processes in transforming one rock type into another. The concept of the rock cycle, which may be considered as a basic outline of physical geology, was initially proposed by James Hutton. This rock cycle, shown in Figure 1.16, indicates processes by arrows and materials in boxes.

The first rock type, **igneous rock,** originates when molten material called **magma** cools and solidifies. This process, called **crystallization,** may occur either beneath the earth's surface or, following a volcanic eruption, at the surface. Initially, or shortly after forming, the earth's outer shell is believed to have been molten. As this molten material gradually cooled and crystallized, it generated a primitive crust that consisted entirely of igneous rocks.

If igneous rocks are exposed at the surface of the earth, they will undergo **weathering,** in which the day-in-and-day-out influences of the atmosphere slowly disintegrate and decompose rocks. The materials that result will be picked up, transported, and deposited by any of a number of erosional agents—gravity, running water, glaciers, wind, or waves. Once these particles and dissolved substances, called **sediment,** are deposited, usually as horizontal beds in the ocean, they will undergo **lithification,** a term meaning "conversion into rock." Sediment is lithified when compacted by the weight of overlying layers or when cemented as percolating water fills the pores with mineral matter. If the resulting **sedimentary rock** is buried deep within the earth or involved in the dynamics of mountain building, it will be subjected to great pressures and heat. The sedimentary rock will react to the changing environment and turn into the third rock type, **metamorphic rock.** When metamorphic rock is subjected to still greater heat and pressure, it will melt, creating

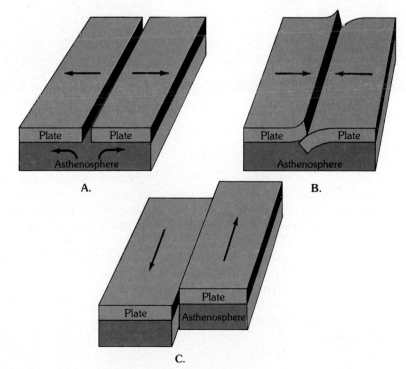

FIGURE 1.14
Schematic of plate boundaries.
A. Divergent boundary.
B. Convergent boundary.
C. Transform fault boundary.

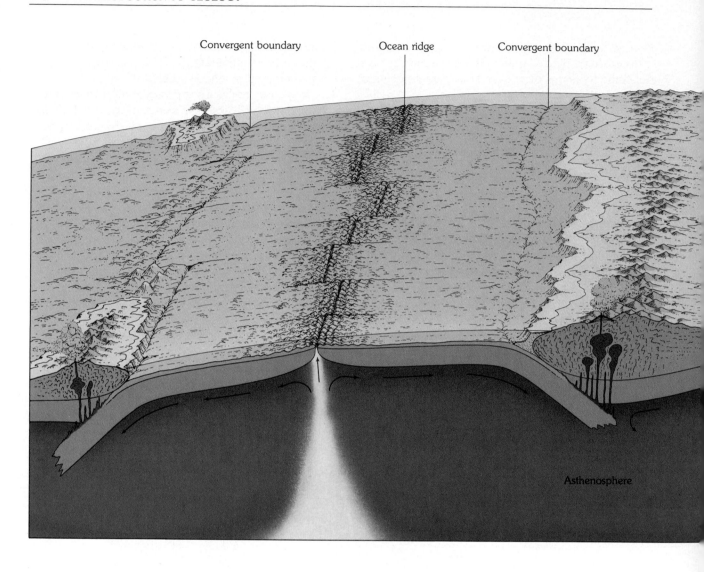

Convergent boundary   Ocean ridge   Convergent boundary

Asthenosphere

magma, which will eventually solidify as igneous rock.

The full cycle just described does not always take place. "Shortcuts" in the cycle are indicated by dashed lines in Figure 1.16. Igneous rock, for example, rather than being exposed to weathering and erosion at the earth's surface, may be subjected to the heat and pressure found far below and change to metamorphic rock. On the other hand, metamorphic and sedimentary rocks, as well as sediment, may be exposed at the surface and turned into new raw materials for sedimentary rock.

When the rock cycle was first proposed by James Hutton, very little was actually known about the processes by which one rock was transformed into another; only evidence for the transformation existed. In fact, it was not until very recently with the development of the theory of plate tectonics that a complete picture of the rock cycle became clear.

Figure 1.17 on page 25 illustrates the rock cycle in terms of the plate tectonics model. According to this model, weathered material from elevated landmasses is transported to the continental margins where it is deposited in layers that collectively are thousands of

Divergent boundary

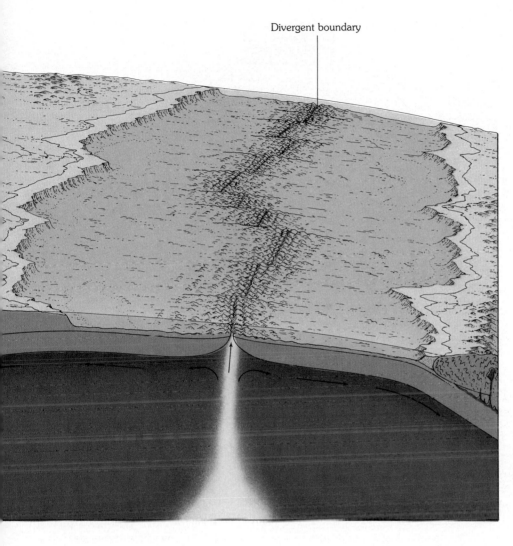

FIGURE 1.15
View of the earth showing the relationship between divergent and convergent plate boundaries.

meters thick. Once lithified, these sediments create a thick wedge of sedimentary rocks flanking the continents.

Eventually the relatively quiescent activity of sedimentation along a continental margin may be interrupted if the region becomes a convergent plate boundary. When this occurs, the oceanic lithosphere adjacent to the continent begins to inch downward into the asthenosphere beneath the continent. Along active continental margins such as this, the converging plate deforms the margin's sedimentary rocks and transforms them into linear belts of metamorphic rocks. Further, as the oceanic plate descends, some of the overlying sediments that were not crumpled into mountains are carried downward into the hot asthenosphere where they too undergo metamorphism. Eventually some of this metamorphic material will be transported to depths where the temperatures and pressures are sufficiently great to initiate melting. This newly formed magma will then migrate upward and occasionally erupt at the surface. Crystallization of this magma generates igneous rocks that are immediately attacked by the processes of weathering. Thus, the rock cycle is ready to begin anew.

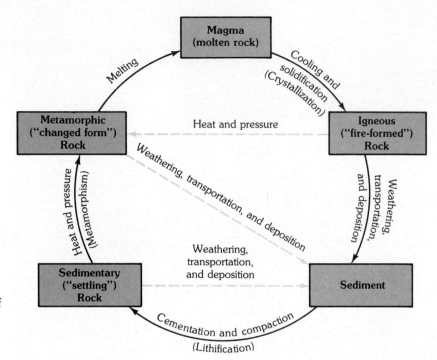

FIGURE 1.16
The rock cycle. Originally proposed by James Hutton, the rock cycle illustrates the role of the various geologic processes which act to transform one rock type into another.

REVIEW QUESTIONS

1 Geology is traditionally divided into two broad areas. Name and describe these two subdivisions.

2 Briefly describe Aristotle's influence on the science of geology.

3 How did the proponents of catastrophism perceive the age of the earth?

4 Describe the doctrine of uniformitarianism. How did the advocates of this idea view the age of the earth?

5 Briefly describe the contributions of Hutton, Playfair, and Lyell.

6 How old is the earth currently thought to be?

7 The geologic calendar was established without the aid of radiometric dating. What principles were used to develop the calendar?

8 Contrast the asthenosphere and the lithosphere.

9 The present shoreline is not the boundary between the continents and the ocean basins. Explain.

10 With which type of plate boundary is each of the following associated: subduction zone, San Andreas fault, sea-floor spreading, and Mount St. Helens?

11 Using the rock cycle, explain the statement "one rock is the raw material for another."

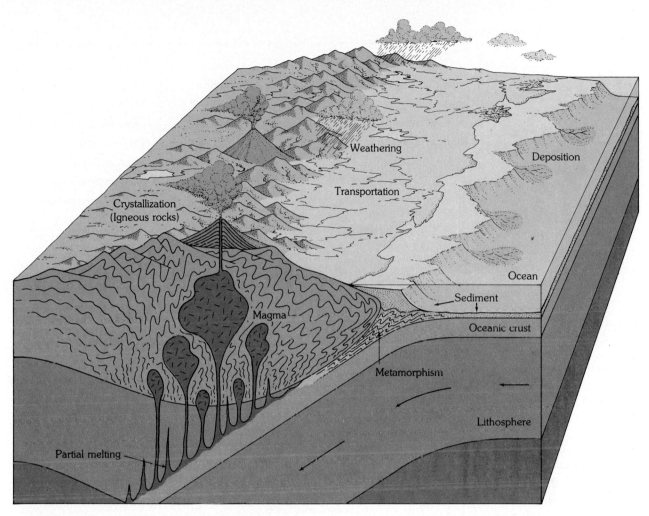

FIGURE 1.17
The rock cycle as it relates to the plate tectonics model.

KEY TERMS

asthenosphere (p. 11)

atmosphere (p. 11)

catastrophism (p. 3)

Cenozoic era (p. 8)

continental shelf
(p. 14)

continental slope
(p. 14)

convergent boundary
(p. 18)

crust (p. 11)

crystallization (p. 21)

divergent boundary
(p. 18)

epoch (p. 8)

era (p. 7)

faunal succession (p. 7)

fossil (p. 7)

geology (p. 3)

historical geology
(p. 3)

hydrosphere (p. 11)

igneous rock (p. 21)

inner core (p. 11)

lithification (p. 21)

lithosphere (p. 11)

magma (p. 21)

mantle (p. 11)

Mesozoic era (p. 8)

metamorphic rock
(p. 21)

nebular hypothesis
(p. 10)

oceanic ridge system
(p. 15)

outer core (p. 11)

Paleozoic era (p. 7)

period (p. 8)

physical geology (p. 3)

plate (p. 15)

plate tectonics (p. 15)

Precambrian (p. 8)

relative dating (p. 7)

rock cycle (p. 21)

sea-floor spreading
(p. 18)

sediment (p. 21)

sedimentary rock
(p. 21)

subduction zone (p. 20)

superposition (p. 7)

transform fault
boundary (p. 18)

trench (p. 15)

uniformitarianism
(p. 4)

weathering (p. 21)

2

MATTER
AND
MINERALS

The outer layer of the earth, which we call the crust, is only as thick when compared to the remainder of the earth as a peach skin is to a peach, yet it is of supreme importance to us. We depend on it for fossil fuels and as a source of such diverse minerals as talc for baby powder, salt to flavor food, and gold for world trade. In fact, on occasion, the availability or absence of certain earth materials has altered the course of history.

In addition to the economic uses of rocks and minerals, all of the processes studied by geologists are in some way dependent upon the properties of these basic earth materials. Events such as volcanic eruptions, mountain building, weathering and erosion, and even earthquakes involve rocks and minerals. Consequently, a basic knowledge of earth materials is essential to the understanding of all geologic phenomena.

ROCKS VERSUS MINERALS

Many people consider rocks to be rather nondescript objects that are hard and often dirty. Minerals are

Quartz crystals exhibit a characteristic external form. (Photo by E. J. Tarbuck)

considered by many to be dietary supplements, or possibly rare ores or precious gems that are mined for their economic value. However, these common perceptions are far from the actual situation.

A **rock** can be defined simply as an aggregate of one or more minerals. Here, the term *aggregate* implies that the minerals are found together as a *mixture* in which the properties of the individual minerals are retained. Although most rocks are composed of more than one mineral, certain minerals are commonly found by themselves in large quantities. In these instances they are considered to be both a mineral and a rock. A common example is the mineral calcite, which frequently is the dominant constituent in large rock units, where it is given the name *limestone*.

By contrast, **minerals** are defined as naturally occurring inorganic solids, which possess a definite internal structure and a specific chemical composition. Although this definition is quite precise, it is not without shortcomings. For example, this definition excludes organic compounds; however, many geologists would classify coal, and occasionally even petroleum, as minerals. Further, the chemical composition of many minerals actually varies over a wide range.

This chapter deals primarily with the nature of minerals. However, keep in mind that rocks are simply aggregates of minerals. Thus, the properties of rocks are determined solely by the chemical composition and internal structure of those minerals which compose them.

THE COMPOSITION OF MATTER

Minerals, like all matter, are made of **elements.** At present, over 100 elements are known, a dozen and a half of which have been produced only in the laboratory. Some minerals such as gold and sulfur are made entirely of one element, but most are a combination of two or more elements joined to form a chemically stable **compound.** In order to better understand how elements combine to form compounds, we must first consider the **atom,** the smallest part of matter that still retains the characteristics of an element, because it is this extremely small particle that does the combining.

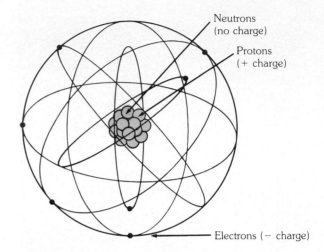

Neutrons (no charge)

Protons (+ charge)

Electrons (− charge)

FIGURE 2.1
Simplified model of an atom. Atoms consist of a central nucleus composed of protons and neutrons which is encircled by electrons.

ATOMIC STRUCTURE

Individual atoms are far too small to be observed directly; therefore, our concept of atomic structure has come from experimental evidence and mathematical models. A simplified model of the structure of an atom is shown in Figure 2.1. Each atom has a central region, called the **nucleus,** which contains very dense positively charged **protons** and equally dense neutral particles called **neutrons.** Orbiting the nucleus are negatively charged particles known as **electrons.** Unlike the orderly orbiting of the planets around the sun, electrons move so rapidly that their positions cannot be pinpointed. Hence, a more realistic picture of the positions of electrons can be obtained by envisioning a cloud of electrons surrounding the nucleus. It is also known that individual electrons are located at given distances from the nucleus in regions called **energy-level shells.** As we shall see, an important fact about these shells is that each can hold only a specific number of electrons.

The number of protons found in the nucleus determines the **atomic number** and name of the element. For example, all atoms with six protons are carbon atoms, all those with eight protons are oxygen atoms, and so forth. Since all atoms have the same number of electrons as protons, the atomic number

TABLE 2.1
Atomic number and distribution of electrons in the main shells.

Element	Symbol	Atomic Number	Number of Electrons in Each Shell			
			1	2	3	4
Hydrogen	H	1	1			
Helium	He	2	2			
Lithium	Li	3	2	1		
Beryllium	Be	4	2	2		
Boron	B	5	2	3		
Carbon	C	6	2	4		
Nitrogen	N	7	2	5		
Oxygen	O	8	2	6		
Fluorine	F	9	2	7		
Neon	Ne	10	2	8		
Sodium	Na	11	2	8	1	
Magnesium	Mg	12	2	8	2	
Aluminum	Al	13	2	8	3	
Silicon	Si	14	2	8	4	
Phosphorus	P	15	2	8	5	
Sulfur	S	16	2	8	6	
Chlorine	Cl	17	2	8	7	
Argon	Ar	18	2	8	8	
Potassium	K	19	2	8	8	1
Calcium	Ca	20	2	8	8	2

also equals the number of electrons surrounding the nucleus. Moreover since neutrons have no charge, the positive charge of the protons is exactly balanced by the negative charge of the electrons. Consequently, every atom is an electrically neutral particle. Elements can be considered to be a large collection of electrically neutral atoms, all having the same atomic number.

The simplest element, hydrogen, is composed of atoms that have only one proton in the nucleus and one electron surrounding the nucleus. Each successively heavier atom has one more proton and one more electron, in addition to a certain number of neutrons (Table 2.1). Studies of electron configurations have shown that each electron is added in a systematic fashion to a particular energy level or shell. In general, electrons enter higher energy levels only after lower energy levels have been filled to capacity. The first principle shell holds a maximum of two electrons, while each of the higher shells holds eight or more electrons. However, any shell that is an outermost shell (other than the first shell which is filled with

two electrons) will contain a maximum of only eight electrons. As we shall see, it is the electrons found in outermost shells which are generally involved in chemical bonding.

BONDING

Chemical bonds occur when atoms of two or more elements join to form a compound. When the atoms separate, the bonds are broken and the compound is destroyed. Through experimentation it has been learned that the forces bonding the atoms together are electrical in nature. Further, it is known that chemical bonding results in a change in the electronic structures of the bonded atoms. Hence, the electron configurations of the atoms involved are important in determining the strength and nature of the chemical bonds that are produced.

As we noted earlier, the outermost electrons are generally involved in chemical bonding. Further, the atoms of most elements have less than the maximum number of electrons in their outermost shell. Only the noble gases such as neon and argon have a complete outer shell, which accounts for their chemical stability and the fact they do not readily react with other elements. However, *every atom seeks a full outer shell* to become chemically stable like the noble gases. The octet rule, literally meaning "a set of eight," refers to the concept of a completely filled outermost energy level. Simply, the **octet rule** states that atoms combine in order that each may have the electron arrangement of a noble gas, with the outer energy level containing eight electrons.

In order to satisfy the octet rule, an atom can either gain, lose, or share electrons with one or more atoms. The result of this process is the formation of an elec-

trical "glue" that bonds the atoms. The electrons involved in the bonding process are commonly called **valence electrons.** The number of valence electrons that an element has determines the number of bonds it will form. For example, the element silicon has four valence electrons and forms four bonds in the process of completing its outer shell. On the other hand, oxygen forms two bonds, and hydrogen forms only one.

Ionic Bonds   Perhaps the easiest type of bond to visualize is an **ionic bond.** In ionic bonding, one or more valence electrons are transferred from one

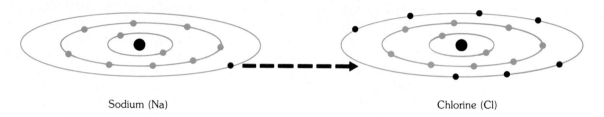

Sodium (Na)   Chlorine (Cl)

FIGURE 2.2
Chemical bonding of sodium and chlorine to produce sodium chloride. Through the transfer of one electron from sodium to chlorine, sodium becomes a positive ion and chlorine a negative ion.

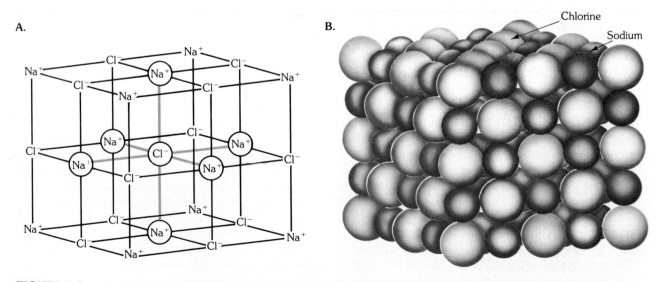

FIGURE 2.3
Schematic illustrating the arrangement of sodium and chloride ions in table salt.
A. Structure has been opened up to show arrangement of ions. **B.** Actual ions are closely packed.

atom to another. One atom becomes stable by giving up its valence electrons and the other uses them to complete its outer shell. An example of ionic bonding using sodium (Na) and chlorine (Cl) to produce sodium chloride (common table salt) is shown in Figure 2.2. Notice that sodium loses its single outer electron to chlorine. As a result, sodium acquires the electron configuration of the noble gas neon, which has two electrons in the first shell and eight in its outermost shell. By adding an electron, a chlorine atom fills its outermost shell to acquire the arrangement of argon. However, these atoms are no longer electrically neutral because neither contains an equal number of protons and electrons. Atoms such as these, which have an unequal charge because of a gain or loss of electrons, are called **ions.** Sodium becomes a positively charged ion and chlorine becomes a negatively charged ion. An ionic bond results from the attraction of these oppositely charged particles. Restated simply, an ionic bond is one in which oppositely charged ions attract one another to produce a neutral chemical compound. Figure 2.3 illustrates the arrangement of sodium and chloride ions in ordinary table salt. Notice that salt consists of alternating sodium and chloride ions, positioned such that each positive ion is attracted to and surrounded on all sides by negative ions, and vice versa. Ionic compounds therefore consist of an orderly arrangement of oppositely charged ions assembled in a definite ratio that provides overall electrical neutrality.

This is an appropriate place in our discussion to point out that the properties of a chemical compound are dramatically different from the properties of the elements composing it. For example, chlorine is a green, poisonous gas that is so toxic it was used as a weapon during World War I. Sodium is a soft, silvery metal that reacts vigorously with water and, if held in

your hand, could burn it severely. Together, however, these atoms produce the compound sodium chloride (table salt), which is a clear crystalline solid that is essential for human life. This example also illustrates an important difference between a rock and a mineral. A *mineral* is a *chemical compound* with unique properties that are very different from the elements which make it up. A *rock,* on the other hand, is a *mixture* of minerals, with each mineral retaining its own identity.

Covalent Bonds   Not all atoms combine by forming ions. For example, the gaseous elements oxygen (O_2), hydrogen (H_2), and chlorine (Cl_2) exist as stable molecules consisting of two atoms bonded together without the complete transfer of electrons. This is necessary because even if one of the atoms in each pair did accept one or more electrons to form a stable octet, the other atom would move farther away from such a stable condition. Instead, a stable octet is obtained when some of the outer electrons of both atoms are shared. Figure 2.4 illustrates the sharing of a pair of electrons between two chlorine atoms to form a molecule of chlorine gas. By overlapping the outer shells, one electron in each chlorine atom, which has seven electrons in its outer shell, has acquired through cooperative action, the needed electron to complete the octet. The bond produced by the sharing of electrons to acquire the stable noble gas arrangement is called a **covalent bond.** The most common mineral group, the silicates, contains the element silicon that readily forms covalent bonds with oxygen.

A common analogy may help you visualize a covalent bond. Imagine two people at opposite ends of a dimly lit room, both reading under a lamp. By moving the lamps to the center of the room, they are

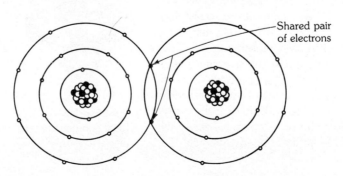

FIGURE 2.4
Schematic drawing showing the sharing of a pair of electrons between two chlorine nuclei to form a chlorine molecule.

Shared pair of electrons

able to combine their light sources so each can see better. Just as it is impossible to determine the source of the overlapping light, shared electrons are indistinguishable from each other.

It should be pointed out that most chemical bonds are actually a blend, consisting to some degree of electron sharing, as in covalent bonding, and to some degree of electron transfer, as in ionic bonding. In addition, there also exists an extreme type of electron sharing, in which the electrons move freely from atom to atom. This type of bonding is found in metals such as copper, gold, aluminum, and silver. The term **metallic bonding** is applied to this type of electron sharing. Metallic bonding accounts for the high electrical conductivity of metals, the ease with which metals are reshaped, and numerous other special properties of metals.

ATOMIC MASS

Subatomic particles, such as protons, are so incredibly small that a special unit was devised to express their mass. A proton or a neutron has a mass just slightly more then one **atomic mass unit,** whereas an electron is only about one two-thousandths of an atomic mass unit. Thus, although electrons play an active role in chemical reactions, they do not contribute significantly to the mass of an atom. Because of this, the **mass number** of an atom is obtained simply by totaling the number of neutrons and the number of protons in the nucleus. Atoms of the same element commonly have varying numbers of neutrons, and therefore, different mass numbers. Such atoms are called **isotopes** of that element. For example, carbon has two well-known isotopes, one having a mass number of 12 (carbon-12), the other a mass number of 14 (carbon-14). Recall that all atoms of the same element must have the same number of protons (atomic number) and that carbon always has six. Hence, carbon-12 must have six neutrons to give it a mass number of 12, whereas carbon-14 must have eight neutrons to give it a mass number of 14. The term commonly used to express the average of the atomic masses of isotopes for a given element is **atomic weight.** The atomic weight of carbon is much closer to 12 than 14, because carbon-12 is the more common isotope. Note that in a chemical sense all isotopes of the same element are nearly identical. To

distinguish among them would be like trying to differentiate individual members from a group of similar objects, all having the same shape, size, and color, with only some being slightly heavier.

Although the vast majority of atoms are stable, many elements do have isotopes that are unstable. Unstable isotopes such as carbon-14 go through a process of natural disintegration called **radioactivity,** which occurs when the forces that bind the nucleus are not strong enough. The rate at which the unstable nuclei break apart (decay) is measurable and makes such elements useful "clocks" in dating the events of earth history. A discussion of radioactivity and its application in dating events of the geologic past can be found in Chapter 19.

THE STRUCTURE OF MINERALS

A mineral is composed of an ordered array of atoms chemically bonded together to form a particular crystalline structure. This orderly stacking of atoms is reflected in the regularly-shaped objects we call crystals (see chapter-opening photo).

What determines the particular crystalline structure a mineral will exhibit? For those compounds formed by ions, the internal atomic arrangement is

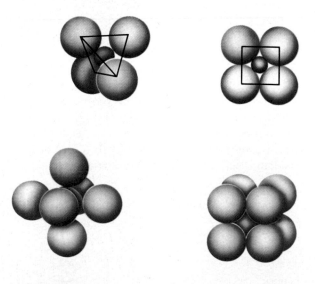

FIGURE 2.5
Ideal geometrical packing for various-sized positive and negative ions.

determined partly by the charges on the ions, but more importantly by the size of the ions involved. In order to form stable ionic compounds, each positively charged ion is surrounded by the largest number of negative ions that will fit, while maintaining overall electrical neutrality, and vice versa. Figure 2.5 shows some ideal geometries for various-sized ions. We have already examined the geometric arrangement of sodium and chloride ions in the mineral halite. In Figure 2.6 we see that on a large scale the

A.

B.

FIGURE 2.6
The structure of sodium chloride. **A.** The coordination of sodium and chloride ions in the mineral halite. **B.** The orderly arrangement at the atomic level produces regularly-shaped crystals. (Courtesy of the American Museum of Natural History)

orderly packing of sodium and chloride ions produces cubic halite crystals. Like halite, all samples of a particular mineral contain the same elements, joined together in the same orderly arrangement.

Although it is true that every mineral has a particular internal structure, some elements are able to join together in more than one way. Thus, two minerals with totally different properties may have exactly the same chemical composition. Minerals of this type are said to be **polymorphs** (many forms). Graphite and diamond are particularly good examples because they consist exclusively of carbon yet are drastically different. Graphite is the soft gray material of which pencil lead is made, while diamond is the hardest known mineral. The differences between these minerals can be attributed to the conditions under which they formed. Diamonds are believed to form at depths approaching 200 kilometers, where extreme pressures produce the compact structure shown in Figure 2.7A. Graphite, on the other hand, consists of sheets of carbon atoms that are widely spaced and weakly held together (Figure 2.7B). Since these carbon sheets will easily slide past one another, graphite makes an excellent lubricant.

PHYSICAL PROPERTIES OF MINERALS

Minerals are solids formed by inorganic processes. Each mineral has an orderly arrangement of atoms (crystalline structure) and a definite chemical composition which give it a unique set of physical properties. Since the internal structure and chemical composition of a mineral are difficult to determine without the aid of sophisticated tests and apparatus, the more easily recognized physical properties are frequently used in identification. A discussion of some diagnostic physical properties follows.

CRYSTAL FORM

Most people think of a crystal as a rare commodity, when in fact most inorganic solid objects are composed of crystals. The reason for this misconception is that most crystals do not exhibit their crystal form. The **crystal form** is the external expression of a mineral that reflects the orderly internal arrangement of atoms. Figure 2.8A illustrates the characteristic form of the iron-bearing mineral pyrite. Any time a mineral is permitted to form without space restrictions, it will

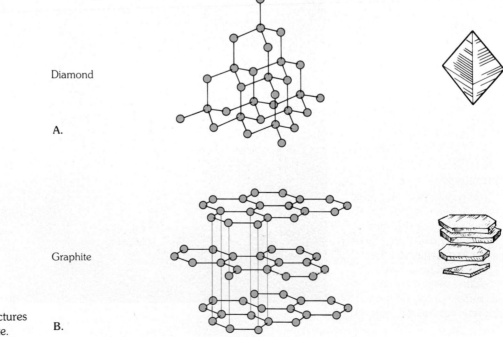

Diamond

A.

Graphite

FIGURE 2.7
Comparison of the structures of diamond and graphite.

B.

develop individual crystals with well-formed crystal faces. Some crystals such as those of the mineral quartz have a very distinctive crystal form that can be helpful in identification (Figure 2.8B). However, most of the time crystal growth is interrupted because of competition for space, resulting in an intergrown mass of crystals, none of which exhibits its crystal form.

LUSTER

Luster is the appearance or quality of light reflected from the surface of a mineral. Minerals that have the appearance of metals, regardless of color, are said to have a *metallic luster*. Minerals with a *nonmetallic luster* are described by various adjectives, including vitreous (glassy), pearly, silky, resinous, and earthy (dull). Some minerals appear partially metallic in luster and are said to be *submetallic*.

COLOR

Although **color** is the most obvious feature of a mineral, it is often an unreliable diagnostic property. Slight impurities in the common mineral quartz, for example, give it a variety of colors, including pink, purple (amethyst), white, and even black. When a mineral, such as quartz, exhibits a variety of colors, it is said to possess *exotic coloration*. Other minerals, for example, sulfur, which is generally yellow, and malachite, which is bright green, are said to have *inherent coloration* because their color does not vary significantly.

STREAK

Streak is the color of a mineral in its powdered form and is obtained by rubbing the mineral across a piece of unglazed porcelain termed a *streak plate*. Although the color of a mineral may vary from sample to sam-

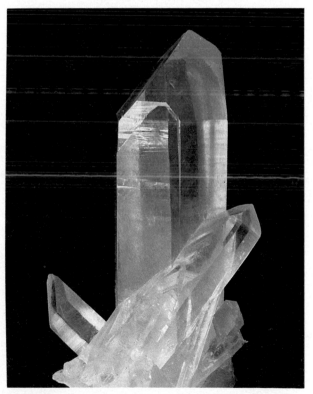

A.

B.

FIGURE 2.8
Crystal form is the external expression of a mineral's orderly internal structure. **A.** Pyrite crystals. (Courtesy of JLM Visuals). **B.** Quartz crystals. (Photo by E. J. Tarbuck)

ple, the streak usually does not, and is therefore the more reliable property. Streak can also be an aid in distinguishing minerals with metallic lusters from those having nonmetallic lusters. Metallic minerals generally have a dense, dark streak, whereas minerals with nonmetallic lusters do not.

HARDNESS

One of the most useful diagnostic properties is **hardness,** the resistance of a mineral to abrasion or scratching. This is a relative property that is determined by rubbing a mineral of unknown hardness against one of known hardness, or vice versa. A numerical value can be obtained by using **Mohs scale** of hardness, which consists of ten minerals arranged in order from 1 (softest) to 10 (hardest) as follows:

Hardness	Mineral
1	Talc
2	Gypsum
3	Calcite
4	Fluorite
5	Apatite
6	Orthoclase
7	Quartz
8	Topaz
9	Corundum
10	Diamond

Any mineral of unknown hardness can be compared to these or to other objects of known hardness. For example, a fingernail has a hardness of 2.5, a copper penny 3, and a piece of glass 5.5. The mineral gypsum, which has a hardness of 2, can be easily scratched with your fingernail. On the other hand, the mineral calcite which has a hardness of 3, will scratch your fingernail but will not scratch glass. Quartz, the hardest of the common minerals, will scratch a glass plate with ease.

CLEAVAGE

Cleavage is the tendency of a mineral to break along planes of weak bonding. Minerals that possess cleavage are identified by the smooth surfaces which are produced when the mineral is broken. The simplest

FIGURE 2.9
Sheet-type cleavage common to the micas. (Courtesy of Ward's Natural Science Establishment, Inc., Rochester, N.Y.)

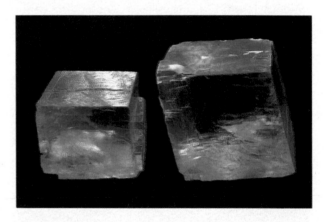

FIGURE 2.10
Smooth surfaces produced when a mineral with cleavage is broken. These samples exhibit three planes of cleavage (six sides.) The mineral on the left has cleavage planes which meet at 90 degree angles, whereas the mineral on the right has cleavage planes which meet at 75 degree angles. (Photo by E. J. Tarbuck)

type of cleavage is exhibited by the micas (Figure 2.9). Because the micas have excellent cleavage in one direction, they break to form thin, flat sheets. Some minerals have several cleavage planes which produce smooth surfaces when broken, while others exhibit poor cleavage, and still others have no cleavage at all. When minerals break evenly in more than one direction, cleavage is described by the number of planes exhibited and the angles at which they meet (Figure 2.10).

Cleavage should not be confused with crystal form. When a mineral exhibits cleavage, it will break into pieces that have the same configuration as the original sample. By contrast, the quartz crystals shown in the chapter-opening photo do not have cleavage, and if broken, would shatter into shapes that do not resemble each other or the original crystals.

FRACTURE

Such minerals as the quartz just described do not exhibit cleavage and are therefore said to **fracture** when broken. Those that break into smooth curved surfaces resembling broken glass have a *conchoidal fracture* (Figure 2.11). Others break into splinters or fibers, but most minerals fracture irregularly.

SPECIFIC GRAVITY

Specific gravity is a number representing the ratio of the weight of a mineral to the weight of an equal

FIGURE 2.11
Conchoidal fracture. The smooth curved surfaces result when minerals break in a glasslike manner. (Photo by E. J. Tarbuck)

volume of water. For example, if a mineral weighs three times as much as an equal volume of water, its specific gravity is 3. With a little practice, you can estimate the specific gravity of minerals by hefting them in your hand. For example, if a mineral feels as heavy as the common rocks you have handled, its specific gravity will probably be somewhere between 2.5 and 3. Some metallic minerals have a specific gravity two or three times the average. Galena, which is an ore of lead, has a specific gravity of roughly 7.5, while the specific gravity of 24 carat gold is approximately 20.

MINERAL GROUPS

Over two thousand minerals are presently known to exist and new ones are still being discovered. Fortunately for those of us who study minerals, no more than two dozen are abundant. Collectively, these few make up most of the rocks of the earth's crust and as such, are classified as the *rock-forming minerals*. It is also interesting to note that only eight elements compose the bulk of these minerals and represent over 98 percent (by weight) of the continental crust (Table 2.2). The two most abundant elements are silicon and oxygen, which combine to form the framework of the most common mineral group, the **silicates.**

TABLE 2.2
Relative abundance of the most common elements in the earth's crust.

Element	Approximate Percentage by Weight
Oxygen (O)	46.6
Silicon (Si)	27.7
Aluminum (Al)	8.1
Iron (Fe)	5.0
Calcium (Ca)	3.6
Sodium (Na)	2.8
Potassium (K)	2.6
Magnesium (Mg)	2.1
All others	1.5
Total	100

SOURCE: Data from Brian Mason.

Every silicate mineral contains oxygen and silicon, and except for quartz, one or more additional elements are needed to acquire electrical neutrality. Perhaps the next most common mineral group is the carbonates, of which calcite is the most prominent member. Other common rock-forming minerals include gypsum and halite (table salt).

In addition to the rock-forming minerals, a number of minerals are prized for their economic value. Included in this group are the ores of metals such as hematite (iron), sphalerite (zinc), and galena (lead); the native elements including gold, silver, and carbon (diamonds); and a host of others such as fluorite, corundum, and uraninite. Note that the rock-forming minerals themselves are not without economic value. For instance, quartz is used in the production of glass, calcite is the main constituent in portland cement, and plaster is composed of the mineral gypsum.

SILICATE STRUCTURES

All silicate minerals have the same fundamental building block, the **silicon-oxygen tetrahedron.** This structure consists of four oxygen atoms surrounding a much smaller silicon atom positioned in the space between them (Figure 2.12). The silicon-oxygen tetrahedron is not, however, a stable compound; rather it is a complex ion with a charge of -4. This excess negative charge results because each of the four oxygen atoms contributes a charge of -2, whereas the one silicon atom has a charge of $+4$. In nature, one of the simplest ways in which these tetrahedra are neu-

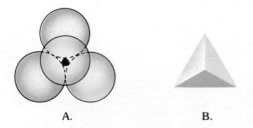

A.   B.

FIGURE 2.12
Top view of the silicon-oxygen tetrahedron. **A.** The four large spheres represent oxygen atoms, and the dark sphere represents a silicon atom. **B.** Diagrammatic representation of the tetrahedron using four points to represent the positions of the oxygen atoms.

tralized is through the addition of positively charged ions. In this way a chemically stable structure consisting of individual tetrahedra linked together by positively charged ions is produced.

In addition to positive ions acting as the "glue" to bind the tetrahedra, the tetrahedra themselves may be linked in a variety of configurations. For example, the tetrahedra may join to form single chains, double chains, or sheet structures a shown in Figure 2.13. The joining of tetrahedra in each of these configurations results from the sharing of oxygen atoms by pairs of silicon atoms. In order to better understand how this sharing takes place, select one of the silicon atoms (small spheres) near the middle of the single chain structure shown in Figure 2.13. Notice that this silicon atom is completely surrounded by four larger oxygen atoms. Also notice that two of the four oxygen atoms are joined to two silicon atoms, while the other two are not shared in this manner. It is the linkage across the shared oxygen atoms that joins the tetrahedra into a chain structure. Now examine a silicon atom near the middle of the sheet structure and count the number of shared and unshared oxygen atoms surrounding it. The increase in the degree of sharing accounts for the sheet structure. Although they are not shown, other silicate structures exist. The most common silicate structure has all of the oxygen atoms shared to produce a complex three-dimensional framework.

By now we can see that the ratio of oxygen atoms to silicon atoms differs in each of the silicate structures. In the isolated tetrahedron there are 4 oxygen atoms for every silicon atom, in the single chain the oxygen to silicon ratio is 3 to 1, and in the three-dimensional framework this ratio is 2 to 1. Consequently, as more of the oxygen atoms are shared, the percentage of silicon in the structure increases. The silicate minerals are therefore described as having a high or low silicon content based on their ratio of oxygen to silicon. This difference in silicon content is quite important as we shall see later when we consider the formation of igneous rocks.

These silicate structures, with the exception of the three-dimensional framework, are not neutral chemical compounds themselves. Thus, like the individual tetrahedra, they all are neutralized by the inclusion of positively charged metallic ions that bond them together into a variety of crystalline configurations. The ions that most often link silicate structures are those of the elements iron (Fe), magnesium (Mg), potassium (K), sodium (Na), aluminum (Al), and calcium (Ca). Notice in Figure 2.14 that each of these positive ions has a particular atomic size and a particular charge. Generally, ions of approximately the same size are able to freely substitute for one another. For instance, the ions of iron (Fe^{2+}) and magnesium (Mg^{2+}) are nearly the same size, and substitute for each other without altering the mineral structure. This also holds true for calcium and sodium, which can occupy the same site in a crystalline structure, and aluminum (Al), which substitutes for silicon in the silicon-oxygen tetrahedron.

Because of the ability of mineral structures to readily accommodate different ions at a given bond-ing site, individual specimens of a particular mineral may contain varying amounts of certain elements. A mineral of this type is often expressed by a chemical formula that uses parentheses to set apart the vari-able component. A good example is the mineral oliv-ine, $(Fe,Mg)_2SiO_4$. As we can see from the formula, it is the iron (Fe^{2+}) and magnesium (Mg^{2+}) ions in oliv-ine that freely substitute for each other. At one extreme, olivine may contain iron without any mag-nesium (Fe_2SiO_4) and at the other, iron is totally lack-ing (Mg_2SiO_4). Between these end members any ratio of iron to magnesium is possible. Thus olivine, as well as many other silicate minerals, is actually a family of minerals that has a range of composition between the two end members.

In certain substitutions the ions that interchange do not have the same electrical charge. For instance,

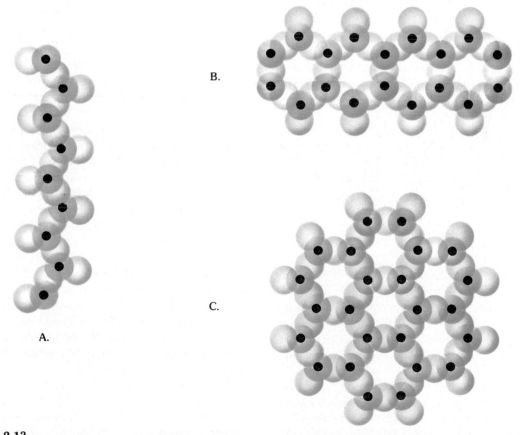

B.

C.

A.

FIGURE 2.13
Three types of silicate structures. **A.** Single chains. **B.** Double chains. **C.** Sheet struc-tures.

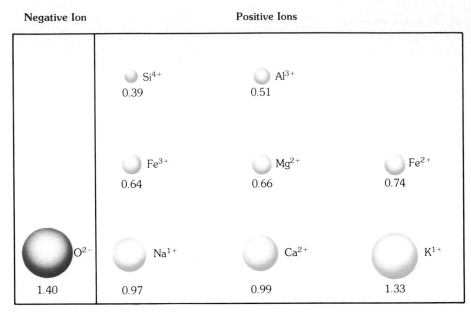

Negative Ion

Positive Ions

Si^{4+}
0.39

Al^{3+}
0.51

Fe^{3+}
0.64

Mg^{2+}
0.66

Fe^{2+}
0.74

O^{2-}
1.40

Na^{1+}
0.97

Ca^{2+}
0.99

K^{1+}
1.33

FIGURE 2.14
Relative sizes and electrical charges of ions commonly found in rock-forming minerals. Ionic radii are expressed in Angstroms (one Angstrom equals 10^{-8} cm).

when calcium (Ca^{2+}) substitutes for sodium (Na^{1+}), the structure gains a positive charge. In nature, one way in which this substitution is accomplished, while still maintaining overall electrical neutrality, is that a simultaneous substitution of aluminum (Al^{3+}) for silicon (Si^{4+}) takes place. This particular double substitution occurs in the feldspar group, which is the most abundant family of minerals found in the crust. The end members of this particular feldspar series are anorthite, $CaAl_2Si_2O_8$, and albite, $NaAlSi_3O_8$.

SILICATE MINERALS

The main groups of silicate minerals and common examples of each are given in Figure 2.15. The feldspars are by far the most abundant group, comprising over 50 percent of the earth's crust. Quartz, the second most common mineral in the continental crust, is the only one made completely of silicon and oxygen.

Notice in Figure 2.15 that each group has a particular silicate structure. A relationship exists between the internal structure of a mineral and the cleavage it exhibits. Because the silicon-oxygen bonds are strong, silicate minerals tend to cleave between the silicon-oxygen structures rather than across them. For example, the micas have a sheet structure and

tend to cleave into flat plates (Figure 2.9). Quartz, which has equally strong silicon-oxygen bonds in all directions, has no cleavage.

Most silicate minerals form when molten rock cools. This cooling can occur at or near the earth's surface, or at great depths where temperatures and pressures are very high. The environment during crystallization and the chemical composition of the molten rock to a large degree determine the minerals that are produced. For example, the silicate mineral olivine crystallizes at high temperatures and possesses a chemical structure that is stable at high temperatures. Quartz, on the other hand, crystallizes at much lower temperatures. In addition, some silicate minerals are stable at the earth's surface and represent the weathered products of pre-existing silicate minerals. Still other silicate minerals are formed under the extreme pressures associated with metamorphism. Each silicate mineral therefore has a structure and a chemical composition that indicate the conditions under which it formed.

The various silicate minerals can be divided on the basis of chemical makeup. The *ferromagnesian silicates* are those minerals containing ions of iron and/or magnesium in their structure. Those minerals that do not contain these ions are simply called *nonferromagnesians*. Usually, ferromagnesian minerals are dark in color and have a specific gravity between

Mineral		Idealized Formula	Cleavage	Silicate Structure	
Olivine		$(Mg,Fe)_2SiO_4$	None	Single tetrahedron	
Pyroxene		$(Mg,Fe)SiO_3$	Two planes at right angles	Chains 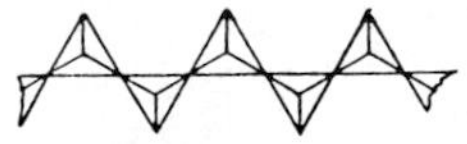	
Amphibole		$(Ca_2Mg_5)Si_8O_{22}(OH)_2$	Two planes at 60° and 120°	Double chains 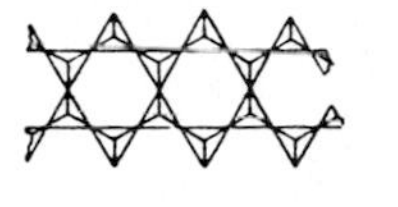	
Micas	Muscovite	$KAl_3Si_3O_{10}(OH)_2$	One plane	Sheets 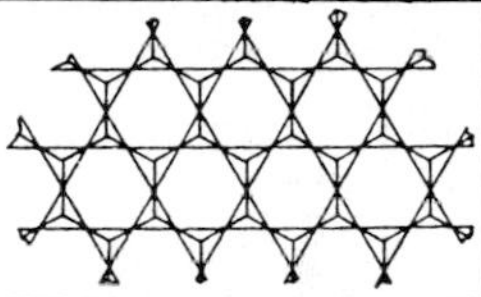	
	Biotite	$K(Mg,Fe)_3Si_3O_{10}(OH)_2$			
Feld-spars	Orthoclase	$KAlSi_3O_8$	Two planes at 90°	Three-dimensional networks 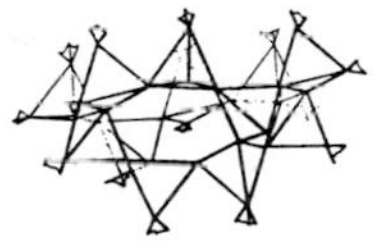	
	Plagioclase	$(Ca,Na)AlSi_3O_8$			
Quartz		SiO_2	None		

FIGURE 2.15
Common silicate minerals. Note that the complexity of the silicate structure increases down the chart.

3.2 and 3.6. By comparison, nonferromagnesian silicates are generally light in color and have an average specific gravity of 2.7. These observed differences are mainly attributable to the presence or absence of iron.

FERROMAGNESIAN SILICATES

Olivine is a high-temperature silicate mineral that is black to olive green in color, has a glassy luster, and a conchoidal fracture. Rather than developing large crystals, olivine commonly forms small, rounded crystals that give the mineral a granular appearance. Olivine is composed of individual tetrahedra which are bonded together by a mixture of iron and magnesium ions positioned so as to link the oxygen atoms together. Since the three-dimensional network generated in this fashion does not have its weak bonds aligned, olivine does not possess cleavage.

Pyroxene is a black, opaque mineral with two planes of cleavage that meet at nearly a 90 degree angle. Its crystalline structure consists of single chains of tetrahedra bonded together by ions of iron and magnesium. Since the silicon-oxygen bonds are stronger than the bonds joining the silicate structures, pyroxene cleaves parallel to the silicate chains. Pyroxene is one of the dominant minerals in basalt, a common igneous rock of the oceanic crust which is also prevalent in volcanic areas on the continents.

Hornblende is the most common member of a chemically complex group of minerals called *amphiboles*. Hornblende is usually dark green to black in color and except for its cleavage angles, which are about 60 degrees and 120 degrees, it is very similar in appearance to pyroxene (Figure 2.16). The double chains of tetrahedra in the hornblende structure account for its particular cleavage. In a rock, hornblende often forms elongated crystals. This helps distinguish it from pyroxene, which forms rather blocky crystals. Hornblende is predominantly found in continental rocks, where it often makes up the dark portion of an otherwise light-colored rock.

Biotite is the dark iron-rich member of the mica family. Like other micas, biotite possesses a sheet structure which gives it excellent cleavage in one direction. Biotite also has a very shiny black appearance that helps distinguish it from the other dark fer-

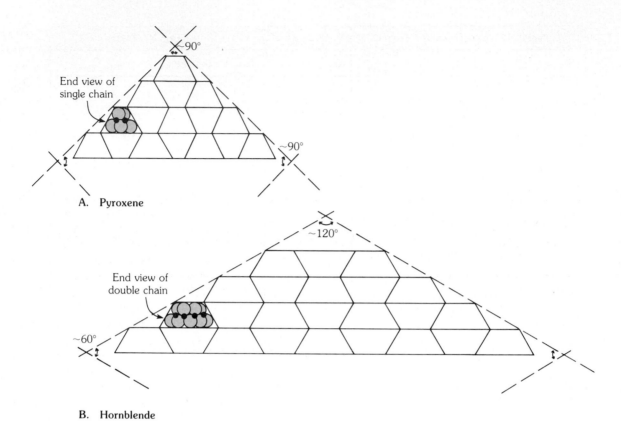

A. Pyroxene

B. Hornblende

FIGURE 2.16
Cleavage angles for pyroxene and hornblende.

FIGURE 2.17
Garnet crystals in a common metamorphic rock. (Courtesy of the American Museum of Natural History)

romagnesian minerals. Like hornblende, biotite is a common constituent of continental rocks, including the igneous rock granite.

Garnet is similar to olivine in that its structure is composed of individual tetrahedra linked by metallic ions. Also like olivine, garnet has a glassy luster, lacks cleavage, and possesses conchoidal fracture. Although the colors of garnet are varied, this mineral is most often brown to deep red. Garnet readily forms equidimensional crystals that are most commonly found in metamorphic rocks (Figure 2.17). When garnets are transparent, they may be used as gemstones.

NONFERROMAGNESIAN SILICATES

Muscovite is a common member of the mica family. It is light in color and has a pearly luster. Like other micas, muscovite has excellent cleavage in one direction. In thin sheets muscovite is clear, a property which accounts for its use as window "glass" during the Middle Ages. Since muscovite is very shiny, it can often be identified by the sparkle it gives a rock. If you have ever looked closely at beach sand, you may have seen the glimmering brilliance of the mica flakes scattered among the other sand grains.

Feldspar, the most common mineral group, can form under a very wide range of temperatures and pressures, a fact that partially accounts for its abundance. All of the feldspars have similar physical properties. They have two planes of cleavage meeting at or near 90 degree angles, are relatively hard (6 on Mohs scale), and have a luster which ranges from glassy to pearly. As one component in a rock, feldspar crystals can be identified by their rectangular shape and rather smooth shiny faces (Figure 2.18).

The structure of the feldspar minerals is a three-dimensional framework formed when oxygen atoms are shared by adjacent silicon atoms. In addition, one-fourth to one-half of the silicon atoms in the feldspar structure are replaced by aluminum atoms. The difference in charge between aluminum (+3) and silicon (+4) is made up by the inclusion of one or more of the following ions into the crystal lattice: potassium (+1), sodium (+1), and calcium (+2). Because of the large size of the potassium ion as compared to the size of the sodium and calcium ions, two different feldspar structures exist. *Orthoclase feldspar* is a common member of a group of feldspar minerals that contains potassium ions in its structure. The other group, called *plagioclase feldspar,* contains both sodium and calcium ions that freely substitute for one another depending on the environment during crystallization.

Orthoclase feldspar is usually light cream to salmon pink in color. The plagioclase feldspars, on the other hand, range in color from white to medium gray. However, color should not be used to distinguish these groups. The only sure way to physically distinguish the feldspars is to look for a multitude of fine parallel lines, called *striations*. Striations are found on some cleavage faces of plagioclase feldspar, but are not present on orthoclase feldspar (Figure 2.19).

Quartz is the only silicate mineral consisting entirely of silicon and oxygen. As such, the term *silica* is applied to quartz, which has the chemical formula SiO_2. Since the structure of quartz contains a ratio of two oxygen ions (O^{2-}) for every one silicon ion (Si^{4+}), no other positive ions are needed to attain neutrality. In quartz, a three-dimensional framework is developed through the complete sharing of oxygen by adjacent silicon atoms. Thus, all of the bonds in

FIGURE 2.18
Samples of the mineral feldspar. (Photo by E. J. Tarbuck)

FIGURE 2.19
These parallel lines, called striations, are a distinguishing characteristic of the plagioclase feldspars. (Photo by E. J. Tarbuck)

quartz are of the strong silicon-oxygen type. Consequently, quartz is hard, very resistant to weathering, and does not have cleavage. When broken, quartz generally exhibits conchoidal fracture. In a pure form, quartz is clear and if allowed to solidify without interference, will form hexagonal crystals which develop pyramidal-shaped ends (see chapter-opening photo). However, like most other clear minerals, quartz is often colored by the inclusion of various ions (impurities) and forms without developing good crystal faces. The most common varieties of quartz are milky (white), smokey (gray), rose (pink), amethyst (purple), and rock crystal (clear).

Clay is a term used to describe a variety of complex minerals which, like the micas, have a sheet structure. The clay minerals are generally very fine grained and can only be studied microscopically. Most clay minerals originate as products of the chemical weathering of the other silicate minerals. Thus, clay minerals make up a large percentage of the surface material we call soil. Because of the importance of soil in agriculture, and because of its role as a supporting material for buildings, clay minerals are extremely important to humans. One of the most common clay minerals is *kaolinite,* which is used in the manufacture of fine chinaware and occasionally, pottery.

NONSILICATE MINERALS

Although many are important from an economic standpoint, other mineral groups can be considered to be scarce when compared to the silicates. Table 2.3 lists examples of oxides, sulfides, sulfates, halides, and native elements of economic value. A discussion of a few of the more common nonsilicate, rock-forming minerals follows.

The carbonate minerals are much simpler structurally than the silicates. This mineral group is composed of the complex carbonate ion, (CO_3^{2-}), and one or

TABLE 2.3
Common nonsilicate mineral groups.

Group	Member	Formula	Economic Use
Oxides	Hematite	Fe_2O_3	Ore of iron
	Magnetite	Fe_3O_4	Ore of iron
	Corundum	Al_2O_3	Used as an abrasive
	Ice	H_2O	Solid form of water
Sulfides	Galena	PbS	Ore of lead
	Sphalerite	ZnS	Ore of zinc
	Pyrite	FeS_2	Fool's gold
	Chalcopyrite	$CuFeS_2$	Ore of copper
Sulfates	Gypsum	$CaSO_4 \cdot 2H_2O$	Used for plaster
	Anhydrite	$CaSO_4$	Used for plaster
Native elements	Gold	Au	Used for trade
	Copper	Cu	Used as an electrical conductor
	Diamond	C	Gemstone
	Sulfur	S	Used in numerous chemicals
	Graphite	C	Pencil lead and dry lubricant
Halides	Halite	$NaCl$	Common salt
	Fluorite	CaF_2	Used in steel making, chemicals, ceramics
Carbonates	Calcite	$CaCO_3$	Portland cement
	Dolomite	$CaMg(CO_3)_2$	Portland cement
	Malachite	$Cu_2(OH)_2CO_3$	Ore of copper

more positive ions. The two most common carbonate minerals are *calcite,* $CaCO_3$, and *dolomite,* $CaMg(CO_3)_2$. Because these minerals are quite similar both physically and chemically, they are difficult to distinguish from one another. Both have a vitreous luster, a hardness between 3 and 4, and nearly perfect rhombic cleavage. They can, however, be distinguished by using dilute hydrochloric acid. Calcite reacts vigorously with this acid, whereas dolomite will react only when powdered. Calcite and dolomite are usually found together as the primary constituents in the sedimentary rocks limestone and dolostone. When calcite is the dominant mineral, the rock is called limestone, whereas dolostone results from a predominance of dolomite. Limestone has numerous economic uses, including road aggregate, building stone, and as the main ingredient in portland cement.

Two other nonsilicate minerals frequently found in sedimentary rocks are *halite* and *gypsum.* Both minerals are commonly found in thick layers, which are the last vestiges of ancient seas that have long since evaporated. Halite is the mineral name for common table salt (NaCl). Gypsum ($CaSO_4 \cdot 2H_2O$) is the mineral from which plaster and other similar building materials are composed.

REVIEW QUESTIONS

1   Define the term *rock.*

2   List the three main particles of an atom and explain how they differ from one another.

3   If the number of electrons in an atom is 35 and its mass number is 80, calculate the following:
   (a)   The number of protons.
   (b)   The atomic number.
   (c)   The number of neutrons.

4   What is the octet rule? What is the significance of valence electrons?

5   Briefly distinguish between ionic and covalent bonding.

6   What occurs in an atom to produce an ion?

7   What is an isotope?

8   Although all minerals have an orderly internal arrangement of atoms (crystalline structure), most mineral samples do not demonstrate their crystal form. Why?

9   Why might it be difficult to identify a mineral by its color?

10   If you found a glassy-appearing mineral while rock hunting and had hopes that it was a diamond, what simple test might help you make a determination?

11   Explain the use of corundum as given in Table 2.3 in terms of Mohs hardness scale.

12   Gold has a specific gravity of almost 20. If a 25-liter pail of water weighs about 25 kilograms, how much would a 25-liter pail of gold weigh?

13   Explain the difference between the terms *silicon* and *silicate.*

14   What do ferromagnesian minerals have in common? List examples of ferromagnesian minerals.

15   What do muscovite and biotite have in common? How do they differ?

16   Should color be used to distinguish between orthoclase and plagioclase feldspar? What is the best means of distinguishing between the two types of feldspar?

17 Each of the following statements describes a silicate mineral or mineral group. In each case, provide the appropriate name.
   (a) The most common member of the amphibole group.
   (b) The most common nonferromagnesian member of the mica family.
   (c) The only silicate mineral made entirely of silicon and oxygen.
   (d) A high-temperature silicate with a name that is based on its color.
   (e) Characterized by striations.
   (f) Originates as a product of chemical weathering.

18 What simple test can be used to distinguish calcite from dolomite?

KEY TERMS

atom (p. 30)

atomic mass unit (p. 34)

atomic number (p. 30)

atomic weight (p. 34)

cleavage (p. 38)

color (p. 37)

compound (p. 30)

covalent bond (p. 33)

crystal form (p. 36)

electron (p. 30)

element (p. 30)

energy-level shell (p. 30)

fracture (p. 39)

hardness (p. 38)

ion (p. 33)

ionic bond (p. 32)

isotope (p. 34)

luster (p. 37)

mass number (p. 34)

metallic bond (p. 34)

mineral (p. 30)

Mohs scale (p. 38)

neutron (p. 30)

nucleus (p. 30)

octet rule (p. 32)

polymorph (p. 36)

proton (p. 30)

radioactivity (p. 34)

rock (p. 30)

silicate mineral (p. 39)

silicon-oxygen tetrahedron (p. 40)

specific gravity (p. 39)

streak (p. 37)

valence electron (p. 32)

3

IGNEOUS ROCKS

In our discussion of the rock cycle, it was pointed out that igneous rocks form when **magma** cools and crystallizes. This molten rock, which originates at depths as great as 200 kilometers within the earth, consists primarily of the elements found in silicate minerals, along with some gases, particularly water vapor, which are confined within the magma by the pressure of the surrounding rocks. Because the magma body is lighter than the surrounding rocks, it works its way toward the surface, and on occasion breaks through, producing a volcanic eruption (Figure 3.1). The spectacular explosions that sometimes accompany an eruption are produced by the gases (volatiles) escaping as the confining pressure lessens near the surface. Sometimes blockage of the vent coupled with surface water seepage into the magma chamber can produce catastrophic explosions. Along with ejected rock fragments, a volcanic eruption often generates extensive lava flows. **Lava** is similar to magma, except that most of the gaseous component has escaped. The rocks which result when lava solidifies are classified as **extrusive,** or **volcanic.** The magma not able to reach the surface eventually crystallizes at depth. Igneous rocks produced in this manner are termed **intrusive,** or **plutonic,** and would never be observed if not for the processes of erosion stripping away the overlying rocks.

Igneous rocks exposed in Yosemite National Park, California. (Photograph used by permission of Dennis Tasa)

CRYSTALLIZATION OF MAGMA

Because magma is a hot liquid, the ions that compose it move about freely and are said to be unordered. However, as magma cools, the random movements of the ions slow and the ions begin to arrange themselves into orderly patterns. This process is called **crystallization.** Before we examine crystallization in more detail, let us first examine how a simple crystalline solid melts. In a crystalline solid, the ions are arranged in a closely packed regular pattern. However, they are not without some motion. They exhibit a sort of restricted vibration about a fixed point. As the temperature rises, the ions vibrate more and more rapidly, and consequently collide with ever-increasing vigor with their neighbors. Continued heating causes the ions to occupy additional space. This results in expansion of the solid and greater distance between ions. When the melting point is reached, the ions are far enough apart and are vibrating rapidly enough to overcome the force of the chemical bonds which had joined them. At this stage, the ions are able to slide past one another, destroying their orderly crystalline structure. Thus, what was once a solid has become a liquid composed of unordered ions moving randomly about.

In the process of crystallization, cooling reverses the events of melting. As the temperature of the liquid drops, the ions pack closer together and begin to lose their freedom of movement. When cooling is suffi-

FIGURE 3.1
The volcano Parícutin a few months after its inception. (Photo courtesy of Tad Nichols)

cient, the force of the chemical bonds will again confine the atoms to an orderly crystalline arrangement. Usually, all of the molten material does not solidify at the same time. Rather, as it cools, numerous embryo crystals develop. In a systematic fashion, ions are added to these centers of crystal growth. When the crystals grow large enough that their edges meet, their growth ceases and crystallization continues elsewhere. Eventually, all of the liquid is transformed into a solid mass of interlocking crystals (Figure 3.2).

The rate of cooling strongly influences the crystallization process, in particular the size of the crystals. When a magma cools very slow, relatively few centers of crystal growth develop. Slow cooling also allows ions to migrate over relatively great distances. Consequently, slow cooling results in the formation of rather large crystals. On the other hand, when cooling occurs quite rapidly, the ions quickly lose their motion and readily combine. This results in the development of large numbers of nuclei which all compete for the available ions. The outcome is the formation of a solid mass formed of very small intergrown crystals.

When the molten material is quenched instantly, there is not sufficient time for the ions to arrange themselves into a crystalline network. Therefore, the solids produced in this manner consist of randomly distributed ions. Rocks that consist of unordered

FIGURE 3.2
Photomicrograph of interlocking crystals in a coarse-grained igneous rock. (Photo by A. H. Koschmann, U.S. Geological Survey)

atoms are referred to as **glass** and are quite similar to ordinary manmade glass.

The crystallization of a magma, although more complex, occurs in a manner similar to that just described. Rather than being composed of only one or two different elements, most magma consists of the eight elements that are the primary constituents of the silicate minerals. These include silicon, oxygen, aluminum, sodium, potassium, calcium, iron, and magnesium. In addition, trace amounts of many other elements, as well as volatiles, particularly water and carbon dioxide, are also found in magma. A *volatile* is a material that is commonly a gas at temperatures and pressures existing at the earth's surface.

When magma cools, it is generally the silicon and oxygen atoms that link together first to form silicon-oxygen tetrahedra. As cooling continues, the tetrahedra join with each other and with other ions to form crystal nuclei of the various silicate minerals. Each crystal nucleus grows as ion after ion is added to the crystalline network in an unchanging pattern. However, the minerals which compose a magma do not all form at the same time or under the same conditions. As we shall see, certain minerals crystallize at much higher temperatures than others. Consequently, magmas often consist of solid crystals surrounded by molten material.

In addition to the rate of cooling, the mineral composition of a magma and the amount of volatile material influence the crystallization process. Since magmas differ in each of these aspects, the physical appearance and mineral composition of igneous rocks vary widely. Nevertheless, it is possible to classify igneous rocks based on their mode of origin and mineral constituents. The environment during crystallization can be inferred from the size and arrangement of the mineral grains, a property called texture. Consequently, igneous rocks are most often classified by their texture and mineral composition. We will consider both of these rock characteristics in the following sections.

IGNEOUS TEXTURES

The term *texture,* when applied to an igneous rock, is used to describe the overall appearance of the rock

A.

B.

C.

D.

FIGURE 3.3
Igneous rock textures. **A.** Aphanitic. **B.** Phaneritic. **C.** Porphyritic. **D.** Glassy. (Photos by E. J. Tarbuck)

based on the size and arrangement of its interlocking crystals (Figure 3.3). Texture is a very important characteristic since it reveals a great deal about the environment in which the rock formed. This fact allows geologists to make inferences about a rock's origin while working in the field where sophisticated equipment is not available.

The most important factor affecting the texture of a rock is the rate at which the magma cooled. From our discussion of crystallization, we learned that rapid cooling produces small crystals, whereas very slow cooling results in the formation of much larger crystals. As we might expect, the rate of cooling is quite slow in magma chambers lying deep within the crust, while a thin layer of lava extruded upon the earth's surface may chill in a matter of hours, and small molten blobs ejected into the air during a violent eruption can solidify almost instantly.

Igneous rocks that form at the earth's surface or as small masses within the upper crust possess a very fine-grained texture termed **aphanitic.** By definition, the grains of aphanitic rocks are too small for individual minerals to be distinguished with the unaided eye (Figure 3.3A). Although mineral identification is not possible, fine-grained rocks are commonly characterized as being light, intermediate, or dark in color.

Using this system of grouping, light-colored aphanitic rocks are those composed primarily of light-colored nonferromagnesian silicate minerals, and so forth.

A common feature in many aphanitic rocks are the voids left by escaping gases (Figure 3.4). These spherical or elongated openings are called **vesicles** and are limited to the outer portion of lava flows (Figure 3.5). It is in the outer zone of a lava flow that

FIGURE 3.4
Vesicular texture. Vesicles form as gas bubbles escape near the top of a lava flow. (Photo by E. J. Tarbuck)

cooling occurs rapidly enough to "freeze" the lava, thereby preserving the openings produced by the escaping gas.

When large masses of magma solidify far below the surface, they form igneous rocks that exhibit a coarse-grained texture described as **phaneritic.** These coarse-grained rocks have the appearance of a mass of intergrown crystals, which are roughly equal in size and large enough so that the individual minerals can be identified with the unaided eye (Figure 3.3B). Because phaneritic rocks form deep within the crust, their exposure at the surface results only after erosion removes the overlying rocks that once surrounded the magma chamber.

A large mass of magma located at depth may require tens of thousands, even millions, of years to solidify. Since all minerals within a magma do not crystallize at the same rate or at the same time during cooling, it is possible for some to become quite large before others even start to form. If magma containing some large crystals should change environments, by erupting at the surface, for example, the molten portion of the lava would cool quickly. The resulting rock, which has large crystals embedded in a matrix of smaller crystals, is said to have a **porphyritic texture** (Figure 3.3C). The large crystals in such a rock

FIGURE 3.5
A vesicular lava which flowed from the base of Sunset Crater, Arizona. (Photo by E. J. Tarbuck)

are referred to as **phenocrysts,** while the matrix of smaller crystals is called **groundmass.** A rock which has such a texture is called a **porphyry.**

During some volcanic eruptions, molten rock is ejected into the atmosphere where it is quenched very quickly. Rapid cooling of this type may generate rock with a **glassy texture.** As was indicated earlier, glass results when the ions have not been permitted the time to unite into an orderly crystalline structure. *Obsidian,* a common type of natural glass, is similar in appearance to a dark chunk of manmade glass (Figure 3.3D).

Although the rate of cooling is the major factor determining the texture of an igneous rock, other factors are also important. In particular, the composition of the magma influences the resulting texture. For example, basaltic magma, which is very fluid, will usually generate crystalline rocks when cooled quickly in a thin lava flow. Under the same conditions, granitic magma, which is quite viscous (resists flow), is much more likely to produce a rock with a glassy texture. Consequently, most of the lava flows which are composed of volcanic glass are granitic in composition. However, when basaltic lava flows into the sea, its surface may be quenched rapidly enough to form a thin, glassy skin. Moreover, small ash fragments of basaltic composition are usually cooled rapidly enough to produce a glassy texture.

Some igneous rocks are formed from the consolidation of individual rock fragments that are ejected during a violent eruption. The ejected particles may be very fine ash, molten blobs, or large angular blocks which are torn from the walls of the vent during the eruption. Igneous rocks composed of these rock fragments are said to have a **pyroclastic texture.**

A common type of pyroclastic rock is composed of glass shards (thin strands) which remained hot enough during their flight to fuse together upon impact. Other pyroclastic rocks are composed of fragments that solidified before impact and became cemented together at some later time. Because pyroclastic rocks are made of individual rock fragments rather than interlocking crystals, their overall textures are often more similar to sedimentary rocks than to igneous rocks.

MINERAL COMPOSITION

The mineral makeup of an igneous rock is ultimately determined by the chemical composition of the magma from which it crystallized. Such a large variety of igneous rocks exists that it is logical to assume an equally large variety of magmas must also exist. However, geologists have found that various eruptive stages of the same volcano often extrude lavas exhibiting somewhat different mineral compositions, particularly if an extensive period of time separated the eruptions. Evidence of this type led them to look into the possibility that a single magma might produce rocks of varying mineral content.

A pioneering investigation into the crystallization of magma was carried out by N. L. Bowen in the first quarter of this century. Bowen discovered that as magma cools in the laboratory, certain minerals crystallize first. At successively lower temperatures, other minerals begin to crystallize as shown in Figure 3.6. As the crystallization process continues, the composition of the melt (liquid portion of a magma, excluding any solid material) continually changes. For example, at the stage when about 50 percent of the magma has solidified, the melt will be greatly depleted in iron, magnesium, and calcium, because these elements are found in the earliest-formed minerals. But at the same time, it will be enriched in the elements contained in the later-forming minerals, namely sodium and potassium. Further, the silicon content of the melt becomes enriched toward the latter stages of crystallization.

Bowen also demonstrated that if a mineral remained in the melt after it had crystallized, it would react with the remaining melt and produce the next mineral in the sequence shown in Figure 3.6. For this reason, this arrangement of minerals became known as **Bowen's reaction series.** On the upper left branch of this reaction series, olivine, the first mineral to form, will react with the remaining melt to become pyroxene. This reaction will continue until the last mineral in the series, biotite, is formed. This left branch is called a *discontinuous reaction series* because each mineral has a different crystalline structure. Recall that olivine is composed of single tetrahedra and that the other minerals in this sequence are composed of single chains, double chains, and sheet structures, respectively. Ordinarily, these reactions are not complete so that various amounts of each of these minerals may exist at any given time.

The right branch of the reaction series is a continuum in which the earliest formed calcium-rich feldspar crystals react with the sodium ions contained in the melt to become progressively more sodium rich. Oftentimes the rate of cooling occurs rapidly enough to prohibit the complete transformation of calcium-rich feldspar into sodium-rich feldspar. In these instances, the feldspar crystals will have calcium-rich interiors surrounded by zones that are progressively richer in sodium.

During the last stage of crystallization, after most of the magma has solidified, the remaining melt will form the minerals quartz, muscovite, and potassium feldspar. Although these minerals crystallize in the order shown, this sequence is not a true reaction series.

Bowen demonstrated that minerals crystallize from magma in a systematic fashion. But how does Bowen's reaction series account for the great diversity of igneous rocks? It appears that at one or more stages in the crystallization process, a separation of the solid and liquid components of a magma frequently occurs. This can happen, for example, if the earlier-formed minerals are heavier than the liquid portion and settle to the bottom of the magma chamber as shown in Figure 3.7A. This settling is thought to occur frequently with the dark silicates, such as olivine. When the remaining melt crystallizes, either in place or in a new location if it migrates out of the chamber, it will form a rock with a chemical composition much different from the original magma (Figure 3.7B). In many instances the melt which has migrated from the initial magma chamber will undergo further segregation. As crystallization progresses in the "new" magma, the solid particles may accumulate into rocklike masses surrounded by pockets of the still

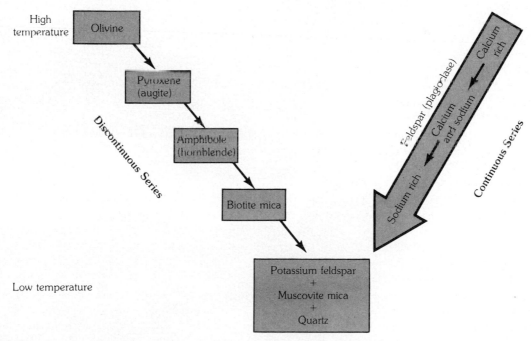

FIGURE 3.6
Bowen's reaction series shows the sequence in which minerals crystallize from a magma. Compare this figure to the mineral composition of the rock groups in Table 3.1. Note that each rock group consists of minerals that crystallize at the same time.

molten material. It is very likely that some of this melt will be squeezed from the mixture into the cracks which develop in the surrounding rock. This process will generate an igneous rock of yet another composition.

The process involving the segregation of minerals by differential crystallization and separation is called **fractional crystallization.** At any stage in the crystallization process the melt might be separated from the solid portion of the magma. Consequently, fractional crystallization can produce igneous rocks having a wide range of compositions.

Bowen successfully demonstrated that through fractional crystallization one magma can generate several different igneous rocks. However, more re-

cent work has indicated that this process cannot account for the relative quantities of the various rock types known to exist. Although more than one rock type can be generated from a single magma, apparently other mechanisms also exist to generate magmas of quite varied chemical compositions. We will examine some of these mechanisms at the end of the next chapter.

NAMING IGNEOUS ROCKS

As was stated previously, igneous rocks are most often classified, or grouped, on the basis of their texture and mineral composition. The various igneous

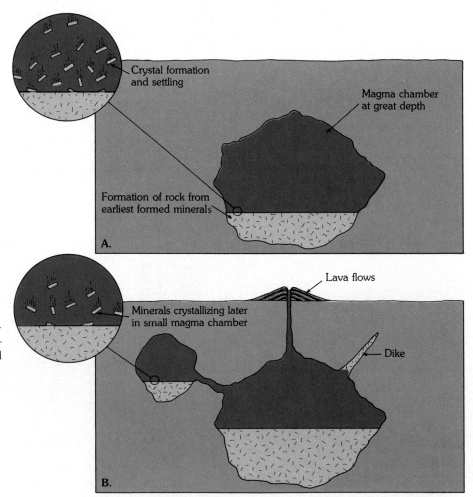

FIGURE 3.7
Separation of minerals by fractional crystallization. **A.** Illustration of how the earliest-formed minerals can be separated from a magma by settling. **B.** The remaining melt could migrate to a number of different locations and, upon further crystallization, generate rocks having a composition much different from the parent magma.

textures result from different cooling histories, while the mineral composition of an igneous rock is the consequence of the chemical makeup of the parent magma and the environment of crystallization. As we might expect from the results of Bowen's work, minerals that crystallize under similar conditions are most often found together composing the same igneous rock. Hence, the classification of igneous rocks closely corresponds to Bowen's reaction series (Figure 3.6).

The first minerals to crystallize—calcium feldspar, pyroxene, and olivine—are high in iron, magnesium, and calcium, and low in silicon. Basalt is a common extrusive rock of this composition; thus, the term *basaltic* is used to denote rocks of this type. Because of their iron content, basaltic rocks are typically darker in color and slightly heavier than other igneous rocks commonly found at the earth's surface.

Among the last minerals to crystallize are potassium feldspar and quartz. Igneous rocks in which these two minerals predominate are referred to as having a *granitic* composition. Intermediate igneous rocks are made up of minerals found near the middle of Bowen's reaction series. Amphibole along with the intermediate plagioclase feldspars are the main constituents of this rock group. We will refer to those rocks that have a composition between that of granite and basalt as being *andesitic.*

Although each of the basic rock groups is composed mainly of minerals located in a specific region of Bowen's reaction series, other constituents are usually present in lesser amounts. For example, granitic rocks are composed mainly of quartz and potassium feldspar (K feldspar), but may also contain muscovite, biotite, amphibole, and sodium feldspar (Na feldspar). See Table 3.1 on page 62.

This discussion has concentrated on only three mineral compositions, yet it is important to note that gradations among these types also exist (Figure 3.8). For example, an abundant intrusive igneous rock called *granodiorite* has a mineral composition between that of rocks with a granitic composition and those with an andesitic composition. Another important igneous rock called *peridotite* contains mostly olivine and thus falls near the very beginning of Bowen's reaction series. Peridotite is believed to be a major constituent of the upper mantle.

An important aspect of the mineral composition of igneous rocks is silica (SiO_2) content. Recall that most of the minerals in igneous rocks contain some silica. Typically, the silica content of crustal rocks ranges from a low of 50 percent in basaltic rocks to a high of about 70 percent in granitic rocks. The percentage of silica in igneous rocks actually varies in a systematic

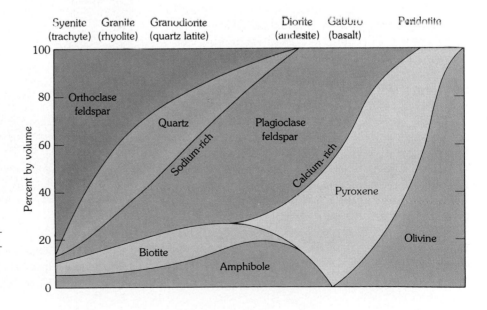

FIGURE 3.8
Mineralogy of the common igneous rocks. Parentheses indicate the name of equivalent extrusive rock. (After Turekian).

manner which parallels the abundance of the other elements. For example, rocks low in silica contain large amounts of calcium, iron, and magnesium. Consequently, the chemical makeup of an igneous rock can be inferred directly from its silica content. Further, the amount of silica present in magma strongly influences its behavior. Granitic magma, which has a high silica content, is quite viscous and exists as a fluid at temperatures as low as 800°C. On the other hand, basaltic magmas are low in silica and generally quite fluid. Basaltic magmas are also extruded at much higher temperatures, often 1200°C or higher.

It is not at all uncommon for two rocks to have the same mineral constituents and yet have different names. This resulted, in part, from the fact that many igneous rocks have ancient names which were given to them on the basis of their overall appearance rather than on their mineral composition. For example, the coarse-grained intrusive rock granite has a fine-grained volcanic equivalent called rhyolite. Although these rocks are mineralogically the same, they have different textures and do not look at all alike. Therefore the rocks were given different names (Figure 3.9).

GRANITIC ROCKS

Granite is perhaps the best known of all the igneous rocks (Figure 3.9A). This is partly because of its natural beauty, which is enhanced when it is polished, and partly because of its abundance. Slabs of polished granite are commonly used for tombstones, monuments, and as building stones.

Granite is a phaneritic rock composed of up to 25 percent quartz and over 50 percent potassium feldspar and sodium-rich feldspar. The quartz crystals, which are roughly spherical in shape, are most often clear to light gray in color. In contrast to quartz, the feldspar crystals in granite are not as glassy, but rectangular in shape and generally salmon pink to white in color. Other common constituents of granite are muscovite and the dark silicates, particularly biotite and amphibole. Although the dark components of granite make up less than 20 percent of most samples, dark minerals appear to be more prominent than their percentage would indicate. In some granites, K feldspar is dominant and very dark pink in

color, so that the rock appears almost reddish. This variety is very popular as a building stone. However, most often the feldspar grains are white, so that when viewed at a distance granite appears light gray in color. Granite may also have a porphyritic texture, in which feldspar crystals a centimeter or more in length are scattered among a coarse-grained groundmass of quartz and amphibole.

Granite is often produced by the processes which

A.

B.

FIGURE 3.9
A. Granite, one of the most common coarse-grained igneous rocks. B. Rhyolite, the fine-grained equivalent of granite is far less abundant. (Photos by E. J. Tarbuck)

generate mountains. Because granite is a by-product of mountain building and is very resistant to weathering and erosion, it frequently forms the core of eroded mountains. For example, Pikes Peak in the Rockies, Mount Rushmore in the Black Hills, the White Mountains of New Hamsphire, Stone Mountain in Georgia, and Yosemite National Park in the Sierra Nevada are all areas where large quantities of granite are exposed at the surface (Figure 3.10). As we can see from these examples, granite is a very abundant rock. However, it has become common practice among geologists to apply the term *granite* to any coarse-grained intrusive rock composed predominately of light silicate minerals. We will follow this practice for the sake of simplicity. The student should keep in mind that this use of the term *granite* covers rock having a range of mineral compositions.

Magma having a granitic composition contains up to 5 percent water. Because water will not crystallize in the magma chamber, it can make up a much higher percentage of the melt during the final phase of solidification. Crystallization in a water-rich environment, where ion migration is enhanced, is believed to result in the formation of crystals several centimeters, or even a few meters, in length. The resulting rocks, called **pegmatites,** are composed of unusually large crystals.

Some of the largest crystals ever uncovered have been found in pegmatites. Feldspar masses the size of houses have been quarried from a pegmatite located in North Carolina. Gigantic hexagonal crystals of muscovite measuring a few meters across have been found in Ontario, Canada. In the Black Hills, crystals as large as telephone poles of the lithium-bearing

FIGURE 3.10
Yosemite National Park, located in the Sierra Nevada, is one of many areas where vast amounts of granite are exposed at the surface. (Photograph used by permission of Dennis Tasa)

TABLE 3.1
Common igneous rocks.

	Granitic	Andesitic	Basaltic
Intrusive	Granite	Diorite	Gabbro
Extrusive	Rhyolite	Andesite	Basalt
Mineral Composition	Quartz Potassium feldspar Sodium feldspar	Amphibole Intermediate plagioclase feldspar Biotite	Calcium feldspar Pyroxene
Minor Mineral Constituents	Muscovite Biotite Amphibole	Pyroxene	Olivine Amphibole

mineral spodumene have been mined. The largest of these was more than 12 meters long. Not all pegmatites contain such large crystals, but these examples emphasize the special conditions that must exist during the formation of pegmatites.

Although most pegmatites are granitic in composition and consist of unusually large crystals of quartz, feldspar, and muscovite, pegmatites of other compositions also exist. Some granitic pegmatites are commercially valuable. The feldspar is used in the production of ceramics and the muscovite is used for isinglass, electrical insulation, and glitter. Further, because pegmatites form at the end of the crystallization process, they often contain some of the least abundant elements. Thus, some rare minerals may also be found in pegmatites. In addition to the common silicates, some pegmatites contain semiprecious gems such as beryl, topaz, and tourmaline. Also, minerals containing the elements lithium, cesium, uranium, and the rare earths[1] are occasionally found. Most pegmatites are located within large igneous masses or as veins which cut into the rock that surrounds the magma chamber. In the latter case, hydrothermal (hot water) solutions are thought to have deposited the minerals in cracks that penetrated into the country rock.

Rhyolite is the volcanic equivalent of granite. Like granite, rhyolite is composed primarily of the light-

colored silicates (Figure 3.9B). This fact accounts for its color, which is usually buff to pink or occasionally very light gray. Rhyolite is usually aphanitic and frequently contains glassy fragments and voids indicating rapid cooling in a surface environment. In those instances when rhyolite contains phenocrysts, they are usually small and composed of either quartz or potassium feldspar. In contrast to granite, rhyolite is rather uncommon. Yellowstone Park is one well-known exception. Here rhyolitic lava flows and ash deposits of similar composition are widespread.

Obsidian is a dark-colored, glassy rock which forms when lava is quenched very quickly (Figure 3.11A). In contrast to the orderly arrangement of ions that is characteristic of minerals, the ions in glass are unordered. Consequently, glassy rocks like obsidian are not composed of minerals in the same sense as most other rocks.

Although usually black or reddish-brown in color, obsidian has a high silica content. Thus, its composition is more akin to the light igneous rocks such as granite than to the dark rocks of basaltic composition. By itself, silica is clear like window glass; the dark color results from the presence of metallic ions. If you examine a thin edge of a piece of obsidian, it will be nearly transparent. Because of its excellent conchoidal fracture, obsidian was a prized material from which the American Indians made arrowheads and cutting tools.

Pumice is a volcanic rock which, like obsidian, has a glassy texture. Usually found with obsidian, pumice forms when large amounts of gas escape through

---

[1]The rare earths are a group of fifteen elements (atomic numbers 57 through 71) that possess similar properties. They are useful catalysts in petroleum refining and are used to improve color retention in television picture tubes.

A.

B.

FIGURE 3.11
A. Obsidian, a glassy volcanic rock. (Courtesy of Ward's Natural Science Establishment, Inc., Rochester, N.Y.). **B.** Pumice, a glassy rock containing numerous tiny voids. (Photo by E. J. Tarbuck)

FIGURE 3.12
Andesite porphyry, a common volcanic rock. (Photo by E. J. Tarbuck)

lava to generate a gray, frothy mass (Figure 3.11B). This material is similar to the foam which flows from a newly opened bottle of champagne. In some samples, the voids are quite noticeable, while in others, the pumice resembles fine shards of intertwined glass. Because of the large percentage of voids, many samples of pumice will float when placed in water. Oftentimes flow lines are visible in pumice, indicating some movement before solidification was complete. Moreover, pumice and obsidian often form in the same rock mass, where they exist in alternating layers.

ANDESITIC ROCKS

Andesite is a medium gray, fine-grained rock of volcanic origin. Its name comes from the Andes Mountains where numerous volcanoes are composed of this rock type. In addition to the volcanoes of the Andes, many of the volcanic structures encircling the Pacific Ocean are of andesitic composition. Andesite quite commonly exhibits a porphyritic texture (Figure 3.12). In these cases, the phenocrysts are often light, rectangular crystals of plagioclase feldspar or black, elongated hornblende crystals.

Diorite is a coarse-grained intrusive rock that looks somewhat similar to gray granite. However, it can be distinguished from granite by the absence of visible quartz crystals. The mineral makeup of diorite is primarily sodium-rich plagioclase and amphibole, with lesser amounts of biotite. Because the white feldspar grains and dark amphibole crystals are roughly equal

in abundance, diorite has a "salt and pepper" appearance.

BASALTIC ROCKS

Basalt is a very dark green to black, fine-grained volcanic rock composed primarily of pyroxene and calcium-rich feldspar, with lesser amounts of olivine and amphibole present (Figure 3.13). When porphyritic, basalt commonly contains small, light-colored calcium feldspar phenocrysts or glassy-appearing olivine phenocrysts embedded in a dark groundmass.

Basalt is the most common extrusive igneous rock. Many volcanic islands, such as the Hawaiian Islands and Iceland, are composed mainly of basalt. Further, the upper layers of the oceanic crust consist of basalt. In the United States, large portions of central Oregon and Washington were the sites of extensive basaltic outpourings (see Figure 4.21). At some locations these once fluid basaltic flows have accumulated to thicknesses approaching 2 kilometers.

Gabbro is the intrusive equivalent of basalt. Like basalt, it is very dark green to black in color and composed primarily of pyroxene and calcium-rich plagioclase. Although gabbro is not a common constituent of the continental crust, it undoubtedly makes up a significant percentage of the oceanic crust. Here large portions of the magma found in underground reservoirs that once fed basalt flows eventually solidified at depth to form gabbro.

PYROCLASTIC ROCKS

Pyroclastic rocks are those which form from fragments ejected during a volcanic eruption. One of the most common pyroclastic rocks, called *tuff,* is composed of tiny ash-sized fragments which were later cemented together. In situations where the ash particles remained hot enough to fuse, the rock is generally called *welded tuff.* Since welded tuffs consist of glass shards, their appearance may closely resemble pumice. Deposits of partially welded tuffs are easily quarried and used as a durable building material. Several villages in Cappadocia in central Turkey, which date as far back as the fourth century, have

FIGURE 3.13

Lava of basaltic composition from Kilauea Caldera, Hawaii. (Photo by R. B. Moore, U.S. Geological Survey)

been carved into vertical cliffs composed of this material.

Pyroclastic rocks composed of particles larger than ash are called *volcanic breccia* (Figure 3.14). The particles in volcanic breccia can consist of streamlined fragments that solidified in air, blocks broken from the walls of the vent, crystals, and glass fragments. Unlike the other igneous rock names, the terms *tuff* and *volcanic breccia* do not denote mineral composition.

FIGURE 3.14

Volcanic breccia. Fragments of solidified lava are torn from the wall of the vent and incorporated into the erupting molten mass. (Photo by E. J. Tarbuck)

OCCURRENCE OF IGNEOUS ROCKS

Although volcanic eruptions can be among the most violent and spectacular events in nature and therefore worthy of detailed study, most magma is believed to be emplaced at depth. Thus, an understanding of intrusive igneous activity is as important to geologists as the study of volcanic events. The structures that result from the emplacement of igneous material at depth are called **plutons.** Since all plutons form out of our view beneath the earth's surface, they can be studied only after uplifting and erosion have exposed them. The challenge lies in reconstructing the events that generated these structures millions or even hundreds of millions of years ago.

For the sake of clarity, we have separated our discussions of volcanism and plutonic activity. Volcanism will be treated in the following chapter; here we will concentrate on plutonic activity. Keep in mind, however, that these diverse processes occur simultaneously and involve basically the same earth materials.

NATURE OF PLUTONS

Plutons are known to occur in a great variety of sizes and shapes. Some of the most common types are illustrated in Figure 3.15. Notice that some of these structures have a tabular shape, while others are quite massive. Also, observe that some of these bodies cut across existing structures, such as the layering of sedimentary beds, while others form when magma is injected between sedimentary layers. Because of these differences, intrusive igneous bodies are generally classified according to their shape as either **tabular** or **massive,** and by their orientation with respect to the country (host) rock. Plutons are said to be **discordant** if they cut across existing structures and **concordant** if they form parallel to the existing structures. Further, as we can see in Figure 3.15, plutons are closely associated with volcanic activity. The largest intrusive bodies are thought to be the remnants of magma chambers which possibly fed volcanoes.

Dikes Dikes are discordant masses that are produced when magma is injected into fractures. The force exerted by the emplaced magma can be great enough to further separate the walls of the fracture. Once crystallized, these tabular structures have thicknesses ranging from less than a centimeter to more than a kilometer. The largest have lengths of a hundred kilometers or more. Most dikes, however, are a few meters thick and extend laterally for no more than a few kilometers. Dikes are often oriented vertically and represent pathways followed by molten rock which fed ancient lava flows. Some dikes end abruptly at depth; still others terminate at plutons.

Dikes may weather more slowly than the surrounding rock. When exposed, these dikes have the appearance of a wall as shown in Figure 3.16. Dikes are often found radiating, like spokes on a wheel, from an eroded volcanic neck (see Figure 4.18). In these situations the active ascent of magma is thought to have generated stress fractures in the volcanic cone.

Sills Sills are tabular plutons formed when magma is injected along sedimentary bedding surfaces (Figure 3.17). Horizontal sills are the most common,

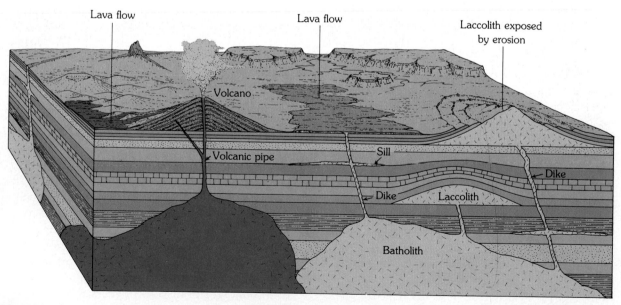

FIGURE 3.15
Cross-sectional view illustrating the basic intrusive igneous structures.

although all orientations, even vertical, are known to exist. Because of their relatively uniform thickness and large extent, sills are believed to form from very fluid lava. As we may expect, sills most often are composed of basaltic magma, which is typically quite fluid.

The emplacement of a sill requires that the overlying sedimentary rock be lifted to a height equal to the thickness of the sill. Although this seems to be a formidable task, it may require less energy than forcing the magma up the remaining distance to the surface. Consequently, sills form only at rather shallow depths where the pressure exerted by the weight of overlying strata is relatively low. Although sills are intruded between existing layers, they need not be concordant along their entire extent. Large sills frequently cut across sedimentary layers and resume their concordant nature at a higher level.

FIGURE 3.16
The tabular structure in the foreground is a dike. In the background is Shiprock, New Mexico, the remnant of a pipe which once fed a volcano that has long since eroded away. (Courtesy of Ward's Natural Science Establishment, Inc., Rochester, N.Y.)

FIGURE 3.17
Banks Island, Canada. The dark, essentially horizontal band is a sill of basaltic composition that intruded into flat-lying sedimentary rock. (Courtesy of the Geological Survey of Canada, photo no. 131185)

One of the largest and best-known sills in the United States is the Palisades Sill, which is exposed along the west shore of the Hudson River in southeastern New York and northeastern New Jersey. This sill is about 300 meters thick and, due to its resistant nature, has formed an imposing cliff that can be seen easily from the opposite side of the Hudson. Because of its great thickness and subsequent slow rate of crystallization, the Palisades Sill provides geologists with an excellent example of magmatic differentiation. The sill formed from magma rich in the minerals olivine, pyroxene, and plagioclase. Olivine, the first and heaviest of these minerals to crystallize, sank toward the bottom and makes up about 25 percent of the lower portion of the sill. By contrast, near the top of the sill, olivine represents only about one percent of the rock mass. Conversely, the lightest mineral of this group, plagioclase, floated toward the top and comprises nearly two-thirds of the upper portion of the sill. Examples such as the Palisades are important to geologists because they confirm the results obtained in the laboratory, where the actual conditions found in nature can only be approximated.

In many respects, sills closely resemble buried lava flows. Both are tabular and often exhibit *columnar jointing* (Figure 3.18). Further, because sills form in near-surface environments and may only be a few meters thick, the emplaced magma is often chilled quickly enough to generate a fine-grained texture. When attempts are made to reconstruct the geologic history of a region, it becomes important to differentiate between sills and buried lava flows. Fortunately, under close examination these two structures can be readily distinguished. The upper portion of a buried lava flow usually contains voids produced by escaping gases and only the rocks beneath a lava flow show evidence of metamorphic alteration. Sills, on the other hand, form when magma has been forcefully intruded between sedimentary layers. Inclusions of adjacent country rock found within the upper and lower zones of the structure and "baked" zones above and below are trademarks of a sill.

Laccoliths   Laccoliths are similar to sills because they form when magma is intruded between sedimentary layers in a near-surface environment. However, unlike sills, the magma that generates laccoliths is believed to be quite viscous. This thick, nonfluid magma collects as a lens-shaped mass that arches the overlying strata upward (see Figure 3.15). Consequently, a laccolith can be detected because of the dome it creates at the surface even before the overlying rock is stripped away by erosional forces.

Most large laccoliths are probably not much wider than a few kilometers. The Henry Mountains in southeastern Utah are composed of several large laccoliths believed to have been fed by a much larger magma body emplaced nearby. Some geologists also consider the well-known structure called Devil's Tower, located in eastern Wyoming, to be the remnant of a laccolith.

Batholiths   By far the largest intrusive igneous bodies are **batholiths.** The largest batholiths are linear structures several hundred kilometers long and nearly one hundred kilometers wide as shown in Figure 3.19. The Idaho batholith, for example, encompasses an area of more than 40,000 square kilometers. Indirect evidence gathered from gravitational studies indicates that batholiths are also very thick, possibly even extending through most of the crust. Based on the amount exposed by erosion, some batholiths are at least several kilometers thick. By definition, a plutonic mass must have an aerial extent of over 80 square kilometers (30 square miles) to be considered a batholith. Smaller plutons of this type are termed **stocks.** Many stocks appear to be portions of batholiths that are not yet fully exposed. Other stocks are believed to be smaller plutons formed separately from the main magma body.

Batholiths are usually composed of rock types having chemical compositions near the granitic end of the spectrum, although diorite is also found. Small batholiths can be rather simple structures composed almost entirely of one rock type. However, studies of large batholiths have shown that they resulted from several distinct events that occurred over a period of millions of years. The plutonic activity which created the Sierra Nevada batholith, for example, is thought to have occurred as five separate events over a 130-million-year period which ended about 80 million years ago (Figure 3.20).

Batholiths frequently compose the core of mountain systems. Here uplifting and erosion have removed the surrounding rock, thereby exposing the resistant igneous body. Some of the highest mountain peaks, such as Mount Whitney in the Sierra Nevada, are carved from such a granitic mass. Large expanses of granitic rock are also exposed in the stable interiors of the continents, such as the Canadian Shield of North America. These relatively flat outcrops are believed to be the remnants of ancient mountains that have long since been leveled by erosion. Thus, the rocks composing the batholiths of

youthful mountain ranges were generated near the top of a magma chamber, whereas in shield areas, the roots of former mountains and the lower portions of batholiths are exposed. We will consider the role of igneous activity as it relates to mountain building further in Chapter 18.

EMPLACEMENT OF BATHOLITHS

One ongoing and interesting debate in geology concerns the emplacement of granitic batholiths. One

FIGURE 3.18
Devil's Post Pile National Monument, California, exhibits columnar joints and the columns that result. These five- to seven-sided columns are the consequence of contraction that occurs as a relatively thin layer of molten rock cools. (Courtesy of the National Park Service, U.S. Department of the Interior)

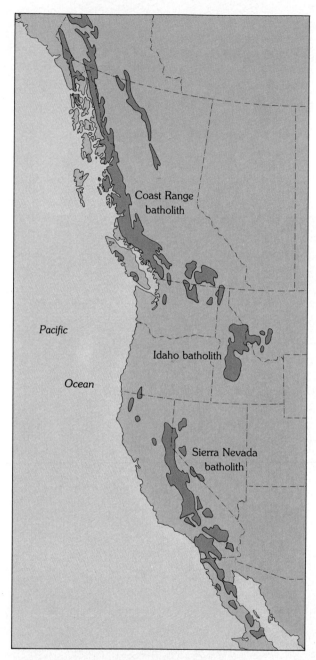

FIGURE 3.19
Location of granitic batholiths that occur along the western margin of North America. These gigantic, elongated bodies were emplaced during the last 100 million years of earth history.

Coast Range batholith

Pacific

Ocean

Idaho batholith

Sierra Nevada batholith

group of geologists supports the idea that batholiths formed from magma that migrated upward from great depths. This idea, however, presents a space problem. What happened to the rock originally in the location now occupied by these igneous masses? Further, the problem of explaining how magma is able to force its way through several kilometers of solid rock also plagued those supporting the magmatic origin of batholiths. The group opposing the magmatic origin hypothesis has suggested that the granite in batholiths originated when hot, ion-rich fluids and gases migrated through the rock and chemically altered the rock's composition. The process of converting country rock into granite is called **granitization.** Although granitization undoubtedly generates small quantities of granite, the strongest evidence points to a magmatic origin for the largest intrusive bodies.

This controversy was resolved, for many at least, when careful studies were made of structures called *salt domes.* These structures are of economic importance as they are found in close association with major oil-producing areas in the Gulf Coast states and the Persian Gulf. Salt domes are produced in regions where extensive salt deposits were subsequently buried by thousands of meters of sediment. The salt, which is less dense than the overlying sediments, migrates very slowly upward. This is possible because salt behaves like a mobile fluid when it is subjected to differential stress over a long period of time. Since salt beds are not perfectly uniform, the zone of upward movement is thought to originate at a high spot along the layer. As the salt moves slowly upward the stress exerted on the overlying sediments causes them to mobilize and be pushed aside. Some of the displaced sediment will move downward to occupy the space made available by the rising salt dome as shown in Figure 3.21A on page 72. Occasionally the salt breaches the surface, where it begins to flow outward not unlike a very thick lava flow.

It is now generally accepted that batholiths are emplaced in a manner similar to the formation of salt domes (Figure 3.21B). Because magma is less dense than the overlying rock, its buoyancy propels it upward. Also, like a salt dome, the mobile magma forcibly makes room for itself by pushing aside the country rocks. As the magma moves upward, some of the country rock which was shouldered aside will fill in the space left by the magma body as it passes. An analogous situation occurs when a can of oil-based paint is left in storage. The oil in the paint is less dense than the pigments used for coloration; thus, oil collects into drops that slowly migrate upward while the heavier pigments settle to the bottom. In the case of ascending granitic magma, gradual cooling results in a loss of mobility. Thus, much of the magma crystallizes at depth to form granite batholiths rather than extruding at the surface as a volcanic eruption.

The upper portions of batholiths often contain unmelted remnants of the country rock that are called **xenoliths.** These inclusions indicate that yet another process may operate during the emplacement of a batholith, at least in a near-surface environment where rocks are brittle. As the magma buoys upward, stress is believed to cause numerous cracks in the overlying rock. The force of the injected magma is strong enough to dislodge blocks of the surrounding rock and incorporate them into the magma body. However, this process of assimilating country rock is only minor compared to the earlier mentioned activity of mobilizing and displacing country rock.

FIGURE 3.20
North Dome and Basket Dome in Yosemite National Park are only two of many dome-shaped structures that comprise the Sierra Nevada Batholith. (Courtesy of the National Park Service)

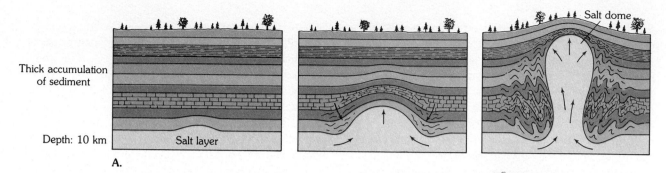

Thick accumulation of sediment

Depth: 10 km — Salt layer

A.

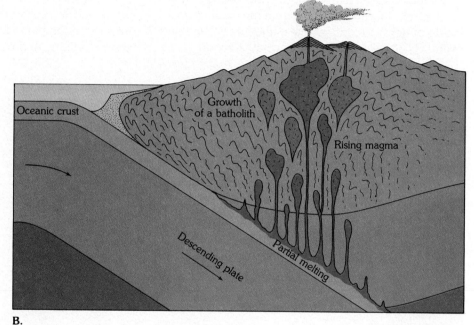

Oceanic crust

Growth of a batholith

Rising magma

Descending plate

Partial melting

B.

FIGURE 3.21
Many geologists believe that the emplacement of a salt dome (Part A) is analogous to the processes involved in the emplacement of large magma bodies (Part B).

REVIEW QUESTIONS

1 How does lava differ from magma?

2 How does the rate of cooling influence the crystallization process?

3 In addition to the rate of cooling, what other factors influence the crystallization process?

4 The classification of igneous rocks is based largely upon two criteria. Name these criteria.

5 The statements that follow relate to terms describing igneous rock textures. For each statement, identify the appropriate term.
   (a) Openings produced by escaping gases.
   (b) Obsidian exhibits this texture.
   (c) A matrix of fine crystals surrounding phenocrysts.
   (d) Crystals are too small to be seen with the unaided eye.
   (e) A texture characterized by two distinctively different crystal sizes.
   (f) Coarse grained, with crystals of roughly equal size.

6 What does a porphyritic texture indicate about an igneous rock?

7 What is fractional crystallization? How might fractional crystallization lead to the formation of several different igneous rocks from a single magma?

8 Relate the classification of igneous rocks to Bowen's reaction series.

9 How are granite and rhyolite different? In what way are they similar?

10 Why are the crystals in pegmatites so large?

11 Compare and contrast each of the following pairs of rocks:
   (a) Granite and diorite.
   (b) Basalt and gabbro.
   (c) Andesite and rhyolite.

12 How do tuff and volcanic breccia differ from other igneous rocks such as granite and basalt?

13 What name is given to a tabular, discordant pluton?

14 Why might a laccolith be detected at the earth's surface before being exposed by erosion?

15 What is the largest of all plutons? Is it tabular or massive? Concordant or discordant?

16 Relate the mechanism of batholith emplacement to the formation of salt domes.

KEY TERMS

aphanitic texture (p. 54)

batholith (p. 68)

Bowen's reaction series (p. 56)

concordant (p. 66)

crystallization (p. 52)

dike (p. 66)

discordant (p. 66)

extrusive (p. 51)

fractional crystallization (p. 58)

granitization (p. 70)

glass (p. 53)

glassy texture (p. 56)

groundmass (p. 56)

intrusive (p. 51)

laccolith (p. 68)

lava (p. 51)

magma (p. 51)

massive (p. 66)

pegmatite (p. 61)

phaneritic texture (p. 55)

phenocryst (p. 56)

pluton (p. 65)

plutonic (p. 51)

porphyritic texture (p. 55)

porphyry (p. 56)

pyroclastic texture (p. 56)

sill (p. 66)

stock (p. 68)

tabular (p. 66)

vesicle (p. 55)

volcanic (p. 51)

xenolith (p. 71)

4

VOLCANIC
ACTIVITY

A t 8:32 A.M. on Sunday, May 18, 1980, one of the largest volcanic eruptions to occur in North America in recent times transformed a picturesque volcano into a decapitated remnant (Figure 4.1). On this date in southwestern Washington state, Mount St. Helens erupted with a force hundreds of times greater than that of the atomic bombs dropped on Japan during World War II. The blast blew out the entire north flank of the volcano, leaving a gaping hole. A once prominent volcano that had grown to more than 2900 meters had, in one brief moment, been lowered by about 410 meters.

The early morning blast totally devastated a wide swath of timber-rich land on the north side of the mountain (see Figure 4.2 on page 80). Trees within a 400 square kilometer area lay intertwined and flattened, stripped of their branches and appearing from the air like toothpicks strewn about. The immense force caused trees as far away as 25 kilometers to topple. The gases and ash unleashed from the volcano had temperatures that probably exceeded 800°C! Thirty-six persons were killed and 23 others were listed as missing. Some died from the intense heat and the suffocating cloud of ash and gases. Others perished as they were hurled from the mountain by the force of the blast. Still others were trapped by debris-laden mudflows.

Vicinity of Mount St. Helens. Mount Rainier appears on the horizon. (Photo by Jim Hughes, USDA Forest Service)

The blast and accompanying mudflows carried ash, trees, and water-saturated rock debris 29 kilometers down the Toutle River. The river quickly became a mud-filled torrent and reached depths of 60 meters in some places. Further, a debris dam was deposited at the outlet of Spirit Lake, causing its level to rise by more than 30 meters. For several days, the threat of pent-up waters breaching the dam posed another potential hazard.

The eruption of May 18th ejected an estimated three to four cubic kilometers of ash and rock debris. By comparison, this is roughly equal to the quantity of ash that buried the city of Pompeii during the historic eruption of Mount Vesuvius in 79 A.D.

Following the devastating explosion, Mount St. Helens continued to emit great quantities of hot gases and ash. Only minutes after the eruption began, a dark plume rose from the volcano. The force of the blast was so strong that some of the ash was propelled high into the stratosphere, more than 18,000 meters above the ground. During the next hours and days, this very fine-grained material was carried great distances by the strong upper air winds. Measurable deposits were reported from as far away as Okla-

FIGURE 4.1
Before and after photographs showing the transformation of Mount St. Helens caused by the May 18, 1980 eruption. ("Before" photo by Roland V. Emetaz, courtesy of USDA Forest Services; "after" photo by Jim Hughes).

homa and Minnesota. Meanwhile, the ash fallout in the immediate vicinity accumulated to depths exceeding 2 meters, and the air over Yakima, Washington, 130 kilometers to the east, was so filled with ash that residents experienced midnight-like darkness at noon. Crop damage from the volcanic fallout was reported as far away as central Montana.

The events leading to the May 18th eruption began about two months earlier, on the 20th of March, as a series of minor earth tremors centered beneath the awakening mountain. The first volcanic activity took place on March 27th, when a small amount of ash and steam rose from the summit. Over the next several weeks, sporadic eruptions of varied intensity occurred.

Prior to the main eruption, the primary concern had been the potential hazard of mudflows. These moving lobes of saturated debris were created when ice and snow were melted by heat from the magma within the volcano. The only sign of a potentially hazardous eruption was a bulge on the volcano's north flank. Careful monitoring of this dome-shaped structure indicated a very slow but steady growth rate of a few meters per day. Geologists monitoring the

FIGURE 4.2
Forest lands and logging truck, 12 kilometers (7.5 miles) northwest of Mount St. Helens destroyed by the lateral blast of May 18, 1980. (Photo by R. P. Hoblitt, U.S. Geological Survey)

FIGURE 4.3
Sequence of events in the May 18, 1980 eruption of Mount St. Helens. **A.** Prior to May 18, minor eruptions covered the winter's snow and made the volcano appear dark. **B.** The eruption began when a minor earthquake caused the bulge on the north slope to slide down toward Spirit Lake. Some of the landslide material is visible in the lower right.

activity suggested that if the growth rate of the bulge changed appreciably, an eruption might quickly follow. Unfortunately, no such variation was detected prior to the explosion. In fact, the seismic activity decreased during the two days preceding the huge blast.

"Vancouver, Vancouver, this is it!" was the only warning to precede the unleashing of tremendous quantities of pent-up gases. The trigger was an earth tremor with a rating of 5.1 on the Richter scale. The vibrations sent the north slope of the cone plummeting into the Toutle River, effectively removing the overburden which had trapped the magma below (Figure 4.3). With the pressure reduced, the water-rich magma is thought to have ruptured like an overheated steam boiler. Since the eruption originated in the vicinity of the bulge, which was several hundred meters below the summit, the main impact of the eruption was directed laterally rather than vertically. Had the full force of the eruption been upward, far less destruction would have occurred.

Mount St. Helens is only one of the 15 large volcanoes and enumerable smaller ones extending from British Columbia to northern California. Eight of the largest cones have been active in the past few hundred years, while the last eruptive phase of Mount St. Helens came to an end in 1857. Of the remaining seven "active" volcanoes, Mount Baker, Mount Shasta, Lassen Peak, and Mount Rainer are believed most likely to erupt again. It is hoped that the eruptions of Mount St. Helens will provide geologists with enough data to more effectively evaluate the potential hazards of future volcanic eruptions.

THE NATURE OF VOLCANIC ACTIVITY

Volcanic activity is generally perceived as a process that produces a picturesque, cone-shaped structure which periodically erupts in a violent manner. However, although some eruptions may be cataclysmic,

C. In a matter of seconds the ash cloud begins to expand vertically (darker material) but with even greater lateral force (lighter material). **D.** The lateral blast races out, destroying nearly every living thing unfortunate enough to be in its path. (Photos courtesy of Keith Ronnholm)

TABLE 4.1
Variations in properties among magmas of differing compositions.

Property	Basaltic	Andesitic	Granitic
Silica content	Least (about 50%)	Intermediate (about 60%)	Most (about 70%)
Typical minerals	Ca feldspar Pyroxene Olivine	Na feldspar Amphibole Pyroxene Mica	K feldspar Quartz Mica Amphibole
Viscosity	Least	Intermediate	Highest
Tendency to form lavas	Highest	Intermediate	Least
Tendency to form pyroclastics	Least	Intermediate	Highest
Density	Highest	Intermediate	Lowest
Melting point	Highest	Intermediate	Lowest

SOURCE: Modified from Peter Francis.

many are relatively quiescent. The primary factors which determine the nature of volcanic eruptions include the magma's composition, its temperature, and the amount of dissolved gases it contains. These factors affect the magma's mobility, or **viscosity.** The more viscous the material, the greater its resistance to flow. For example, molasses is more viscous than water. The effect of temperature on viscosity is easily seen. Just as heating molasses makes it more fluid, the mobility of lava is also influenced by temperature changes. As a lava flow cools and begins to congeal, its mobility decreases and eventually the flowing halts.

The chemical composition of magmas was discussed in Chapter 3 with the classification of igneous rocks. One major difference between various igneous rocks and therefore between the magmas from which they originate is their silica (SiO_2) content (Table 4.1). Magmas that produce basaltic rocks contain about 50 percent silica, whereas rocks of granitic composition (granite and its extrusive equivalent, rhyolite) contain over 70 percent silica. The intermediate rock types, andesite and diorite, contain around 60 percent silica. It is important to note that a magma's viscosity is directly related to its silica content. In general, the higher the percentage of silica in magma, the greater its viscosity. It is believed that the flow of magma is impeded because the silica molecules link into long chains even before crystallization begins. Consequently, because of their low silica content,

basaltic lavas tend to be quite fluid, whereas rhyolitic lavas are very viscous and incapable of flow over appreciable distances even at relatively high temperatures (Figure 4.4).

The gas content of a magma also affects its mobility. Dissolved gases tend to increase the fluidity of magma. Of far greater consequence is the fact that escaping gases provide enough force to propel molten rock from a volcanic vent. As magma moves into a near-surface environment, such as within a volcano, the confining pressure in the uppermost portion of the magma body is greatly reduced. This reduction in confining pressure allows the gases, which had been dissolved when they were at greater depths, to be released suddenly. At temperatures of 1000°C and low, near-surface pressures, these gases will expand to occupy hundreds of times their original volume. Very fluid basaltic magmas allow the expanding gases to migrate upward and escape from the vent with relative ease. As they escape, the gases will often carry incandescent lava hundreds of meters into the air, producing lava fountains. Although spectacular, such fountains are not generally associated with major explosive events of the type which cause great loss of life and property. Rather, eruptions of fluid basaltic lavas, such as those that occur in Hawaii, are relatively quiescent.

At the other extreme, highly viscous magmas impede the upward migration of gases. As a consequence, gases collect as bubbles and pockets that increase in size and pressure until they explosively eject the semimolten rock from the volcano.

Once magma in the upper portion of the vent is ejected, the reduced pressure on the molten rock directly below is believed to cause it to be blown out also. Thus, rather than a single "bang," volcanic eruptions are really a series of explosions. This process might logically continue until the magma chamber is emptied, much like a geyser empties itself of water (see Chapter 10). However, this is generally not the case. The soluble gases in a viscous magma migrate quite slowly. Hence, only within the uppermost portion of the magma body, where the confining pressure is low, does the gas pressure build to explosive levels. Thus, an explosive event is commonly followed by the quiet emission of gas-free lavas. However, once this eruptive phase ceases, the process of gas buildup begins anew. This time lag may partially explain the sporadic eruptive patterns of volcanoes that eject viscous lavas.

To summarize, we have seen that the quantity of dissolved gases, as well as the ease with which the gases can escape, largely determine the nature of a volcanic eruption. We can now understand why the volcanic eruptions on Hawaii are relatively quiet, whereas the volcanoes bordering the Pacific are explosive and pose the greatest threat to people, because these latter volcanoes generally contain great quantities of gas and emit viscous lavas.

MATERIALS EXTRUDED DURING AN ERUPTION

Many people believe that lava is the primary material extruded from a volcano. However, this is not always true. Explosive eruptions that eject huge quantities of broken rock, lava bombs, and fine ash and dust, occur just as frequently. Moreover, all volcanic eruptions emit large amounts of gas into the atmosphere. In this section we will examine each of these materials associated with a volcanic eruption.

LAVA FLOWS

Due to their low silica content, basaltic lavas are usually very fluid and flow in thin, broad sheets or

FIGURE 4.4
Aerial view of Big Glass Mountain in northern California. Here high-silica lavas moved as viscous blocks of obsidian and rhyolite. These slow-moving lava flows were very thick (note the roads for scale). (Photo courtesy of Ron Greeley)

tongues. On the island of Hawaii such lavas have been clocked at speeds of 30 kilometers per hour on steep slopes. These velocities are rare however, and flow rates of 10 to 300 meters per hour are more common. Further, basaltic lavas have been known to travel distances of 150 kilometers or more before congealing. In contrast, the movement of silica-rich lava is often too slow to be perceptible.

When fluid basaltic lavas of the Hawaiian type congeal, they often form a relatively smooth skin that sometimes wrinkles as the still-molten subsurface lava continues to advance (Figure 4.5A). These are known as **pahoehoe flows** and resemble the twisting braids in ropes. Another common type of basaltic lava has a surface of rough, jagged blocks with dangerously sharp edges and spiny projections (Figure 4.5B). The name **aa** (pronounced "ah ah") is given to these flows. Active aa flows are relatively cool and thick and, depending upon the slope, advance at rates of from 5 to 50 meters per hour. Further, escaping gases fragment the cool surface and produce numerous voids and sharp spines in the congealing lava. As the molten interior advances, the outer crust is broken further, giving the flow the appearance of an advancing mass of lava rubble.

The lava which flowed from the famous Mexican volcano Parícutin and buried the city of San Juan Parangaricutiro was of the aa type (see Figure 4.12). At times one of the flows from Parícutin moved only one meter per day, but continued to advance day in and day out for more than three months.

Hardened lava flows commonly contain tunnels that once were horizontal conduits carrying lava from the vent to the flow's leading edge. These lava tubes are found near the interior of a flow where temperatures remained high long after the surface had congealed. Under these conditions, the still molten lava within the conduits continued its forward motion, leaving behind the voids called *lava tunnels*. Lava tunnels can play an important role in allowing fluid lavas to advance great distances from their source. The rocks that surround the tunnels act as excellent insulation. Therefore, the lava flowing through the tunnels cools very slowly and can travel far before congealing.

When lava flows enter the ocean, or when lava outpourings actually originate within an ocean basin,

A.

B.

FIGURE 4.5
A. Typical pahoehoe (ropy) lava flow, Kilauea, Hawaii.
B. Typical slow moving aa flow. (Photos by D. W. Peterson, U.S. Geological Survey)

the flows' outer zones quickly congeal. The lava within the flows is usually able to move forward by breaking through the hardened surface. This process occurs over and over, generating a lava flow composed of elongated structures resembling large bed pillows stacked one upon the other. **Pillow lavas** help interpret earth history since, when they are identified, they indicate that deposition occurred in an underwater environment.

GASES

Magmas contain varied amounts of dissolved gases held in the molten rock by confining pressure, just as carbon dioxide is held in soft drinks. As with soft drinks, as soon as the pressure is reduced, the gases begin to escape. Because obtaining samples from an erupting volcano is very difficult and dangerous, geologists usually only estimate the amount of gas originally contained within the magma.

The gaseous portion of most magmas is believed to compose from 1 to 5 percent of the total weight, and most of this is in the form of water vapor. Although the percentage may be small, the actual quantity of emitted gas can exceed thousands of tons per day. The composition of the gases is also of interest to scientists, since much evidence points to these as the source of the earth's atmosphere and the oceans. The analysis of samples taken during Hawaiian eruptions indicated that the gases emitted there consist of about 70 percent water vapor, 15 percent carbon dioxide, 5 percent each of nitrogen and sulfur compounds, and lesser amounts of chlorine, hydrogen, and argon. Sulfur compounds are easily recognized by their pungent odor and because they readily form sulfuric acid, which when inhaled produces a burning sensation.

In addition to propelling magma from a volcano, gases are thought to create the narrow conduit that connects the magma chamber to the surface. First, the intense heat from the magma body cracks the rock above. Then, hot streams of high-pressure gases expand the cracks and develop a passageway to the surface. Once completed, the hot gases armed with rock fragments erode the walls of the passageway, producing a larger conduit. Because these erosive forces will be concentrated on any protrusion along the pathway, the volcanic pipes that are produced have a circular shape. As the conduit enlarges, magma moves upward to produce surface activity. Following an eruptive phase the volcanic pipe often becomes choked with debris that was not thrown clear of the vent. Before the next eruption, a new surge of explosive gases may then clear the conduit.

PYROCLASTIC MATERIALS

When basaltic lava is extruded, the dissolved gases escape quite freely and continually. As stated earlier, these gases often carry incandescent blobs of lava to great heights, thereby producing spectacular lava fountains. Some ejected material may land near the vent and produce a cone structure, while smaller particles will be carried great distances by the wind. The gases in highly viscous magmas, on the other hand, are less able to escape and may build up an internal pressure capable of producing a violent eruption. Upon release, these superheated gases expand a thousandfold as they blow pulverized rock and lava from the vent. The particles produced by these processes are called **pyroclastic materials.** These ejected lava fragments range in size from very fine dust and sand-sized volcanic ash, to large volcanic bombs and blocks.

The fine *ash* and *dust* particles are produced when the extruded lava contains so many gas bubbles that it resembles the froth flowing from a newly opened bottle of champagne. As the hot gases expand explosively, the lava is disseminated into very fine fragments. When the hot ash falls, the glassy shards often fuse to form welded tuff. Sheets of this material, as well as ash deposits that consolidate later, cover vast portions of the western United States. In some instances the froth-like lava is ejected in larger pieces called *pumice*. This material has so many voids that it is often light enough to float in water.

Pyroclastics the size of walnuts, called *lapilli* ("little stones"), and pea-sized particles called *cinders* are also very common. Cinders contain numerous voids and form when ejected lava blobs are pulverized by the escaping gases. Particles larger than lapilli are called *blocks* when they are made of hardened lava and *bombs* when they are ejected as incandescent lava. Since volcanic bombs are semimolten upon ejection, they often take on a streamlined shape as shown in Figure 4.6. Due to their size, bombs and blocks usually fall on the slopes of a cone; however, bombs may occasionally be propelled like rockets far from the volcano by the force of escaping gases (Figure 4.7).

Fine volcanic debris can be scattered great distances from its source. Dust in particular may be blasted high into the atmosphere, where it can

FIGURE 4.6
Volcanic bombs. Ejected lava fragments acquire a
streamlined shape as they sail through the air. (Photo by
E. J. Tarbuck)

remain for extended periods. While present, dust
produces brilliant sunsets and has on occasion slightly
lowered the earth's average temperature. The possible effects of volcanic eruptions on climate are discussed more thoroughly in an upcoming section.
Table 4.2 provides data on the amount of ejected
pyroclastic material for some well-known eruptions.

VOLCANOES AND VOLCANIC ERUPTIONS

Successive eruptions from a central vent result in the
formation of a mountainous accumulation of material
known as a **volcano.** Located at the summit of many
volcanoes is a steep-walled depression, a **crater,**
which is connected to a magma chamber via a pipe-
like conduit, or vent. As was indicated earlier, the
conduit as well as the crater is produced by the erosive force of gases and effervescent magma. Some
volcanoes have unusually large summit depressions
that exceed one kilometer in diameter and are known
as **calderas.**

When fluid lava leaves a conduit, it is often stored
in the crater or caldera, until it overflows. On the
other hand, lava that is very viscous forms a plug in
the pipe which rises slowly or is blown out, often
enlarging the crater. However, lava does not always

issue from a central crater. Sometimes it is easier for
the magma or escaping gases to push through fissures located on the volano's flanks. Continued activity from a flank eruption may build a so-called **parasitic cone.** Mount Etna, for example, has more than
200 secondary vents. Some of these secondary vents
extrude only gases and are appropriately called **fu-
maroles.**

The eruptive history of each volcano is unique;
consequently, all volcanoes are somewhat different
in form and size. Nevertheless, volcanologists have
recognized that volcanoes exhibiting somewhat similar eruptive styles can be grouped. Based on their
"typical" eruptive patterns and characteristic form,
three groups of volcanoes are generally recognized:
shield volcanoes, cinder cones, and composite cones
(stratovolcanoes) (Figure 4.8).

SHIELD VOLCANOES

When fluid lava is extruded, the volcano takes the
shape of a broad, slightly domed structure called a
shield volcano (Figure 4.8C). Shield volcanoes are
built primarily of basaltic lava flows and contain only
a small percentage of pyroclastic material. Typically
they have a slope of a few degrees at their flanks and
generally do not exceed 15 degrees near their summit, as exemplified by the volcanoes of the Hawaiian
Islands. Mauna Loa, probably the largest volcano on
earth, is one of the five shield volcanoes that together
make up the island of Hawaii (Figure 4.9). Its base
rests on the ocean floor 5000 meters below sea level,
while its summit reaches a height of 4170 meters
above the water. Nearly one million years and numerous eruptive cycles were required to build this
truly gigantic pile of volcanic rock. Many other volcanic structures, including Midway Island and the Galapagos Islands, have been built in a similar manner
from the ocean's depths.

The 1959–60 eruption of the volcano Kilauea
illustrates the "typical" cycle of a Hawaiian-type
eruption. Kilauea is located on the island of Hawaii
on the southeast flank of the larger volcano Mauna
Loa (Figure 4.10). The summit caldera of Kilauea is
an oval-shaped depression about 100 meters deep
and roughly 5 kilometers long by 3 kilometers wide.
Located within this crater is an even deeper depression called Halemaumau, or "firepit," which from

FIGURE 4.7
Parícutin Volcano in eruption at night, 1943. (Photo by K. Segerstrom, U.S. Geological Survey)

TABLE 4.2
Approximate volume of volcanic debris extruded for some well-known eruptions.

Eruption	Amount Extruded (mostly pyroclastics, km^3)
Tambora, Indonesia (1815)	80–100
Crater Lake, Oregon (prehistoric)	50–70
Krakatoa, Indonesia (1883)	18
Parícutin, Mexico (1943–52)	1.3
Mt. St. Helens (1980)	4
Mt. Vesuvius, Italy (79 A.D.)	3
Yellowstone, Wyoming (prehistoric)	2400

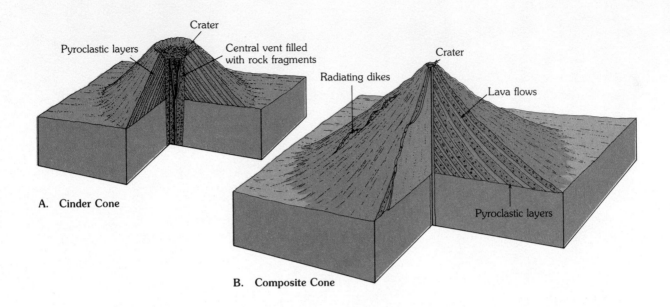

A. **Cinder Cone**

B. **Composite Cone**

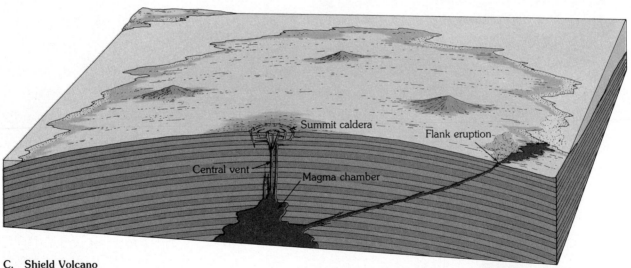

C. **Shield Volcano**

FIGURE 4.8
Comparison of the three basic types of volcanic structures. **A.** Cinder cone. **B.** Composite cone. **C.** Shield volcano.

time to time contains a ''boiling'' lava lake reaching depths upwards of 400 meters (1300 feet). In addition to the summit caldera, numerous smaller craters dot the landscape.

Kilauea has erupted over 50 times in recorded history. A testimonial to the quiescent nature of these eruptions is the location of the Hawaiian Volcano Observatory on the very rim of the summit caldera. Two years prior to the 1959–60 eruption, volcanologists equipped with tiltmeters discovered swelling in the summit area. Earth tremors produced by magma migrating toward the surface indicated that the

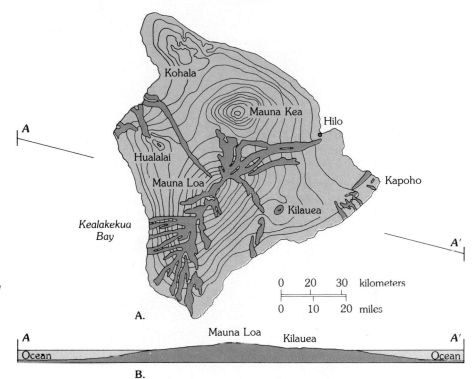

FIGURE 4.9
Map of the island of Hawaii.
A. Five volcanoes collectively
make up the island. Contour
interval is 300 meters (1000
feet). **B.** Illustration of the very
gentle slope which is characteristic of a shield volcano (no
vertical exaggeration). (After
H. T. Stearns and G. A.
MacDonald, U.S. Geological
Survey)

FIGURE 4.10
Aerial view of Kilauea Caldera with the volcano Mauna Loa in the background. This
caldera is about 5 kilometers long and 2 kilometers wide. Located within the caldera is
a depression called the "fire pit." Notice the road traversing the caldera for scale.
(Photo by Dan Ozrisin, U.S. Geological Survey)

material originated between 60 and 100 kilometers below the surface. This molten rock slowly worked its way upward and accumulated in smaller reservoirs located about 3 to 5 kilometers below the summit.

The first phase of the 1959–60 eruption began in November of 1959 in Kilauea Iki, a crater located just east of the summit caldera. The initial event consisted of lava fountaining from a one-kilometer-long fissure along the south wall of the crater. Within a day the eruption was restricted to a single vent, occasionally spraying lava to heights of more than 300 meters (Figure 4.11). Most of the ejected lava fell back into the crater where it formed a lava lake. Around the vent, falling cinders built a small cone that frequently collapsed as the crater filled with lava. This phase of the eruption lasted about a month, during which a lava lake filled the crater to a depth of 125 meters. Immediately after this eruptive phase ceased, some of this lava drained back into the vent. The draining of this incandescent material resembled water spiraling into a drain.

The earth tremors persisted long after the activity at Kilauea Iki ceased. This seismic activity indicated that the magma was still flowing underground, but had been diverted toward the volcano's flank. Then abruptly on January 13, 1960, ground cracks were observed as a one-kilometer-long fissure opened near the village of Kapoho, located 45 kilometers east of Kilauea Iki. Here lava fountains played continuously day and night for several weeks. Within the first few days, molten lava collected into flows and advanced to the sea, 5 kilometers away. As the hot lava invaded the sea, boiling clouds of steam, occasionally darkened by explosive blasts of black ash, rose thousands of meters into the air. During the next few days the lava flows spread laterally, engulfing more of the countryside. Earthen dams were constructed in vain to protect the village of Kapoho. During this phase several developments were destroyed, while the island was enlarged by one-half kilometer in one location.

The activity at Kapoho was accompanied by gradual subsidence of the summit area and collapse of the

FIGURE 4.11
Kilauea Iki erupting in 1959, with lava fountaining high above the vent. (Courtesy of U.S. Geological Survey)

floor of Halemaumau. About three months after the initial event, the fountain near Kapoho died, marking the end of this eruptive phase. However, geologists continued to monitor ground subsidence and to sample molten rock from beneath the crust of the lava lake in Kilauea Iki. From this type of information a better understanding of the formation of the Hawaiian Islands and volcanism in general has emerged.

There is now general agreement that the early stages of a shield volcano's formation consist of frequent eruptions of thin flows of very fluid basalts. As the structure enlarges, flank eruptions occur along with the summit eruptions. Collapse of the summit area frequently follows each eruptive phase. In the latter stages of growth, the activity is more sporadic and pyroclastic ejections are more prevalent. Further, the lavas increase in viscosity, resulting in thicker, shorter flows. These activities tend to steepen the slope of the summit area. This explains why Mauna Kea, an older and inactive volcano to the north, has a steeper summit than Mauna Loa.

CINDER CONES

As the name suggests, **cinder cones** are built from ejected lava fragments. Because unconsolidated pyroclastic material maintains a high angle of repose (between 30 and 40 degrees), volcanoes of this type have very steep slopes. Cinder cones are rather small, usually less than 300 meters (1000 feet) high, and often form as parasitic cones on or near larger volcanoes. In addition, they frequently occur in groups, where the cones apparently represent the last phase of activity in a region of older basaltic flows. This may result because the contributing magma has cooled and become more viscous.

One of the very few volcanoes whose formation has been observed by geologists from beginning to end is a cinder cone called Parícutin. This volcano's history serves to illustrate the formation and structure of a larger-than-usual cinder cone.

In 1943, about 200 miles west of Mexico City, the volcano Parícutin was born (see Figure 3.1). The eruption site was a cornfield owned by Dionisio Pulido, who with his wife Paula witnessed the event as they were preparing the field for planting. For two weeks prior to the first eruption, numerous earth tremors caused apprehension in the village of Parícutin about 3.5 kilometers away. Then around 4:00 P.M. February 20th, smoke with a sulfurous odor arose from a small hole that had been in the cornfield for as long as Dionisio could remember. During the night, hot, glowing rock fragments thrown into the air from the hole produced a spectacular fireworks display. By the next day the cone had grown to a height of 40 meters and by the fifth day it was over 100 meters high. At this time explosive eruptions were throwing hot fragments 1000 meters above the crater rim. The larger fragments fell near the crater, some remaining incandescent as they rolled down the slope. These fragments built an aesthetically pleasing cone, while finer ash fell over a much larger area, burning and eventually covering the village of Parícutin. Within two years the cone had grown to 400 meters high and would rise only a few tens of meters more.

The first lava flow came from a fissure that had opened just north of the cone, but after a few months of activity, flows began to emerge from the base of the cone itself. In June of 1944, a clinkery flow 10 meters thick moved over the village of San Juan Parangaricutiro, leaving only the church steeple exposed (Figure 4.12). After nine years the activity

FIGURE 4.12
The village of San Juan Parangaricutiro engulfed by lava from Parícutin, shown in the background. Only the church towers remain. (Photo by Tad Nichols)

ceased almost as quickly as it began. Now Parícutin is just another one of the numerous cinder cones dotting the landscape in this region of Mexico. Like the others, it will probably not erupt again.

COMPOSITE CONES

The earth's most picturesque volcanoes are **composite cones,** or **stratovolcanoes.** Just as shield volcanoes owe their shape to the fluid nature of the extruded lavas, so too do composite cones reflect the nature of the erupted material. Composite cones are produced when relatively viscous lavas of andesitic composition are extruded. For long periods a composite cone may extrude viscous lava. Then suddenly, the eruptive style will change and the volcano will violently eject pyroclastic material. Most of the ejected pyroclastic material falls near the summit, building a steep-sided mound of cinders. In time this debris will be covered by lava. Occasionally both activities occur simultaneously. The resulting structure consists of alternating layers of lava and pyroclastics. Two of the most perfect cones, Mount Mayon in the Philippines and Fujiyama in Japan, exhibit the classic form of the stratovolcano with its steep summit area and rather gently sloping flanks.

Although composite cones are the most picturesque, they also represent the most violent type of volcanic activity. Their eruption can be unexpected and devastating as was the 79 A.D. eruption of the Italian volcano we now call Vesuvius. Prior to this eruption, Vesuvius was dormant for centuries. Although minor earthquakes probably warned of the events to follow, Vesuvius was covered with a heavy coat of vegetation and hardly looked threatening. On August 24th, however, the tranquility ended, and in the next three days the city of Pompeii (near Naples) and more than 2000 of its 20,000 residents were buried. They remained so for nearly seventeen centuries, until the city was rediscovered and excavated.

Although the destruction of Pompeii was truly catastrophic, eruptions of a more devastating nature occur when hot gases infused with incandescent ash are ejected, producing a fiery cloud called **nuée ardente.** Also referred to as *glowing avalanches,* these flows, which are black in daylight and glow red at night, move down steep volcanic slopes at speeds exceeding 150 kilometers per hour (Figure. 4.13).

Although very dense, these glowing avalanches are supported by the expanding gases emitted from the hot lava particles. Thus, this material, which can be composed of rather large lava fragments, flows downslope in an almost frictionless environment cushioned by the expanding gases.

In 1902, a nuée ardente from Mount Pelée, a small volcano on the Caribbean island of Martinique, destroyed the port town of St. Pierre. The destruction was instantaneous and so devastating that almost all of St. Pierre's 28,000 inhabitants were killed. Only a prisoner protected in a dungeon, a shoemaker, and a few people on ships in the harbor were spared (Figure 4.14). Satis N. Coleman, in *Volcanoes, New and Old*, relates a vivid account of this event, which lasted less than five minutes:

FIGURE 4.13
Fiery clouds (nuée ardentes) race down the slope of Mount Mayon, Philippines, 1968. The steeple in the foreground is the remnant of a church in which several hundred persons who had sought refuge during an eruption in 1814 were killed by a mudflow. (Courtesy of the Smithsonian Center for Short-Lived Phenomena)

I saw St. Pierre destroyed. The city was blotted out by one great flash of fire. Nearly 40,000 people were killed at once. Of eighteen vessels lying in the roads, only one, the British steamship *Roddam* escaped and she, I hear, lost more than half of those on board. It was a dying crew that took her out. Our boat, the *Roraima*, arrived at St. Pierre early Thursday morning. For hours before entering the roadstead we could see flames and smoke rising from Mt. Pelée. . . . The spectacle was magnificent. As we approached St. Pierre we could distinguish the rolling and leaping of red flames that belched from the mountain in huge volumes and gushed into the sky. Enormous clouds of black smoke hung over the volcano. There was a constant muffled roar. It was like the biggest oil refinery in the world burning up on the mountain top. There was a tremendous explosion about 7:45, soon after we got in. The mountain was blown to pieces. There was no warning. The side of the volcano was ripped out and there was hurled straight toward us a solid wall of flame. It sounded like a thousand cannons.

The wave of fire was on us and over us like a flash of lightning. It was like a hurricane of fire. I saw it strike the cable steamship *Grappler* broadside on, and capsize her. From end to end she burst into flames and then sank. The fire rolled in mass straight down upon St. Pierre and the shipping. The town vanished before our eyes.

The air grew stifling hot and we were in the thick of it. Wherever the mass of fire struck the sea, the water boiled and sent up vast columns of steam. . . . The blast of fire from the volcano lasted only a few minutes. It shrivelled and set fire to everything it touched. Thousands of casks of rum were stored in St. Pierre, and these were exploded by the terrific heat. The burning rum ran in streams down every street and out into the sea. This blazing rum set fire to the *Roraima* several times. . . . Before the volcano burst, the landings of St. Pierre were covered with people. After the explosion, not one living soul was seen on land.[1]

When we compare the destruction of St. Pierre with that of Pompeii, several differences are noticed. Pompeii was totally buried by an event lasting three days, whereas St. Pierre was destroyed in a brief instant and its remains were only mantled by a thin

[1]New York: John Day, 1946, pp. 80–81.

FIGURE 4.14
St. Pierre as it appeared shortly after the eruption of Mount Pelée, 1902. (Reproduced from the collection of the Library of Congress)

FIGURE 4.15
Following the May 18, 1980 eruption of Mount St. Helens, a large lava dome began to develop. (Photo by Robert Krimmels, U.S. Geological Survey)

layer of volcanic debris. Also, the structures of Pompeii remained intact except for the roofs that collapsed under the weight of the ash. In St. Pierre masonary walls nearly one meter thick were knocked over like dominoes; large trees were uprooted and cannons were torn from their mounts.

Nuée ardentes are usually associated with highly viscous magmas near the granitic end of the compositional spectrum (see Table 4.1). Bulbous masses called **lava domes** are another feature associated with volcanoes that extrude highly viscous material. These structures are produced when thick, viscous lava is slowly "squeezed" out of the vent (Figure 4.15). Lava domes often act as plugs, which deflect subsequent gaseous eruptions. Occasionally lava domes will form on a volcano's flanks. As these structures enlarge they pose a potential threat to inhabitants in the surrounding area. If the upper part of the dome slips downslope in response to the pull of gravity, the loss of confining pressure may trigger an explosive eruption of the magma below. These laterally directed eruptions, such as the May 18, 1980 eruption of Mount St. Helens, can be devastating.

CALDERAS

Earlier it was pointed out that some volcanoes have unusually large craters known as calderas. Some calderas are thought to form when the summit of a volcano collapses into the partially emptied magma chamber below (Figure. 4.16). Crater Lake in Oregon, which is 8–10 kilometers wide and 1300 meters deep, is located in such a depression (Figure 4.17). The creation of Crater Lake began about 7000 years ago when the volcano, later to be named Mount Mazama, put forth a violent ash eruption much like that of Vesuvius. However, this ancient eruption was on a much larger scale, extruding an estimated 50–70 cubic kilometers of volcanic material. With the loss of support, 1500 meters of this once prominent 3600-meter cone collapsed. After the collapse, rainwater filled the caldera. Later activity built a small cinder cone called Wizard Island, which today provides a mute reminder of past activity.

The spectacular eruption of the Indonesian volcano Krakatoa in 1883 also resulted in the formation of a caldera. This explosion, which has not been equalled since, was heard 4800 kilometers away in

southern Australia as pumice and ash rose thousands of meters into the atmosphere. After nearly two days of total darkness the sky began to clear, revealing that two-thirds of this island chain had disappeared. A portion of the volcano probably collapsed into the

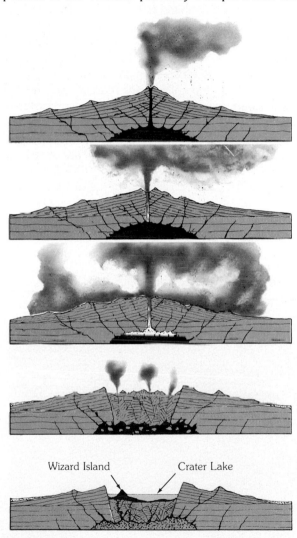

FIGURE 4.16
Sequence of events that formed Crater Lake, Oregon. About 7000 years ago, the summit of former Mount Mazama collapsed following a violent eruption which partly emptied the magma chamber. Subsequent eruptions produced the cinder cone called Wizard Island. Rainfall and groundwater contributed to form the lake. (After H. Williams, *The Ancient Volcanoes of Oregon,* p. 47, courtesy of the University of Oregon)

space left when the 18 cubic kilometers of magma were extruded. The ejected material was mostly in the form of pumice. Following the blast, such large quantities remained floating that navigation in the nearby Sunda Straits was disrupted. Although Krakatoa was an uninhabited island, the destructive, 30-meter (100-foot) high sea wave[1] generated by the explosion took a toll of 36,000 lives in nearby Java and Sumatra. Calderas of varying sizes are known to exist, the largest more than 20 kilometers across. Some, such as the calderas of Mauna Loa and Kilauea, clearly result from subsidence that occurred as supporting magma was diverted to a flank eruption. Others were formed by an explosive event that blasted the upper portion of the volcano away, as occurred at Mount St. Helens in 1980. However,

since most large calderas are prehistoric, the mechanism primarily reponsible for their creation is often unknown. Although attempts have been made to reconstruct many of the ancient eruptions that created these features, it is difficult to determine precisely how much of the volcano was blown away and what portion subsided into the partially emptied magma chamber.

VOLCANIC NECKS AND PIPES

Volcanoes, like all land areas, are continually being lowered by the forces of weathering and erosion. Cinder cones are easily eroded, because they are composed of unconsolidated materials. However, all volcanic structures will eventually be worn away. As erosion progresses, the rock occupying the vent is often more resistant, and may remain standing above the terrain long after most of the cone has vanished.

---

[1]Such waves, called *tsunami,* are discussed in Chapter 14.

FIGURE 4.17
Crater Lake occupies a caldera about 10 kilometers (6 miles) in diameter. (Courtesy of the National Park Service)

Shiprock, New Mexico, is thought to be such a feature, called a **volcanic neck** (Figure 4.18). This structure, higher than many skyscrapers, is but one of many that protrude conspicuously from the red desert landscape of the southwestern United States.

The conduits feeding most volcanoes are thought to be connected to a magma source emplaced near the surface. By contrast, some ferromagnesian-rich pipes are thought to extend in a tubelike fashion directly into the asthenosphere, a distance of 200 kilometers. Consequently, the materials found in these **pipes** are thought to be samples of the asthenosphere which have undergone very little alteration during their ascent. Geologists consider pipes, there-fore, to be "windows" into the earth since they allow us to view rocks found at great depths.

Occasionally a volcanic pipe will reach the surface, but the eruptive phase will cease before any lava is extruded. The upper portions of these pipes often contain a jumble of lava fragments and fragments that were torn from the walls of the vent by the violently escaping gases. The best known of these structures are the diamond-bearing pipes of South Africa. Here the rocks filling the pipes are thought to have originated at depths of about 200 kilometers, where pressure is great enough to generate diamonds and other high-pressure minerals. In these instances the diamonds are phenocrysts that crystallized at depth

FIGURE 4.18
Shiprock, New Mexico, is the remnant neck of a volcano which erosion has almost completely removed. (Photo by John S. Shelton)

and were carried upward by the still-molten portion of the magma.

THE LOST CONTINENT OF ATLANTIS

One of the most popular legends of all time concerns the disappearance of the so-called continent of Atlantis. According to the accounts provided by Plato, an island empire named Atlantis disappeared beneath the sea in a single day and night. This event, which reportedly took place between 1500 B.C. and 1400 B.C. and caused the collapse of the Minoan civilization, is thought by many to have been the result of a cataclysmic volcanic eruption.

Research efforts in the eastern Mediterranean have provided evidence apparently linking Atlantis to the volcanic island of Santorin. Although once a majestic volcano, Santorin now consists of five islands located roughly midway between the island of Crete and Greece. Evidence collected from core samples taken around the remnants of Santorin revealed that a violent eruption occurred about 1500 B.C. This eruption generated great volumes of ash and pumice that reached a maximum depth of 60 meters. Many nearby Minoan cities were buried and their ruins preserved beneath the ash.[1] Even as far away as Crete, enough ash fell to kill crops and possibly livestock grazing on the ash-laden plants. Following the ejection of this large quantity of material, Santorin collapsed, producing a caldera 14 kilometers across. The eruption and collapse of Santorin undoubtedly generated large destructive sea waves that caused widespread destruction to the coastal villages of Crete as well as those on the nearby islands to the north.

The connection between the eruption of Santorin and the disappearance of Atlantis is further supported by the fact that the Minoan civilization also disappeared between 1500 B.C. and 1400 B.C. About 1400 B.C. some of the Minoan traditions began to appear in the culture of Greece.

Most scholars agree that the eruption of Santorin contributed to the collapse of the Minoan civilization.

Was this eruption the main cause of the dispersal of this great civilization or only one of many contributing factors? Was Santorin the island continent of Atlantis described by Plato? Whatever the answers to these questions, it seems evident that volcanism can dramatically change the course of human events.

VOLCANOES AND CLIMATE

The idea that explosive volcanic eruptions may cause changes in the earth's climate was first proposed many years ago and is still regarded as a plausible explanation for some aspects of climatic variability. Explosive eruptions emit huge quantities of gases and fine-grained debris into the atmosphere. The greatest eruptions are sufficiently powerful to inject material high into the stratosphere, where it spreads around the globe and remains for many months or even years. The theory's basic premise is that this suspended volcanic material will filter out a portion of the incoming solar radiation which, in turn, will lower air temperatures.

Perhaps the most notable cool period linked to a volcanic event is the "year without a summer" that followed the 1815 eruption of Mount Tambora in Indonesia. In many northern hemisphere locations, including New England, the abnormally cold spring and summer of 1816 were believed to be caused by the cloud of volcanic debris ejected from Tambora.

When Mount St. Helens erupted on May 18, 1980, there was almost immediate speculation about the possible effects of this event on our climate. Can an eruption such as this cause our climate to change? Although spectacular, a single explosive volcanic eruption of the magnitude of Mount St. Helens occurs somewhere in the world every 2 to 3 years. Studies of these events indicate that a very slight cooling of the lower atmosphere does occur. However, it is believed that the cooling is so slight, less than one-tenth of one degree Celsius, as to be inconsequential. On April 4, 1982, El Chichón, a little-known volcano on Mexico's Yucatan peninsula, erupted. The cloud of debris and sulfur gases lofted into the atmosphere was huge, probably 20 times greater than the cloud from Mount St. Helens. Following the blast, scientists predicted a gradual lower-

---

[1]One of these buried cities, Akrotiri, on the island of Thera, has been excavated.

ing of temperatures in the northern hemisphere, perhaps as great as 0.3–0.5°C. Such a change is large enough to be distinguishable from normal temperature fluctuations but is probably too small to affect our life styles. Nevertheless, many scientists agree that such a hemispheric cooling could alter the general pattern of atmospheric circulation for a limited period. Such a change, in turn, could have an effect on the weather in some regions. However, the prediction of specific regional effects still presents a considerable challenge to atmospheric scientists.

The preceding examples illustrate that the impact on climate of a single volcanic eruption, no matter how great, is relatively small and short-lived. Therefore, if volcanism is to have a pronounced impact over an extended period, many great eruptions, closely spaced in time, would have to occur. If this happened, the stratosphere could be loaded with enough volcanic dust to seriously diminish the amount of solar radiation reaching the surface. Since no such period of explosive volcanism is known to have occurred in historic times, the volcanic dust theory is most often mentioned as a possible cause for such prehistoric climatic shifts as the Ice Age.

The Ice Age, however, was not a single period of continuous glaciation. Instead it was characterized by alternating periods of glacial advance and retreat. The warm periods between advances are termed *interglacial periods*. Thus, if volcanic activity was the primary forcing mechanism for the Ice Age, periods of explosive volcanism must have alternated with relatively quiet conditions. At present, even those who believe that volcanic activity may have triggered the Ice Age believe the data are not sufficiently clear to deal with these glacial-interglacial episodes. Hence, the full importance of volcanic activity as a possible cause for the Ice Age is still something of a mystery and a matter for continued speculation.[1] Stephen H. Schneider, an active researcher and writer in the field of climatic change, summarizes the current uncertainty of many scientists regarding the volcanic dust theory as follows:

> **Although the role of volcanoes in forcing climatic change may prove critical—in which case I should**

[1]Other theories on the causes of the Ice Age are discussed near the end of Chapter 11.

> **be trying to learn to predict volcanic eruptions and not spend most of my time building atmospheric models—as yet we cannot be sure the extent to which the potential climatic effects of volcanic blasts are detectable in climatic records.[1]**

FISSURE ERUPTIONS AND PYROCLASTIC FLOW DEPOSITS

Although volcanic eruptions from a central vent are the most familiar, larger amounts of volcanic material are probably extruded from cracks or fractures in the crust called *fissures*. Rather than building an isolated cone, lava is extruded from several vents along these narrow cracks, resulting in the distribution of volcanic materials over a wide area (Figure 4.19). An extensive region in the northwestern United States known as the Columbia Plateau was formed in this manner (Figure 4.20). Here, numerous **fissure eruptions** extruded very fluid basaltic lava. Successive flows as thick as 50 meters buried the old landscape and built a lava plain, which in some places is 2 to 3 kilometers

[1]*The Genesis Strategy* (New York: Plenum Press, 1976), p. 134.

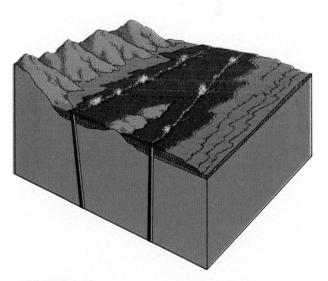

FIGURE 4.19

Flood basalts. Magma erupts from fissures cut through earlier crystallized flows to produce a thick accumulation of basaltic rock. (After Richard S. Fiske, U.S. Geological Survey)

thick (Figure 4.21). The lava's fluidity is evident, since some flows remained molten long enough to travel 150 kilometers from their source. The term **flood basalts** appropriately describes these flows.

Although a few large continental areas are covered with basalt flows, the greatest activity of this type occurs on the ocean floor hidden from view. Along oceanic ridges, where sea-floor spreading is active, fissure eruptions generate new sea floor. Iceland, which is located astride the Mid-Atlantic Ridge, has experienced numerous fissure eruptions of the type believed to occur at spreading centers. The largest Icelandic eruptions in historic times occurred in 1783. A rift 25 kilometers long generated over 20 separate vents which initially extruded sulfurous gases and ash deposits that built several small cinder cones. This activity was followed by huge outpourings of very fluid basalt. The total volume of lava extruded by this so-called Laki eruption was in excess of 10 cubic kilometers. The fluid lava first collected in the nearby valleys and after repeated eruptive phases led to the creation of a flat plateau-like topography.

When silica-rich magma is extruded from fissures, **pyroclastic flows** consisting largely of ash and pum-ice fragments are the rule. When these pyroclastic materials are ejected, they move away from the vent at high speeds and may blanket extensive areas before coming to rest. Once deposited, the pyroclastic materials closely resemble lava flows.

Extensive pyroclastic flow deposits are found in many parts of the world and are most often associated with large calderas. The first phase in the origin of these flows is thought to occur when a highly viscous magma body is emplaced near the surface, upwarping the overlying rocks. Next, fracturing of the roof allows the gas-rich magma to reach the surface where it produces short-lived, explosive eruptions. Finally, the loss of magma from beneath the surface allows the roof to collapse.

Perhaps the best-known region of pyroclastic flows is the Yellowstone Plateau in northwestern Wyoming. Here a large magma body, rich in silica, exists a few kilometers below the surface. Several times over the past two million years, fracturing of the

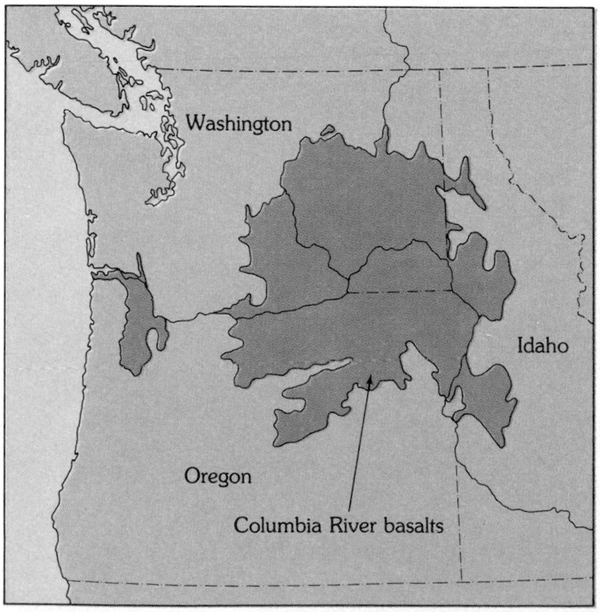

FIGURE 4.20
Map of the area covered by the Columbia River basalts. (After U.S. Geological Survey)

FIGURE 4.21
Basalt flows of the Columbia Plateau. (Photo by E. T. Jones, U.S. Geological Survey)

rocks overlying the magma chamber has resulted in huge eruptions accompanied by the formation of calderas. In the northwestern portion of Yellowstone National Park, 27 fossil forests have been discovered, one resting upon another. During periods of inactivity, a forest developed upon the newly-formed volcanic surface, only to be covered by ash from the next eruptive phase. An examination of Table 4.2 on page 87 reveals that very large quantities of pyroclastic minerals have been extruded in the Yellowstone area. Fortunately, no eruption of this type has occurred in modern times.

VOLCANISM AND PLATE TECTONICS

The origin of magma has been a controversial topic in geology almost from the very beginning of the science. How do magmas of different compositions arise? Why do volcanoes located in the deep-ocean basins primarily extrude basaltic lava, whereas those adjacent to oceanic trenches extrude mainly andesitic lava? Why are basaltic lavas common at the earth's surface, while most granitic magma is emplaced at depth? Why does an area of igneous activity commonly called the "Ring of Fire" surround the Pacific Ocean? New insights gained from the theory of plate tectonics are providing some answers to these questions. We will first examine the origin of magma and then look at the global distribution of volcanic activity as viewed from the model provided by plate tectonics.

ORIGIN OF MAGMA

We know that magma can be produced when rock is heated to its melting point. In a surface environment, rocks of granitic composition begin to melt at temperatures near 750°C, whereas basaltic rocks must reach temperatures above 1000°C before melting will begin. One important difference exists between the melting of a substance that consists of a single compound, such as ice, and the melting of igneous rocks, which are mixtures of several different minerals. Whereas ice melts at 0°C, most igneous rocks melt over a temperature range of a few hundred degrees. As a rock is heated, the first liquid to form will contain a higher percentage of the low-melting-point miner-

als than the original rock. Should melting continue, the composition of the melt will steadily approach the overall composition of the rock from which it is derived. Most often, however, melting is not complete. This process, known as **partial melting,** produces most, if not all, magma.

A significant result of partial melting is the production of a melt with a higher silica content than the parent rock. Recall that basaltic rocks have a relatively low silica content and that granitic rocks have a much higher silica content. Consequently, magmas generated by partial melting are nearer the granitic end of the compositional spectrum than the parent material from which they formed. As we shall see, this idea will help us to understand the global distribution of the various types of volcanic activity.

What is the heat source to melt rock? One source is the heat liberated during the decay of radioactive elements that are thought to be concentrated in the upper mantle and crust. Workers in underground mines have long recognized that temperatures increase with depth. Although the rate of increase varies from place to place, it is thought to average about 30°C per kilometer in the upper crust. This gradual increase in temperature with depth is known as the **geothermal gradient.**

If temperature were the only factor to determine whether or not a rock melts, the earth would be a molten ball covered with only a thin, solid outer shell. However, pressure also increases with depth. Since rock expands when heated, extra heat is needed to melt buried rocks, in order to overcome the effect of confining pressure. In general, an increase in the confining pressure causes an increase in the rock's melting point. Another important factor affecting the melting point of rock is its water content. Up to a point, the more water present, the lower the melting point. The effect of water on lowering the melting point is magnified by increased pressure. Consequently, "wet" rock under pressure has a much lower melting temperature than "dry" rock of the same composition. Whereas dry rocks melt at lower temperatures in a low-pressure environment, water-rich rocks melt most readily in a high-pressure environment (Figure 4.22). Therefore, in addition to a rock's composition, its temperature, confining pressure, and water content determine whether the rock exists as a solid or liquid.

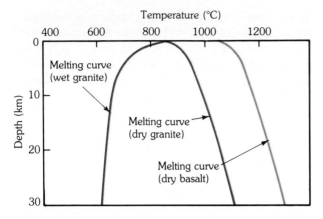

FIGURE 4.22
Idealized melting point curves for basalt and granite. Notice that while dry granite and dry basalt melt at higher temperatures with increasing depth, the melting point of wet granite decreases as the confining pressure increases.

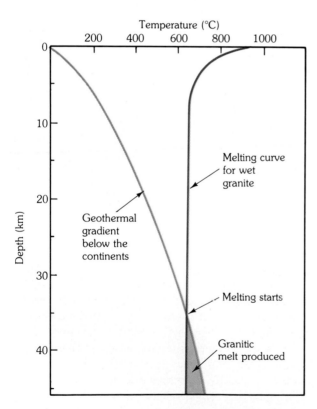

FIGURE 4.23
Proposed melting curve and geothermal gradient which accounts for the fact that burial of wet granitic material to a depth of about 35 kilometers will produce some melting.

In nature, deeply buried rocks melt for one of two reasons. First, rocks melt when they are heated to their melting points. Second, without increasing temperature, a reduction in the confining pressure can lower the melting temperature sufficiently to trigger melting. Both processes are thought to play significant roles in magma formation.

Most basaltic magmas are believed to originate from the partial melting of the rock peridotite, the major constituent of the upper mantle. Laboratory studies confirm that partial melting of this dry, silica-poor rock produces magma having a basaltic composition. Since mantle rocks exist in environments that are characterized by high temperatures and pressures, melting most often results from a reduction in confining pressure. This can occur, for example, where mantle rock ascends as part of a slow-moving convection cell.

Due to the fact that magmas form many kilometers below the surface, we might expect that most of this material would cool and crystallize before reaching the surface. However, as dry basaltic magma moves upward, the confining pressure steadily diminishes and further reduces the melting point. Basaltic magmas appear to ascend rapidly enough so that as they enter cooler environments the heat loss is offset by a drop in the melting point. Consequently, large outpourings of basaltic magmas are common on the earth's surface.

Conversely, granitic magmas are thought to be generated by partial melting of water-rich rocks that were subjected to increased pressure and temperature. Recall that an increase in pressure significantly lowers the melting point of rock that contains appreciable amounts of water. Therefore, burial of wet quartz-rich material to relatively shallow depths is thought to be sufficient to trigger melting (Figure 4.23). Thus, in contrast to some basaltic magmas, where melting is generated by a reduction in pressure, granitic magma is expected to form in a compressional environment characterized by rising pressures.

As a wet granitic melt rises, the confining pressure decreases, which in turn reduces the effect of water on lowering the melting temperature. Thus, in contrast to dry basaltic magmas that produce vast outpourings of lava, most granitic magmas lose their mobility before reaching the surface and therefore tend to produce large intrusive features such as batholiths. On those occasions when silica-rich magmas reach the surface, explosive pyroclastic flows, such as those that produced the Yellowstone Plateau, are the rule.

Since andesitic magma is intermediate in composition between basaltic and granitic magma, its properties are intermediate as well. Although outpourings of andesitic magma are relatively common, the flows are more viscous and consequently thicker and less extensive than those produced when more fluid basaltic magma is extruded. Moreover, a high percentage of andesitic magma is extruded as pyroclastic debris.

Even though most magma is thought to be generated by partial melting, once formed the composition of a magma body can change dramatically with time. For example, as a magma body migrates upward, it may incorporate some of the surrounding country rock. As the country rock is assimilated, the composition of the magma is altered. Further, recall from our discussion of Bowen's reaction series that during solidification, magma often undergoes fractional crystallization. This produces a magma quite unlike the parent material. These processes may account, at least in part, for the fact that a single volcano can extrude lavas with a wide range of chemical compositions.

DISTRIBUTION OF IGNEOUS ACTIVITY

Most of the more than 600 active volcanoes that have been identified are located in the vicinity of convergent plate margins (Figure 4.24). Further, extensive volcanic activity occurs out of view along spreading centers of the oceanic ridge system. In this section we will examine three zones of volcanic activity and

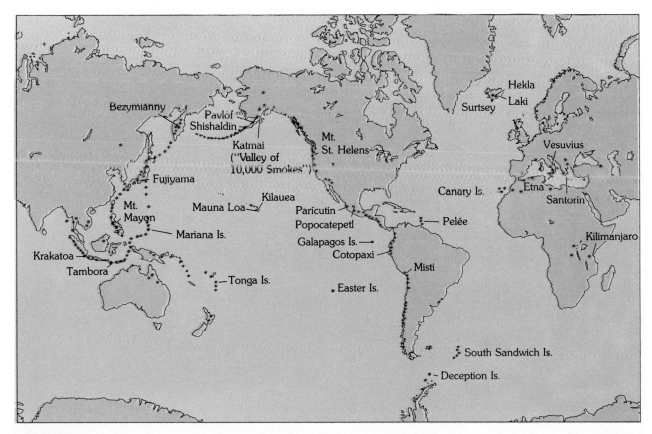

FIGURE 4.24
Locations of some of the most recently formed volcanoes.

relate them to global tectonic activity. These active areas are found along the oceanic ridges, adjacent to ocean trenches, and within the plates themselves (Figure 4.25).

Spreading Center Volcanism As stated earlier, the greatest volume of volcanic rock is produced along the oceanic ridge system where sea-floor spreading is active (Figure 4.25). As the rigid lithosphere pulls apart, the pressure on the underlying rocks is lessened. This reduced pressure, in turn, lowers the melting point of the mantle rocks. Partial melting of these rocks (primarily peridotite) generates large quantities of basaltic magma that move upward to fill the newly formed cracks.

Some of the molten basalt reaches the ocean floor where it produces extensive lava flows or occasionally grows into a volcanic pile. Sometimes this activity

produces a volcanic cone that rises above sea level as the island of Surtsey did in 1963 (Figure 4.26). Numerous submerged volcanic cones also dot the flanks of the ridge system and the adjacent deep-ocean floor. Many of these formed along the ridge crests and were moved away as new oceanic crust was created by the seemingly unending process of sea-floor spreading.

Subduction Zone Volcanism Rocks having an andesitic to granitic composition are confined to the continents and to volcanic island chains, such as the Aleutians, which lie along oceanic margins. Only very small amounts are found as part of the volcanoes in the deep-ocean basins. Further, most active volcanoes that extrude andesitic magma are found on continental areas or island arcs located adjacent to deep-ocean trenches. Recall that ocean trenches are sites

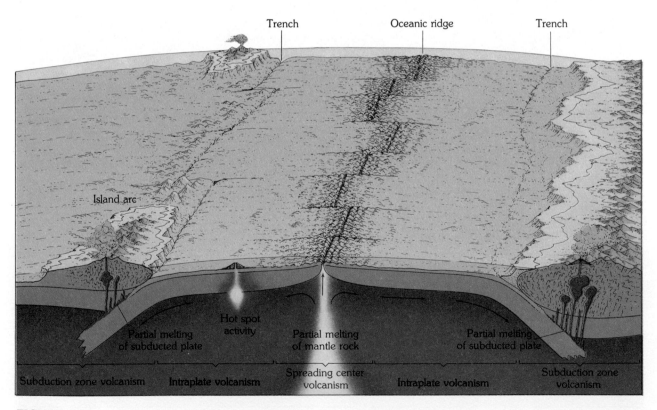

FIGURE 4.25
Three zones of volcanism. Two of these zones are plate boundaries, and the third includes areas within the plates themselves.

FIGURE 4.26
Surtsey emerged from the ocean just south of Iceland in 1963. (Courtesy of Icelandic Photo and Press Service)

where slabs of oceanic crust are bent and move downward into the upper mantle.

When cold oceanic lithosphere reaches depths of about 125 kilometers, melting is believed to take place. The partial melting of these wet, sediment-laden basalts yields a magma of andesitic composition. After a sufficient quantity has melted, this magma buoys upward, because it is less dense than the surrounding rock. The volcanoes of the Andes Mountains, from which andesite obtains its name, are examples of this mechanism at work.

Similarly, some granitic magma may be formed in subduction zones. The parent material is thought to be sediments that were weathered from the continents and transported to the subduction zones. These sediments, which are carried along on subducting oceanic plates, have a high silica content and when melted generate a magma of granitic composition.

The "Ring of Fire," is associated with subduction and melting of the Pacific plate. The volcanoes in this very active zone primarily extrude magma having an intermediate silica content. The volcanoes of the Cascade Range in the northwestern United States, including Mounts St. Helens, Rainier, and Shasta, are all of this type (Figure 4.27).

Intraplate Volcanism The processes that actually trigger volcanic activity within a rigid plate are difficult to establish. Activity such as in the Yellowstone

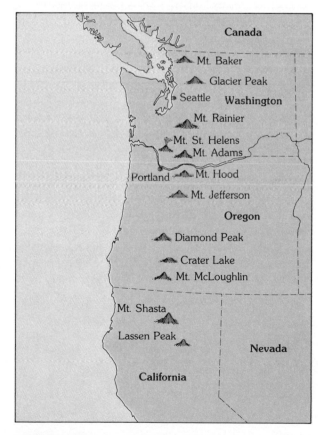

FIGURE 4.27
Locations of several of the larger composite cones that comprise the Cascade Range.

region and other nearby areas produced rhyolitic pumice and ash flows, while extensive basaltic flows cover vast portions of our Northwest. Yet these rocks of greatly varying compositions actually overlie one another in several locations.

Since basaltic extrusions occur on the continents as well as within the ocean basins, the partial melting of upper mantle rocks is the most probable source for this activity. Recall that earth tremors indicate that the island of Hawaii does in fact tap the upper mantle. One proposal is that a small percentage of the rocks of the asthenosphere exists in the molten state. The geothermal gradient and melting curve that would account for this zone of partial melting is illustrated in Figure 4.28. If this environment actually exists, the zone between the depths of 100 and 250 kilometers would contain some melt. From these molten pockets, called **hot spots,** plumes of magma are thought to migrate upward where they often penetrate to the surface. Hot spots are believed to be located beneath Hawaii and Iceland, and may have formerly existed beneath the Columbia Plateau.

Generally lavas and ash of granitic composition are extruded from vents located landward of the continental margins. This suggests that remelting of the continental crust may be one of the mechanisms responsible for the formation of these silica-rich magmas. But what mechanism causes large quantities of continental material to be melted? One proposal suggests that a thick segment of continental crust occasionally becomes situated over a plume of rising magma; that is, a hot spot. Rather than producing vast outpourings of basaltic lava as occurs at oceanic sites such as Hawaii, the magma from the rising plume is emplaced at depth. Here the incorporation and melting of the surrounding country rock, coupled with fractional crystallization, result in the formation

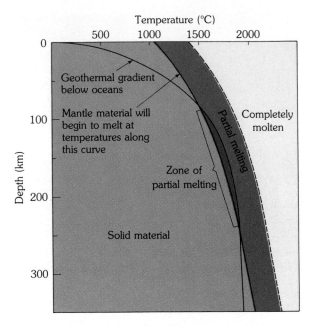

FIGURE 4.28
Idealized geothermal gradient and melting point curve to illustrate the cause of partial melting in the asthenosphere.

of a secondary, silica-rich magma which slowly migrates upward. Continued hot spot activity supplies heat to the rising mass, thereby aiding its ascent. The activity in the Yellowstone region may have resulted from just this type of activity.

Although the plate tectonics theory has answered many of the questions which have plagued volcanologists for decades, many new questions have arisen; for example, Why does sea-floor spreading occur in some areas and not others? How do hot spots originate? These are just two of the many unanswered questions.

REVIEW QUESTIONS

1 List three factors that determine the nature of a volcanic eruption. What role does each play?

2 Why is a volcano fed by highly viscous magma likely to be a greater threat than a volcano supplied with very fluid magma?

3 Contrast pahoehoe and aa lava.

4 List the main gases released during a volcanic eruption.

5 How do volcanic bombs differ from blocks of pyroclastic debris?

6 Compare and contrast the main types of volcanoes as to size, shape, and eruptive style.

7 Name a prominent volcano for each of the three types.

8 Briefly compare the eruptions of Kilauea and Parícutin.

9 Contrast the destruction of Pompeii with the destruction of St. Pierre.

10 Describe the formation of Crater Lake. Compare it to the caldera formed during the eruption of Krakatoa.

11 What is Shiprock, New Mexico, and how did it form?

12 How do the eruptions that created the Columbia Plateau differ from eruptions that create volcanic peaks?

13 Where are fissure eruptions most common?

14 What is partial melting? What factors influence the melting points of rocks?

15 Since basaltic magma forms at great depths, why does it often reach the surface rather than crystallize below the surface?

16 Under what circumstances does andesitic magma originate? Granitic magma?

17 Where does most basaltic magma originate?

KEY TERMS

aa lava (p. 84)

caldera (p. 86)

cinder cone (p. 91)

composite cone (p. 92)

crater (p. 86)

fissure eruption (p. 99)

flood basalt (p. 100)

fumarole (p. 86)

geothermal gradient (p. 101)

hot spot (p. 106)

lava dome (p. 95)

nuée ardente (p. 92)

pahoehoe lava (p. 84)

parasitic cone (p. 86)

partial melting (p. 101)

pillow lava (p. 84)

pipe (p. 97)

pyroclastic flow (p. 100)

pyroclastic material (p. 85)

shield volcano (p. 86)

stratovolcano (p. 92)

viscosity (p. 82)

volcanic neck (p. 97)

volcano (p. 86)

5

WEATHERING
AND
SOIL

To the casual observer the face of the earth may appear to be without change, unaffected by time. For that matter, less than 200 years ago most people believed that mountains, lakes, and deserts were permanent features of an earth that was thought to be no more than a few thousand years old. Today, however, we know that mountains eventually succumb to weathering and erosion and are washed into the sea, lakes fill with sediment and vegetation or are drained by streams, and deserts come and go as relatively minor climatic changes occur.

The earth is indeed a dynamic body. Volcanic and tectonic activities are elevating parts of the earth's surface, while opposing processes are continually removing materials from higher elevations and moving them to lower elevations. The latter processes include:

(1) **Weathering**—the disintegration and decomposition of rock at or near the earth's surface.

(2) **Erosion**—the incorporation and transportation of material by mobile agents such as water, wind, or ice.

(3) **Mass wasting**—the transfer of rock material downslope under the influence of gravity.

The primary focus of this chapter is on rock weathering and the products generated by this activity. However, weathering cannot be easily separated from the other two processes because as weathering breaks rocks apart, it encourages the movement of rock debris by erosion and mass wasting. On the other hand, the transport of material by erosion and mass wasting furthers the disintegration and decomposition of rock.

Delicate Arch, Arches National Park, a monument to the forces of rock weathering. (Photograph used by permission of Dennis Tasa)

111

WEATHERING

All materials are susceptible to weathering. Consider, for example, the fabricated product concrete, which closely resembles a sedimentary rock called conglomerate. A newly-poured concrete sidewalk has a smooth, fresh, unweathered look. However, not many years later the same sidewalk will appear chipped, cracked, and rough, with pebbles exposed at the surface. If a tree is nearby, its roots may heave and buckle the concrete as well. The same natural processes which eventually destroy a concrete sidewalk also act to disintegrate rock.

Weathering occurs when rock is mechanically fragmented (disintegrated) and chemically altered (decomposed). Mechanical weathering is accomplished by physical forces which break rock into smaller and smaller pieces without changing the rock's mineral composition. Chemical weathering, on the other hand, involves a chemical transformation of the rock into one or more new compounds. A simple example of these two concepts can be made using a piece of paper. Disintegration of the paper is accomplished by tearing it into smaller and smaller pieces, whereas decomposition occurs when the paper is set afire and burned.

Why does rock weather? Simply, weathering is the response of earth materials to a changing environment. For instance, after millions of years of uplift and erosion, the rocks overlying a large intrusive igneous body may be removed, exposing it at the surface. This mass of crystalline rock, which formed in a high-temperature, high-pressure environment perhaps several kilometers below ground, is now subjected to a very different and comparatively hostile surface environment. In response, this rock mass will gradually change until it is once again in equilibrium, or balance, with its new environment. This transformation of rock is what we call weathering.

In the following sections we will discuss the various modes of mechanical and chemical weathering. Although we will consider these two processes separately, keep in mind that they usually work simultaneously in nature.

MECHANICAL WEATHERING

When a rock undergoes **mechanical weathering** it is broken into smaller and smaller pieces, each retaining the characteristics of the original material. The end result is many small pieces from a single large one. Figure 5.1 shows that breaking a rock into smaller pieces increases the surface area available for chemical attack. An analogous situation occurs when sugar is added to a liquid. In this situation, a cube of sugar will dissolve much slower than an equal volume of granules because of the vast difference in surface area. Hence, by breaking rocks into smaller pieces, mechanical weathering increases the amount of surface area available for chemical weathering.

In nature four important physical processes lead to the fragmentation of rock: frost wedging, expansion resulting from unloading, thermal expansion, and organic activity.

Frost Wedging   Alternate freezing and thawing is one of the most important processes of mechanical weathering. Water has the unique property of expanding about 9 percent as it freezes. This increase in volume occurs because as water solidifies, the water molecules arrange themselves into a very open crystalline structure. As a result, when water freezes it expands and exerts a tremendous outward force. This can be verified by filling a container with water and freezing it. If sufficient volume does not exist in the container, it will shatter.

In nature, water works its way into cracks or voids in rock and, upon freezing, expands and wedges the rock apart. This process is appropriately called **frost wedging** (Figure 5.2). Frost wedging is most pro-

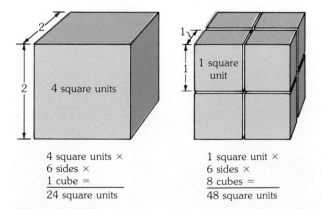

4 square units × 6 sides × 1 cube =
24 square units

1 square unit × 6 sides × 8 cubes =
48 square units

FIGURE 5.1

Mechanical weathering increases the surface area available for chemical attack.

nounced in mountainous regions in the middle latitudes where a daily freeze-thaw cycle often exists. Here, sections of rock are wedged loose and may tumble into large piles called **talus slopes** that often form at the base of steep rock outcrops (Figure 5.3).

Frost wedging also causes great destruction to the highways in the northern United States, particularly in the early spring when the freeze-thaw cycle is well established. Roadways acquire numerous potholes and are occasionally heaved and buckled by this destructive force.

Unloading   When large igneous bodies, particularly those composed of granite, are exposed by erosion, concentric slabs begin to break loose. The process generating these onionlike layers is called **sheeting** and is thought to occur, at least in part, because of the great reduction in pressure when the overlying rock is stripped away. Accompanying the unloading, the outer layers expand more than the rock below, and thus separate from the rock body (Figure 5.4). The

fractures separating the individual slabs are usually more closely spaced near the earth's surface and therefore layers that result are generally less than a meter thick. Further, the fractures typically develop parallel to the surface topography and give the exhumed igneous body a domed shape. Continued weathering eventually causes the slabs produced by sheeting to separate and spall[1] off these large structures which are known as **exfoliation domes.** Excellent examples of exfoliation domes are Stone Mountain, Georgia, and Half Dome and Liberty Cap in Yosemite National Park (Figure 5.5).

Mine shafts provide us with a view of how rocks behave once the confining pressure is removed. Large rock slabs have been known to explode off the walls of newly cut mine shafts because of the reduced pressure. Evidence of this type, plus the fact that fracturing occurs parallel to the floor of a rock quarry when large blocks are removed, strongly supports the process of unloading as the cause of sheeting.

---

[1]Spalling means to break off into chips, scales, or slabs.

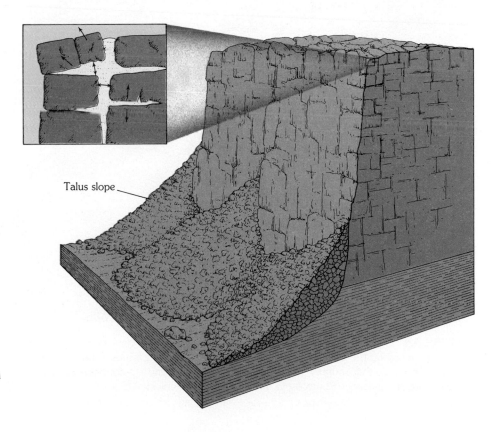

Talus slope

FIGURE 5.2
Frost wedging. As water freezes it expands, exerting a force great enough to break rock.

FIGURE 5.3
Talus slope made of weathered debris at the base of a cliff. (Photo by H. E. Malde, U.S. Geological Survey)

FIGURE 5.4
Sheeting caused by the expansion of crystalline rock as erosion removes the overlying material. Fractures that roughly parallel the surface topography are common in large, intrusive granitic masses. (Photo by G. K. Gilbert, U.S. Geological Survey)

Although many fractures are created by expansion, others are produced by contraction during the crystallization of magma, and still others by tectonic forces during mountain building. Fractures produced by these activities generally form a definite pattern and are called **joints** (Figure 5.6). Joints are important rock structures which allow water to penetrate to depth and start the process of weathering long before the rock reaches the surface.

Thermal Expansion   The daily cycle of temperature change is thought to weaken rocks, particularly in hot, dry regions where daily variations may exceed 30°C. Heating a rock causes expansion and cooling causes contraction. Repeated swelling and shrinking

of minerals with different expansion rates should logically exert some stress on the rock's outer shell.

Although this process was once thought to be of major importance in the disintegration of rock, laboratory experiments have not substantiated this. In one test, unweathered rocks were heated to temperatures much higher than those normally experienced on the earth's surface and then cooled. This procedure was repeated many times to simulate hundreds of years of weathering, but the rocks showed little apparent change.

Nevertheless, in desert areas pebbles do exhibit unmistakable evidence of shattering from what appears to be temperature changes (Figure 5.7). A proposed solution to this dilemma suggests that rocks

FIGURE 5.5
Half Dome, an exfoliation dome in Yosemite National Park. (Photo by F. C. Calkins, U.S. Geological Survey)

FIGURE 5.6
Aerial view of an extensively jointed area in the Canyonlands of southeastern Utah.
(Photo by John S. Shelton)

must first be weakened by chemical weathering before they can be broken down by thermal activity. Further, this process may be aided by the rapid cooling of a desert rainstorm. Additional data are needed before a definite answer can be given as to the impact of temperature variation on rock disintegration.

Organic Activity Weathering is also accomplished by the activities of organisms, including plants, burrowing animals, and man. Plant roots in search of minerals and water grow into fractures, and as the roots grow, they wedge the rock apart (Figure 5.8). Burrowing animals further break down rock by moving fresh material to the surface, where physical and chemical processes can more effectively attack it. Further, decayed organisms produce acids which contribute to chemical weathering. Where rock has been blasted in search of minerals or for road construction, the impact of man is quite noticeable, but on a worldwide scale, humans probably rank behind burrowing animals in earth-moving accomplishments.

Although usually considered separately from mechanical weathering, the activities of the erosional agents—wind, running water, and glaciers—are nonetheless important. For as these mobile agents move rock debris, they relentlessly disintegrate the earth materials they carry.

CHEMICAL WEATHERING

Chemical weathering involves the complex processes that alter the internal structures of minerals by removing and/or adding elements. During this transformation, the original rock decomposes into substances that are in equilibrium, or balance, with the surface environment. Consequently, the products of chemical weathering will remain essentially unchanged as long as they remain in their new environment.

Water is by far the most important agent of chemical weathering. Although pure water is nonreactive, a small amount of dissolved material is generally all that is needed to activate it. The major processes by which water decomposes rock are solution, oxidation, and hydrolysis.

Solution   Perhaps the easiest type of decomposition to envision is the process of **solution.** Just as sugar readily dissolves in water, so too do certain minerals. One of the most water-soluble minerals is halite (table salt), which as you may recall, is composed of sodium and chloride ions. The reason halite readily dissolves in water has to do with the fact that, although this compound maintains overall electrical neutrality, the individual ions retain their respective

FIGURE 5.7
These stones were once rounded stream gravels; however, long exposure in a hot desert climate disintegrated them. (Photo by C. B. Hunt, U.S. Geological Survey)

FIGURE 5.8
Root wedging widens fractures in rock and aids the process of mechanical weathering. (Photo by G. K. Gilbert, U.S. Geological Survey)

charge. Moreover, the surrounding water molecules are polar; that is, the oxygen end of the molecule has a small residual negative charge whereas the end with hydrogen has a small positive charge. As the water molecules collide with the halite crystal, their negative ends contact and disrupt the oppositely-charged sodium ions (Figure 5.9). The attractive force of the water pulls the sodium ions from the crystalline lat-

tice. The chloride ions are similarly removed layer after layer by the positive end of the water molecules.

Although most minerals are, for all practical purposes, insoluble in pure water, the presence of even a small amount of acid dramatically increases the corrosive force of water. (An acidic solution contains the reactive hydrogen ion, H^+.) For instance, the mineral calcite, $CaCO_3$, which composes the common building stones marble and limestone, is easily attacked by even a weakly acidic solution:

$$CaCO_3 \quad + 2[H^+(H_2O)] \longrightarrow$$

calcium carbonate   aqueous acid
(insoluble)

$$Ca^{2+} \quad + CO_2 \uparrow + 3H_2O$$

calcium ion   carbon   water
(soluble)   dioxide

During this process, the insoluble calcium carbonate is transformed into soluble products. In nature, over periods of thousands of years, large quantities of limestone are dissolved and carried away by groundwater. This activity is clearly evidenced by the large number of subsurface caverns found in every one of the contiguous forty-eight states. Monuments and buildings made of limestone or marble are also subjected to the corrosive work of acids, particularly in industrial areas that have smoggy, polluted air.

The soluble ions from reactions of this type are retained in our underground water supply. It is these dissolved ions that are responsible for the so-called "hard water" found in many locales. Simply, hard water is undesirable because the active ions react with soap to produce an insoluble material that renders soap nearly useless in removing soil. To solve this problem a water softener can be used to remove these ions, generally by replacing them with others that do not chemically react with soap.

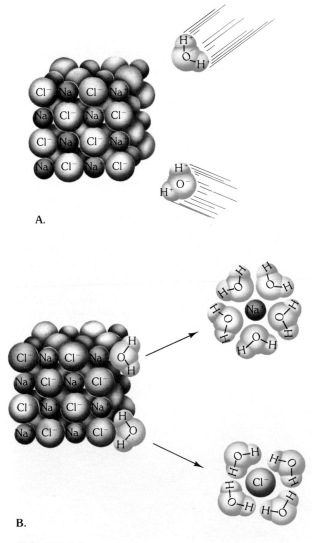

A.

B.

FIGURE 5.9
Illustration of halite dissolving in water. **A.** Sodium and chloride ions are attacked by the polar water molecules. **B.** Once removed, these ions are surrounded and held by a number of water molecules as shown.

Oxidation The process of rusting occurs when oxygen combines with iron to form iron oxide as follows:

$$4Fe + 3O_2 \longrightarrow 2Fe_2O_3$$

iron   oxygen   iron oxide
(hematite)

This type of chemical reaction, called **oxidation**[1], occurs when electrons are lost from one element during the reaction. In this case, we say that iron was oxidized because it lost electrons to oxygen. Although the oxidation of iron progresses very slowly in a dry environment, the addition of water greatly speeds the reaction.

Oxidation is very important in decomposing such ferromagnesian minerals as olivine, pyroxene, and amphibole. Oxygen readily combines with the iron in these minerals to form the reddish-brown iron oxide called hematite (Fe_2O_3) or in more extreme cases a yellowish-colored rust called limonite [$FeO(OH)$]. These products are responsible for the rusty color on the surfaces of dark igneous rocks, such as basalt, as they begin to weather. However, oxidation can only occur after iron is freed from the silicate structure by another process called hydrolysis.

Hydrolysis The most common mineral group, the silicates, is decomposed primarily by the processes of **hydrolysis,** which in the broadest sense is the reaction of any substance with water. Ideally, the hydrolysis of a mineral could take place in pure water since some of the water molecules dissociate to form the very reactive hydrogen (H^+) and hydroxyl (OH^-) ions. It is the hydrogen ion which attacks and replaces other positive ions found in the crystal lattice. With the introduction of hydrogen ions into the crystalline structure, the original orderly arrangement of atoms is destroyed and the mineral decomposes.

In nature, water usually contains other substances that contribute additional hydrogen ions, thereby greatly accelerating hydrolysis. The most common of these substances is carbon dioxide, CO_2, which dissolves in water to form carbonic acid, H_2CO_3. Generally, rain dissolves some carbon dioxide as it falls through the atmosphere, and additional amounts, released by decaying organic matter, are acquired as the water percolates through the soil. In water, carbonic acid ionizes to form hydrogen ions (H^+) and bicarbonate ions (HCO_3^-). To illustrate how a rock undergoes hydrolysis in the presence of carbonic acid, we will examine the chemical weathering of

granite, the most abundant continental rock. Recall that granite consists mainly of quartz and potassium feldspar. The weathering of the potassium feldspar component of granite is as follows:

$$2KAlSi_3O_8 + 2(H^+ + HCO_3^-) + H_2O \longrightarrow$$
potassium   carbonic acid   water
feldspar

$$Al_2Si_2O_5(OH)_4 + 2K^+ + 2HCO_3^- + 4SiO_2$$
kaolinite   potassium   bicarbonate   silica
(residual clay)   ion   ion
in solution

In this reaction, the hydrogen ions (H^+) attack and replace potassium ions (K^+) in the feldspar structure, thereby disrupting the crystalline network. Once removed, the potassium is available as a nutrient for plants or becomes the soluble salt potassium bicarbonate ($KHCO_3$), which may be incorporated into other minerals or carried to the ocean.

The most abundant by-product of the chemical breakdown of potassium feldspar is the clay mineral kaolinite, which forms through the addition of hydroxyl ions to the crystal lattice. Clay minerals are the end product of weathering and are very stable under surface conditions. Consequently, clay minerals make up a high percentage of the inorganic material in soils. Further, the most abundant sedimentary rock, shale, is also composed of clay minerals. Also during this reaction some silica is removed from the feldspar structure and carried away by groundwater. This dissolved silica will eventually precipitate, producing nodules of chert or flint, or fill in the pore spaces of such things as buried wood to produce petrified wood, or be carried to the ocean, where microscopic animals will remove it to build hard silica shells.

To summarize, the weathering of potassium feldspar generates a residual clay mineral, a soluble salt (potassium bicarbonate), and some silica which enters into solution.

Quartz, the other main component of granite, is very resistant to chemical weathering; hence it remains substantially unaltered when attacked by weakly acidic solutions. As a result, when granite weathers, the feldspar crystals dull and slowly turn to

---

[1]The reader should note that *oxidation* is a term referring to any chemical reaction in which a compound or radical loses electrons. The element oxygen is not necessarily present.

clay, releasing the once-interlocked quartz grains, which still retain their fresh, glassy appearance. Although some of the quartz remains in the soil, much is transported to the sea, where it becomes the main constituent of sandy beaches and in time is often converted to the sedimentary rock sandstone.

Table 5.1 lists the weathered products of some of the most common silicate minerals. Remember that silicate minerals make up most of the earth's crust and that these minerals are essentially composed of only eight elements. When chemically weathered, the silicate minerals yield sodium, calcium, potassium, and magnesium ions that form soluble products which may be removed by groundwater. The element iron combines with oxygen, producing relatively insoluble iron oxides, most notably hematite and limonite, which give soil a reddish-brown or yellowish color. Under most conditions the three remaining elements, aluminum, silicon, and oxygen, join with water to produce residual clay minerals. However, even the highly insoluble clay minerals are very slowly removed (leached) by subsurface water.

Alterations Caused by Chemical Weathering   As noted earlier the most significant result of chemical weathering is the decomposition of unstable minerals and the generation or retention of those materials which are in equilibrium at the earth's surface. This accounts for the predominance of certain minerals in the surface material we call soil.

TABLE 5.1
Products of weathering.

Mineral	Residual Products	Material in Solution
Quartz	Quartz grains	Silica
Feldspars	Clay minerals	Silica K^+, Na^+, Ca^{2+}
Amphibole (hornblende)	Clay minerals Limonite Hematite	Silica Ca^{2+}, Mg^{2+}
Olivine	Limonite Hematite	Silica Mg^{2+}

A.

B.

C.

FIGURE 5.10
Spheroidal weathering of extensively jointed rock. Water moving through the joints begins to enlarge them. Since the rocks are attacked more on the corners and edges, they take on a spherical shape.

In addition to altering the internal structure of minerals, chemical weathering causes physical changes. For instance, when angular rock fragments are attacked by water flowing through joints, the fragments tend to take on a spherical shape. The gradual rounding of the corners and edges of angular blocks is illustrated in Figure 5.10. The corners are attacked most readily because of the greater surface area for their volume as compared to the edges and faces. This process gives the weathered rock a spherical shape and is called **spheroidal weathering** (Figure 5.11).

Quite commonly during the formation of spheroidal boulders, successive shells separate from the rock's main body (Figure 5.12). Eventually the outer shells spall off, allowing the chemical weathering activity to penetrate deeper into the boulder. This spherical scaling results because, as the minerals in the rock weather to clay, they increase in size through the addition of water to their structure. This increased bulk exerts an outward force that causes concentric layers of rock to break loose and fall off. Hence, chemical weathering does produce forces great enough to cause mechanical weathering. This type of spheroidal weathering in which shells spall off should not be confused with the phenomenon of sheeting discussed earlier. In sheeting, the fracturing occurs as a result of unloading and the rock layers which separate from the main body are largely unaltered at the time of separation.

RATES OF WEATHERING

Several factors influence the type and rate of rock weathering. Most important of these are rock structure, climate, and topography.

Rock Structure Rock structure encompasses all of the chemical characteristics of rocks, including mineral composition and solubility, as well as any physical features that may be present, such as fractures, bedding planes, and voids. The variations in weathering rates, attributable to the mineral constituents, can be demonstrated by comparing old headstones carved from different rock types. Headstones made of granite, which is composed of silicate minerals, are relatively resistant to chemical weathering as we can see by examining the inscriptions on the headstones

FIGURE 5.11
Spheroidal weathering of a massive granite outcrop in eastern Arizona. (Photo by Stephen Trimble)

FIGURE 5.12
Successive shells are loosened as the weathering process continues to penetrate ever deeper into the rock. (Photo by Kenneth Hasson)

shown in Figure 5.13. This is not true of the marble headstone which shows signs of extensive chemical alteration over a relatively short period. Recall that marble is composed of calcium carbonate, which readily dissolves even in a weakly acidic solution.

The most abundant mineral group, the silicates, weathers in the order shown in Figure 5.14. This arrangement of minerals is identical to that of Bowen's reaction series. The order in which the silicate minerals weather is essentially the same as their order of crystallization. The minerals that crystallize first form under much higher temperatures than those that crystallize last. Consequently, the early-formed minerals are not as stable at the earth's surface, where the temperature and pressure are drastically different from the environment in which they formed. By examining Figure 5.14, we see that olivine crys-

tallizes first and is therefore the least resistant to chemical weathering, while quartz, which crystallizes last, is the most resistant.

Climate Climatic factors, particularly temperature and moisture, are of primary significance to the rate of rock weathering. These climatic elements largely determine the weathering rate and indirectly determine the kind and amount of vegetation present. Regions with lush vegetation generally have a thick mantle of soil rich in decayed organic matter from which chemically active fluids such as carbonic and humic acids are derived.

The optimum environment for chemical weathering is a combination of warm temperatures and abundant moisture. In polar regions chemical weathering is ineffective because frigid temperatures keep the

A. B.

FIGURE 5.13
An examination of headstones reveals the rate of chemical weathering on diverse rock types. The granite headstone (left) was erected six years before the marble headstone (right), whose inscription date, 1894, is nearly illegible. (Photos by E. J. Tarbuck)

FIGURE 5.14
The stability of common silicate minerals in relation to chemical weathering. Those minerals which crystallize at high temperatures decompose most rapidly and vice versa.

Olivine
Pyroxene
Amphibole
Biotite

Calcium feldspar

Sodium feldspar

**Increasing
Resistance
to Chemical
Weathering**

Potassium feldspar
Muscovite
Quartz

available moisture locked up as ice, whereas in arid regions there is insufficient moisture to foster rapid chemical weathering. A classic example of how climate affects the rate of weathering was provided when Cleopatra's Needle, a granite obelisk, was moved from Egypt to New York City. After withstanding approximately 3500 years of exposure in the dry climate of Egypt, the hieroglyphics were almost completely removed from the windward side in less than 75 years in the wet and chemical-laden air of New York City (Figure 5.15).

Topography Topography greatly influences the amount of rock exposed to the forces of weathering. In addition, the topographic setting may indirectly determine the amount of precipitation as well as influence the kind and amount of vegetation present.

Angular topography with large rock outcrops is most prevalent in arid regions whereas more subdued topography mantled with soil and vegetation is found in humid areas. These differences are often attributed to the predominance of chemical weathering in a humid region and mechanical weathering in

A.

B.

FIGURE 5.15
Chemical weathering of Cleopatra's Needle, a granite obelisk. **A.** Before it was removed from Egypt. (Courtesy of the Metropolitan Museum of Art) **B.** After a span of 75 years in New York City's Central Park. After surviving intact for about 35 centuries in Egypt, the windward side has been almost completely defaced in less than a century. (Courtesy of New York City Parks)

FIGURE 5.16
Bryce Canyon National Park. Weathering accentuates differences in rocks to produce some of our most spectacular scenery. (Courtesy of G. Marie Conover)

an arid region. It is probably more nearly correct to say that chemical weathering is most important in both environments, while the effects of mechanical weathering are relatively more significant and thus more obvious in an arid setting.

The sum of these factors determines the type and rate of rock weathering for a given region. However, there is generally enough variation, even within a relatively small area, for the rocks to exhibit some differential weathering. Differential weathering and subsequent erosion are responsible for most of the unusual and often spectacular landforms. Included are features such as natural bridges like those found in Arches National Park (see chapter-opening photo) and sculptured rock pinnacles such as those found in Bryce Canyon National Park (Figure 5.16).

SOIL

Soil has accurately been called "the bridge between life and the inanimate world." All life owes its existence to the dozen or so elements that ultimately come from the earth's crust. First, weathering disintegrates and decomposes rock. Then, plants execute their intermediary role by assimilating the necessary elements from this weathered debris and making them available to both animals and humans.

With few exceptions, the earth's land surface is covered by **regolith,** the layer of rock and mineral fragments produced by weathering. Some would call this material soil. However, true soil is more than an accumulation of loose rock debris. **Soil** is a combination of mineral and organic matter, water, and air. It is the portion of the regolith that supports plant growth. Although the proportions may vary, the major components do not (Figure 5.17). About one-half of the total volume of a good quality surface soil is a mixture of disintegrated and decomposed rock (mineral matter) and the decayed remains of animal and plant life (organic matter). The remaining half consists of pore spaces, where air and water circulate.

THE SOIL PROFILE

If you were to dig a trench, you would see that its walls consisted of a series of essentially horizontal layers. These layers, called **horizons,** collectively make up the **soil profile** (Figure 5.18). Three basic horizons

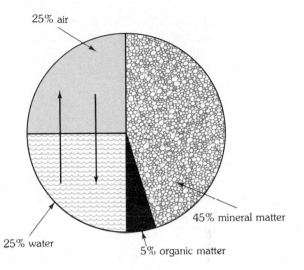

FIGURE 5.17
Composition (by volume) of a soil in good condition for plant growth. Although the percentages vary, each soil is composed of mineral and organic matter, water, and air.

are identified and from top to bottom are labeled A, B, and C.

The uppermost layer in a soil profile, the A horizon, is often called the **surface soil.** This is the part of the soil with the greatest biological activity and is therefore the horizon where organic matter is most plentiful. Since it lies at the surface, the A horizon is also the part of the soil that rainwater penetrates first. As a consequence, soluble materials and tiny particles such as clay are *leached* (washed out) by the percolating water.

Lying immediately below the surface soil is the B horizon, or **subsoil.** In this zone much of the material removed from the A horizon is deposited. Because of this, the B horizon is also referred to as the *zone of accumulation.* Since the B horizon has an intermediate position in the soil profile, it may be considered, at least in part, a transitional zone. Living organisms and organic matter are more abundant in the B than in the C horizon, but considerably less so than in the A horizon. The A and B horizons together constitute the **solum,** or "true soil." It is in the solum that the soil-forming processes are active and that living roots and other plant and animal life are largely confined.

Below the solum is the C horizon, a layer characterized by partially altered rock debris and little if any organic matter (Figure 5.18). Thus, while the material from which the soil formed may be so dramatically altered in the solum that its original character is not recognizable, it is easily identifiable in the *C* horizon.

The boundaries between soil horizons may be very sharp, or the horizons may blend gradually from one to another. Furthermore, some soils lack horizons altogether. Such soils are called **immature,** because soil building has been going on for only a short time. Immature soils are also characteristic of steep slopes where erosion continually strips away the soil, preventing full development.

CONTROLS OF SOIL FORMATION

Soil is the product of the complex interplay of several factors, including parent material, time, climate, plants and animals, and slope. Although all of these factors are interdependent, it will be helpful to examine their roles separately.

Parent Material　The **parent material** from which a soil has evolved may be either the underlying bedrock or a layer of unconsolidated deposits. Soils formed on bedrock are termed **residual soils,** whereas those developed on unconsolidated deposits are called **transported soils** (Figure 5.19).

FIGURE 5.18
Soil profile. Mature soils are characterized by a series of horizontal layers called horizons, which comprise the soil profile.

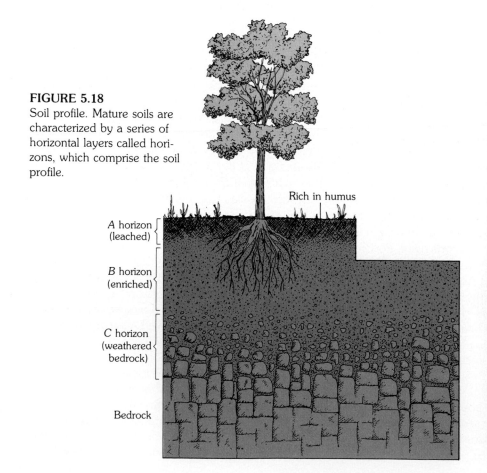

Rich in humus

A horizon (leached)

B horizon (enriched)

C horizon (weathered bedrock)

Bedrock

The nature of the parent material influences soils in two ways. First, the type of parent material to some degree will affect the rate of weathering, and thus the rate of soil formation. Also, since unconsolidated deposits are already partly weathered, soil development on such material will likely progress more rapidly than when bedrock is the parent material. Second, the chemical makeup of the parent material will affect the soil's fertility. For instance, if it lacks the elements necessary for plant growth, its usefulness is obviously diminished.

At one time the parent material was believed to be the primary factor causing differences among soils. Today soil scientists realize that other factors, especially climate, are more important. In fact, it has been found that similar soils are often produced from different parent materials and that dissimilar soils have developed from the same parent material. Such discoveries reinforce the importance of the other soil-forming factors.

Time If weathering has been going on for a comparatively short time, the character of the parent material determines to a large extent the characteristics of the soil. As the weathering process continues, the influence of parent material on soil is overshadowed by the other soil-forming factors. The amount of time required for various soils to evolve cannot be listed because the soil-forming processes act at varying rates under different circumstances. However, as a rule the longer a soil has been forming, the thicker it becomes and the less it resembles the parent material.

Climate Climate is considered to be the most important control of soil formation, since it determines whether chemical or mechanical weathering will predominate and also greatly influences the rate and depth of weathering. For instance, a hot, wet climate may produce a thick layer of chemically weathered soil in the same amount of time that a cold, dry

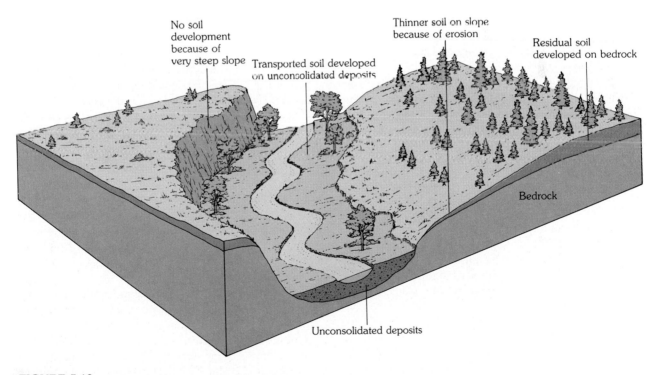

FIGURE 5.19
Parent material. The parent material for residual soils is the underlying bedrock, while transported soils form on unconsolidated deposits. Also note that soils are thinner, or nonexistent, on the slopes.

climate produces a thin mantle of mechanically weathered debris. Also, the amount of precipitation influences the degree to which various materials are leached from the soil, thereby affecting soil fertility. Finally, climatic conditions are an important control on the type of plant and animal life present.

Plants and Animals The chief function of plants and animals is to furnish organic matter to the soil. Certain bog soils are composed almost entirely of organic matter, while desert soils may contain as little as a small fraction of one percent. Although the quantity of organic matter varies substantially among soils, no soil completely lacks it.

The primary source of organic matter is plants, although animals and an infinite number of microorganisms also contribute. When organic matter is decomposed, important nutrients are supplied to plants, as well as to animals and microorganisms living in the soil. Consequently, soil fertility is in part related to the amount of organic matter present. Furthermore, the decay of plant and animal remains causes the formation of various organic acids. These complex acids hasten the weathering process. Organic matter also has a high water-holding ability and thus aids water retention in a soil.

Microorganisms, including fungi, bacteria, and single-celled protozoa, play an active role in the decay of plant and animal remains. The end product is **humus,** a jellylike material that no longer resembles the plants and animals from which it formed. In addition, certain microorganisms aid soil fertility because they have the ability to *fix* (change) atmospheric nitrogen into soil nitrogen.

Earthworms and other burrowing animals act to mix the mineral and organic portions of a soil. Earthworms, for example, feed on organic matter and thoroughly mix soils in which they live, often moving and enriching many tons per acre each year. Burrows and holes also aid the passage of water and air through the soil.

Slope Slope has a significant impact on the amount of erosion and the water content of soil. On steep slopes soils are often poorly developed. In such situations the quantity of water soaking in is slight, and as

a result, the moisture content of the soil may not be sufficient for vigorous plant growth. Further, because of accelerated erosion on steep slopes, the soils are thin, or in some cases nonexistent (Figure 5.19). On the other hand, poorly drained and waterlogged soils found in bottomlands have a much different character. Such soils are usually very thick and very dark. The dark color results from the large quantity of organic matter that accumulates because saturated conditions retard the decay of vegetation. The optimum slope for soil development is a flat-to-undulating upland surface. Here we find good drainage, minimum erosion, and sufficient infiltration of water into the soil.

Slope orientation, the direction the slope is facing, is another aspect worthy of mention. In the mid-latitudes, a south-facing slope will receive a great deal more sunlight than a north-facing slope. In fact, a steep north-facing slope may receive no direct sunlight at all. The difference in the amount of solar radiation received will cause differences in soil temperature and moisture, which in turn may influence the nature of the vegetation and the character of the soil.

Although this section dealt separately with each of the soil-forming factors, remember that all work together to form soil. No single factor is responsible for a soil being as it is, but rather it is the combined influence of parent material, time, climate, plants and animals, and slope, that determines a soil's character.

SOIL TYPES

In the following discussion, we will briefly examine some common soil types. As you read, notice that the characteristics of each soil type are primarily manifestations of the prevailing climatic conditions. A summary of the characteristics of the soils discussed in this section is provided in Table 5.2.

The term **pedalfer** gives a clue to the basic characteristic of this soil type. The word is derived from the Greek **ped***on,* meaning ''soil,'' and the chemical symbols **Al** (aluminum) and **Fe** (iron). Pedalfers are characterized by an accumulation of iron oxides and aluminum-rich clays in the *B* horizon. In mid-latitude areas where the annual rainfall exceeds 63 centimeters (25 inches) most of the soluble materials, such as calcium carbonate, are leached from the soil and carried away by underground water. The less soluble iron oxides and clays are carried from the *A* horizon and deposited in the *B* horizon, giving it a brown to red-brown color. These soils are best developed under forest vegetation where large quantities of decomposing organic matter provide the acid conditions necessary for leaching. In the United States pedalfers are found east of a line extending from northwestern Minnesota to south-central Texas.

Pedocal is derived from the Greek **ped**on, meaning "soil," and the first three letters of **cal**cite (calcium carbonate). As the name implies, pedocals are characterized by an accumulation of calcium carbonate. This soil type is found in the drier western United States in association with grassland and brush vegetation. Since chemical weathering is less intense in

drier areas, pedocals generally contain a smaller percentage of clay minerals than pedalfers.

In the arid and semiarid western states a lime-enriched layer called **caliche** often develops in the soils. In these areas little of the rain that falls penetrates to great depths. Rather it is held by the soil particles near the surface until it evaporates. As a result, the soluble materials, chiefly calcium carbonate, are removed from the uppermost layer and redeposited below, forming the caliche layer.

In the hot, wet climates of the tropics, soils called **laterites** develop. Since chemical weathering is intense under such climatic conditions, these soils are usually deeper than soils developing over a similar period in the mid-latitudes. Not only does leaching remove the soluble materials such as calcite, but the great quantities of percolating water also remove much of the silica, with the result that oxides of iron

TABLE 5.2
Summary of soil types.

Climate	Temperate humid (>63 cm rainfall)	Temperate dry (<63 cm rainfall)	Tropical (heavy rainfall)	Extreme arctic or desert
Vegetation	Forest	Grass and brush	Grass and trees	Almost none, so no humus develops
Typical Area	Eastern U.S.	Western U.S.		
Soil Type	Pedalfer	Pedocal	Laterite	
Topsoil	Sandy; light colored; acid	Commonly enriched in calcite; whitish color	Enriched in iron (and aluminum); brick red color. All other elements removed by leaching	No real soil forms because there is no organic material. Chemical weathering is very slow
Subsoil	Enriched in aluminum, iron, and clay; brown color	Enriched in calcite; whitish color	*(Zones not developed)*	
Remarks	Extreme development in conifer forests, because abundant humus makes groundwater very acid. Produces light gray soil because of removal of iron	*Caliche* is name applied to the accumulation of calcite	Apparently bacteria destroy humus, so no acid is available to remove iron	

and aluminum become concentrated in the soil. The iron gives the soil a distinctive red color (Figure 5.20). When dried, laterites are very hard. In fact, some people use this soil for making bricks. If the parent rock contained little iron, the product of weathering is an aluminum-rich accumulation called *bauxite*. Bauxite is the primary ore of aluminum.

Since bacterial activity is very high in the tropics, laterites contain practically no humus. This fact, coupled with the highly leached and bricklike nature of these soils, makes laterites poor for growing crops. The infertility of these soils has been borne out repeatedly in tropical countries where cultivation has been expanded into such areas.

In cold or dry climates soils are generally very thin and poorly developed. The reasons for this are fairly obvious. Chemical weathering progresses very slowly in such climates, and the scanty plant life yields very little organic matter.

FIGURE 5.20
Characteristic red color of a well-developed lateritic soil.

REVIEW QUESTIONS

1 Differentiate between the products of mechanical weathering and chemical weathering.

2 In what type of environment is frost wedging most effective?

3 Describe the processes of sheeting and spheroidal weathering. How are they different and how are they similar?

4 How does mechanical weathering add to the effectiveness of chemical weathering?

5 Granite and basalt are exposed at the surface in a hot, wet region.
   (a) Which type of weathering will predominate?
   (b) Which of these rocks will weather most rapidly? Why?

6 Heat speeds up a chemical reaction. Why then does chemical weathering proceed slowly in a hot desert?

7 How is carbonic acid (H_2CO_3) formed in nature? What results when this acid reacts with potassium feldspar?

8 What is the difference between soil and regolith?

9 List the characteristics associated with each of the horizons in a well-developed soil profile. Which of the horizons constitute the solum? Under what circumstances do soils lack horizons?

10 What factors might cause different soils to develop from the same parent material, or similar soils to form from different parent materials?

11 Which of the controls of soil formation is most important? Explain.

12 How can slope affect the development of soil? What is meant by the term *slope orientation?*

13 Distinguish between pedalfers and pedocals.

14 Soils formed in the humid tropics and the Arctic both contain little organic matter. Do both lack humus for the same reasons?

KEY TERMS

caliche (p. 129)

chemical weathering (p. 117)

erosion (p. 111)

exfoliation dome (p. 113)

frost wedging (p. 112)

horizon (p. 125)

humus (p. 128)

hydrolysis (p. 119)

immature soil (p. 126)

joints (p. 115)

laterite (p. 129)

mass wasting (p. 111)

mechanical weathering (p. 112)

oxidation (p. 119)

parent material (p. 126)

pedalfer (p. 128)

pedocal (p. 129)

regolith (p. 125)

residual soil (p. 126)

sheeting (p. 113)

soil (p. 125)

soil profile (p. 125)

solum (p. 125)

solution (p. 117)

spheroidal weathering (p. 121)

subsoil (p. 125)

surface soil (p. 125)

talus slope (p. 113)

transported soil (p. 126)

weathering (p. 111)

6

SEDIMENTARY ROCKS

6

The products of mechanical and chemical weathering constitute the raw materials for sedimentary rocks. The word *sedimentary* indicates the nature of these rocks, for it is derived from the Latin *sedimentum,* which means "settling," a reference to solid material settling out of a fluid. Most but not all sediment is deposited in this fashion. Weathered debris is constantly being swept from bedrock, carried away, and eventually deposited in lakes, river valleys, seas, and countless other places. The particles in a desert sand dune, the mud on the floor of a swamp, the gravels in a stream bed, and even household dust are examples of this never-ending process. Since the weathering of bedrock and the transport and deposition of the weathering products are continuous, sediment is found almost everywhere. As piles of sediment accumulate, the materials near the bottom are compacted. Over long periods, these sediments are cemented together by mineral matter deposited in the spaces between particles to form solid rock.

Geologists estimate that sedimentary rocks account for only about 5 percent (by volume) of the earth's outer 16 kilometers (10 miles). However, the importance of this group of rocks is far greater than this percentage would imply. If we were to sample the rocks exposed at the earth's surface, we would find that the great majority are sedimentary. Indeed, about 75 percent of all rock outcrops on the continents are sedimentary. Therefore, we may think of sedimentary rocks as comprising a relatively thin and somewhat discontinuous layer in the uppermost portion of the crust. This fact is

Tapeats Sandstone outcropping in the Grand Canyon. (Photo by E. J. Tarbuck)

135

readily understood when we consider that sediment accumulates at the surface of the earth.

Since sediments accumulate at the earth's surface, the rock layers that they eventually form contain evidence of past events at the surface. By their very nature, sedimentary rocks contain within them indications of past environments in which their particles were deposited and in some cases, clues to the mechanisms involved in their transport. Furthermore, it is sedimentary rocks that contain fossils, which are vital tools in the study of the geologic past. Thus, it is largely from this group of rocks that geologists must reconstruct the details of earth history.

Finally, it should be mentioned that many sedimentary rocks are very important economically. Coal, for example, is classified as a sedimentary rock, whereas our other major energy resources, petroleum and natural gas, are found in association with sedimentary rocks. Still others represent major sources of iron, aluminum, manganese, and fertilizer as well as numerous materials essential to the construction industry.

TYPES OF SEDIMENTARY ROCKS

Materials accumulating as sediment have two principle sources. First, sediments may be accumulations of materials that originate and are transported as solid particles derived from both mechanical and chemical weathering. Deposits of this type are termed *detrital* and the sedimentary rocks that they form are called **detrital sedimentary rocks.** The second major source of sediment is soluble material produced largely by chemical weathering. When these dissolved substances are precipitated by either inorganic or organic processes, the material is known as chemical sediment and the rocks formed from it are called **chemical sedimentary rocks.**

DETRITAL SEDIMENTARY ROCKS

Though a wide variety of minerals and rock fragments may be found in detrital rocks, clay minerals and quartz are the chief constituents of most sedimentary rocks in this category. Recall that clay minerals are the most abundant product of the chemical weathering of silicate minerals, especially the feld-

spars. Clays are fine-grained minerals with sheetlike crystalline structures similar to the micas. The other common mineral, quartz, is abundant because it is very resistant to chemical weathering. Thus, when igneous rocks such as granite are attacked by weathering processes, individual quartz grains are freed.

Other common minerals in detrital rocks are the feldspars and micas. Since chemical weathering rapidly transforms these minerals into new substances, their presence in a sedimentary rock indicates that mechanical rather than chemical weathering was responsible for creating the sediment.

Particle size is the primary basis for distinguishing among various detrital sedimentary rocks. Table 6.1 presents the size categories for particles making up detrital rocks. Note that in this context the term *clay* refers only to a particular size and not to the minerals of the same name. Although most clay minerals are of clay size, not all clay-sized sediment consists of clay minerals.

The size of the particles in a detrital rock can often be related to the energy of the transporting medium. Currents of water or air sort the particles by size; the stronger the current, the larger the particle size carried. Gravels, for example, are moved by swiftly flowing rivers as well as by landslides and glaciers. Less energy is required to transport sand, thus it is common to such features as windblown dunes, as well as some river deposits and beaches. Since silts and clays settle very slowly, accumulations of these materials are generally associated with the quiet waters of a lake, lagoon, swamp, or marine environment.

TABLE 6.1
Particle size classification for detrital rocks.

Size Range (millimeters)	Particle Name	Common Sediment Name	Detrital Rock
>256 64–256 4–64 2–4	Boulder Cobble Pebble Granule	Gravel	Conglomerate or breccia
1/16–2	Sand	Sand	Sandstone
1/256–1/16 <1/256	Silt Clay	Mud	Shale or mudstone

Shale *Shale* is a sedimentary rock consisting of silt- and clay-sized particles (Figure 6.1A). These fine-grained detrital rocks account for an estimated 70 percent of all sedimentary rocks. The particles in these rocks are so small that they cannot be readily identified without great magnification (Figure 6.2). Since the sediments are not only microscopic but tend to be flat or tabular as well, they usually become tightly packed. As a result there is very little space through which solutions containing cementing material can circulate. Therefore, shales are commonly not well cemented and readily crumble. Although shale is very abundant, it is still the least well known common sedimentary rock. Shale does not form prominent outcrops as do sandstone and other common sedimentary rocks. In addition, shale weathers easily and usually forms a cover of soil that hides the unweathered rock below.

The word *shale* is often applied to all fine-grained detrital rocks, yet many geologists have a more restricted use of the term. In this more restricted usage, shale must exhibit the ability to split into thin

A.

B.

C.

D.

FIGURE 6.1
Common detrital sedimentary rocks. **A.** Shale. **B.** Sandstone. **C.** Conglomerate.
D. Breccia. (Photos by E. J. Tarbuck)

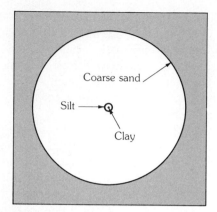

FIGURE 6.2
Relative sizes of three kinds of detrital sediments. Even though the sediments have been enlarged about 60 times, the clay particle can barely be seen.

layers along well-developed, closely spaced planes. If the rock breaks into chunks or blocks, the name *mudstone* is applied.

Sandstone   *Sandstone* is the name given rocks when sand-sized grains predominate (Figure 6.1B). After shale, sandstone is the most abundant sedimentary rock, accounting for approximately 20 percent of the total group. In most sandstones, quartz is the predominant mineral. When this is the case, the rock may simply be called *quartz sandstone*. If the sandstone contains appreciable quantities of feldspar, the rock is called *arkose*. The presence of abundant feldspar is an indication that the sediment was subjected to little chemical weathering. Finally, a third type of sandstone is known as *graywacke*. In addition to containing quartz and feldspar, this dark-colored rock contains abundant angular rock fragments and clay. Because its particles are characteristically poorly sorted, graywacke is often referred to as ''dirty'' sandstone.

Conglomerate   *Conglomerate* consists largely of gravels (Figure 6.1C). As Table 6.1 indicates, these particles may range in size from large boulders to particles as small as garden peas. The large particles in a conglomerate are commonly rock fragments. Usually

the openings between the gravel particles are filled with mud and sand, then the entire mass in cemented into a hard rock. If the large particles are angular rather than rounded, the rock is called *breccia* (Figure 6.1D).

CHEMICAL SEDIMENTARY ROCKS

In contrast to detrital rocks, which form from the solid products of weathering, chemical sediments derive from material that is carried in solution to lakes and seas. This material does not remain dissolved in the water indefinitely, however. Rather, some of it precipitates to form chemical sediments. This precipitation of material may occur directly as the result of inorganic processes or indirectly as the result of the life processes of water-dwelling organisms. Sediment formed in this second way is said to have a biochemical origin (Figure 6.3).

An example of a deposit resulting from inorganic chemical processes is the salt left behind as a body of salt water evaporates. In contrast, many water-dwelling animals and plants extract dissolved mineral matter to form shells and other hard parts. After the organisms die, their skeletons collect on the floor of a lake or ocean.

Limestone   Representing about 10 percent of the total volume of all sedimentary rocks, *limestone* is the most abundant chemical sedimentary rock. It is composed chiefly of the mineral calcite ($CaCO_3$) and forms by either inorganic means or as the result of biochemical processes. Limestones having a biochemical origin are by far the most common. As much as 90 percent of the world's limestone may have originated as accumulations of biochemical sediment.

Although most limestone is the product of biological processes, this origin is not always evident because shells and skeletons may undergo considerable change before being converted to rock. However, one easily identified biochemical limestone is *coquina,* a coarse rock composed of poorly cemented shells and shell fragments. Another less obvious, but nevertheless familiar example is *chalk*, a rock made up almost entirely of the hard parts of foraminifera,

FIGURE 6.3
This sedimentary rock composed largely of shells and shell fragments has an obvious biochemical origin. (Photo by E. J. Tarbuck)

microscopic organisms no larger than the head of a pin (Figure 6.4).

Limestones having an inorganic origin form when evaporation and/or high water temperatures increase the concentration of calcium carbonate to the point that it precipitates. *Travertine,* the type of limestone commonly seen in caves, is an example, as is *oolitic limestone.* Travertine is deposited when groundwater containing calcium carbonate evaporates. The formation of oolitic limestone, a rock composed of small spherical grains called *oolites,* is somewhat more complex. Oolites form in shallow marine waters as tiny "seed" particles are moved back and forth by currents and become coated with layer upon layer of calcium carbonate as they roll on the ocean floor.

Dolomite   Closely related to limestone is *dolomite,* a rock composed of the calcium-magnesium carbonate mineral of the same name. Dolomite is one of the rare cases where the same term is used to designate both a mineral and a rock. To avoid confusion, some

FIGURE 6.4
The White Chalk Cliffs of Dover. (Photo by Robert Ryan)

geologists refer to the rock as *dolostone.* Although dolomite can form by direct precipitation from seawater, it is thought that most originates when magnesium in seawater replaces some of the calcium in limestone. The latter theory is reinforced by the fact that there are practically no young dolomite rocks. Rather, most dolomites are ancient rocks in which there was ample time for magnesium to replace calcium.

Chert *Chert* is a name used for a variety of very dense and hard rocks made of microcrystalline silica (SiO_2). One well-known form is *flint,* whose dark color results from the organic matter it contains. *Jasper,* a red variety, gets its bright color from the iron oxide it contains.

Chert deposits are commonly found in one of two situations: as irregularly shaped nodules in limestone and as layers of rock. The silica composing most chert nodules is believed to have been deposited directly from water. Thus, these nodules have an inorganic origin. However, it is unlikely that a very high percentage of chert layers was precipitated directly from seawater, because seawater is not generally saturated with silica. Hence, beds of chert are thought to have originated largely as biochemical sediment. Although most water-dwelling organisms that produce hard parts secrete shells made of calcium carbonate, some, such as diatoms and radiolarians, produce glasslike silica skeletons. These tiny organisms are able to extract silica from very unsaturated solutions. It is from these remains that beds of chert are made. Note that when a chert specimen is being examined, there are few reliable criteria by which the mode of origin (inorganic versus biochemical) can be determined.

Rock Salt and Rock Gypsum Very often evaporation is the mechanism triggering deposition of chemical precipitates. Minerals commonly precipitated in this fashion include halite (sodium chloride), the chief component of *rock salt,* and gypsum (hydrous calcium sulfate), the main ingredient of *rock gypsum.*

In the geologic past, many places that are now dry land were covered by shallow arms of the sea that eventually became cut off from the main body of water (much as the Caspian Sea is today). As the water evaporated, the salts were left behind as **evaporite deposits.** Today these deposits serve as an important source of many chemicals. Similar deposits may be seen in such places as Death Valley, California. Here, following rains or periods of snowmelt in the mountains, streams flow from the surrounding mountains into an enclosed basin. As the water evaporates, **salt flats** form from dissolved materials left behind as a white crust on the ground (Figure 6.5).

TURNING SEDIMENT INTO SEDIMENTARY ROCK

Lithification refers to the processes by which unconsolidated sediments are transformed into solid sedimentary rocks. One of the most common processes affecting sediments is **compaction.** As sediments accumulate through time, the weight of overlying material compresses the deeper sediments. As the grains are pressed closer and closer, there is a considerable reduction in pore space. For example, when clays are buried beneath several thousand meters of material, the volume of the clay may be reduced by as much as 40 percent. Since sands and other coarse sediments are only slightly compressible, compaction is most significant as a lithification process in fine-grained sedimentary rocks such as shale.

Cementation is another important means by which sediments are converted to sedimentary rocks. The cementing materials are carried in solution by water percolating through the open spaces between particles. Through time, the cement precipitates onto the sediment grains, fills the open spaces, and joins the particles. Calcite, silica, and iron oxide are the most common cements. The identification of the cementing material is a relatively simple matter. Calcite cement will effervesce with dilute hydrochloric acid. Silica is the hardest cement and thus produces the hardest sedimentary rocks. When a sedimentary rock has an orange or dark red color, this usually means that iron oxide is present. In some instances, cements can even be economically significant. For example, the iron oxide cement in the Clinton Formation in the

Appalachians was rich enough to make this sandstone into an important iron ore.

Although most sedimentary rocks are lithified by compaction, cementation, or a combination of both, some are made of interlocking crystals. This type of lithification is confined largely to certain chemical sedimentary rocks.

CLASSIFICATION OF SEDIMENTARY ROCKS

The classification scheme in Table 6.2 on page 142 divides sedimentary rocks into two major groups: detrital and chemical. Further, we can see that the main criterion for subdividing the detrital rocks is particle size, whereas the primary basis for distinguishing among different rocks in the chemical group is their mineral composition.

As is the case with many (perhaps most) classifications of natural phenomena, the categories presented in Table 6.2 are more rigid than the actual state of nature. In reality, many of the sedimentary rocks classified into the chemical group also contain at least small quantities of detrital sediment. Many limestones, for example, contain varying amounts of mud or sand, giving them a "sandy" or "shaly" quality. On the other hand, since practically all detrital rocks are cemented with material that was originally dissolved in water, they too are far from being "pure."

As was the case with the igneous rocks examined in Chapter 3, texture is an important aspect of sedimentary rock classification. There are two major textures used in the classification of sedimentary rocks: clastic and nonclastic. The term **clastic** is taken from a Greek word meaning "broken." Thus clastic rocks are made of broken fragments. An examination of Table 6.2 reveals that all detrital rocks have a clastic texture. The table also shows that some chemical sedimentary rocks may also exhibit this texture. For example, coquina, the limestone composed of shells and shell fragments, is obviously as clastic as a con-

FIGURE 6.5
These salt flats (composed of gypsum and rock salt) in Death Valley, California, are examples of evaporite deposits. (Photo by James E. Patterson)

TABLE 6.2
Classification of sedimentary rocks.

DETRITAL ROCKS

Texture	Sediment Name and Particle Size	Comments	Rock Name
Clastic	Gravel >2 mm	Rounded rock fragments	Conglomerate
		Angular rock fragments	Breccia
	Sand 1/16–2 mm	Quartz predominates	Quartz sandstone
		Quartz with considerable feldspar	Arkose
		Dark color; quartz with considerable feldspar, clay, and rock fragments	Graywacke
	Mud <1/16 mm	Splits into thin layers	Shale
		Breaks into clumps or blocks	Mudstone

CHEMICAL ROCKS

Group	Texture	Composition	Rock Name
Inorganic	Clastic or nonclastic	Calcite, $CaCO_3$	Limestone
	Clastic or nonclastic	Dolomite, $CaMg(CO_3)_2$	Dolomite (dolostone)
	Nonclastic	Microcrystalline quartz, SiO_2	Chert
	Nonclastic	Halite, $NaCl$	Rock salt
	Nonclastic	Gypsum, $CaSO_4 \cdot 2H_2O$	Rock gypsum
Biochemical	Clastic or nonclastic	Calcite, $CaCO_3$	Limestone
	Nonclastic	Microcrystalline quartz, SiO_2	Chert
	Nonclastic	Altered plant remains	Coal

glomerate or sandstone. The same applies for some varieties of oolitic limestone.

Some chemical sedimentary rocks have a **nonclastic** texture in which the minerals form a pattern of interlocking crystals. Because of this, some nonclastic sedimentary rocks may look more like igneous rocks, which are also composed of intergrown crystals. However, the two are usually easy to distinguish between because the minerals that compose nonclastic sedimentary rocks are quite different from the minerals found in igneous rocks.

SEDIMENTARY STRUCTURES

As stated earlier, sedimentary rocks are particularly important in the interpretation of earth history. These rocks form at the earth's surface and as layer upon layer of sediment accumulates, each records the nature of the environment at the time the sediment was deposited. These layers, called **strata** or **beds,** are probably the single most characteristic feature of sedimentary rocks (Figure 6.6).

The thickness of beds ranges from microscopically thin to tens of meters thick. Separating the strata are **bedding planes,** flat surfaces along which rocks tend to separate or break. Changes in the grain size or in the composition of the sediment being deposited can create bedding planes. Pauses in deposition can also lead to layering because chances are slight that newly deposited material will be exactly the same as previously deposited sediment. Generally each bedding plane marks the end of one deposit and the beginning of another.

Since sediments usually accumulate as particles settle from a fluid, most strata were originally deposited as horizontal layers. There are circumstances, however, when sediments do not accumulate in horizontal beds. Sometimes when a bed of sedimentary rock is examined, we see layers inclined at a steep angle to the horizontal. Such layering is termed **cross-bedding** and is most characteristic of sand dunes and river deltas (Figure 6.7).

Graded beds represent another special type of bedding. In this case the particles within a single sedimentary layer gradually change from coarse at the bottom to fine at the top. Graded beds are most characteristic of rapid deposition from water containing sediment of varying sizes. When a current experiences a rapid energy loss, the largest particles settle first, followed by successively smaller grains. The deposition of a graded bed is most often associated with a turbidity current, a mass of sediment-choked water that is denser than clear water and moves downslope along the bottom of a lake or the ocean.

As geologists examine sedimentary rocks, much can be deduced. A conglomerate, for example, may indicate a high-energy environment, such as a rushing stream, where only course materials can settle out (Figure 6.8). If the rock is arkose, it may signify a dry climate where little chemical alteration of feldspar is possible. Carbonaceous shale is a sign of a low-energy, organic-rich environment, such as a swamp or lagoon.

Other features found in some sedimentary rocks also give clues to past environments. Ripple marks are such a feature. **Ripple marks** are small waves of sand that develop on the surface of a sediment layer by the action of moving water or air (Figure 6.9A, page 146). The ridges form at right angles to the direction of motion. If the ripple marks were formed by air or water moving in essentially one direction, their form will be asymmetrical. These *current ripple marks* will have steeper sides in the downcurrent direction and more gradual slopes on the upcurrent side. Ripple marks produced by a stream flowing across a sandy channel or by wind blowing over a sand dune are two common examples of current ripples. When present in solid rock, they may be used to determine the direction of movement of ancient wind or water currents. Other ripple marks have a symmetrical form. These features, called *oscillation ripple marks*, result from the back and forth movement of surface waves in a shallow nearshore environment.

Mud cracks (Figure 6.9B) indicate that the sediment in which they formed was alternately wet and dry. When exposed to air, wet mud dries out and shrinks, producing cracks. Mud cracks are associated with such environments as shallow lakes and desert basins.

A.

B.

FIGURE 6.7
The cross-bedding in this sandstone indicates that it was once a sand dune. (Photograph used by permission of Dennis Tasa)

FIGURE 6.8
Cross-section of an ancient stream channel filled with conglomerate. In a high-energy environment, such as a rushing stream, only the coarse materials can settle out. (Photo by W. R. Hansen, U.S. Geological Survey)

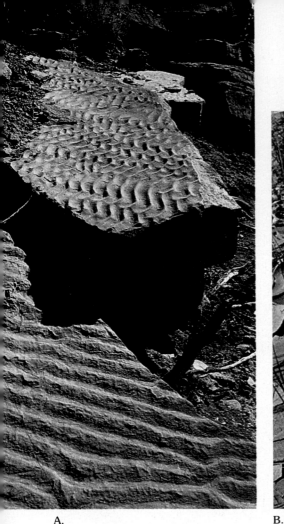

A.   B.

FIGURE 6.9
A. Asymmetrical ripple marks such as these are produced by currents of water or wind. (Photo by Stephen Trimble). **B.** Mud cracks form when wet mud or clay dries out and shrinks. (Photo by Garrett Deckert)

FOSSILS

Fossils, the remains or traces of prehistoric life, are perhaps the most important inclusions in sedimentary rocks. Fossils are important tools used to interpret the geologic past. Knowing the nature of the life forms that existed at a particular time helps researchers understand past environmental conditions. Furthermore, fossils are important time indicators and play a key role in correlating rocks of similar ages but from different places.[1]

Fossils can be of many types. The remains of relatively recent organisms may not be altered at all. However, given enough time, actual remains are

[1]Chapter 19 contains a more detailed discussion of the role of fossils in the interpretation of earth history.

likely to be modified. Often fossils become petrified (literally "turned into stone"), meaning that the original substance, such as wood or bone, has been replaced by mineral matter from circulating solutions or that open spaces have been filled by minerals (Figure 6.10A). Impressions constitute another common class of fossils (Figure 6.10B). Sometimes impressions serve as molds that may be filled with sediment that subsequently hardens. When the rock is split, both the impression (mold) and cast may faithfully reflect the shape and surface markings of the organism (Figure 6.10C). A type of fossilization in which a thin carbon film is left behind is a common way in which leaves and delicate animals are preserved (Figure 6.10D). In addition, there are numerous other fossil types, many of them only traces of prehistoric life.

A.

B.

C.

D.

FIGURE 6.10
There are many types of fossilization. Four examples are shown here. **A.** Petrified wood. **B.** Impressions are common fossils and often show considerable detail. **C.** Natural casts of some shelled invertebrates. **D.** A fossil bee preserved as a thin carbon film. (Photos A, B, and C by E. J. Tarbuck; Photo D courtesy of the National Park Service)

Only a tiny fraction of the organisms that lived during the geologic past has been preserved as fossils. Normally the remains of an animal or plant are totally destroyed. Under what circumstances are they preserved? Two special conditions appear to be necessary: rapid burial and the possession of hard parts.

Usually when an organism perishes, its remains are quickly eaten by scavengers or decomposed by bacteria. Occasionally, however, the remains are buried by sediment. When this occurs the remains are removed from the environment where destructive forces operate most effectively. Rapid burial therefore is an important condition favoring preservation.

In addition, organisms have a much better chance of being preserved as part of the fossil record if they have hard parts. Although traces and imprints of soft-bodied animals such as jellyfish, worms, and insects exist, they are rare, to say the least. Flesh usually decays so rapidly that preservation is exceedingly remote. Shells, bones, and teeth as well as similar hard parts predominate in the record of past life.

Because preservation is contingent on special conditions, the record of life in the geologic past is biased. The fossil remains of those organisms with hard parts that lived in areas of sedimentation is quite abundant. However, we only get an occasional glimpse of the vast array of other life forms that did not meet the special conditions favoring preservation.

ENERGY RESOURCES FROM SEDIMENTARY ROCKS

Coal, oil, and natural gas are the primary fuels of our complex industrial economy. In recent years, as the "oil crisis" became headline news, so too did interest in these vital energy resources. In addition to spurring an increase in the production and use of coal, the "crisis" led to renewed interest in the development of alternative energy sources. One of these, oil shale, has been mentioned often as part of the possible solution to our nation's energy problems. In the following section, we will briefly examine these important resources associated with sedimentary rocks.

COAL

Coal is difficult to classify because it is quite different from other sedimentary rocks. Nevertheless, it is often grouped with biochemical sedimentary rocks. However, unlike other rocks in this category, which are calcite- or silica-rich, coal is made of organic matter. Close examination of a piece of coal under a microscope or magnifying glass often reveals the presence of various plant structures such as leaves, bark, and wood that have been chemically altered but are nevertheless still identifiable. This supports the conclusion that coal is the end product of the burial of large amounts of plant material over extended periods.

Along with oil and natural gas, coal is commonly called a fossil fuel. Such a designation is certainly appropriate since each time we burn coal we are using energy from the sun that was stored by plants many millions of years ago. We are indeed burning a "fossil."

The initial stage in coal formation is the accumulation of large quantities of plant remains. However, special conditions are required for such accumulations, because dead plants readily decompose when exposed to the atmosphere or other oxygen-rich environments. One important environment that allows for the buildup of plant material is a swamp. Since stagnant swamp water is oxygen deficient, complete decay (oxidation) of the plant material is not possible. Rather, the plants are attacked by certain bacteria that partly decompose the organic material and liberate oxygen and hydrogen. As these elements escape, the percentage of carbon gradually increases. The bacteria are not able to finish the job of decomposition because they are themselves destroyed by acids liberated from the plants.

The partial decomposition of plant remains in an oxygen-poor swamp creates a layer of *peat,* a soft, brown material in which plant structures are still easily recognized. With shallow burial, peat is changed to *lignite,* a soft, brown coal. Burial increases the temperature of sediments as well as the pressure on them. The higher temperatures bring about chemical reactions within the plant materials and yield water and organic gases (volatiles). As the load increases, the water and volatiles are pressed out and the proportion of fixed carbon (solid combustible material) increases. The greater the carbon content, the higher the coal will rank as a fuel. During burial the coal also becomes increasingly compact. For example, deeper burial transforms lignite into a harder, more compacted black coal called *bituminous*. Compared to the peat from which it formed, a bed of bituminous coal may be only one-tenth as thick.

Lignite and bituminous coals are sedimentary rocks, but a later product, called *anthracite* (a very hard, black coal) is a metamorphic rock. Anthracite forms when sedimentary layers are subjected to the folding and deformation associated with mountain building. The heat and pressure of mountain building cause a further loss of volatiles and water, thus increasing the concentration of fixed carbon. Although anthracite is a clean-burning fuel, only a relatively small amount is mined. This stems from the fact that anthracite is not widespread and is more difficult and expensive to extract than the flat-lying layers of bituminous coal.

Coal has been an important fuel for centuries. In the nineteenth and early twentieth centuries, cheap and plentiful coal powered the industrial revolution.

By 1900 coal was providing 90 percent of the energy produced in the United States. Although still important, coal currently provides less than 20 percent of the Unites States' energy needs and ranks third as a source of fuel behind oil and natural gas. More than 70 percent of present-day coal usage is for the generation of electricity. Although coal has not been the premiere fuel for many decades, its use in the future will no doubt increase. The resurgence in coal production is prompted by diminishing and sometimes unreliable oil supplies coupled with increasing oil prices. Expanded coal production is possible because the world has enormous reserves. In the United States, coal fields are widespread and contain supplies that should last for hundreds of years (Figure 6.11).

Although coal enjoys the advantage of being plentiful, its recovery and use present a number of problems. Strip mining can turn the countryside into a

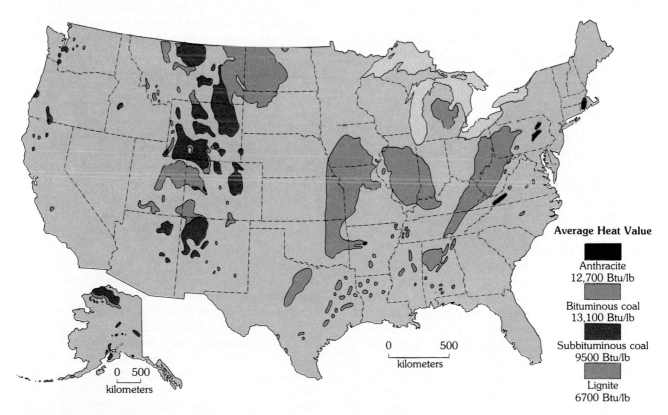

Average Heat Value

Anthracite
12,700 Btu/lb

Bituminous coal
13,100 Btu/lb

Subbituminous coal
9500 Btu/lb

Lignite
6700 Btu/lb

FIGURE 6.11
Coal fields of the United States. (Courtesy of the Bureau of Mines, U.S. Department of the Interior)

scarred wasteland if careful (and costly) reclamation is not carried out to restore the land. Although underground mining does not scar the landscape to the same degree, it is more costly in terms of human life and health. Mining accidents and ''black lung'' disease plague those who mine coal below the surface.

Air pollution is a major problem associated with the burning of coal. Much coal contains significant quantities of sulfur, which when burned is converted into noxious sulfur oxide gases. Through a series of complex chemical reactions in the atmosphere, the sulfur oxides are converted to sulfuric acid, which then falls to the earth's surface as rain or snow. This acid precipitation can have severe ecological effects over widespread areas. Although steps have been taken and will continue to be, the air pollution produced by burning coal is still a significant threat to our environment.

Since none of the problems just mentioned is likely to prevent the increased use of this important and abundant fuel, stronger efforts must be made to correct the problems associated with the mining and use of coal.

OIL AND NATURAL GAS

Petroleum and natural gas are found in similar environments and typically occur together. Both are mixtures of hydrocarbon compounds, that is, compounds consisting of hydrogen and carbon, and may also contain small quantities of other elements such as sulfur, nitrogen, and oxygen. Like coal, petroleum and natural gas are biological products derived from the remains of organisms. However, coal is formed mostly from plant material that accumulated in a swampy environment above sea level. Oil and gas, on the other hand, are derived from the remains of both plants and animals having a marine origin.

Petroleum formation is complex and not completely understood. Nevertheless, we know that it begins with the accumulation of sediment in ocean areas that are rich in plant and animal remains. These accumulations must occur where biological activity is high, such as in nearshore areas. However, most marine environments are oxygen-rich, which leads to the decay of organic remains before they can be bur-

ied by other sediments. Therefore, accumulations of oil and gas are not as widespread as the marine environments that support abundant biological activity. This limiting factor notwithstanding, large quantities of organic matter are buried and protected from oxidation in many offshore sedimentary basins. With increasing burial over millions of years, chemical reactions gradually transform some of the original organic matter into the liquid and gaseous hydrocarbons we call petroleum and natural gas.

Unlike the organic matter from which they formed, the newly-created petroleum and natural gas are mobile. These fluids are gradually squeezed from the compacting, mud-rich layers where they originated into adjacent permeable beds such as sandstone, where openings between sediment grains are larger. The rock layers containing the oil and gas are saturated with water and, since they are not as dense as water, oil and gas migrate upward through the water-filled pore spaces of the enclosing rocks. Unless something acts to halt this upward migration, the fluids will eventually reach the surface, at which point the volatile components will evaporate.

A geologic environment that allows for economically significant amounts of oil and gas to accumulate is termed an **oil trap.** Although, as we shall see, several geologic structures may act as oil traps, all have two basic conditions in common: a porous, permeable **reservoir rock** that will yield petroleum and natural gas in sufficient quantities to make drilling worthwhile; and a **cap rock** such as shale, that is virtually impermeable to oil and gas. The cap rock keeps the upwardly mobile oil and gas from escaping at the surface.

Figure 6.12 illustrates some common oil and natural gas traps. One of the simplest traps is an anticline, an uparched series of sedimentary strata (Figure 6.12A). As the strata are folded, the rising oil and gas collect at the apex of the fold. Because of its lower density, the natural gas collects above the oil. Both rest upon the denser water that saturates the reservoir rock. One of the world's largest oil fields, El Nala in Saudi Arabia, is the result of an anticlinal trap as is the famous Teapot Dome in Wyoming. *Fault traps* form when strata are displaced in such a manner as to bring a dipping reservoir rock opposite an impermeable bed as shown in Figure 6.12B. In this case the

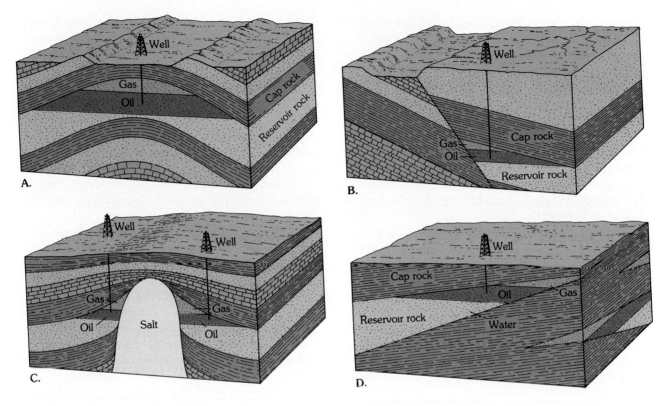

FIGURE 6.12
Common oil traps. **A.** Anticline. **B.** Fault trap. **C.** Salt dome. **D.** Stratigraphic (pinch out) trap.

upward migration of the oil and gas is halted at the fault zone. In the Gulf coastal plain region of the United States important accumulations of oil occur in association with *salt domes*. In such areas, which are characterized by thick accumulations of sedimentary strata, layers of rock salt occurring at great depth have been forced to rise in columns by the pressure of overlying beds. These rising salt columns gradually deform the overlying strata. Since oil and gas migrate to the highest level possible, they accumulate in the upturned sandstone beds adjacent to the salt column (Figure 6.12C). Yet another important geologic circumstance that may lead to significant accumulations of oil and gas is termed a *stratigraphic trap*. These oil-bearing structures result primarily from the original pattern of sedimentation rather than structural deformation. The stratigraphic trap illustrated in Figure 6.12D exists because a sloping bed of sandstone

thins to the point of disappearance.

When the lid created by the cap rock is punctured by drilling, the oil and natural gas, which are under pressure, migrate from the pore spaces of the reservoir rock to the drill hole. On rare occasions the fluid pressure may force oil up the drill hole to the surface. Usually, however, a pump is required to lift the oil out.

A drill hole is not the only means by which oil and gas can escape from a trap. Traps can be broken by natural forces as well. For example, earth movements may create fractures that allow the hydrocarbon fluids to escape, or surface erosion may breach a trap with similar results. The older the rock strata, the greater the chance that deformation or erosion have affected a trap. Indeed, not all ages of rock yield oil and gas in the same proportions. The greatest production comes from the youngest rocks, those of

Cenozoic age. Older Mesozoic rocks produce considerably less, followed by even smaller yields from the still older Paleozoic strata. There is virtually no oil produced from the most ancient rocks, those of Precambrian age.

OIL SHALE

In recent years, the development of our oil shale resources has been suggested as a partial solution to the problem of dwindling fuel supplies. Indeed, an enormous amount of untapped oil is locked in the rock called oil shale. Worldwide, the U.S. Geological Survey estimates that there are more than 3000 billion barrels of oil contained in shales that would yield more than 38 liters (10 gallons) of oil per ton of shale. This figure, however, is deceptively high because less than 200 billion barrels are known to be recoverable with present technology. Roughly half of the worldwide supply is in the Green River Formation that encompasses portions of Colorado, Utah, and Wyoming (Figure 6.13). Within this region the oil shales are part of sedimentary layers that accumulated at the bottoms of two vast, shallow lakes about 50 million years ago.

Oil shale is a very fine-grained sedimentary rock that contains enough organic matter to yield at least 38 liters of oil per ton. Although many other organic-rich shales contain some oil, these shales are such a low grade that they usually are not called oil shale. Oil shale does not actually contain oil but rather a waxy hydrocarbon material called *kerogen*. When heated to about 480°C (900°F), this complex solid mixture of carbon compounds decomposes into hydrocarbons and a carbonaceous residue. When cooled, the hydrocarbons condense into a liquid called shale oil. The oil from shale produces a fuel that can be burned in boilers or with proper (but expensive) conditioning can be upgraded for other uses.

Although pilot projects to extract shale oil have been in operation for years, there have not yet been any serious attempts to commercially exploit this vast resource. The primary reason for this lack of development is the fact that oil from shale has not been economically competitive with oil from traditional sources. In addition, shale oil production requires significant quantities of water, a scarce resource in the semiarid states where the richest oil shale reserves are located. The disposal of huge quantities of waste shale left from the extraction process presents another serious, unanswered problem.

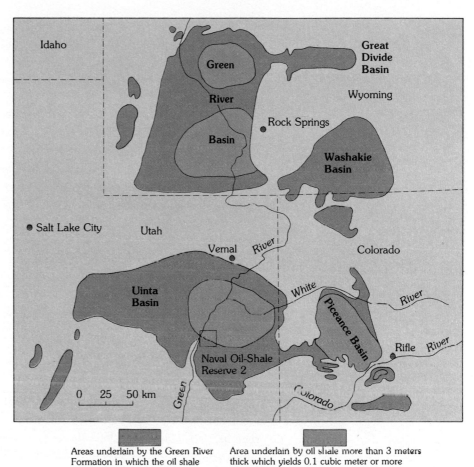

FIGURE 6.13
Distribution of oil shale in the Green River Formation of Colorado, Utah, and Wyoming. (After D. C. Duncan and V. E. Swanson, U.S. Geological Survey Circular 523, 1965)

Areas underlain by the Green River Formation in which the oil shale is unappraised or of low grade

Area underlain by oil shale more than 3 meters thick which yields 0.1 cubic meter or more oil per ton of shale

REVIEW QUESTIONS

1 How does the volume of sedimentary rocks in the earth's crust compare with the volume of igneous rocks in the crust? Are sedimentary rocks evenly distributed throughout the crust?

2 What minerals are most common in detrital sedimentary rocks? Why are these minerals so abundant?

3 What is the primary basis for distinguishing among various detrital sedimentary rocks?

4 The term *clay* can be used in two different ways. Describe the two meanings of this term.

5 Why does shale usually crumble quite easily?

6 Distinguish between conglomerate and breccia.

7 Distinguish between the two categories of chemical sedimentary rocks.

8 What are evaporite deposits? Name a rock that is an evaporite.

9 Compaction is an important lithification process with which sediment size?

10 List three common cements for sedimentary rocks. How might each be identified?

11 What is the primary basis for distinguishing among different chemical sedimentary rocks?

12 Distinguish between clastic and nonclastic textures. What type of texture is common to all detrital sedimentary rocks?

13 What is probably the single most characteristic feature of sedimentary rocks?

14 Distinguish between cross-bedding and graded bedding.

15 How is coal different from other biochemical sedimentary rocks?

16 How is bituminous coal different from lignite? How is anthracite different from bituminous?

17 Coal enjoys the advantage of being plentiful. What are some disadvantages associated with the production and use of coal?

18 What is an oil trap? List two conditions common to all oil traps.

19 Many organic-rich shales are not called oil shale. Why?

20 Although the United States has huge oil shale deposits, shale oil is not produced commercially. Explain.

KEY TERMS

beds (strata) (p. 143)

bedding plane (p. 143)

cap rock (p. 150)

cementation (p. 140)

chemical sedimentary rock (p. 136)

clastic (p. 141)

compaction (p. 140)

cross-bedding (p. 143)

detrital sedimentary rock (p. 136)

evaporite deposit (p. 140)

fossil (p. 146)

graded bed (p. 143)

lithification (p. 140)

mud crack (p. 143)

nonclastic (p. 143)

oil trap (p. 150)

reservoir rock (p. 150)

ripple mark (p. 143)

salt flat (p. 140)

strata (bed) (p. 143)

7

METAMORPHIC ROCKS

7

METAMORPHISM

The process of metamorphism involves the transformation of pre-existing rock. Metamorphic rocks can form from igneous, sedimentary, or even from other metamorphic rocks. The term for this process is very appropriate because it literally means to "change form." The agents of change include heat, pressure, and chemically active fluids, while the changes that occur are textural as well as mineralogical.

In some instances metamorphic rocks are only slightly changed, becoming more compact. In other cases the transformation is so complete that the identity of the original rock cannot be determined. In high-grade metamorphism, such features as bedding planes, fossils, and vesicles that may have existed in the parent rock are completely destroyed. Further, when subjected to intense heat and directional pressure, these rocks behave plastically and bend into intricate folds (Figure 7.1). In the most extreme metamorphic environments, the temperatures approach those at which rocks melt. However, during metamorphism the deformed material must remain solid, for once melting occurs, we have entered the realm of igneous activity.

The process of metamorphism takes place when rock is subjected to conditions unlike those in which it formed; the rock becomes unstable and gradually

Vishnu Schist. Precambrian-aged metamorphic rocks exposed in the inner gorge of the Grand Canyon, Arizona. (Photo by E. J. Tarbuck)

FIGURE 7.1
Intricately folded rock of Cabbage Island, Maine. (Photo by W. B. Hamilton, U.S. Geological Survey)

changes until a state of equilibrium with the new environment is reached. The changes occur at the temperatures and pressures existing in the region extending from a few kilometers below the earth's surface to the crust-mantle boundary. Since the formation of metamorphic rocks is completely hidden from view (which is not the case for many sedimentary and some igneous rocks), metamorphism is undoubtedly one of the most difficult processes for geologists to study.

Metamorphism most often occurs in one of three settings. First, during mountain building great quantities of rock are subjected to the intense stresses and temperatures associated with large-scale deformation. The end result may be extensive areas of metamorphic rocks that are said to have undergone **regional metamorphism.** The greatest volume of metamorphic rock is produced in this fashion. Second, when rock is in contact or close proximity to a mass of magma, **contact metamorphism** takes place. In this circumstance the changes are caused primarily by the high temperatures of the molten material, which in effect "bake" the surrounding

rock. The third and least common type of metamorphism occurs along fault zones. Here rock is broken and distorted as crustal blocks on opposite sides of a fault grind past one another.

AGENTS OF METAMORPHISM

As stated earlier, the agents of metamorphism include heat, pressure, and chemically active fluids. During metamorphism, rocks are often subjected to all three metamorphic agents simultaneously. However, the degree of metamorphism and the contribution of each agent varies greatly from one environment to another. In low-grade metamorphism, rocks are subjected to temperatures and pressures only slightly greater than those associated with the lithification of sediments. High-grade metamorphism, on the other hand, involves extreme conditions closer to those at which rocks melt.

HEAT AS A METAMORPHIC AGENT

Perhaps the most important agent of metamorphism is heat. Rocks formed near the earth's surface may be subjected to intense heat when they are intruded by

158

molten material rising from below. The effects of contact metamorphism are most apparent when it occurs at or near the surface where the temperature contrast between the molten, intrusive rock and the country rock is most pronounced. Here the adjacent country rock is "baked" by the emplaced magma. In this high-temperature and high-pressure environment, the boundary that forms between the intrusive igneous body and the altered rocks is usually quite distinct.

Rocks that originate in a surface environment may also be subjected to extreme temperatures if they are subsequently buried deep within the earth. Recall that temperatures increase with depth at a rate known as the thermal gradient. In the upper crust, this increase in temperature averages about 30°C per kilometer. As we discussed earlier, earth materials are continually being transported to great depths at convergent plate boundaries. When buried to a depth of only a few kilometers, certain minerals, such as clay, become unstable and begin to recrystallize into minerals that are stable in this environment. Other minerals, particularly those found in crystalline igneous rocks, are stable at relatively high temperatures and pressures and therefore require burial to 20 kilometers or more before metamorphism will occur.

PRESSURE AS A METAMORPHIC AGENT

Pressure, like temperature, also increases with depth. Buried rocks are subjected to the force exerted by the load above. This confining pressure is analogous to air pressure where the force is applied equally in all directions.

In addition to the pressure exerted by the load of material above, rocks are subjected to **stress** during the process of mountain building. Here, the applied force is directional, squeezing the material as if it had been placed in a vise. Rock located at great depth is quite warm and behaves plastically during deformation. This accounts for its ability to flow and bend into intricate folds (Figure 7.2). By contrast, cooler, near-surface rocks will usually **shear** during deformation. Shearing results when relatively brittle rock is broken into thin slabs that are able to slide past each other. This phenomenon can be demonstrated using a deck of playing cards. Shearing is similar to the slippages that occur between individual cards when the deck is held between your hands and the top of the deck is moved relative to the bottom.

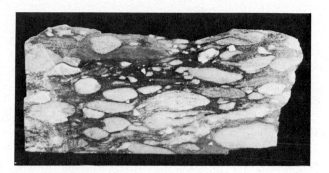

FIGURE 7.2
Metaconglomerate. These once-rounded rock fragments have been elongated as if squeezed in a gigantic vise. (Photo by E. J. Tarbuck)

CHEMICAL ACTIVITY AND METAMORPHISM

Chemically active fluids, most commonly water containing ions in solution, also enhance the metamorphic process. Some water is contained in the pore spaces of virtually every rock. In addition, many minerals are hydrated, and thus contain water within their crystalline structures. When deep burial occurs, water is forced out of the mineral structures and is then available to aid in chemical reactions. Water that surrounds the crystals acts as a catalyst by aiding ion migration. In some instances the minerals recrystallize to form more stable configurations. In other cases, ion exchange among minerals results in the formation of completely new minerals. Complete alteration of rock by hot, mineral-rich water has been observed in the near-surface environment of Yellowstone National Park. Further, along oceanic ridge systems seawater circulating through the still-hot rock helps transform dark basaltic minerals into metamorphic minerals such as serpentine and talc.

TEXTURAL AND MINERALOGICAL CHANGES

The degree of metamorphism is reflected in the texture and mineralogy of metamorphic rocks. When rocks are subjected to very low-grade metamorphism, they become more compact and thus more dense. A common example is the metamorphic rock slate, formed from the further compaction of shale. Under more extreme conditions, pressure causes certain minerals to recrystallize. As described earlier, water is believed to play a very important role in the recrystallization process by aiding the migration of ions. In general, recrystallization encourages the growth of larger crystals. Consequently, many metamorphic rocks consist of visible crystals, much like phaneritic igneous rocks. The crystals of some minerals, such as micas, which have a sheet structure, and hornblende, which has an elongated structure, will recrystallize with a preferred orientation. The new orientations will be essentially perpendicular to the direction of the compressional force. The resulting mineral alignment usually gives the rock a layered or banded appearance termed **foliation** (Figure 7.3).

Various types of foliation exist, depending to a large extent upon the degree of metamorphism. For example, during the transformation of shale to slate, clay minerals, which are stable at the surface, recrystallize into minute mica flakes, which are stable at much higher temperatures and pressures. Further, during recrystallization these fine-grained mica crystals become aligned so that their flat surfaces are nearly parallel. Consequently, slate can be split easily along these layers of mica grains into rather flat slabs. This property is called **rock cleavage,** to differentiate it from the type of cleavage exhibited by minerals (Figure 7.4). Since the mica flakes composing slate are tiny, slate is not usually visibly foliated. However, because it exhibits excellent rock cleavage, which is evidence that its minerals are aligned, slate is considered to be foliated.

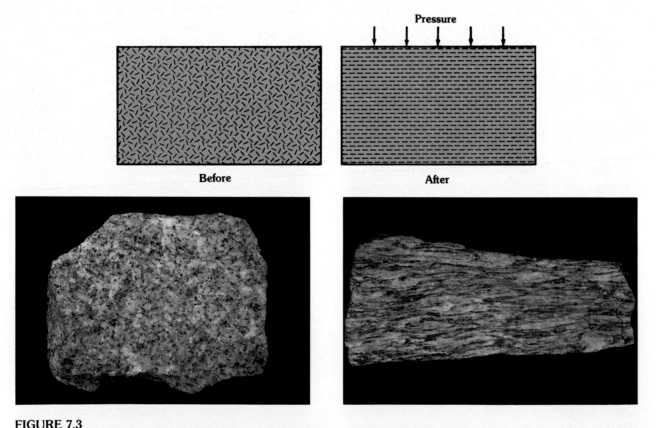

Before   **After**

FIGURE 7.3
Under the pressures of metamorphism, some mineral grains become reoriented and aligned at right angles to the pressure. The resulting linear orientation of mineral grains gives the rock a foliated texture. If the coarse-grained igneous rock (granite) on the left underwent intense metamorphism, it could end up closely resembling the metamorphic rock on the right (gneiss). (Photos by E. J. Tarbuck)

FIGURE 7.4
Rock cleavage in schist, northeastern California. (Photo by E. J. Tarbuck)

Under more extreme temperature-pressure regimes, the very fine mica grains found in slate will grow many times larger. These mica crystals, which are about a centimeter in diameter, will give the rock a platy or scaly appearance. This type of foliation is called **schistosity,** and a rock having this texture is called *schist.* Many types of schist exist and are named according to their mineral constituents. By far the most abundant are the mica schists, which are usually composed of either muscovite or biotite, or both (Figure 7.5).

During high-grade metamorphism, ion migrations can be extreme enough to cause minerals to segregate. An example of a metamorphic rock that exhibits mineral segregation is shown in Figure 7.3. Notice that the dark and light silicate minerals have sepa-

rated, giving the rock a banded appearance. Metamorphic rocks with this texture are called *gneiss* (pronounced "nice") and are quite common. Gneiss usually forms from the metamorphism of granite or diorite, but can form from gabbro or even by the high-grade metamorphism of shale. Although banded, gneiss will not usually split parallel to the crystals as easily as slate.

Not all metamorphic rocks have a foliated texture. Such rocks are said to be **nonfoliated.** Metamorphic rocks composed of only one mineral which forms equidimensional crystals are as a rule not visibly foliated. For example, when a fine-grained limestone is metamorphosed, the small calcite crystals combine to form relatively large interlocking crystals. The resulting rock has an appearance similar to a coarse-

161

FIGURE 7.5
Mica schist, a common metamorphic rock composed of shiny mica flakes. (Photo by E. J. Tarbuck)

grained igneous rock. This metamorphic equivalent of limestone is *marble*. Although it is considered to be nonfoliated, microscopic investigation of marble may reveal some flattening and parallelism of the grains. Further, some limestones contain thin layers of clay minerals which may become distorted during metamorphism. These "impurities" will often appear as curved bands of dark material flowing through the marble.

In the metamorphism of shale to slate we saw that clay minerals recrystallized to form mica crystals. In most instances, including this example, the chemical composition of the rock does not change during recrystallization. Rather, the existing minerals and available ions in the water will recombine to form minerals which are stable in the new environment. A common example is the formation of the metamorphic mineral *wollastonite*. Wollastonite is generated when limestone ($CaCO_3$), which contains abundant sandy material in the form of quartz (SiO_2), is subjected to high temperatures during contact metamorphism. The calcite and quartz crystals chemically react to form wollastonite ($CaSiO_3$), while carbon dioxide is liberated.

In some environments, however, new materials are actually introduced during the metamorphic process. For example, country rock adjacent to a large magma body would be altered by ion-rich **hydrothermal** (hot water) **solutions** released during the latter stages of crystallization. Many metallic ore deposits were formed by the deposition of minerals from hydrothermal solutions. Further, the seawater percolating through newly formed oceanic crust contains numerous active ions that chemically react with existing rocks. Some of the earth's richest copper ores have been formed in this manner.

In summary, the metamorphic process causes many changes in rocks, including increased density, growth of larger crystals, reorientation of the mineral grains into a layered or banded appearance known as foliation, and the transformation of low-temperature minerals into high-temperature minerals. Further, the introduction of ions generates new minerals, some of which are economically important.

COMMON METAMORPHIC ROCKS

FOLIATED ROCKS

Slate is a very fine-grained foliated rock composed of minute mica flakes. The most noteworthy characteristic of slate is its excellent rock cleavage. This property has made slate a most useful rock for roof and floor tile, blackboards, and billiard tables. Slate is most often generated by the low-grade metamorphism of shale, although less frequently it forms from the metamorphism of volcanic ash. Slate can be almost any color depending on its mineral constituents. Black (carbonaceous) slate contains organic material, red slate gets its color from iron oxide, and green slate is usually composed of chlorite, a mica-like mineral formed by the metamorphism of iron-rich silicates. Because slate forms during low-grade metamorphism, evidence of the original bedding planes is often preserved. However, the orientation of slate's rock cleavage generally trends across the original sedimentary layering (Figure 7.6). Thus, unlike shale, which splits along bedding planes, slate splits across them.

Phyllite represents a gradation in metamorphism between slate and schist. Its constituent platy miner-als are larger than those in slate, but not yet large enough to be clearly identifiable. Although phyllite appears similar to slate, it can be easily distinguished from slate by its glossy sheen (Figure 7.7). Phyllite usually exhibits rock cleavage and is composed mainly of very fine crystals of either muscovite or chlorite.

Schists are distinctive metamorphic rocks and are almost as common as gneisses. By definition, schists contain more than 50 percent platy minerals, most commonly muscovite and biotite. Like slate, the parent material from which most schists originate is shale, but in the case of schist, the metamorphism is more intense. If the parent rock contained abundant silica, schist will often contain thin layers of quartz and possibly feldspar as well.

Schists are named according to their mineral com-

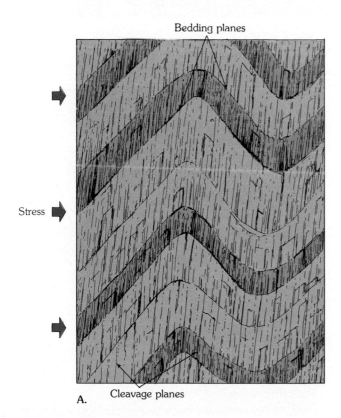

Bedding planes

Stress

A.   Cleavage planes

B.

FIGURE 7.6
Illustration and photograph showing the relationship between slate cleavage and original bedding planes. (Photo by G. K. Gilbert, U.S. Geological Survey)

FIGURE 7.7
Phyllite (left) can be distinguished from slate by its glossy sheen. (Photo by E. J. Tarbuck)

position. Those composed primarily of muscovite and biotite with lesser amounts of quartz and feldspar are called *mica schist.* Depending upon the degree of metamorphism, mica schists often contain accessory minerals quite unique to metamorphic rocks. Some common accessory minerals include garnet, staurolite, and sillimanite (Figure 7.8). Some schists contain graphite which is recovered for use as pencil lead, graphite fibers, and lubricant. In addition, schists may be composed largely of the minerals chlorite or talc, in which case they are called *chlorite schist* and *talc schist,* respectively. Both chlorite and talc schists form when rocks with a basaltic composition undergo metamorphism.

Gneiss is the term applied to foliated metamorphic rocks that contain mostly granular, as opposed to platy, minerals. The most common minerals found in gneisses are quartz, potassium feldspar, and sodium feldspar. In addition, lesser amounts of muscovite, biotite, and hornblende are common. The segregation of light and dark silicates gives gneisses a characteristic foliated texture. Thus, most gneisses consist of alternating bands of white or reddish feldspar-rich zones and layers of dark ferromagnesian minerals. These banded gneisses are often deformed while in a plastic state into rather intricate folds (Figure 7.9).

FIGURE 7.8
Garnet-mica schist. (Courtesy of the American Museum of Natural History)

Some gneisses will split readily along the layers of platy minerals, but most break in an irregular fashion, like other crystalline rocks.

Gneisses generally have a composition similar to granite and are probably derived from granite or its aphanitic equivalent. However, they may also form from the high-grade metamorphism of shale. In this

FIGURE 7.9
A contorted gneiss found at the bottom of the Grand Canyon of the Colorado River.
(Photo by E. J. Tarbuck)

instance, gneiss represents the last rock in the sequence of shale, slate, phyllite, schist, and gneiss. Like schists, gneisses may also include large crystals of accessory minerals such as garnet and staurolite. When banded rocks are made up predominately of dark minerals, such as those common in basalt, they are called *amphibolites,* after the mineral amphibole.

NONFOLIATED ROCKS

Marble is a coarse, crystalline rock whose parent rock was limestone or dolomite (Figure 7.10). When a hand sample is examined, marble closely resembles crystalline limestone. Pure marble is white and composed essentially of the mineral calcite. Because of its

color and relative softness (hardness of 3), marble is a popular building stone. White marble is particularly prized as a stone from which to carve monuments and statues, such as the famous statue of David by Michelangelo.

Often the limestone from which marble forms contains impurities that color the marble. Thus, marble can be pink, gray, green, or even black. Also, when impure limestone is metamorphosed, it may yield a variety of accessory minerals including chlorite, mica, garnet, and commonly wollastonite. When marble forms from limestone interbedded with shales, it will appear banded. Occasionally, marble will split along these bands and reveal the mica minerals that crystallized from clay minerals. Under extreme deforma-

FIGURE 7.10
Marble, a crystalline rock formed by the metamorphism of limestone. (Photo by E. J. Tarbuck)

tion, the bands in marble may become highly contorted and give the rock a rather artistic design.

Quartzite is a very hard metamorphic rock most often formed from quartz sandstone. Under moderate-to-high-grade metamorphism, the quartz grains in sandstone fuse. The recrystallization is so complete that when broken, quartzite will split across the original quartz grains, rather than between them. In some instances, such sedimentary structures as crossbedding are preserved and give the rock a banded appearance.

Although pure quartzite is white, it commonly contains some iron oxide stain and thus appears pink to red. Occasionally quartzite may contain a small percentage of dark minerals which give it a gray color.

Quartzite, like marble, is made up of only one mineral, which develops equidimensional crystals. Consequently, it does not develop any parallelism of mineral grains and therefore lacks foliation.

OCCURRENCES OF METAMORPHIC ROCKS

Recall that metamorphic rocks commonly form in one of three environments: along fault zones, in contact with igneous bodies, or during dynamic episodes associated with mountain building.

METAMORPHISM ALONG FAULT ZONES

When faulting occurs near the surface, the stress and frictional heat produced along the fault zone generate a loosely coherent rock composed of broken or distorted rock fragments. When these rocks are composed of angular granules, they are called *fault breccia*. Metamorphic rocks produced along fault zones which are located at depth often have elongated grains and more closely resemble the rocks formed by the other metamorphic processes. Consequently, their origin may not be discernible when hand samples are examined.

The quantity of metamorphic rock generated solely by faulting is relatively insignificant when compared to the amount generated by the other two processes. Nevertheless, in some areas these granulated rocks are quite abundant. For example, movements along California's San Andreas fault have created a zone of fault breccia and related rock types nearly 1000 kilometers long and up to 3 kilometers wide.

CONTACT METAMORPHISM

Contact metamorphism occurs when molten rock comes into contact with cooler rock. Contact metamorphism is clearly distinguishable only when it occurs at the surface or in a near-surface environment where the temperature contrast between the magma

FIGURE 7.11
Close-up view of a contact zone where light-colored igneous rock has invaded and metamorphosed the dark-colored host rock. (Photo by John S. Shelton)

and host rock is great (Figure 7.11). Undoubtedly, contact metamorphism is also an active process at great depth. However, its effect is blurred due to the general alteration of regional metamorphism.

During contact metamorphism, a zone of alteration called an **aureole** (or halo) forms around the emplaced magma. Small intrusive bodies such as dikes and sills have aureoles only a few centimeters thick. On the other hand, large igneous bodies such as batholiths and laccoliths may form zones of metamorphic rocks a few kilometers or more in thickness (Figure 7.12). These large aureoles often consist of definite zones of metamorphism. Near the magma body, high-temperature minerals such as garnet may form, while farther away such low-grade minerals as chlorite are produced. In addition to the size of the intrusive magma body, the mineral composition of the country rock and the availability of water greatly affect the size of the aureole produced. In chemically active rock such as limestone, the zone of alteration

FIGURE 7.12
The dark layer, called a roof pendant, consists of metamorphosed country rock adjacent to the upper part of an igneous pluton. The term *roof pendant* infers that the rock was once the roof of a magma chamber. This roof pendant is in the Sierra Nevada of eastern California, near Split Mountain. (Photo by John S. Shelton)

can extend to distances of 10 kilometers or more from the igneous body. Here the occurrence of minerals such as garnet and wollastonite marks the extent of metamorphism.

Most contact metamorphic rocks are fine-grained, dense, tough rocks of various chemical compositions. For example, during contact metamorphism, clay minerals are baked, as if placed in a kiln, and can generate a very hard, fine-grained rock not unlike porcelain. Since directional pressure is not a major factor in the formation of these rocks, they generally are not foliated. The name applied to a wide variety of rather hard, nonfoliated metamorphic rocks is *hornfels*.

When larger igneous masses are involved in contact metamorphism, hydrothermal solutions which originate within the magma can migrate great distances. As these solutions percolate through the host rock, chemical reactions with this rock greatly enhance the metamorphic process. Further, ores of numerous metals are thought to result from the emplacement of metallic ions whose source is hydro-

thermal solutions. These deposits include ores of copper, zinc, lead, iron, and gold.

REGIONAL METAMORPHISM

By far the greatest quantity of metamorphic rock is produced during regional metamorphism. As stated earlier, regional metamorphism takes place at considerable depths over an extensive area and is associated with the process of mountain building. During mountain building, a large segment of the earth's crust is intensely squeezed into a highly deformed mass. As the rocks are folded and faulted, the crust is shortened and thickened. This general thickening of the crust results in mountain terrains that stand high above sea level. Although material is obviously elevated to great heights during mountain building, an equally large quantity of rock is forced downward where it is exposed to high temperatures and pressures. Here in the "roots" of mountains, the most intense metamorphic activity occurs. Some of the deformed rock is thought to be heated enough to melt and thereby generate magma. As we shall see in

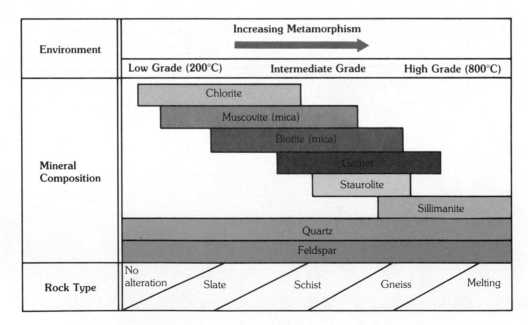

FIGURE 7.13
The typical transition in mineralogy that results from progressive metamorphism of shale.

the next section magma is also generated by other mountain-building processes. This magma, being less dense than the surrounding rock, will migrate upward. Magmas emplaced in a near-surface environment will cause contact metamorphism within the zone of regional metamorphism. Consequently, the cores of many mountain ranges consist of intrusive igneous bodies surrounded by high-grade metamorphic rocks. As these deformed rock masses are uplifted, erosion removes the overlying material to expose the igneous and metamorphic rocks composing the central core of the mountain range.

Since metamorphic rocks that form during regional metamorphism are deformed by directional stress, they are usually foliated. Further, in regional metamorphism, there usually exists a gradation in the intensity of metamorphism. As we progress from areas of low-grade metamorphism to areas of high-grade metamorphism, changes in mineralogy and rock texture can be observed.

A somewhat oversimplified example of progressive metamorphism can be made using the sedimentary rock shale, which under low-grade metamorphism yields the metamorphic rock slate. In high-temperature, high-pressure environments, slate will turn into mica schist. Under more extreme conditions, the micas in schist will recrystallize into minerals such as feldspar and hornblende, and eventually generate a gneiss. We can see certain aspects of this transition as we approach the Appalachian Mountains from the west. Beds of shale, which once extended over large areas of the eastern United States, can still be found as nearly flat-lying strata in Ohio. However, in the broadly folded Appalachians of central Pennsylvania, these beds are inclined and composed of low-grade slate. As we progress farther eastward to the intensely deformed crystalline Appalachians, we find large outcrops of schist and gneiss, some of which are perhaps remnants of once flat-lying shale beds. The most intense zones of metamorphism are found in regions such as Vermont and New Hampshire, often in close association with igneous intrusions.

In addition to the textural changes already considered, changes in mineralogy will be encountered as we progress from areas of low-grade metamorphism to areas of high-grade metamorphism. The typical transition in mineralogy that would result from the regional metamorphism of shale is shown in Figure 7.13. The first new mineral to be produced in the formation of slate is chlorite, which as we move toward the region of high-grade metamorphism would be replaced by ever greater amounts of muscovite and biotite. Mica schists are formed under more extreme conditions and may contain garnet and staurolite crystals as well. At temperatures and pressures approaching the melting point of rock, sillimanite forms. Sillimanite is a high-temperature metamorphic mineral used to make refractory porcelains such as those used in the casings of spark plugs.

Through the study of metamorphic rocks and laboratory experiments, researchers have learned that certain minerals are good indicators of the metamorphic environment in which they formed. Using these **index minerals,** geologists distinguish among different zones of regional metamorphism. For example, the mineral chlorite is produced when temperatures are relatively low, about 200°C (Figure 7.13). The mineral sillimanite on the other hand forms in very extreme environments where temperatures exceed 600°C. By mapping the occurrences of index minerals, geologists in effect map zones of varying metamorphic intensities. Figure 7.14 outlines the zones of metamorphic intensities that outcrop in New England and the Maritime Provinces of Canada.

In the most extreme environments even the highest-grade metamorphic rocks are subjected to change. In a low-pressure environment where temperatures exceed 800°C, a schist or gneiss having a chemical composition similar to granite will begin to melt. However, recall from our discussion of igneous rocks that not all minerals melt at the same temperature. The light-colored silicates, usually quartz and potassium feldspar, will melt first, whereas the dark silicates, such as amphibole and biotite, will remain solid. If this partially melted rock cools, the light bands will be made of crystalline igneous rock while the dark bands will consist of unmelted metamorphic material. Rocks of this type fall into a transitional zone somewhere between ''true'' igneous rocks and ''true'' metamorphic rocks and are called **migmatites** (Figure 7.15).

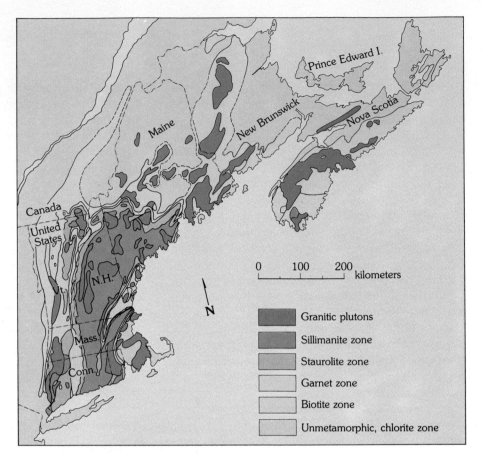

FIGURE 7.14
Zones of metamorphic intensities in New England and the Maritime Provinces. (After Donald W. Hyndman, *Petrology of Igneous and Metamorphic Rocks,* New York: McGraw-Hill, 1972)

Granitic plutons
Sillimanite zone
Staurolite zone
Garnet zone
Biotite zone
Unmetamorphic, chlorite zone

FIGURE 7.15
Migmatite. The lightest colored layers are igneous rock composed of quartz and feldspar, while the darker layers have a metamorphic origin. (Photo by J. B. Hadley, U.S. Geological Survey)

170

METAMORPHISM AND PLATE TECTONICS

Most of the present knowledge of metamorphism appears to conform well to the dynamics of the earth as proposed by the plate tectonics theory. In this model, mountain building and associated metamorphism occur along convergent zones where slabs of lithosphere are moving toward one another (Figure 7.16). It is at these locations that compressional forces squeeze and generally deform the converging plates and the sediments that have accumulated along the margins of continents. The plate tectonics model also accounts for the igneous activity associated with mountain building. At convergent zones, material is being thrust to depths where temperatures and pressures are high. The eventual melting of some subducted material creates magma that migrates upward to crystallize in the core of mountain masses.

A close examination of Figure 7.16 shows that more than one type of metamorphic environment exists along convergent boundaries. Near ocean trenches, slabs of cold lithosphere are being subducted to great depths. As the lithosphere and associated sediment descend, the pressure increases more rapidly than the temperature. This occurs because rock is a rather poor conductor of heat; thus, heating of the cold, thick slabs progresses slowly. Rock formed in this environment of high pressure relative to temperature is called *blueschist,* after the presence of the blue-colored amphibole glaucophane, which forms under these conditions. The Coast Range of California was once a subduction zone of this type. Here highly deformed rocks that were once deeply buried have been elevated as a result of a change in the plate boundary.

In near-surface zones landward of trench areas, the metamorphic environment consists of high tem-

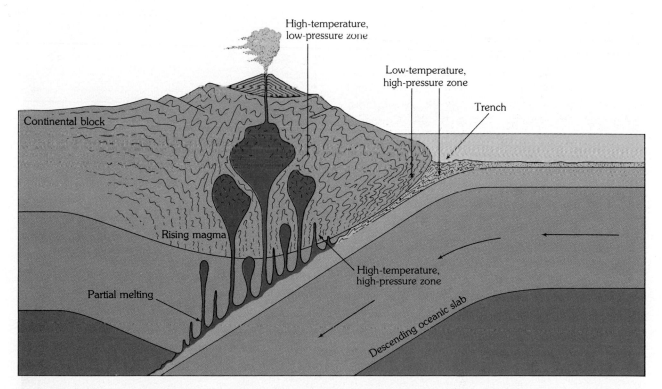

FIGURE 7.16
Metamorphic environments according to the plate tectonics model.

peratures relative to pressures. Here the emplacement of molten rock from below alters the existing rocks in an environment characterized by low to moderate pressures. The Sierra Nevada, which consist of igneous intrusions and associated metamorphic rocks, exemplifies this type of environment.

Apparently, most of the deformed material found adjacent to oceanic trenches consists of two distinctive linear belts of metamorphic rocks (Figure 7.17). Nearest to the trench we find a high-pressure, low-temperature metamorphic regime similar to that of the Coast Range of California. Farther inland, in the region of plutonic emplacement, metamorphism is dominated by high temperatures and low to moderate pressures; that is, environments similar to those that generated the Sierra Nevada Batholith. Between these two areas rocks such as those of the Great Valley of California are essentially unaltered by metamorphism.

In addition to the linear belts of metamorphic rocks that are found in the axes of most mountain belts, even larger expanses of metamorphic rocks exist within the stable continental interiors. These relatively flat expanses of metamorphic rocks and igneous plutons are called **shields.** One such structure, the Canadian Shield, has very little topographic expression and forms the bedrock over much of central Canada, extending from Hudson Bay to northern Minnesota. Radiometric dating of the Canadian Shield rocks indicates that they are among the oldest rocks on earth. Because shields are old and since their rock structure is similar to that found in the central core of existing mountains, they are believed to be remnants of much earlier periods of mountain building. If this concept is correct, it indicates that the earth has been a dynamic planet throughout a great expanse of its history. Additional study of these vast areas of metamorphism, in the context of the plate

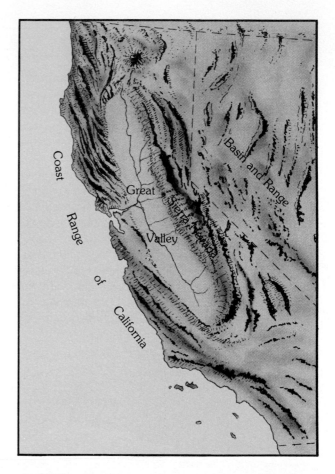

FIGURE 7.17
Map of the California Coast Range and the Sierra Nevada.

tectonics model, should give geologists new insights into the problem of discerning just how the continents came to exist. We will consider that topic further in Chapter 18.

REVIEW QUESTIONS

1 What is metamorphism? What are the agents of change?

2 What is foliation? Distinguish between *rock cleavage* and *schistosity*.

3 List some changes that might occur to a rock in response to metamorphic processes.

4 Slate and phyllite resemble each other. How might you distinguish one from the other?

5 Each of the following statements describes one or more characteristics of a particular metamorphic rock. For each statement, name the metamorphic rock that is being described.
  (a) Calcite-rich and nonfoliated.
  (b) Foliated and composed mainly of granular minerals.
  (c) Represents a grade of metamorphism between slate and schist.
  (d) Very fine-grained and foliated; excellent rock cleavage.
  (e) Foliated and composed of more than 50 percent platy minerals.
  (f) Often composed of alternating bands of light and dark silicate minerals.
  (g) Hard, nonfoliated rock resulting from contact metamorphism.

6 Distinguish between contact metamorphism and regional metamorphism. Which creates the greatest quantity of metamorphic rock?

7 What feature would make schist and gneiss easily distinguished from quartzite and marble?

8 Briefly describe the textural and mineralogical differences between slate, mica schist, and gneiss. Which one of these rocks represents the highest degree of metamorphism?

9 Are migmatites associated with high-grade or low-grade metamorphism?

10 With what type of plate boundary is regional metamorphism associated?

KEY TERMS

aureole (p. 167)

contact metamorphism (p. 158)

foliation (p. 160)

hydrothermal solution (p. 162)

index mineral (p. 169)

migmatite (p. 169)

nonfoliated (p. 161)

regional metamorphism (p. 158)

rock cleavage (p. 160)

schistosity (p. 161)

shear (p. 159)

shield (p. 172)

stress (p. 159)

8

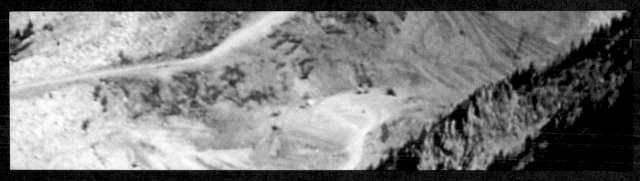

MASS WASTING

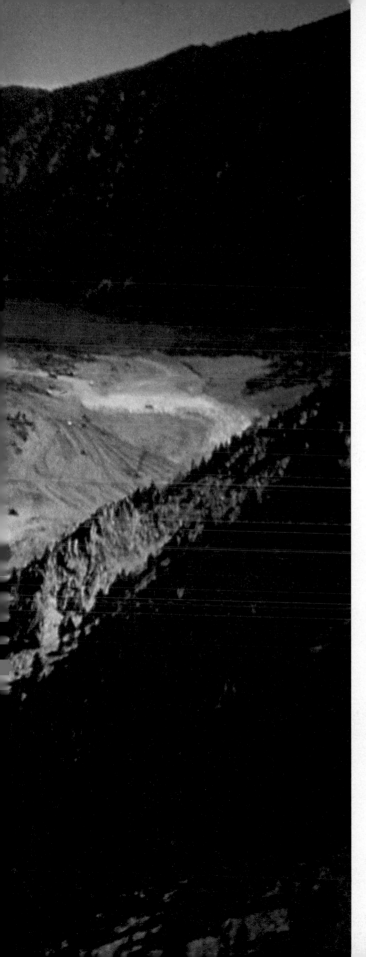

The news media periodically relate the terrifying and often grim details of landslides. On May 31, 1970, one such event occurred when a gigantic rock avalanche buried more than 20,000 people in Yungay and Ranrahirca, Peru (Figure 8.1). There was little warning of the impending disaster; it began and ended in just a matter of a few minutes. The avalanche started 14 kilometers from Yungay, near the summit of 6700 meter high Nevados Huascaran, the loftiest peak in the Peruvian Andes. Triggered by the ground motion from a strong offshore earthquake, a huge mass of rock and ice broke free from the precipitous north face of the mountain. After plunging nearly one kilometer, the material pulverized on impact and immediately began rushing down the mountainside, made fluid by trapped air and melted ice. The initial mass ripped loose additional millions of tons of debris as it roared downhill. The shock waves produced by the event created thunderlike noise and stripped nearby hillsides of vegetation. Although the material followed a previously eroded gorge, a portion of the debris jumped a 200–300 meter high bedrock ridge that had protected Yungay from past rock avalanches and buried the entire city. After inundating another town in its path, Ranrahirca, the mass of debris finally reached the bottom of the valley where its momentum carried it across the Rio Santa and tens of meters up the opposite bank.

The Madison Canyon slide was triggered by an earthquake. (Photo by J. R. Stacy, U.S. Geological Survey)

177

A.

FIGURE 8.1
This Peruvian valley was devastated by the rock avalanche that was triggered by an offshore earthquake in May, 1970. **A.** Before. **B.** After the rock avalanche. (Photos courtesy of Iris Lozier)

B.

178

This was not the first such disaster in the region and will no doubt not be the last. Just eight years earlier, a less spectacular, but nevertheless devastating, rock avalanche took the lives of an estimated 3500 people on the heavily populated valley floor at the base of the mountain.

As with many geologic hazards, the tragic rock avalanche in Peru was triggered by a natural event—in this case, an earthquake. However, disasters also result from the mass movement of surface material triggered by the actions of humans. For example, in 1960 a large dam almost 265 meters tall was built across the Vaiont Canyon in the Italian Alps. Three years later, on the night of October 9, 1963, a violent disaster occurred and the manmade dam was largely responsible.

The bedrock in Vaiont Canyon slanted steeply downward toward the lake impounded behind the dam and was composed of weak, highly fractured limestone strata that contained beds of clay and numerous solution cavities. As the reservoir filled, the lower portions of these rocks became saturated and the clays became swollen and more plastic. The rising water reduced the internal friction that

had kept the rock in place. Measurements made shortly after the reservoir was filled hinted at the problem, because they indicated that a portion of the mountain was slowly creeping downhill at the rate of 1 centimeter per week. In September, 1963, the rate increased to 1 centimeter per day, then 10–20 centimeters per day, and eventually to as much as 80 centimeters on the day of the disaster. Finally, the mountainside let loose. In just an instant, 240 million cubic meters of rock and rubble slid down the face of Mount Toc and filled nearly 2 kilometers of the gorge to heights of 150 meters above the reservoir level (Figure 8.2). The filling of the reservoir pushed the water completely over the dam in a wave more than 90 meters high. More than 1.5 kilometers downstream the wall of water was still 70 meters high and everything in its path was completely destroyed. The entire event lasted less than seven minutes, yet claimed an estimated 2600 lives. This is known as the worst dam disaster in history, but when it was over, the Vaiont Dam was still standing intact. Although the catastrophe was triggered by human interference with the Vaiont River, the slide would have eventually

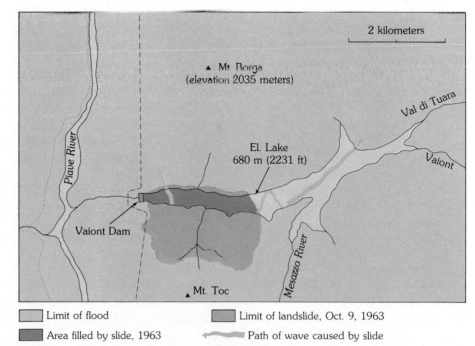

FIGURE 8.2
Sketch map of the Vaiont River area showing the limits of the landslide, the portion of the reservoir that was filled with debris, and the extent of flooding downstream. [After G. A. Kiersch, ''Vaiont Reservoir Disaster,'' *Civil Engineering* 34 (1964) : 32–39]

Limit of flood
Area filled by slide, 1963
Limit of landslide, Oct. 9, 1963
Path of wave caused by slide

occurred on its own; however, the effects would not have been nearly as tragic. Fortunately, mass movements such as those just described are infrequent and only occasionally affect large numbers of people.

CONTROLS OF MASS WASTING

Landslides are spectacular examples of a normal geologic process called **mass wasting.** Mass wasting refers to the downslope movement of rock, regolith, and soil under the direct influence of gravity. Once weathering breaks up rock, gravity pulls the material to lower elevations, where streams usually carry it away, eventually depositing it in the ocean. In this manner the earth's landscape is slowly being shaped.

Although gravity is the controlling force of mass wasting, other factors play an important part in bringing about the downslope movement of material. Water is one of these factors. When the pores in sediment become filled with water, the cohesion among particles is destroyed, allowing them to slide past one another with relative ease. For example, when sand is slightly moist, it sticks together quite well. However, if enough water is added to fill the openings between the grains, the sand will ooze out in all directions. Thus, saturation reduces the internal resistance of materials, which are then easily set in motion by the force of gravity. When clay is wetted, it becomes very slick—another example of the "lubricating" effect of water. Water also adds considerable weight to a mass of material. The added weight in itself may be enough to cause the material to slide or flow downslope.

Oversteepening of slopes is another cause of many mass movements. Loose, undisturbed particles assume a stable slope called the **angle of repose,** the steepest angle at which material remains stable. Depending upon the size and shape of the particles, the angle varies from 25 to 40 degrees. The larger, more angular particles maintain the steepest slopes. If the angle is increased, the rock debris will adjust by moving downslope. There are many situations in nature where this takes place. A stream undercutting a valley wall and waves pounding against the base of a cliff

are but two familiar examples. Furthermore, through their activities, people often create oversteepened and unstable slopes that become prime sites for mass wasting.

CLASSIFICATION OF MASS WASTING PROCESSES

There is a broad complex of activities that geologists call mass wasting. Generally, the different types are divided and described on the basis of the type of material involved, the kind of motion displayed, and by the velocity of the movement.

TYPE OF MATERIAL

The classification of mass wasting processes based on the material involved in the movement depends upon whether the descending mass began as unconsolidated material or as bedrock. If soil and regolith dominate, we typically see terms such as "debris," "mud," or "earth" used in the description. On the other hand, when a mass of bedrock breaks loose and moves downslope, the term "rock" may be part of the description.

TYPE OF MOTION

In addition to characterizing the type of material involved in a mass wasting event, the way in which the material moves may also be important. Generally, the kind of motion is described as either a fall, a slide, or a flow.

When the movement involves the free-fall of detached individual pieces of any size, it is termed a **fall.** Fall is a common form of movement on slopes that are so steep that loose material cannot remain on the surface. The rock may fall directly to the base of the slope or move in a series of leaps and bounds over other rocks along the way. Many falls result when freeze and thaw cycles or the action of plant roots loosen rock to the point that gravity takes over. Although signs along bedrock cuts on highways warn of falling rock, few of us have actually witnessed such an event. However, as Figure 8.3 illustrates, they do indeed occur. In fact, this is the primary way in which talus slopes are built and maintained. Sometimes falls may trigger other forms of movement. For example, recall that the Yungay disaster described at the beginning of the chapter was initiated by a mass of free-falling material that broke from the nearly vertical summit of Nevados Huascaran.

Many mass wasting processes are described as **slides.** Slides occur whenever material remains fairly coherent and moves along a well-defined surface. Sometimes the surface is a joint, a fault, or a bedding plane that is approximately parallel to the slope. However, in the case of the movement called slump, the descending mass moves along a curved surface of rupture. A note of clarification is appropriate at this point. Sometimes the word "slide" is used as a synonym for the word "landslide." It should be pointed out that although many people, including geologists, use the term, the word "landslide" has no specific definition in geology. Rather it should be considered as a popular nontechnical term used to describe all perceptible forms of mass wasting, including those in which sliding does not occur.

The third type of movement common to mass wasting processes is termed **flow.** Flow occurs when material moves downslope as a viscous fluid. Most flows are saturated with water and typically move as lobes or tongues. Figure 8.4 illustrates the distinctive appearance of a flow.

RATE OF MOVEMENT

The events described at the beginning of this chapter clearly involved rapid movements. In both instances, the rock and debris moved downslope at speeds well in excess of 200 kilometers per hour. These most rapid of mass movements are termed **rock avalanches.** Many researchers believe that rock avalanches, such as the one that produced the scene in Figure 8.5, must literally "float on air" as they move downslope. That is, high velocities result when air becomes trapped and compressed beneath the falling mass of debris, allowing it to move as a buoyant, flexible sheet across the surface.

Most mass movements, however, do not move with the speed of a rock avalanche. In fact, a great deal of mass wasting is imperceptibly slow. One process that we will examine later, termed creep, results in particle movements that are usually measured in millimeters or centimeters per year. Thus, as you can see, rates of movement can be spectacularly sudden or exceptionally gradual. Although various types of mass wasting are often classified as either rapid or slow, such a distinction is highly subjective because

FIGURE 8.3
Rockfall blocking a highway. (Photo courtesy of U.S. Geological Survey)

FIGURE 8.4
When debris moves by flowing, it often has a lobelike appearance. (Photo by D. R. Crandell, U.S. Geological Survey)

FIGURE 8.5
Debris deposited atop Sherman Glacier by a rock avalanche. The event was triggered by a tremendous earthquake in March, 1964. (Photo by Austin Post, U.S. Geological Survey)

there is a wide range of rates between the two extremes. Even the velocity of a single process at a particular site can vary considerably from one time to another.

SLUMP

Slump refers to the downward slipping of a mass of rock or unconsolidated material moving as a unit along a curved surface (Figure 8.6). Usually the slumped material does not travel spectacularly fast nor very far. This is a very common form of mass

wasting, especially in thick accumulations of cohesive materials such as clay. The surface of rupture beneath the slump block is characteristically spoon-shaped and concave upward or outward. As the movement occurs, a crescent-shaped scarp (cliff) is created at the head and the block's upper surface is tilted backwards. Although slump may involve a single mass, it often consists of multiple blocks (Figure 8.7). Sometimes water is impounded between the base of the scarp and the top of the tilted block. As this water percolates downward along the surface of rupture, it may promote further instability and additional movement.

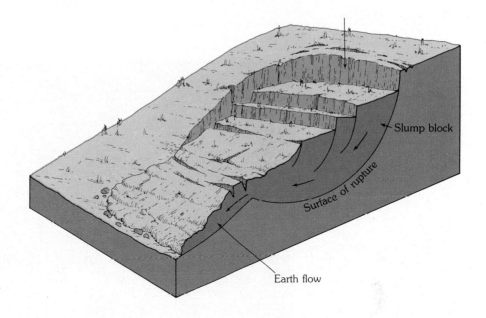

FIGURE 8.6
Slump occurs when material slips downslope en masse along a curved surface of rupture.

Slump block

Surface of rupture

Earth flow

FIGURE 8.7
Like this slump in Madison County, Montana, slumps often consist of multiple blocks. (Photo by J. R. Stacy, U.S. Geological Survey)

183

FIGURE 8.8
Slump at Point Fermin, California. Slump is often triggered when slopes become over-steepened by erosional processes such as wave action. (Photo by John S. Shelton)

Slump commonly occurs because a slope has been oversteepened. The material on the upper portion of a slope is held in place by the material at the bottom of the slope. As this anchoring material at the base is removed, the material above is made unstable and reacts to the pull of gravity. Figure 8.8 provides an example of this process. The cliffed seashore was undercut by wave action at its base. Slumping may also occur when a slope is overloaded, causing internal stress on the material below. This type of slump often occurs where weak, clay-rich material underlies layers of stronger, more resistant rock such as sandstone. The seepage of water through the upper layers reduces the strength of the clay below and slope failure results.

ROCKSLIDE

Rockslides occur when blocks of bedrock break loose and slide down a slope. Such events are among the fastest and most destructive mass movements. Usually rockslides take place in a geologic setting where the rock strata are inclined, or joints and fractures exist parallel to the slope. When such a rock unit is undercut at the base of the slope, it loses support and the rock eventually gives way. Sometimes the rockslide is triggered when rain or melting snow lubricate the underlying surface to the point that friction is

no longer sufficient to hold the rock unit in place. As a result, rockslides tend to be most likely during the spring, when heavy rains and melting snow are most prevalent. Earthquakes are another mechanism which may trigger rockslides and other mass movements. The 1811 earthquake at New Madrid, Missouri, for example, caused slides in an area of more than 13,000 square kilometers (5000 square miles) along the Mississippi River valley. A more recent example occurred on August 17, 1959, when a severe earthquake west of Yellowstone National Park triggered a massive slide in the canyon of the Madison River in Montana (Figure 8.9). In a matter of moments an estimated 27 million cubic meters of rock, soil, and trees slid into the canyon. The debris dammed the river and buried a campground and highway.

The Gros Ventre River flows west from the Wind River Range in northwestern Wyoming, through the Grand Teton National Park, and eventually empties into the Snake River. On June 23, 1925, a classic rockslide took place in its valley, just east of the small town of Kelly. In the span of just a few minutes a great mass of sandstone, shale, and soil crashed down the south side of the valley, carrying with it a dense pine forest. The volume of debris, estimated at 38 million cubic meters (50 million cubic yards), created a 70 meter high dam on the Gros Ventre River (Figure 8.10). The river was completely blocked and created

FIGURE 8.9
This pile of debris (an estimated 27 million cubic meters) damming the Madison River in Montana resulted from an earthquake-triggered rockslide that occurred August 17, 1959. (Photo by J. R. Stacy, U.S. Geological Survey)

FIGURE 8.10
The scar left on the side of Sheep Mountain by the Gros Ventre rockslide. (Photo by Garrett Deckert)

a lake, which filled so quickly that a house that had been 18 meters (60 feet) above the river was floated off its foundation 18 hours after the slide. In 1927 the lake overflowed the dam, partially draining the lake and resulting in a devastating flood downstream.

Why did the Gros Ventre rockslide take place? Figure 8.11 is a diagrammatic cross-sectional view of the geology of the valley. Notice the following points: (1) the sedimentary strata in this area dip (tilt) 15–21 degrees; (2) underlying the bed of sandstone is a relatively thin layer of clay; and (3) at the bottom of the valley the river had cut through much of the sandstone layer. During the spring of 1925 water from heavy rains and melting snow seeped through the sandstone, saturating the clay below. Since much of the sandstone layer had been cut through by the Gros Ventre River, the layer had virtually no support at the bottom of the slope. Eventually the sandstone could no longer hold its position on the wetted clay, and gravity pulled the mass down the side of the valley. The circumstances at this location were such that the event was inevitable.

FIGURE 8.11
Cross-sectional view of the Gros Ventre rockslide. The slide occurred when the tilted and undercut sandstone bed could no longer maintain its position atop the saturated bed of clay. [After W. C. Alden, "Landslide and Flood at Gros Ventre, Wyoming." *Transactions* (AIME) 76 (1928) : 348]

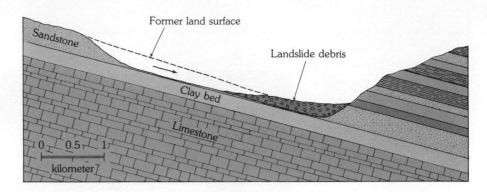

A.

FIGURE 8.12
A. View of the great Slumgullion mudflow from its source to Lake San Cristobal, a vertical distance of about 780 meters (2600 feet). Since it occurred in prehistoric times, there is no direct information about its rate of movement. (Photo by W. Cross, U.S. Geological Survey) B. A house damaged by mudflow along the Toutle River, west-northwest of Mount St. Helens. The end section of the house was torn free and lodged against trees. (Photo by D. R. Crandell, U.S. Geological Survey)

B.

186

MUDFLOW

Mudflow is a relatively rapid type of mass wasting which involves a flowage of debris containing a large amount of water. Mudflows are most characteristic of canyons and gullies in semiarid mountainous regions (Figure 8.12). When such an area experiences a heavy rain, large quantities of sediment are washed into the channel from the valley walls, which usually have little or no vegetation to anchor the loose material. The end product is a rapidly flowing tongue of well-mixed mud, soil, rock, and water. Its consistency may range from that of wet concrete to a soupy mixture not much thicker than muddy water. Because of its high density a mudflow can often carry or push large boulders, trees, and even houses with relative ease. Upon reaching the end of a steep, narrow canyon, the mudflow spreads out, covering the area beyond the mouth of the canyon with a mixture of wet debris. This material contributes to the buildup of fanlike deposits[1] at canyon mouths. Fans are relatively easy to build on, often have nice views, and are close to the mountains; thus, many have become preferred sites for development, especially in California. Since mudflows occur only sporadically, the public is often unaware of this potential hazard. This ignorance has led to many disasters.

Mudflows are also common in some volcanic regions. Here, the layers of fine-grained volcanic ash which mantle the steep volcanic slopes may form large mudflows following sudden heavy rains or periods of snowmelt. Flows such as this were triggered by the period of activity at Mount St. Helen's in 1980. The increased heat flow from the volcano caused rapid melting of great quantities of snow. The ash-rich mudflows that resulted spread over extensive areas.

EARTHFLOW

Unlike mudflows, which are usually confined to channels in semiarid regions, **earthflows** most often form on hillsides in humid areas as the result of excessive rainfall. When water saturates clay-rich regolith on a hillslope, the material may break away and flow

---

[1]These structures are called alluvial fans and will be discussed in greater detail in Chapter 9.

FIGURE 8.13
Earthflow. (Photo by G. K. Gilbert, U.S. Geological Survey)

a short distance downslope, leaving a scar on the hillside (Figure 8.13). Depending upon the steepness of the slope and the material's consistency, the speed of an earthflow may vary from a few meters per hour to several meters per minute. However, since earthflows are quite viscous, they generally move more slowly than the more fluid mudflows described earlier. In addition to occurring as isolated hillside phenomena, earthflows commonly take place in association with large slumps. In this situation, they may be seen as tonguelike flows at the base of the slump block (Figure 8.6).

CREEP

Movements such as rockslides and rock avalanches are certainly the most spectacular and catastrophic forms of mass wasting. Since these events have been known to kill thousands, they deserve intensive study so that, through more effective prediction, timely warnings and better controls can help save lives. However, because of their large size and spectacular nature they give us a false impression of their importance as a mass wasting process. Indeed, sudden movements are responsible for moving less material than the slow and far more subtle action of creep. Whereas rapid types of mass wasting are characteristic of mountains and steep hillsides, creep can take place on gentle slopes and is thus much more widespread.

Creep is a type of mass wasting that involves the gradual downhill movement of soil and regolith. One of the primary causes of creep is the alternate expansion and contraction of surface material caused by freezing and thawing or wetting and drying. As shown in Figure 8.14, freezing or wetting lifts the soil at right angles to the slope (solid lines), and thawing or drying allows the particles to fall back to a slightly lower level (dashed lines). Each cycle therefore moves the material a short distance downhill (Figure 8.15). Creep may also be initiated if the ground becomes saturated with water. Following a heavy rain or snowmelt, a

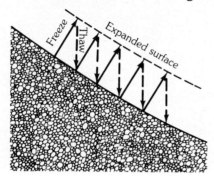

FIGURE 8.14
Path of a particle during creep. Movement results from the repeated expansion and contraction of material on the slope.

FIGURE 8.15
The weathered upper portions of these vertical shale beds are creeping downslope. (Photo by G. W. Stose, U.S. Geological Survey)

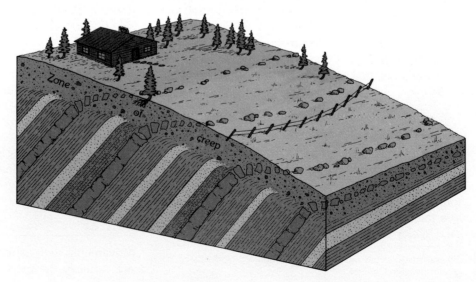

FIGURE 8.16
Although creep is an imperceptibly slow movement, its effects are often visible.

waterlogged soil may lose its internal cohesion, allowing gravity to pull the material downslope. Although the movement is imperceptibly slow, its effects are recognizable. Creep causes fences and telephone poles to tilt, and tree trunks will often be bent as a consequence of this movement (Figure 8.16).

PERMAFROST AND SOLIFLUCTION

When the average annual air temperature is low enough to maintain a continuous surface temperature below 0°C, the depth to which the ground freezes in winter exceeds the depth of summer thawing and a layer of permanently frozen ground is formed. This frozen ground, termed **permafrost,** underlies as much as 20 percent of the earth's land surface. In addition to occurring in Antarctica, permafrost is extensive in the lands surrounding the Arctic Ocean (Figure 8.17).

When changes occur in the surface environment, such as the clearing of the insulating vegetation mat, or the building of roads and other structures, the delicate thermal balance is disturbed and thawing of the permafrost can result. This in turn produces unstable ground that may be susceptible to slides, slumps, subsidence, and severe frost heaving. The many environmental problems stemming from human activities in the Arctic during and following World War II demonstrated the need for a thorough understanding of the nature of permafrost (Figure 8.18).

A form of mass wasting common to regions underlain by permafrost is **solifluction.** This may be regarded as a form of creep in which unconsolidated, water-saturated material gradually moves downslope. Solifluction occurs in a zone above the permafrost called the *active layer,* which thaws in summer and refreezes in winter. During the summer season,

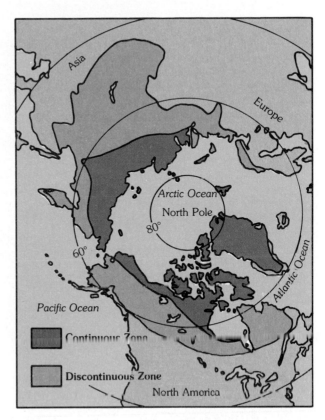

FIGURE 8.17
Distribution of permafrost in the Northern Hemisphere. (Courtesy of U.S. Geological Survey)

meltwaters are unable to percolate into the impervious permafrost layer below. As a result, the active layer becomes supersaturated and slowly flows. The process can occur on slopes as gentle as 2–3 degrees. Where there is a well-developed mat of vegetation, a solifluction sheet may move downward in a series of well-defined lobes or as a series of partially overriding folds (Figure 8.19).

A

FIGURE 8.18
This building located south of Fairbanks, Alaska, subsided because of thawing permafrost. Notice that the right side, which was heated, settled much more than the unheated porch on the left. (Photo courtesy of O. J. Ferrians, Jr., U.S. Geological Survey)

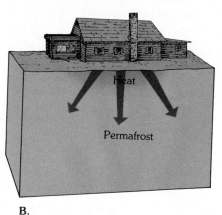

B.

FIGURE 8.19
Solifluction lobes northeast of Fairbanks, Alaska. (Photo by James E. Patterson)

REVIEW QUESTIONS

1 How did a manmade structure contribute to the Vaiont Canyon disaster? Was the disaster avoidable?

2 Describe how mass wasting sculptures the earth's surface.

3 What is the controlling force of mass wasting?

4 How does water affect mass wasting processes?

5 Describe the significance of the angle of repose.

6 Distinguish between fall, slide, and flow.

7 Why can rock avalanches move at such great speeds?

8 Slump and rockslide both move by sliding. In what ways do these processes differ?

9 What factors led to the massive rockslide at Gros Ventre, Wyoming?

10 Compare and contrast mudflow and earthflow.

11 Describe the mass wasting that occurred on the slopes of Mount St. Helen's during its active period in 1980.

12 Since creep is an imperceptibly slow process, what evidence might indicate that this phenomenon is affecting a slope? Describe the mechanism that creates this slow movement.

13 What is permafrost? What portion of the earth's land surface is affected?

14 Why is solifluction only a summertime process?

KEY TERMS

angle of repose (p. 180)

creep (p. 188)

earthflow (p. 187)

fall (p. 180)

flow (p. 181)

mass wasting (p. 180)

mudflow (p. 187)

permafrost (p. 189)

rock avalanche (p. 181)

rockslide (p. 184)

slide (p. 181)

slump (p. 182)

solifluction (p. 189)

9

RUNNING WATER

THE HYDROLOGIC CYCLE

The amount of water on earth is immense: an estimated 1.36 billion cubic kilometers (326 million cubic miles). Of this total, the vast bulk—97.2 percent—is part of the world ocean. Icecaps and glaciers account for another 2.15 percent, leaving only 0.65 percent to be divided among lakes, streams, subsurface water, and the atmosphere. Although the percentage of the earth's total water found in each of the latter sources is but a small fraction of the total inventory, the absolute quantities are great.

An adequate supply of water is vital to life on earth. With increasing demands on this finite resource, science has given a great deal of attention to the continuous exchanges of water among the oceans, the atmosphere, and the continents. This unending circulation of the earth's water supply has come to be called the **hydrologic cycle,** a gigantic system powered by energy from the sun in which the atmosphere provides the vital link between the oceans and continents. Water from the oceans, and to a much lesser extent from the continents, is constantly evaporating into the atmosphere. Wind transports the moisture-laden air, often great distances, until the complex processes of cloud formation are

Waterfall. (Photo by Kenneth Hasson)

set in motion. Precipitation that falls into the ocean has ended its cycle and is ready to begin another. The water that falls on the continents, however, must still make its way back to the ocean.

What happens to precipitation once it has fallen on the land? A portion of the water soaks into the ground, some of it moving downward, then laterally, finally seeping into lakes, streams, or directly into the ocean. When the rate of rainfall is greater than the earth's ability to absorb it, the additional water flows over the surface into lakes and streams. Much of the water which soaks in (**infiltration**) or runs off (**runoff**) eventually finds its way back to the atmosphere because of evaporation from the soil, lakes, and streams. Also, some of the water that infiltrates the ground surface is absorbed by plants, which later release it into the atmosphere. This process is called **transpiration.** Each year a field of crops may transpire an amount of water equivalent to a layer 60 centimeters deep over the entire field, while a forest may pump twice this amount into the atmosphere. Because we cannot clearly distinguish between the amount of water that is evaporated and the amount that is transpired by plants, the term **evapotranspiration** is often used for the combined effect.

When precipitation falls at high elevations or high latitudes, the water may not immediately soak in, run off, or evaporate. Instead it becomes part of a snowfield or glacier. Glaciers store large quantities of water on land, temporarily removing it from the hydrologic cycle. If present-day glaciers were to melt and release their storage of water, sea level would rise by several tens of meters and submerge many heavily populated coastal areas. As we shall see in Chapter 11, over the past two million years, huge continental glaciers have formed and melted on several occasions, each time upsetting the hydrologic cycle.

A diagram of the earth's water balance, a quantitative view of the hydrologic cycle, is shown in Figure 9.1. While the amount of water vapor in the air at any one time is but a minute fraction of the earth's total water supply, the absolute quantities that are cycled through the atmosphere over a one-year period are immense—some 380,000 cubic kilometers.

Since the total amount of water vapor in the atmosphere remains about the same, average annual precipitation over the earth must be equal to the quantity of water evaporated. However, for all of the continents taken together, precipitation exceeds evaporation. Conversely, over the oceans, evaporation exceeds precipitation. Since the level of the world

195

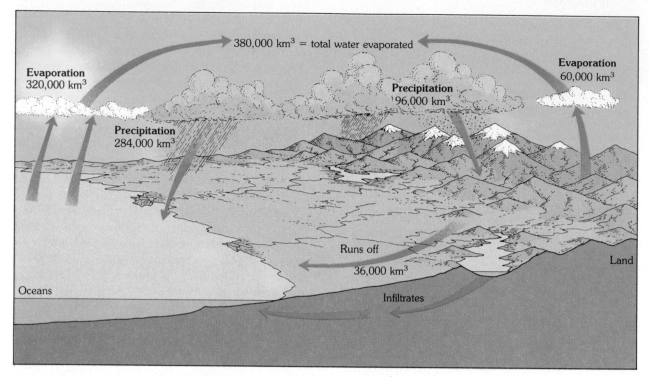

FIGURE 9.1

The earth's water balance. About 320,000 cubic kilometers of water are evaporated each year from the oceans, while evaporation from the land (including lakes and streams) contributes 60,000 cubic kilometers of water. Of this total of 380,000 cubic kilometers of water, about 284,000 cubic kilometers fall back to the ocean, and the remaining 96,000 cubic kilometers fall on the earth's land surface. Since 60,000 cubic kilometers of water evaporate from the land, 36,000 cubic kilometers of water remain to erode the land during the journey back to the oceans.

ocean is not dropping, runoff from land areas must balance the deficit of precipitation over the oceans.

The work accomplished by the 36,000 cubic kilometers of water that flows annually over the land to the sea is enormous. Arthur Bloom effectively described it as follows:

> **The average continental height is 823 meters above sea level. . . . If we assume that the 36,000 cubic kilometers of annual runoff flow downhill an average of 823 meters, the potential mechanical power of the system can be calculated. Potentially, the runoff from all lands would continuously generate almost 9×10^9 kW. If all this power were used to erode the land, it would be comparable to having . . . one horse-drawn scraper or scoop at work on each 3-acre piece of land, day and night, year round. Of course, a large part of the potential energy of runoff is wasted as frictional heat by turbulent flow and splashing of water.[1]**

Although only a small percentage of the energy of running water is used to erode the earth's surface, running water nevertheless represents the single most important agent sculpturing our planet's landscape.

To summarize, the hydrologic cycle represents the continuous movement of water from the oceans to the atmosphere, from the atmosphere to the land, and from the land back to the sea. The wearing down of the earth's land surface is largely attributable to the

---

[1]*Geomorphology: A Systematic Analysis of Late Cenozoic Landforms* (Englewood Cliffs, N.J.: Prentice-Hall, 1978) p. 97.

last of these steps and is the primary focus of the remaining pages of this chapter.

RUNNING WATER

Of all the geologic processes, running water may have the greatest impact on people. We depend upon rivers for energy, travel, and irrigation; their fertile floodplains have fostered human progress since the dawn of civilization. As the dominant agent of landscape alteration, streams have shaped much of our physical environment.

Although people have always depended to a great extent on running water, its source eluded them for centuries. Not until the sixteenth century did they realize that streams were supplied by runoff and underground water, which ultimately had their sources as rain and snow.

Runoff initially flows in broad, thin sheets, appropriately termed **sheet flow.** The amount of water that runs off in this manner rather than sinking into the ground depends upon the **infiltration capacity** of the soil. Infiltration capacity is controlled by many factors including: (1) the intensity and duration of the rainfall, (2) the prior wetted condition of the soil, (3) the soil texture, (4) the slope of the land, and (5) the nature of the vegetative cover. When the soil becomes saturated, sheet flow commences as a layer only a few millimeters thick. After flowing as a thin, unconfined sheet for only a short distance, threads of

current typically develop and tiny channels called **rills** begin to form and carry the water to a stream.

To some, the term *stream* implies relative size. That is to say, streams are thought of as being larger than creeks or brooks but smaller than rivers. In geology, however, this is not the case. Here the word **stream** is used to denote channelized flow of any size, from the smallest trickle to the mightiest river. It should be pointed out, however, that although the terms *river* and *stream* are used interchangeably, the term *river* is often preferred when describing a main stream into which several tributaries flow.

The remainder of this chapter will concentrate on that part of the hydrologic cycle in which the water moves in stream channels. Further, the discussion will deal primarily with the characteristics of streams in humid regions. Streams are also very important in arid landscapes, as we shall see in Chapter 12.

STREAMFLOW

Water may flow in one of two ways, either as **laminar flow** or **turbulent flow.** As Figure 9.2 illustrates, when the movement is laminar, the water particles flow in straight-line paths that are parallel to the channel. The water particles move steadily downstream without mixing. By contrast, when the flow is turbulent, the water moves in a confused and erratic fashion that is often characterized by swirling, whirlpool-like eddies.

The stream's velocity is the primary factor that determines whether the flow is laminar or turbulent.

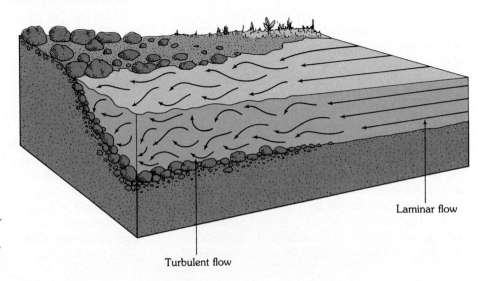

FIGURE 9.2
Laminar flow is only possible when water is moving slowly through a smooth channel. If the velocity increases or the channel becomes rough, laminar flow changes to turbulent flow. Practically all streamflow is turbulent.

Laminar flow

Turbulent flow

Laminar flow is only possible when water is moving very slowly through a smooth channel. If the velocity increases or the channel becomes rough, laminar flow changes to turbulent flow. The movement of water in streams is usually fast enough that most flow is turbulent. The multidirectional movement of turbulent flow is very effective both in eroding a stream's channel and in keeping sediment suspended within the water so that it can be transported downstream.

Flowing water makes its way to the sea under the influence of gravity. The time required for the journey depends upon the velocity of the stream, which is measured in terms of the distance the water travels in a given unit of time. Some sluggish streams travel at less than 1 kilometer per hour, whereas a few rapid ones may reach speeds in excess of 30 kilometers per hour. Velocities are determined at gauging stations where measurements are taken at several locations across the channel and then averaged. This is done because the rate of water movement is not uniform within a stream channel. When the channel is straight, the highest velocities occur in the center just below the surface. It is here that friction is least. Minimum velocities occur along the sides and bottom (bed) of the channel where friction is always greatest. When a stream channel is crooked or curved, the fastest flow is not in the center. Rather, the zone of maximum velocity shifts toward the outside of each bend. As we shall see later, this shift plays an important part in eroding the bank on that side.

The velocity of a stream is directly related to its ability to erode and transport materials; thus, it is a very important characteristic. Even slight variations in velocity can lead to significant changes in the load of sediment transported by the water. Several factors determine the velocity of a stream and therefore control the amount of erosional work a stream may accomplish. These factors include: (1) the gradient, (2) the shape, size, and roughness of the channel, and (3) the discharge.

Certainly one of the most obvious factors controlling stream velocity is the **gradient,** or slope, of a stream channel. Gradient is typically expressed as the vertical drop of a stream over a fixed distance. Gradients may vary considerably from one stream to another as well as along the course of a given stream. Portions of the lower Mississippi River, for example,

have gradients of 10 centimeters per kilometer and less. By way of contrast, some mountain stream channels decrease in elevation at a rate of more than 40 meters per kilometer, 400 times more abruptly than the lower Mississippi. The higher the gradient, the more energy available for streamflow. If two streams were identical in every respect except gradient, the stream with the higher gradient would obviously have the greater velocity.

The cross-sectional shape of a channel determines the amount of water in contact with the channel and hence affects the frictional drag. The most efficient channel is one with the least perimeter for its cross-sectional area. Figure 9.3 compares three types of channels. Although the cross-sectional area of all three is identical, the semicircular shape has less water in contact with the channel than the others and therefore less frictional drag. As a result, if all other factors are equal, the water will flow more rapidly in the semicircular channel.

The size and roughness of the channel also affect the amount of friction. An increase in the size of a channel reduces the ratio of perimeter to cross-sectional area and therefore increases the efficiency of flow. The effect of roughness is obvious. A smooth channel promotes a more uniform flow, while an irregular channel filled with boulders creates enough turbulence to significantly retard the stream's forward motion.

The **discharge** of a stream is the amount of water flowing past a certain point in a given unit of time. This is usually measured in cubic meters per second or cubic feet per second. Discharge is determined by multiplying a stream's cross-sectional area by its velocity:

discharge (m^3/second) =
  channel width (meters) $\times$ channel depth (meters)
  $\times$ velocity (meters/second)

The largest river in North America, the Mississippi, discharges an average of 17,715 cubic meters per second. Although this is a huge quantity of water, it is nevertheless dwarfed by the mighty Amazon, the world's largest river. Draining an area that is nearly three-quarters the size of conterminous United States and averaging about 200 centimeters of rain per year, the Amazon discharges 10 times more water

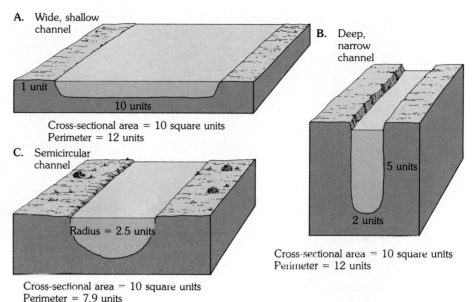

FIGURE 9.3

Influence of channel shape on velocity. Although the cross-sectional area of all three channels is the same, the semicircular channel has less water in contact with the channel, and hence less frictional drag. As a result, the water will flow more rapidly in this channel, all other factors being equal.

A. Wide, shallow channel

Cross-sectional area = 10 square units
Perimeter = 12 units

B. Deep, narrow channel

Cross-sectional area = 10 square units
Perimeter = 12 units

C. Semicircular channel

Radius = 2.5 units

Cross-sectional area = 10 square units
Perimeter = 7.9 units

than the Mississippi (Figure 9.4). In fact, the flow of the Amazon accounts for about 15 percent of all the fresh water discharged into the ocean by all of the world's rivers. Just one day's discharge would supply the water needs of New York City for 9 years!

As Figure 9.5 illustrates, the discharges of the Amazon and Mississippi rivers are far from constant. This is true for most rivers because of such variables as rainfall and snowmelt. If discharge changes, then the factors noted earlier must also change. When discharge increases, the width or depth of the channel must increase or the water must flow faster, or some combination of these factors must change. Indeed, measurements show that when the amount of water in a stream increases, the width, depth, and velocity all increase in an orderly fashion (Figure 9.6). In order to handle the additional water, the stream will increase the size of its channel by widening and deepening it. As we saw earlier, when the size of the channel increases, proportionally less of the water is in contact with the bed and banks of the channel. This means that friction, which acts to retard the flow, is relatively decreased. The less friction, the more swiftly the water will flow.

CHANGES DOWNSTREAM

One useful way of studying a stream is to examine its **longitudinal profile.** Such a profile is simply a cross-

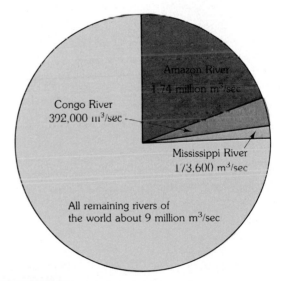

Congo River
392,000 m³/sec

Amazon River
1.74 million m³/sec

Mississippi River
173,600 m³/sec

All remaining rivers of the world about 9 million m³/sec

FIGURE 9.4

The Amazon's average flow is 10 times greater than the flow of the Mississippi, North America's largest river, and represents 15 percent of all the fresh water discharged into the oceans by all the world's rivers. (Courtesy of U.S. Geological Survey)

sectional view of a stream from its source area (called the **head** or **headwaters**) to its **mouth,** the point downstream where the river empties into another water body. By examining Figure 9.7, you can see

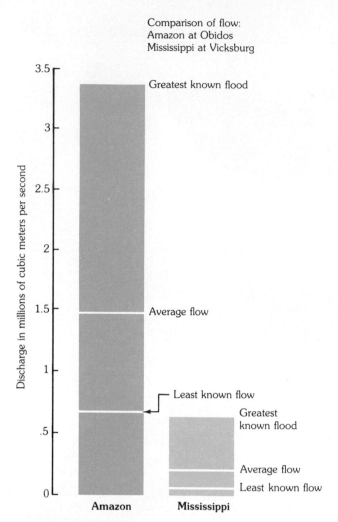

Comparison of flow:
Amazon at Obidos
Mississippi at Vicksburg

Discharge in millions of cubic meters per second

Greatest known flood

Average flow

Least known flow

Greatest known flood

Average flow

Least known flow

Amazon   Mississippi

FIGURE 9.5
As the data for the Amazon and Mississippi rivers illustrate, the discharge of a river can be highly variable. (Courtesy of U.S. Geological Survey)

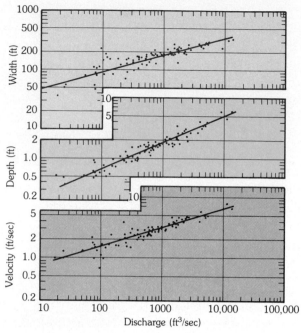

Width (ft)

Depth (ft)

Velocity (ft/sec)

Discharge (ft³/sec)

FIGURE 9.6
Relationship of width, depth, and velocity to discharge of the Powder River at Locate, Montana. As discharge increases, width, depth, and velocity all increase in an orderly fashion. (From L. B. Leopold and Thomas Maddock, Jr., U.S. Geological Survey Professional Paper 252, 1953)

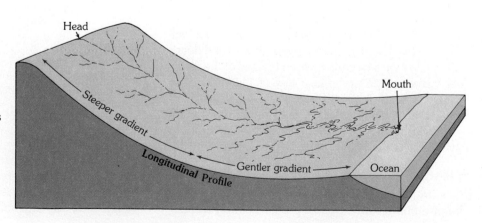

Head

Steeper gradient

Longitudinal Profile

Gentler gradient

Mouth

Ocean

FIGURE 9.7
A longitudinal profile is a cross section along the length of a stream. Note the concave-upward curve of the profile, with a steeper gradient upstream and a gentler gradient downstream.

200

that the most obvious feature of a typical longitudinal profile is a constantly decreasing gradient from the head to the mouth. Although many local irregularities may exist, the overall profile is a smooth concave-upward curve.

The longitudinal profile shows that the gradient decreases downstream. To see how other factors change in a downstream direction, observations and measurements must be made. When data are collected from successive gauging stations along a river, they show that discharge increases toward the mouth. This should come as no surprise since, as we move downstream, more and more tributaries contribute water to the main channel. Furthermore, in most humid regions, additional water is continually being added from the groundwater supply. Since this is the case, the width, depth, and velocity must change in response to the increased volume of water carried by the stream. Indeed, the downstream changes in these variables have been shown to vary in a manner similar to what occurs when discharge increases at one place; that is, width, depth, and velocity all increase systematically. Figure 9.8 depicts these variables along a portion of the Missouri-Mississippi river system to illustrate this point.

The observed increase in velocity that occurs downstream contradicts our impressions about wild, rushing mountain streams and wide, placid rivers. The mental picture that we may have of "old man river just rollin' along" is just not so. Although a mountain stream may have the appearance of a raging torrent, its average velocity is often less than for the river near its mouth. The difference is primarily attributable to the greater efficiency of the larger channel in a downstream direction.

In the headwaters region where the gradient is steep, the water must flow in a relatively small and often boulder-strewn channel. The small channel and rough bed create great drag and inhibit movement by sending water in all directions with almost as much backward motion as forward motion. However, as one progresses downstream, the material on the bed of the stream becomes much smaller, offering less resistance to flow, and the width and depth of the channel increase to accommodate the greater discharge. These factors, especially the wider and deeper channel, permit the water to flow more freely and hence more rapidly.

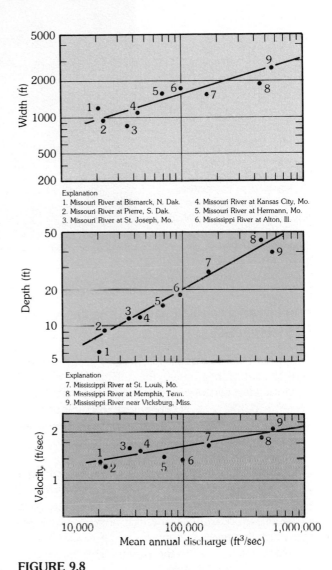

FIGURE 9.8
As illustrated by data from various points along the Missouri-Mississippi system, discharge, velocity, channel depth, and channel width all increase downstream. (From L. B. Leopold and Thomas Maddock, Jr., U.S. Geological Survey Professional Paper 252, 1953)

In summary, we have seen that there is an inverse relationship between gradient and discharge. Where the gradient is high, the discharge is small, and where the discharge is great, the gradient is small. Stated another way, a stream can maintain a higher velocity near its mouth even though it has a lower gradient than upstream because of the greater discharge, larger channel, and smoother bed.

THE EFFECT OF URBANIZATION ON DISCHARGE

When rains occur, stream discharge increases. If the rains are sufficiently heavy, the ability of the channel to contain the discharge is exceeded and water spills over the banks as a flood (Figure 9.9). Floods are natural events that should be expected. However, when cities are built, the magnitude and frequency of flooding increases.

Figure 9.10A is a hypothetical hydrograph that shows the time relationship between a rainstorm and the occurrence of flooding. Notice that the water level in the stream does not rise at the onset of precipitation because time is needed for water to move from the place where it fell to the stream. This time difference is called the **lag time.**

When an area changes from being predominantly rural to largely urban, streamflow is affected. The effect of urbanization on streamflow is illustrated by the hydrograph in Figure 9.10B. Notice that after urbanization the peak discharge during a flood is greater, and that the lag time between precipitation and flood peak is shorter than before urbanization. The explanation for this effect is relatively simple. The construction of streets, parking lots, and buildings covers over the ground that once soaked up water. Thus, less water infiltrates the ground and the rate and amount of runoff increases. Further, since much less water soaks into the ground, the low-water (dry season) flow in urban streams, which is maintained

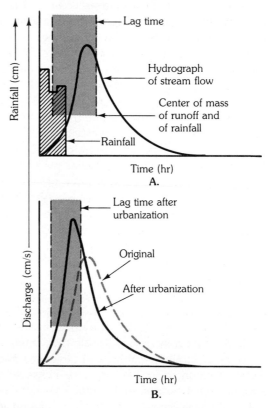

FIGURE 9.10
Generalized hydrographs. **A.** The typical lag time between the time when most of the rainfall occurs and when flooding occurs. **B.** Decreased lag time and higher floodstage due to urbanization. (After L. B. Leopold, U.S. Geological Survey Circular 559, 1968)

by the seepage of groundwater into the channel, is greatly reduced. As one might expect, the magnitude of these effects is a function of the percentage of land that is covered by impermeable surfaces.

Urbanization is just one example of human interference with streams. There are many other ways that land use inadvertently influences the flow of streams and the work they carry out. Moreover there are also many ways by which people intentionally attempt to manipulate and control streams. Some of these will be discussed at appropriate points later in this chapter.

BASE LEVEL AND GRADED STREAMS

In 1875 John Wesley Powell, the pioneering geologist who first explored the Grand Canyon and later headed the U.S. Geological Survey, introduced the concept that there is a downward limit to stream erosion, which he called **base level.** Although the idea is relatively straightforward, it is nevertheless a key concept in the study of stream activity. Base level is defined as the lowest elevation to which a stream can erode its channel. Essentially this is the level at which the mouth of a stream enters the ocean, a lake, or another stream. Base level accounts for the fact that most stream profiles have low gradients near their mouths, because the streams are approaching the elevation below which they cannot lower their beds. Powell recognized that two types of base level exist:

> We may consider the level of the sea to be a grand base level, below which the dry lands cannot be eroded; but we may also have, for local and temporary purposes, other base levels of erosion. . . .[1]

Sea level, which Powell called "grand base level," is now referred to as **ultimate base level. Local** or **temporary base levels** include lakes, resistant layers of rock, and main streams which act as base levels for their tributaries. All have the capacity to limit a stream at a certain level. For example, when a stream enters a lake, its velocity quickly approaches zero and its ability to erode ceases. Thus the lake prevents the

[1]*Exploration of the Colorado River of the West,* (Washington, D.C.: Smithsonian Institution, 1875) p. 203.

stream from eroding below its level at any point upstream from the lake (Figure 9.11A). Although lakes and layers of resistant rock may not seem like temporary features, over the long span of geologic time they are indeed only passing phenomena. Even the largest lakes are eventually drained by the downcutting of their outlets and even the hardest rock layers will ultimately be cut through by the grinding action of sediment in a swiftly moving stream. Thus a lake or a rock layer are only temporary hindrances to a stream's ability to downcut its channel.

Any change in base level will cause a corresponding readjustment of stream activities. When a dam is built along a stream course, the reservoir which forms behind it raises the base level of the stream (Figure 9.11B). Upstream from the dam the stream gradient is reduced, lowering its velocity and, hence, its sediment-transporting ability. The stream, now unable to transport all of its load, will deposit material, thereby building up its channel. This process continues until the stream again has a gradient sufficient to carry its load. The profile of the new channel would be similar to the old, except that it would be somewhat higher.

If, on the other hand, the base level should be lowered, either by uplifting of the land or by a drop in sea level, the stream would again readjust. The stream, now above base level, would have excess energy and downcut its channel to establish a balance with its new base level (Figure 9.11C). Erosion would first progress near the mouth, then work upstream until the stream profile was adjusted along its full length.

The observation that streams adjust their profile for changes in base level led to the concept of a graded stream. A **graded stream** has the correct slope and other channel characteristics necessary to maintain just the velocity required to transport the material supplied to it. On the average, a graded system is not eroding or depositing material but is simply transporting it. Once a stream has reached this state of equilibrium, it becomes a self-regulating system in which a change in one characteristic causes an adjustment in the others to counteract the effect. Referring again to our example of a stream adjusting to a lowering of its base level, the stream would not be graded while it was cutting its new channel but would achieve this state after downcutting had ceased.

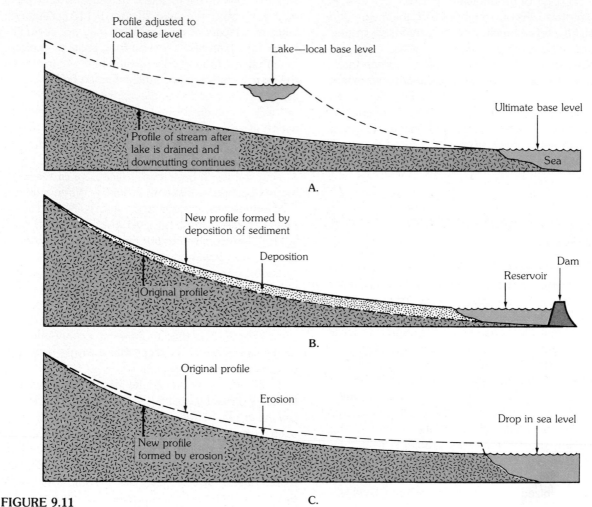

FIGURE 9.11
A. Effect of local base level on a stream profile. **B.** Adjustment of stream profile to a change in base level. **C.** Readjustment of a stream to a lower base level.

STREAM EROSION

Streams may erode their channels in several different ways: by lifting loosely consolidated particles, by abrasion, and by solution activity. The last of these is by far the least significant. Although some erosion results from the solution of soluble bedrock and channel debris, most of the dissolved material in a stream is contributed by groundwater.

As we learned earlier, when the flow of water is turbulent, the water whirls and eddies. When an eddy is sufficiently strong, it can dislodge particles from the channel and lift them into the moving water. In this manner, the force of running water swiftly erodes poorly consolidated materials on the bed and banks of the stream. The stronger the current, the more effectively the stream will lift particles. In some instances water is forced into cracks and bedding planes with sufficient strength to actually pry up pieces of rock from the bed of the channel.

Observing a muddy stream will show that currents of water can lift and carry debris. However, it is not as obvious that a stream is capable of eroding solid rock. However, just as the particles on sandpaper can

A.

FIGURE 9.12
A. Potholes in the bed of the James River exposed during a period of low water. (Photo by C. K. Wentworth, U.S. Geological Survey) **B.** The rotational motion of swirling pebbles acts like a drill to create potholes. (Photo by Kenneth Hasson)

B.

effectively wear down a piece of wood, so too the solid particles carried by a stream (especially sand and gravel) are capable of abrading a bedrock channel. Many steep-sided gorges cut through solid rock by the ceaseless bombardment of particles against the bed and banks of a channel serve as testimony to this erosional strength. In addition, the individual sediment grains are also abraded by the many impacts

with the channel and with one another. Thus, by scraping, rubbing, and bumping, abrasion erodes a bedrock channel and simultaneously smoothes and rounds the abrading particles.

A common feature on some river beds are rounded depressions known as **potholes** (Figure 9.12). They are created by the abrasive action of particles swirling in fast-moving eddies. The rotational motion of the sand and pebbles acts like a drill to bore the holes. As the particles wear down to nothing, they are replaced by new ones that continue to drill the stream bed. Eventually smooth depressions several meters across and just as deep may result.

TRANSPORT OF SEDIMENT BY STREAMS

Streams are the most important erosional agent not only because they have the ability to downcut their channels, but also because they have the capacity to transport the enormous quantities of sediment produced by weathering. Although erosion by running water in the channel contributes significant amounts of material for transport, by far the greatest quantity of sediment carried by a stream is derived from the products of weathering. Weathering produces tremendous amounts of material that are delivered to the stream by sheet flow, mass wasting, and groundwater.

Streams transport their load of sediment in three ways: (1) in solution (**dissolved load**), (2) in suspension (**suspended load**), and (3) along the bottom of the channel (**bed load**).

We have already seen that some of the material carried in solution may be acquired as a stream dissolves bedrock in its channel. However, the greatest portion of the dissolved load transported by most streams is supplied by groundwater. As water percolates through the ground, it first acquires soluble soil compounds. As the water seeps deeper through cracks and pores in the bedrock below, additional mineral matter may be dissolved. Eventually much of this mineral-rich water finds its way into streams.

The velocity of stream flow, which is very important to the transportation of solid particles, has essentially no effect upon a stream's ability to carry its dissolved load. After material is in solution, it goes wherever the stream goes, regardless of velocity. Precipitation occurs only when the chemistry of the water changes.

The quantity of material carried in solution is highly variable and depends upon such factors as climate and the geologic setting. Usually the dissolved load is expressed as parts of dissolved material per million parts of water (parts per million, or ppm). Although some rivers may have a dissolved load of 1000 ppm or more, the average figure for the world's rivers is estimated at between 115 and 120 ppm. Almost 4 billion metric tons of dissolved mineral matter are supplied to the oceans each year by streams.

Most streams (but not all) carry the largest part of their load in suspension (Figure 9.13). Indeed, the visible cloud of sediment suspended in the water is the most obvious portion of a stream's load. Usually only fine sand-, silt-, and clay-sized particles can be carried this way, but during floodstage larger particles are carried as well. Also during floodstage, the total quantity of material carried in suspension increases dramatically, as can be verified by persons whose homes have been sites for the deposition of this material. During floodstage the Hwang Ho (Yellow River) of China is reported to carry an amount of sediment equal in weight to the water that carries it. Rivers like this are appropriately described as "too thick to drink, but too thin to cultivate."

The type and amount of material carried in suspension is controlled by two factors: the velocity of the water and the settling velocity of each sediment grain. **Settling velocity** is defined as the speed at

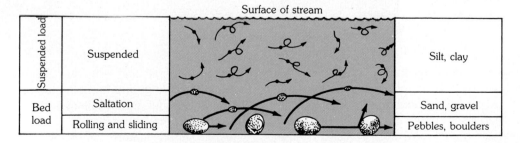

FIGURE 9.13
Transportation of a stream's load.

which a particle falls through a still fluid. The larger the particle, the more rapidly it settles toward the stream bed. In addition to size, the shape and specific gravity of particles also influence settling velocity. Flat grains sink through water more slowly than spherical grains and dense particles fall toward the bottom more rapidly than less dense particles. As long as the velocity of a stream exceeds the settling velocity of a sediment grain, that particle will be carried downstream with the flowing water.

A portion of a stream's load of solid material consists of sediment that is too large to be carried in suspension. These coarser particles move along the bottom of the stream and constitute the bed load. In terms of the erosional work accomplished by a downcutting stream, the grinding action of the bed load is of great importance.

The particles composing the bed load move along the bottom by rolling, sliding, and saltation. Sediment moving by **saltation** appears to jump or skip along the stream bed (Figure 9.13). This occurs as particles are propelled upward by collisions or sucked upward by the current and then carried downstream a short distance until gravity pulls them back to the bed of the stream. Particles that are too large or heavy to move by saltation either roll or slide along the bottom, depending upon their shape.

Unlike the suspended and dissolved loads, which are constantly in motion, the bed load is in motion only intermittently, when the force of the water is sufficient to move the larger particles. Although the bed load may constitute up to 50 percent of the total load of a few streams, it usually does not exceed 10 percent of a stream's total load. For example, consider the distribution of the 750 million tons of material carried to the Gulf of Mexico by the Mississippi River each year. Of this total, it is estimated that approximately 500 million tons are carried in suspension, 200 million tons in solution, and the remaining 50 million tons as bed load. Estimates of a stream's bed load, however, should be viewed cautiously because this fraction of the load is very difficult to measure accurately. Not only is the bed load more inaccessible than the suspended and dissolved loads, but it moves primarily during periods of flooding when the bottom of a stream channel is most difficult to study.

A stream's ability to carry solid particles is typically described using two criteria. First, the maximum load of solid particles that a stream can transport is termed its **capacity.** The capacity of a stream is directly related to its discharge. The greater the amount of water flowing in a stream, the greater the stream's capacity for hauling sediment. Second, the **competence** of a stream is a measure of the maximum size of particles it is capable of transporting. The stream's velocity determines its competence; the stronger the flow, the larger the particles it can carry in suspension or as bed load. It is a general rule that the competence of a stream increases as the square of its velocity. Thus, if the velocity of a stream doubles, the impact force of the water increases four times; if the velocity triples, the force increases nine times, and so forth. Hence, the large boulders that are often visible during a low-water stage and seem immovable can, in fact, be transported during floodstage because of the stream's increased velocity.

Figure 9.14 illustrates the controlling influence of velocity on stream erosion, transportation, and deposition. The velocity necessary to lift a particle of a specific size is shown on the upper curve. Notice that the velocity necessary to lift clay-sized particles is greater than that needed to set sand in motion. This unexpected upturn in the curve is due primarily to strong cohesive forces which cause the tiny clay-sized particles to cling tightly together. However, once these fine particles are set in motion, they can remain suspended at very low velocities. Coarse particles, on the other hand, remain in motion in only a narrow range of velocities.

By now it should be relatively clear why the greatest erosion and transportation of sediment must occur during floods. The increase in discharge not only results in a greater capacity but in an increased velocity. With rising velocity the water becomes more turbulent, and larger and larger particles are set in motion. In the course of just a few days, or perhaps just a few hours, a stream in floodstage can erode and transport more sediment than it does during months of normal flow.

DEPOSITION OF SEDIMENT BY STREAMS

Whenever a stream's velocity subsides, its competence is reduced. The lower curve in Figure 9.14 shows that as the velocity of a river diminish

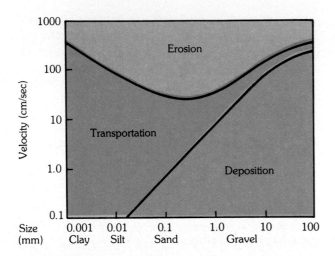

FIGURE 9.14
Relationship of stream velocity to erosion, transportation, and deposition of various particle sizes.

particles of sediment are deposited according to size. As streamflow drops below the critical settling velocity of a certain particle size, sediment in that category begins to settle out. Thus stream transport provides a mechanism by which solid particles of various sizes are separated. This process, called **sorting,** explains why particles of similar size are deposited together.

The well-sorted material typically deposited by a stream is called **alluvium,** a general term applicable to any stream-deposited sediment. Many different depositional features are composed of alluvium. Some of these features may be found within stream channels, some occur on the valley floor adjacent to the channel, and some exist at the mouth of the stream.

CHANNEL DEPOSITS

As a river transports sediment toward the sea, some material may be deposited within the channel. Channel deposits are most often composed of sand and gravel, the coarser components of a stream's load, and are commonly referred to as **bars.** Such features, however, are only temporary, for the material will be picked up again by the running water and be transported farther downstream. Eventually, most of the material will be carried to its ultimate destination, the ocean.

Sand and gravel bars can form in a number of situations. For example, they are common where

streams flow in a series of bends, called meanders. As a stream flows around a bend, the velocity of the water on the outside increases, leading to erosion at that site. At the same time, the water on the inside of the meander slows, which causes some of the sediment load to settle out. Since these deposits occur on the inside "point" of the bend, they are called **point bars.** Actually these deposits would be better described as crescent-shaped accumulations of sand and gravel.

Sometimes a stream deposits materials on the floor of its channel. As these accumulations begin to choke the channel, they force the stream to split and follow several paths. What results is a complex network of converging and diverging channels that thread their way among the bars. Because such channels have an interwoven appearance, the stream is said to be **braided** (Figure 9.15). Braided patterns most often form when the load supplied to a stream exceeds its competency or capacity. For example, if a steeper, more turbulently flowing tributary enters a main stream, its rocky bed load may be deposited at the junction. Excessive load may also be provided when debris from barren slopes is flushed into a channel during a heavy downpour, or at the end of a glacier where ice-eroded sediment is dumped into a meltwater stream flowing away from the glacier. Braided streams also form when there is an abrupt decrease in gradient or a decrease in the stream's discharge. The latter situation could result from a drop in rainfall in the area drained by the stream. It also commonly occurs when a stream leaves a humid area with many tributaries and enters a dry region with few tributaries. In this case the loss of water to evaporation and seepage into the channel results in a diminished discharge.

FLOODPLAIN DEPOSITS

As its name implies, a **floodplain** is that part of a valley that is inundated during a flood. Most streams are bordered by floodplains. Although some are impressive features that are many kilometers across, others are very modest, having widths of just a few meters. If we were to sample the alluvium covering a floodplain, we would find that some of it consists of coarse sand and gravel that was originally deposited as point bars by meanders shifting laterally across the valley floor. Other sediments would be composed of

FIGURE 9.15
Braided stream choked with sediment near the edge of a melting glacier. (Photo by
Bradford Washburn)

fine sands, silts, and clays that were spread across the floodplain whenever water left the channel during floodstage.

Rivers that occupy valleys with broad, flat floors sometimes create a landform called a **natural levee** that parallels the stream channel. Natural levees are built by successive floods over a period of many years. When a stream overflows its banks onto the floodplain, the water moves over the surface as a broad sheet. Since such a flow pattern significantly reduces the water's velocity and turbulence, the coarser portion of the suspended load is deposited in strips bordering the channel. As the water spreads out

over the floodplain, a lesser amount of finer sediment is laid down over the valley floor. This uneven distribution of material produces the very gentle, almost imperceptible slope of the natural levee (Figure 9.16). The natural levees of the lower Mississippi River rise 6 meters above the lower portions of the valley floor. The area behind the levee is characteristically poorly drained for the obvious reason that water cannot flow up the levee and into the river. Marshes called **back swamps** often result. When a tributary stream enters a valley having substantial natural levees, it may not be able to make its way into the main channel. As a consequence the tributary

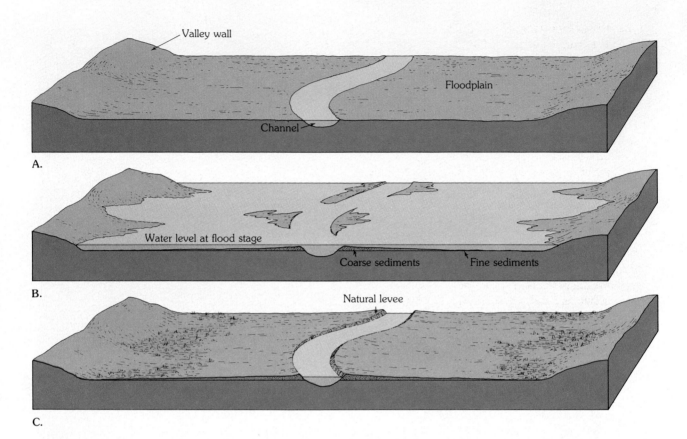

FIGURE 9.16
Formation of natural levees. After repeated flooding, streams may build very gently
sloping levees.

may be forced to flow through the back swamp zone parallel to the main river for many kilometers before it eventually joins the main river. Such streams are called **yazoo tributaries,** after the Yazoo River, which parallels the lower Mississippi River for more than 300 kilometers.

Sometimes artificial levees are built along rivers as a means of flood control. These are usually easy to distinguish from natural levees because their slopes are much steeper. When a river is confined by levees during periods of high water, it deposits material in its channel as the discharge diminishes. This is sediment that otherwise would have been dropped on the floodplain. Thus, each time there is a high flow, deposits are left on the river bed and the bottom of the channel is built up. With the buildup of the bed, less water is required to overflow the original levee.

As a result, the height of the levee must be raised to protect the floodplain. For this reason, many levees along the lower Mississippi River have had to be raised to cope with the increasing height of the water in the channel. As you can see, artificial levees are not a permanent solution to the problem of flooding. If protection is to be maintained, the structures must be heightened periodically, a process that cannot go on indefinitely.

DELTAS AND ALLUVIAL FANS

Two of the most common landforms composed of alluvium are deltas and alluvial fans. They are sometimes similar in shape and are deposited for essentially the same reason: an abrupt loss of competence in a stream. Although similar in those respects, the two features are distinct. Alluvial fans are deposited

FIGURE 9.17
A large alluvial fan in Death Valley, California. These structures develop where the gradient of a stream changes abruptly, such as at the foot of a mountain. (Photo by John S. Shelton)

on land; deltas are deposited in a body of water. In addition deltas are relatively flat, barely protruding above the level surface of the ocean or lake in which they formed. Conversely, alluvial fans can be quite steep.

Alluvial fans typically develop where a high-gradient stream leaves a narrow valley in mountainous terrain and comes out suddenly onto a broad, flat plain or valley floor. Alluvial fans form in response to the abrupt drop in gradient combined with the change from a narrow channel of a mountain stream to the unconfined flow on the slopes of the plain. The sudden drop in velocity causes the stream to dump its load of sediment quickly in a distinctive cone- or fan-shaped accumulation. As illustrated by Figure 9.17,

the surface of the fan slopes outward in a broad arc from an apex at the mouth of the steep valley. Usually, coarse material is dropped near the apex of the fan, while finer material is carried toward the base of the deposit. As we learned earlier, steep canyons in dry regions are prime locations for mud flows. Therefore, it should be expected that many alluvial fans in arid areas have mudflow deposits interbedded with the alluvium.

In contrast to an alluvial fan, a **delta** forms when a stream enters the relatively still waters of the ocean or a lake. The stream's forward motion is checked and the dying current deposits its load of sediment. The finer silts and clays will settle out some distance from the mouth into nearly horizontal layers called

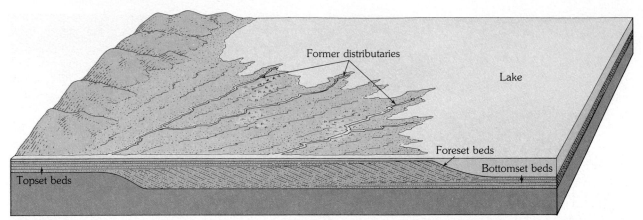

A.

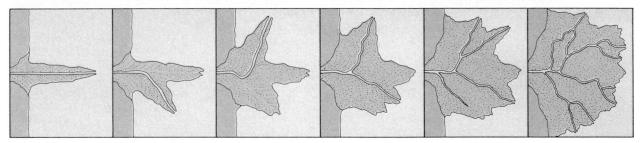

B.

FIGURE 9.18
A. Structure of a simple delta. **B.** Growth of a simple delta. As a stream extends its
channel, the reduced gradient causes it to find a shorter route to its base level. (After
Ward's Natural Science Establishment, Inc., Rochester, N.Y.)

bottomset beds (Figure 9.18A). Prior to the accumulation of bottomset beds, **foreset beds** begin to form. These beds are composed of coarser sediment which drops almost immediately upon entering a lake or ocean to form sloping layers. The foreset beds are usually covered by thin, horizontal **topset beds** deposited during floodstage. As the delta grows outward, the river's gradient is continually lowering, causing the stream to search for a shorter route to base level, a process illustrated in Figure 9.18B. This figure shows how a simple delta grows into the idealized triangular shape of the Greek letter delta (Δ), for which it was named. Note, however, that many deltas do not exhibit this idealized shape. Differences in the nature of shoreline processes and in the configuration of coastlines result in a number of different shapes. Moreover, many of the world's great rivers have created massive deltas, each with its own peculiarities, and none as simple as the delta illustrated in Figure 9.18B.

Many large rivers have deltas extending over thousands of square kilometers. The Mississippi delta began forming millions of years ago near the present-day town of Cairo, Illinois, and has since advanced nearly 1600 kilometers (1000 miles) to the south. New Orleans rests where there was ocean less than 5000 years ago, and the present bird's foot delta shown in Figure 9.19 has been built in the last 500 years. Figure 9.19 also shows the main channel dividing into several smaller ones called **distributaries.** The Mississippi delta, like most others, is characterized by these shifting channels that act in an opposite way to that of tributaries. Rather than carrying water into the main channel, distributaries carry water away

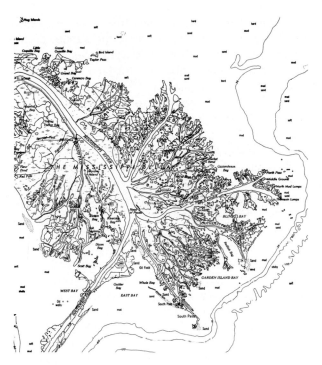

FIGURE 9.19
The bird's foot delta of the Mississippi River as depicted on the Breton Sound, Louisiana, 1:250,000 topographic map. Over the past 400 years, the delta has extended the shoreline by more than 32 kilometers (20 miles). The numerous channels on the delta are distributaries. (Courtesy of U.S. Geological Survey)

from the main channel in various paths to the sea.

Although deltas are deposited by many large rivers, not all rivers create these features. Even streams that transport large loads of sediment may lack deltas because powerful currents and waves quickly redistribute the material as soon as it is deposited. The Columbia River in the Pacific Northwest is one such situation. In other cases, rivers do not carry sufficient quantities of sediment to build up a delta. The St. Lawrence river, for example, has little opportunity to pick up much sediment between Lake Ontario and its mouth in the Gulf of St. Lawrence.

STREAM VALLEYS

Valleys are the most common features on the earth's surface. In fact they exist in such large numbers that they have never been counted except in limited areas

used for study. Prior to the turn of the nineteenth century it was generally believed that valleys were created by catastrophic events that pulled the crust apart and created avenues for streams to follow. Today, however, we know that with a few exceptions, streams create the valleys through which they flow.

Among the first meaningful statements that related streams to the creation of their valleys was one made by the English geologist, John Playfair in 1802. In his well-known work, *Illustrations of the Huttonian Theory of the Earth,* Playfair stated the principle that has come to be called **Playfair's law:**

> Every river appears to consist of a main trunk, fed from a variety of branches, each running in a valley proportioned to its size, and all of them together forming a system of valleys, communicating with one another, and having such a nice adjustment of their declivities, that none of them join the principle valley, either on too high or too low a level; a circumstance that would be indefinitely improbable, if each of these valleys were not the work of the stream that flows in it.

Not only were Playfair's observations essentially correct, they were written in a style that is seldom achieved in scientific prose.

Stream valleys can be divided into two general types. Narrow, V-shaped valleys and wide valleys with flat floors exist as the ideal forms, with many gradations between. In some arid regions, where downcutting is rapid and weathering is slow, and in places where rock is particularly resistant, narrow valleys may not be V-shaped but rather may have nearly vertical walls. However, most valleys, even those that are narrow at their base, are much broader at the top than the width of the channel at the bottom. This would not be the case if the only agent responsible for eroding valleys were the streams flowing through them.

The sides of most valleys are shaped primarily as the result of weathering, sheet flow, and mass wasting. Consider the following example of this process. Figure 9.20 shows the relationship between suspended load and discharge at a gauging station on the Powder River in Wyoming. Notice that as the discharge increases, the quantity of suspended sediment increases. In fact, the increase is exponential; that is to say, if the discharge at the gauging station

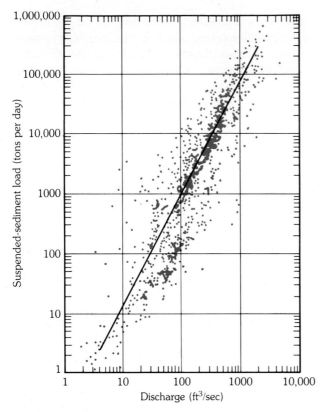

FIGURE 9.20

Relationship between suspended load and discharge on the Powder River at Arvada, Wyoming. (From L. B. Leopold and Thomas Maddock, Jr., U.S. Geological Survey Professional Paper 252, 1953)

increases tenfold, the suspended load may increase by a factor of 100 or more. Measurements and calculations have shown that stream channel erosion during periods of increased discharge can account for only a portion of the additional sediment transported by a stream. Therefore, much of the increased load must be delivered to the stream by sheet flow and mass wasting.

A narrow, V-shaped valley indicates that the primary work of the stream has been downcutting toward base level. The most prominent features in such a valley are **rapids** and **waterfalls** (Figure 9.21). Both occur where the stream profile drops rapidly, a situation usually caused by variations in the erodibility of the bedrock into which the stream channel is cutting. A resistant bed produces a rapids by acting as a temporary base level upstream while allowing

downcutting to continue downstream. Once erosion has eliminated the resistant rock, the stream profile smoothes out again. Waterfalls are places where the stream makes a vertical drop. One type of waterfall is exemplified by Niagara Falls (Figure 9.22A). Here, the falls are supported by a resistant bed of dolomite that is underlain by a less resistant shale (Figure 9.22B). As the water plunges over the lip of the falls it erodes the less resistant shale, undermining a section of dolomite, which eventually breaks off. In this manner the waterfall retains its vertical cliff while slowly but continually retreating upstream. Since its formation Niagara Falls has retreated approximately 11 kilometers (about 7 miles) upstream.

Once a stream has cut its channel closer to base level, it approaches a graded condition, and downward erosion becomes less dominant. At this point more of the stream's energy is directed from side to side. The reason for this change is not fully understood, but the reduced gradient probably is an important factor. Nevertheless it does occur, and the result is a widening of the valley as the river cuts away first at one bank and then the other (Figure 9.23). In this manner the flat valley floor, or floodplain, is produced. This is an appropriate name because the river is confined to its channel except during floodstage, when it overflows its banks and inundates the floodplain.

When a river erodes laterally and creates a floodplain as just described, it is called an *erosional floodplain*. Floodplains can be depositional in nature as well. *Depositional floodplains* are produced by a major fluctuation in conditions, such as a change in base level. The floodplain in Yosemite Valley is one such feature, and was produced when a glacier gouged the former stream valley about 300 meters (1000 feet) deeper than it had been. After the glacial ice melted, the stream readjusted itself to its former base level by refilling the valley with alluvium.

Streams that flow upon floodplains, whether erosional or depositional, move in sweeping bends called **meanders**. The term is derived from a river in western Turkey, the Menderes, which has a very sinuous course. Once a bend in the channel begins to form, it grows larger. Erosion occurs on the outside of the meander where velocity and turbulence are greatest. Commonly the outside bank is undermined, especially during periods of high water. As the bank becomes oversteepened, it fails by slumping into the channel. Because the outside of a meander is a zone of active erosion, it is often referred to as the **cut bank** (Figure 9.24). Much of the debris detached by the stream at the cut bank moves downstream and is soon deposited as point bars in zones of decreased velocity on the insides of meanders. Thus, meanders migrate laterally, while keeping the same cross-sectional area, by eroding on the outside of the bends and depositing on the inside (Figure 9.25). Growth ceases when the meander reaches a critical size that is determined by the size of the stream. The larger the stream, the larger its meanders can be.

Due to the slope of the channel, erosion is more effective on the downstream side of a meander. Therefore, in addition to growing laterally, the bends also gradually migrate down the valley. Sometimes the downstream migration of a meander is slowed when it reaches a more resistant portion of the flood-plain. This allows the next meander upstream to "catch up." Gradually the neck of land between the meanders is narrowed. When they get close enough, the river may erode through the narrow neck of land to the next loop (Figure 9.26). The new, shorter channel segment is called a **cutoff** and, because of its shape, the abandoned bend is called an **oxbow lake.**

FIGURE 9.21
V-shaped valley of the Yellowstone River. The rapids and waterfalls indicate that the river is vigorously downcutting. (Photograph used by permission of Dennis Tasa)

A.

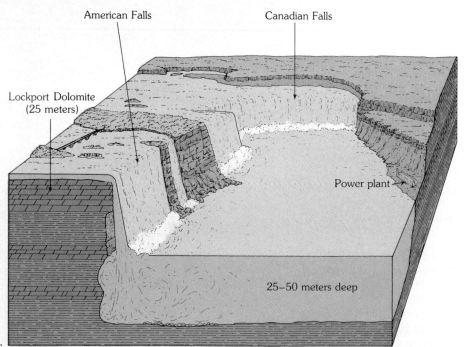

American Falls

Canadian Falls

Lockport Dolomite
(25 meters)

Power plant

25–50 meters deep

FIGURE 9.22
A. Niagara Falls, with the larger Horseshoe Falls at the top and the American Falls on the left. (Courtesy of the Niagara Falls Area Chamber of Commerce) **B.** The river plunges over the falls and erodes the shale beneath the more resistant Lockport Dolomite. As a section of dolomite is undermined, it loses support and breaks off. (After Ward's Natural Science Establishment, Inc., Rochester, N.Y.)

B.

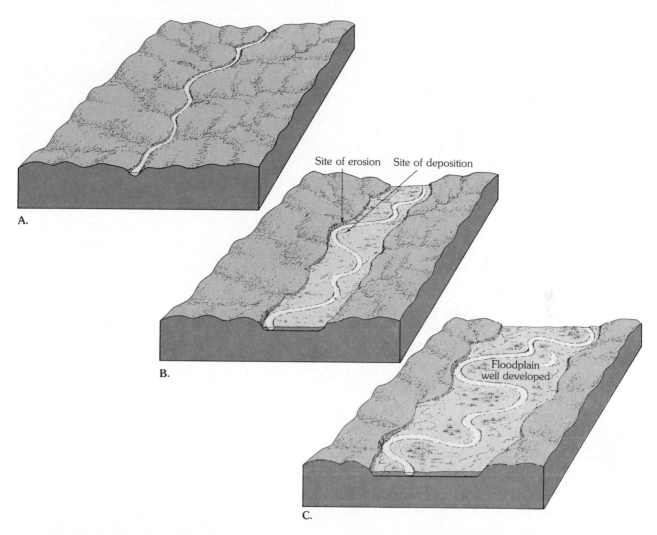

A.

Site of erosion   Site of deposition

B.

Floodplain
well developed

C.

FIGURE 9.23
Stream eroding its floodplain

A.

B.

FIGURE 9.24
Erosion of a cut bank along the Newaukum River, Washington. **A.** January, 1965.
B. March, 1965. (Photos by P. A. Glancy, U.S. Geological Survey)

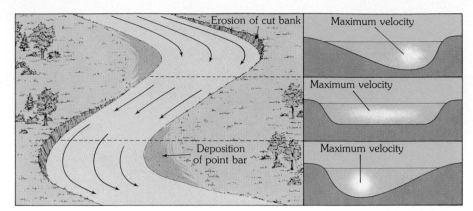

FIGURE 9.25
Lateral movement of meanders. By eroding its outer bank and depositing sediment on the inside of the bend, a stream is able to shift its channel.

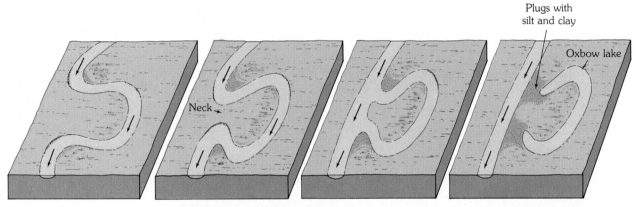

A.

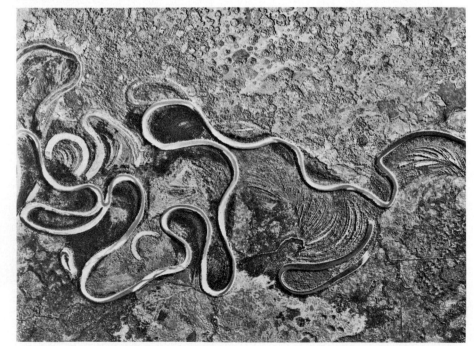

B.

FIGURE 9.26
A. Formation of a cutoff and oxbow lake. **B.** In this vertical aerial view of the meandering Hay River in northern Alberta, oxbow lakes and meander scars are prominent features. (Photo courtesy of the Geological Survey of Canada, catalog no. 203168)

Over a period of time, the oxbow lake fills with sediment to create a **meander scar.**

The process of meander cutoff formation has the effect of shortening the river and was described humorously by Mark Twain in *Life on the Mississippi.*

> **In the space of one hundred and seventy-six years the lower Mississippi has shortened itself two hundred and forty-two miles. This is an average of a trifle over one mile and a third per year. Therefore, any calm person, who is not blind or idiotic, can see that in the Old Oolitic Silurian Period, just a million years ago next November, the Lower Mississippi River was upwards of one million three hundred thousand miles long, and stuck out over the Gulf of Mexico like a fishing rod. And by the same token any person can see that seven hundred and forty-two years from now the Lower Mississippi will be only a mile and three quarters long, and Cairo and New Orleans will have joined their streets together, and be plodding comfortably along under a single mayor and a mutual board of aldermen. There is something fascinating about science. One gets such wholesale returns of conjecture out of such a trifling investment of fact.**

Although the data used by Mark Twain may be reasonably accurate, he purposely forgot to include the fact that the Mississippi created many new meanders, thus lengthening its course by a similar amount. In fact, with the growth of the delta, the Mississippi is actually getting longer, not shorter.

One method of flood control that is sometimes used is to straighten a channel by creating cutoffs artificially. The idea is that by shortening the stream, the gradient and hence the velocity is increased. By increasing velocity, the larger discharge associated with flooding can be dispersed more rapidly. Since 1932 about 16 artificial cutoffs have been constructed on the lower Mississippi for the purpose of increasing the efficiency of the channel and reducing the threat of flooding. The program has been somewhat successful in reducing the height of the river in flood. However, since the river's tendency toward meandering still exists, preventing the river from returning to its previous condition has been difficult.

Artificial cutoffs increase a stream's velocity and may also accelerate erosion of the bed and banks of the channel. A case in point is the Blackwater River in

Missouri, whose meandering course was shortened in 1910. Among the many effects of this project was a dramatic increase in the width of the channel caused by the increased velocity of the stream. One particular bridge over the river collapsed because of bank erosion in 1930. Over the next 17 years the same bridge was replaced on three more occasions, each time with a wider span.

DRAINAGE NETWORKS

A stream is just a small component in a much larger system. Each system consists of a **drainage basin,** the land area that contributes water to the stream. The drainage basin of one stream is separated from another by an imaginary line called a **divide** (Figure 9.27). Divides range in size from a ridge separating two small gullies to continental divides, which split continents into enormous drainage basins. For example, the continental divide that runs somewhat north-south through the Rocky Mountains separates the drainage which flows west to the Pacific Ocean from that which flows to the Atlantic via the Gulf of Mexico. Although divides separate the drainage of two

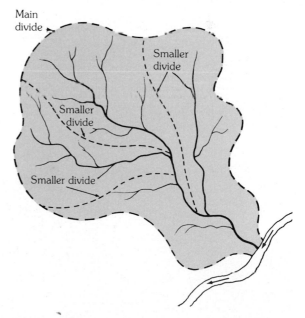

FIGURE 9.27
Drainage basins and divides. Divides separate the drainage basins of each stream. Drainage basins and divides also exist for the smallest tributaries but are not shown.

streams, if they are tributaries of the same river, they are both a part of that larger drainage system.

DRAINAGE PATTERNS

All drainage systems are made up of an interconnected network of streams which together form particular patterns. The nature of a drainage pattern can vary greatly from one type of terrain to another, primarily in response to the kinds of rock on which the streams developed or the structural pattern of faults and folds.

Certainly the most commonly encountered drainage pattern is the **dendritic** pattern (Figure 9.28A). This pattern is characterized by irregular branching of tributary streams that resembles the branching pattern of a deciduous tree. In fact, the word dendritic literally means "treelike." The dendritic pattern forms where underlying bedrock is relatively uniform, such as flat-lying strata or massive igneous rocks. Since the underlying material is essentially uniform in its resistance to erosion, it does not control the pattern of streamflow. Rather, the pattern is determined chiefly by the direction of slope of the land.

Figure 9.28B illustrates a **rectangular** pattern, in which many right-angle bends can be seen. This pattern develops when the bedrock is crisscrossed by a series of joints and faults. Since these structures are eroded more easily than unbroken rock, their geometric pattern guides the directions of valleys.

When streams diverge from a central area like spokes from the hub of a wheel, the pattern is said to be **radial** (Figure 9.28C). This pattern typically develops on isolated volcanic cones and domal uplifts.

Figure 9.28D illustrates a **trellis** drainage pattern, a rectangular pattern in which tributary streams are nearly parallel to one another and have the appearance of a garden trellis. This pattern forms in areas underlain by alternating bands of resistant and less resistant rock and is particularly well displayed in the folded Appalachians, where both weak and strong strata outcrop in nearly parallel belts.

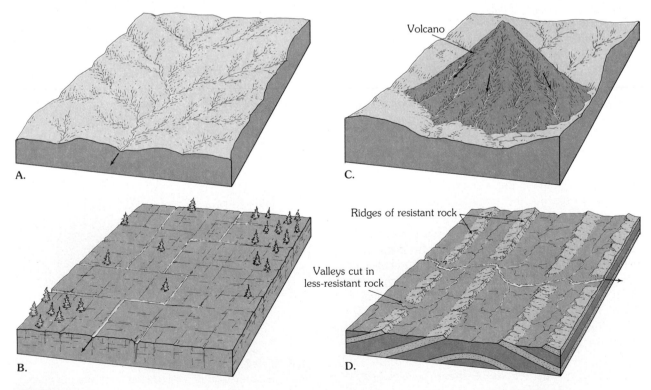

FIGURE 9.28
Drainage patterns. **A.** Dendritic. **B.** Rectangular. **C.** Radial. **D.** Trellis.

ANTECEDENT AND SUPERPOSED STREAMS

Sometimes to fully understand the pattern of streams in an area, the history of the streams must be considered. For example, in many places a river valley can be seen cutting through a ridge or mountain that lies across its path. The steep-walled notch followed by the river through the structure is called a **water gap** (Figure 9.29).

Why does a stream cut across such a structure and not flow around it? One possibility is that the stream existed before the ridge or mountain was formed. In this situation, the stream, called an **antecedent stream,** would have to keep pace downcutting while the uplift progressed. A second possibility is that the stream was **superposed,** or let down, upon the structure. This can occur when a ridge or mountain is buried beneath layers of flat-lying strata. Streams originating on this cover would establish their courses without regard to the structures below. Then, as the valley was deepened and the structure was encountered, the river continued to cut its valley into it. The folded Appalachians again provide some good examples. Here a number of major rivers, such as the Potomac and the Susquehanna, are superposed streams that cut across the folded strata on their way to the Atlantic.

HEADWARD EROSION AND STREAM PIRACY

We have seen that a stream can lengthen its course by building a delta at its mouth. A stream can also lengthen its course by **headward erosion,** that is, by extending the head of its valley upslope. As sheet flow converges, a valley collects water at its headward end. As the water becomes concentrated into a channel, its velocity, and hence its power to erode, increase. The result can be vigorous erosion at the head of the valley. Thus, through headward erosion, the valley extends itself into previously undissected terrain (Figure 9.30).

Headward erosion is not only useful in understanding the dissection of uplands, it also helps explain changes that take place in drainage patterns. One cause for changes that occur in the pattern of streams is **stream piracy,** the diversion of the drainage of one stream because of the headward erosion of another stream. Piracy can occur, for example, if a stream on one side of a divide has a steeper gradient than a stream on the other side. Since the stream with the steeper gradient has more energy, it can extend its valley headward, eventually breaking down the divide and capturing part or all of the drainage of the slower stream. In Figure 9.31, the headwaters of stream A were captured when the more swiftly flowing stream B shifted the divide at its head until the divide intersected and diverted stream A.

Stream piracy also explains the existence of narrow, steep-sided gorges that have no active streams running through them. These abandoned water gaps (called **wind gaps**) form when the stream that cut the notch has its course changed by a pirate stream (Figure 9.32).

FIGURE 9.29
A water gap near Victorville, California. View is upstream where the Mojave River crosses a ridge of massive granitic rock (just above highway bridge). It is speculated that when sediment was deeper and more extensive it completely buried the ridge. On this higher surface the river established its course. As the river cut downward, it cut through the ridge. (Photo by John S. Shelton)

FIGURE 9.30
Valleys extending headward into undissected terrain. (Photo by John S. Shelton)

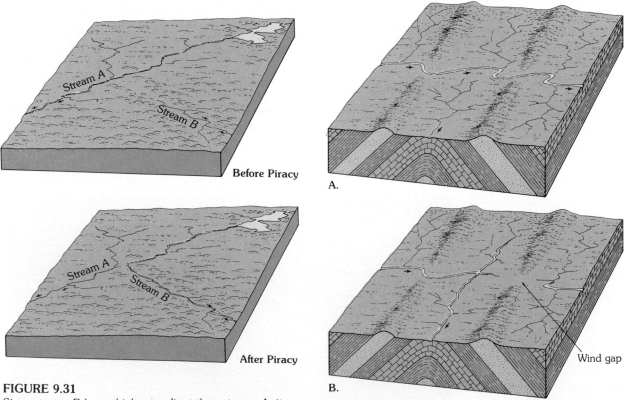

FIGURE 9.31
Since stream *B* has a higher gradient than stream *A*, it extended its valley headward and captured a portion of stream *A*.

FIGURE 9.32
Formation of a wind gap.

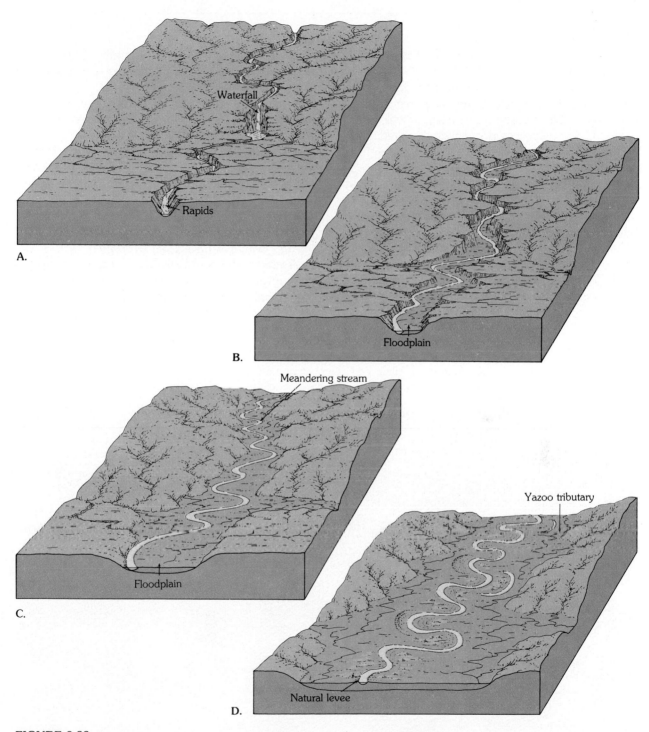

FIGURE 9.33

Stages of valley development. **A.** Youth. The youthful stage is characterized by down-cutting and a V-shaped valley. **B.** and **C.** Maturity. Once a stream has sufficiently lowered its gradient, it begins to erode laterally, producing a wide valley. **D.** Old age. After the valley has been cut several times wider than the width of the meander belt, it has entered old age. (After Ward's Natural Science Establishment, Inc., Rochester, N.Y.)

STAGES OF VALLEY DEVELOPMENT

Contrary to the popular belief of their day, James Hutton and John Playfair proposed that streams were responsible for cutting the valleys in which they flowed. Later geologic work conducted in stream valleys substantiated Hutton's and Playfair's pronouncements and further revealed that the development of stream valleys progresses in a somewhat orderly fashion. The evolution of a valley has been arbitrarily divided into three sequential stages: youth, maturity, and old age.

As long as the stream is downcutting to establish a graded condition with its base level, it is considered youthful. Rapids, an occasional waterfall, and a narrow, V-shaped valley are all visible signs of the vigorous downcutting that is going on. Other features of a youthful stream include a steep gradient, little or no floodplain, and a rather straight course without meanders (see Figure 9.33A on page 223).

When a stream reaches maturity, downward erosion diminishes and lateral erosion dominates. Thus the mature stream begins actively cutting its floodplain and meandering upon it (Figure 9.33B). During the mature stage cutoffs occur, producing oxbows, and a few streams may even produce natural levees (Figure 9.33C). In contrast to the gradient of a youthful stream, the gradient of a mature stream is much lower, and the profile is much smoother, since all rapids and waterfalls have been eliminated.

A stream enters old age after it has cut its floodplain several times wider than its meander belt, which is the width of the meander (Figure 9.33D). When this stage is reached the stream is rarely near the val-

FIGURE 9.34
Entrenched meanders of the San Juan River, Utah. With the uplift of the Colorado Plateau, the San Juan River began downcutting. (Photo by John S. Shelton)

ley walls; hence, it ceases to enlarge the floodplain. For that reason, the primary work of a river in an old-age valley is the reworking of unconsolidated floodplain deposits. Because this task is easier than cutting bedrock, a stream in an old-age valley shifts more rapidly than a stream in a mature valley. For example, some meanders in the lower Mississippi valley move 20 meters a year, and its large floodplain is dotted with oxbow lakes and old cutoffs. Natural levees are also common features of old-age valleys and, when present, are accompanied by back swamps and yazoo tributaries.

Thus far we have assumed that the base level of a stream remains constant as a river progresses from youth to old age. On many occasions, however, the land is uplifted. The effect of uplifting on a youthful stream is to increase its gradient and accelerate its rate of downcutting. However, uplifting of a mature stream would cause it to abandon lateral erosion and revert to downcutting. Rivers of this type are said to be **rejuvenated,** and the meanders are known as **entrenched meanders** (Figure 9.34). Mature streams may eventually readjust to uplift by cutting a new floodplain at a level below the old one. The remnants of the old floodplain are often present in the form of flat surfaces called **terraces.**

Two additional points concerning valley development should be made. First, the time required for a stream to reach any given stage depends on several factors, including the erosive ability of the stream, the nature of the material through which the stream must cut, and its height above base level. Consequently, a stream which starts out very near base level and only has to cut through unconsolidated sediments may reach maturity in a matter of a few hundred years. On the other hand, the Colorado River, where it is ac-tively cutting the Grand Canyon, has retained its youthful nature for an estimated 15 million years. Second, individual portions of a stream reach each stage at different times. Often the lower reaches of a stream attain maturity or old age while the head-waters are still youthful in character.

CYCLE OF LANDSCAPE EVOLUTION

While streams are cutting their valleys they simulta-neously sculpture the land. To describe this unending process we will need a starting point. For this reason only, we will assume the existence of a relatively flat upland area in a humid region. Until a well-established drainage system forms, lakes and ponds will occupy any depressions that exist (Figure 9.35A). As streams form and begin to downcut toward base level they will drain the lakes. During the youthful stage the landscape retains its relatively flat surface, interrupted only by narrow stream valleys (Figure 9.35B). As downcutting continues, relief increases, and the flat, youthful landscape is transformed into one consisting of the hills and valleys which charac-terize the mature stage (Figure 9.35D). Eventually some of the streams will approach base level, and downcutting will give way to lateral erosion. As the cycle nears the old-age stage the effects of sheet flow and mass wasting, coupled with the lateral erosion by streams, will reduce the land to a **peneplain** ("near plain"), an undulating plain near base level (Figure 9.35F). Although no peneplains are known to exist today, there is evidence that they formed in the past and have since been uplifted. Once a peneplain has formed, uplifting starts the cycle over again. Most often, uplifting interrupts the cycle before it reaches old age.

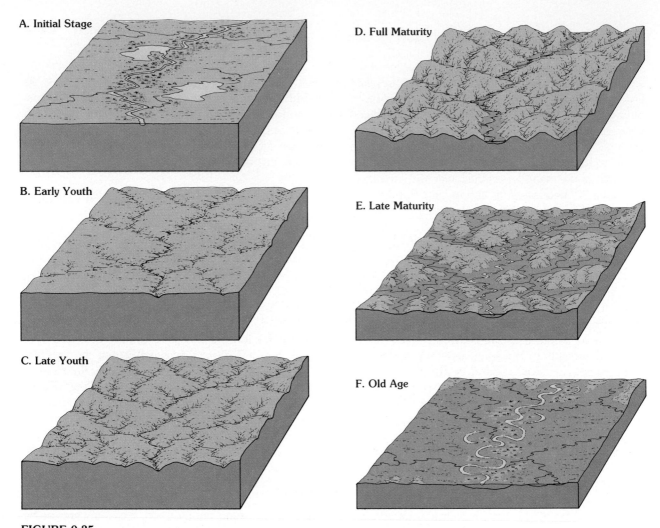

A. Initial Stage

B. Early Youth

C. Late Youth

D. Full Maturity

E. Late Maturity

F. Old Age

FIGURE 9.35

Idealized cycle of landscape evolution. **A.** Initial stage. The land is poorly drained and situated well above base level. **B.** Youthful stage. Streams have cut downward and drained the lakes. The landscape is still relatively flat between stream valleys. **C.** Late youth. **D.** Mature stage. All of the area between the initial streams has been eroded by running water. Most of the landscape is in slope, and maximum relief exists. **E.** Late maturity. **F.** Old-age stage. Lateral erosion, mass wasting, and slope wash have lowered most of the hills to the level of the floodplain. The entire surface has become an undulating plain near base level. (Courtesy of Ward's Natural Science Establishment, Inc., Rochester, N.Y.)

REVIEW QUESTIONS

1 Describe the movement of water through the hydrologic cycle. Once precipitation has fallen on land, what paths are available to it?

2 Over the oceans the quantity of water lost to evaporation is not equaled by the amount gained from precipitation. Why then does the sea level not drop?

3 List several factors that influence infiltration capacity.

4 "Water in streams moves primarily in laminar flow." Briefly explain whether this statement is true or false.

5 A stream originates at 2000 meters above sea level and travels 250 kilometers to the ocean. What is its average gradient in meters per kilometer?

6 Suppose that the stream mentioned in Question 5 developed extensive meanders so that its course was lengthened to 500 kilometers. Calculate its new gradient. How does meandering affect gradient?

7 When the discharge of a stream increases, what happens to the stream's velocity?

8 What typically happens to channel width, channel depth, velocity, and discharge from the point where a stream begins to the point where it ends? You may wish to refer to Figure 9.8. Briefly explain why these changes take place.

9 When an area changes from being predominantly rural to largely urban, how is streamflow affected?

10 Define *base level*. Name the main river in your area. For what streams does it act as base level? What is the base level for the Mississippi River?

11 Why do most streams have low gradients near their mouths?

12 Describe three ways in which a stream may erode its channel. Which one of these is responsible for creating potholes?

13 If you were to collect a jar of water from a stream, what part of the load will settle to the bottom of the jar? What portion will remain in the water? What part of a stream's load was probably not present in your sample?

14 What is settling velocity? What factors influence settling velocity?

15 Distinguish between capacity and competency.

16 The water velocity needed to lift a tiny clay particle is greater than that required to set much larger sand grains in motion. Explain.

17 Describe a situation that might cause a stream channel to become braided.

18 Briefly describe the formation of a natural levee. How is this feature related to back swamps and yazoo tributaries?

19 In what way is a delta similar to an alluvial fan? In what way are they different?

20 For what purpose are artificial cutoffs made?

21 Each of the following statements refers to a particular drainage pattern. Identify the pattern.
    (a) Streams diverging from a central high area such as a dome.
    (b) Branching, "treelike" pattern.
    (c) A pattern that develops when bedrock is crisscrossed by joints and faults.

22 Describe how a water gap might form.

23 Why is it possible for a youthful valley to be older (in years) than a mature valley?

24 Do rivers flowing in mature and old-age valleys make good political boundaries? Explain.

KEY TERMS

alluvial fan (p. 211)

alluvium (p. 208)

antecedent stream (p. 221)

back swamp (p. 209)

bar (p. 208)

base level (p. 203)

bed load (p. 206)

bottomset beds (p. 212)

braided stream (p. 208)

capacity (p. 207)

competence (p. 207)

cut bank (p. 215)

cutoff (p. 215)

delta (p. 211)

dendritic pattern (p. 220)

discharge (p. 198)

dissolved load (p. 206)

distributary (p. 212)

divide (p. 219)

drainage basin (p. 219)

entrenched meander (p. 225)

evapotranspiration (p. 195)

floodplain (p. 208)

foreset beds (p. 212)

graded stream (p. 203)

gradient (p. 198)

head (headwaters) (p. 199)

headward erosion (p. 221)

hydrologic cycle (p. 195)

infiltration (p. 195)

infiltration capacity (p. 197)

lag time (p. 202)

laminar flow (p. 197)

local (temporary) base level (p. 203)

longitudinal profile (p. 199)

meander (p. 214)

meander scar (p. 219)

mouth (p. 199)

natural levee (p. 209)

oxbow lake (p. 215)

peneplain (p. 225)

Playfair's law (p. 213)

point bar (p. 208)

pothole (p. 205)

radial pattern (p. 220)

rapids (p. 214)

rectangular pattern (p. 220)

rejuvenated stream (p. 225)

rills (p. 197)

runoff (p. 195)

saltation (p. 207)

settling velocity (p. 206)

sheet flow (p. 197)

sorting (p. 208)

stream (p. 197)

stream piracy (p. 221)

superposed stream (p. 221)

suspended load (p. 206)

temporary (local) base level (p. 203)

terrace (p. 225)

topset beds (p. 212)

transpiration (p. 195)

trellis pattern (p. 220)

turbulent flow (p. 197)

ultimate base level (p. 203)

waterfall (p. 214)

water gap (p. 221)

wind gap (p. 221)

yazoo tributary (p. 210)

10

GROUNDWATER

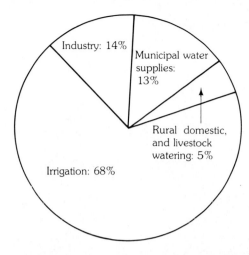

FIGURE 10.1

In 1980 an estimated 334.5 billion liters (88.5 billion gallons) of groundwater were drawn for use each day. Irrigation accounted for 68 percent, industry 14 percent, municipal water supplies 13 percent, and rural domestic and livestock watering 5 percent. (Data from U.S. Water Resources Council)

O f all the world's water only about six-tenths of one percent is found underground. Nevertheless, the amount of water stored in the rocks and sediments beneath the earth's surface is vast. The U.S. Geological Survey estimates that the quantity of water in the upper 800 meters (2600 feet) of the continental crust is about 3000 times greater than the volume of water in all rivers at any one time, and nearly 20 times greater than the combined volume in all lakes and rivers.

In many parts of the world, wells and springs provide the water needs not only for great numbers of people but also for crops, livestock, and industry (Figure 10.1). In the United States, subsurface water supplies about 20 percent of our country's freshwater requirements. In addition, underground water is important as an equalizer of streamflow and as an agent of erosion. The erosional work of subsurface water is responsible for the creation of caverns and many other related features.

Entrance to the Big Room, Carlsbad Caverns, New Mexico. Groundwater is responsible for creating and decorating the caverns. (Photo courtesy of the National Park Service)

DISTRIBUTION OF UNDERGROUND WATER

When rain falls, some of the water runs off, some evaporates, and the remainder soaks into the ground. This last path is the primary source of practically all subsurface water. The amount of water that takes each of these paths, however, varies greatly both in time and space. Several influential factors include steepness of slope, nature of surface material, intensity of rainfall, and type and amount of vegetation. Heavy rains falling upon steep slopes underlain by impervious materials will obviously result in a high percentage of runoff. On the other hand, if rain falls steadily and gently upon more gradual slopes composed of materials that are easily penetrated by the water, a much larger percentage of water soaks into the ground.

Some of the water that soaks in does not travel far, because molecular attraction holds it as a film on the surface of soil particles. A portion of this moisture evaporates back into the atmosphere, while much of the remainder serves as a source of water for use by plants between rains. Water that is not held in this **belt of soil moisture** penetrates downward until it reaches a zone where all of the open spaces in sedi-

231

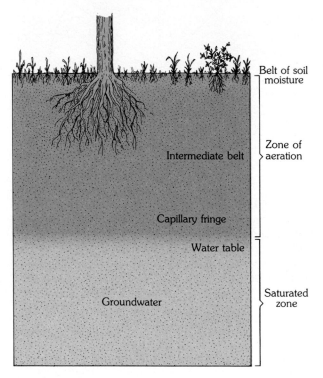

FIGURE 10.2
Distribution of underground water.

ment and rock are completely filled with water (Figure 10.2). The water in this saturated zone is called **groundwater.** The upper limit of this zone is known as the **water table.** Extending upward from the water table is the **capillary fringe.** Here groundwater is lifted against gravity by surface tension in tiny thread-like passages between grains of soil or sediment. Capillary action is easily demonstrated by placing the corner of a paper towel in some water and watching the liquid move upward. The area above the water table that includes the capillary fringe and the belt of soil moisture is called the **zone of aeration.** Unlike the **zone of saturation** below, the spaces in the zone of aeration are unsaturated and filled mainly with air.

THE WATER TABLE

The water table, the upper limit of the zone of saturation, is a very significant feature of the groundwater system. The water table level is important in predicting the productivity of wells, explaining the changes in the flow of springs and streams, and accounting for fluctuations in the levels of lakes.

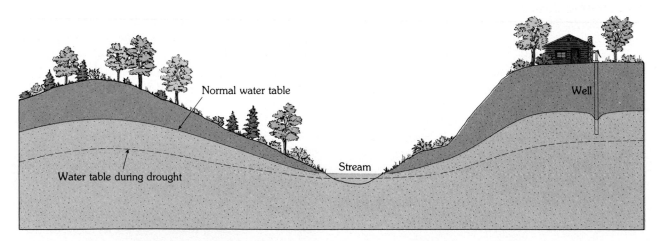

FIGURE 10.3
The shape of the water table is usually a subdued replica of the surface topography. During periods of drought, the water table falls, reducing streamflow and drying up some wells.

Although we cannot observe the water table directly, its position can be mapped and studied in detail in areas where wells are numerous because the water level in wells coincides with the upper boundary of the saturated zone. Such maps reveal that the water table is rarely level as we might expect a table to be. Instead, its shape is usually a subdued replica of the surface topography, reaching its highest elevations beneath hills and then descending toward valleys (Figure 10.3). Where a swamp is encountered, the water table is right at the surface, while lakes and streams generally occupy areas where the water table is above the land surface.

A number of factors contribute to the irregular surface of the water table. For example, variations in rainfall and permeability from place to place can lead to uneven infiltration and thus to differences in the water table level. However, the most important cause is simply the fact that groundwater moves very slowly and at varying rates under different conditions. Because of this, water tends to "pile up" beneath high areas between stream valleys. If rainfall were to cease completely, these water table "hills" would slowly subside and gradually approach the level of the valleys. However, new supplies of rainwater are usually added frequently enough to prevent this. Nevertheless, in times of extended drought, the water table may drop enough to dry up shallow wells (Figure 10.3).

The relationship between the water table and a stream in a humid region is illustrated in Figure 10.4A. Even during dry periods, the movement of groundwater into the channel maintains a flow in the stream. In situations such as this, streams are said to be **effluent.** By contrast, in arid regions, where the water table is far below the surface, groundwater does not contribute to streamflow. Therefore, the only permanent streams in such areas are those that originate in wet regions and then happen to traverse the desert. Under these conditions the zone of saturation below the valley floor is supplied by downward seepage from the stream channel which, in turn, produces an upward bulge in the water table. Streams that provide water to the water table in this manner are called **influent streams** (Figure 10.4B).

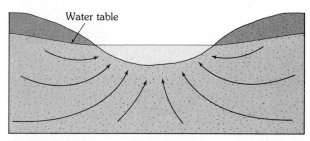

A. Effluent Stream

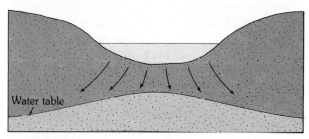

B. Influent Stream

FIGURE 10.4
A. Effluent streams are characteristic of humid areas and are supplied by water from the zone of saturation. **B.** Influent streams are found in deserts. Seepage from such streams produces an upward bulge in the water table.

POROSITY AND PERMEABILITY

Depending upon the nature of the subsurface material, the flow of groundwater and the amount of water that can be stored are highly variable. Water soaks into the ground because bedrock, sediment, and soil contain voids or openings. These openings are similar to those of a sponge and are often called *pore spaces.* The quantity of groundwater that can be stored depends on the **porosity** of the material; that is, the percentage of the total volume of rock or sediment that consists of pore spaces. Although these openings often consist of spaces between particles of sediment or sedimentary rock, such features as vesicles (voids left by gases escaping from lava), joints and faults, and cavities formed by the solution of soluble rocks like limestone are also common.

As one might expect, variations in porosity can be great. Sediment is commonly quite porous, and open spaces may occupy from 10 to 50 percent of the sediment's total volume. The amount of pore space depends on the size and shape of the grains, as well as the packing, degree of sorting, and in the case of sedimentary rocks, the amount of cementing material. For example, clay may have a porosity as high as 50 percent, whereas in some gravels, voids make up only 20 percent of the material's volume. Where sediments of various sizes are mixed, the porosity is reduced because the finer particles tend to fill the openings between the larger grains. Most igneous and metamorphic rocks, as well as some sedimentary rocks, are composed of tightly interlocking crystals. The amount of open space between the grains may be negligible, perhaps as little as one or two percent of the rock's volume. Therefore, if these rocks are to have greater porosity, fractures must provide a significant proportion of the open space.

Porosity alone is not a satisfactory measure of a material's ability or capacity to yield groundwater. Rock or sediment may be very porous and still not allow water to move through it. The **permeability** of a material, its ability to transmit a fluid, is also very important. Groundwater moves by twisting and turning through small openings. The smaller the pore spaces, the slower the water moves. If the spaces between particles are very small, the films of water clinging to the grains will come in contact or overlap. As a result, the force of molecular attraction binding the water to the particles extends across the opening and the water is held firmly in place. Clay exemplifies this circumstance. Although clay's ability to store water is often high, its pore spaces are so small that water is unable to move. Impermeable layers composed of materials such as clay that hinder or prevent water movement are termed **aquicludes.** On the other hand, larger particles, such as sand or gravel, have larger pore spaces. Therefore, the water in the centers of the openings is not bound to the particles by molecular attraction and can move with relative ease. Permeable rock strata or sediment that transmit groundwater freely are called **aquifers.**

In summary, we have seen that porosity is not always a reliable guide to the amount of groundwater that can be produced and the property of permeability is a significant factor in determining the rate of groundwater movement and the quantity of water that might be pumped from a well.

MOVEMENT OF GROUNDWATER

The energy responsible for groundwater movement is provided by the force of gravity. In response to gravity, water moves from areas where the water table is high to zones where the water table is lowest; that is, toward a stream channel, lake, or spring. Although some water takes the most direct path down the slope of the water table, much of the water follows long curving paths toward the zone of discharge. As illustrated in Figure 10.5, water percolates into the stream from all possible directions, with some of the paths turning upward, apparently against the force of gravity, and entering through the bottom of the channel. Such paths are followed because differences in the height of the water table create differences in the groundwater pressure at a particular height. Stated another way, the water at any given height is under greater pressure beneath a hill than beneath a stream channel and the water tends to migrate toward points of lower pressure. Thus, the looping curves followed by water in the saturated zone may be thought of as a compromise between the downward pull of gravity and the tendency of water to move toward areas of reduced pressure.

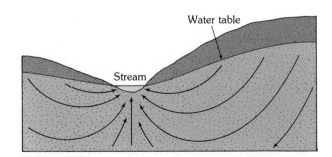

FIGURE 10.5
Arrows indicate groundwater movement through uniformly permeable material. The looping curves may be thought of as a compromise between the downward pull of gravity and the tendency of water to move toward areas of reduced pressure.

The modern concepts of groundwater movement were formulated in the middle of the nineteenth century. During this period Henry Darcy, a French engineer studying the water supply of the city of Dijon in east-central France, formulated a law that now bears his name and is basic to an understanding of groundwater movement. Darcy found that if permeability remains uniform, the velocity of groundwater will increase as the slope of the water table increases. The water table slope, known as the **hydraulic gradient,** is determined by dividing the vertical distance between the recharge and discharge points (a distance known as the **head**) by the length of flow between these points. **Darcy's law** can be expressed by the following formula:

$$V = K\frac{h}{l}$$

where V represents velocity, h the head, l the length of flow, and K a coefficient that accounts for a material's permeability.

Due to the large amount of friction, the movement of groundwater is quite slow. This is fortunate because the slow movement keeps underground reservoirs from being quickly depleted. If groundwater flowed as rapidly as the water in streams, wells would run dry after only a short period of no rain.

Rates of groundwater movement have been determined directly in a number of ways. In one experiment dye is introduced into a well and the time is measured until the coloring agent appears in another well at a known distance from the first. Measurements of groundwater movement have also been made by applying radiometric dating techniques, specifically by using the radioactive isotope of carbon, carbon-14. Upon entering the ground the carbon dioxide dissolved in rainwater contains a characteristic amount of carbon-14. As the water gradually percolates through the ground, the radioactive carbon decays. The rate of movement is determined by measuring the distance between the well where the water was withdrawn and the recharge area, and dividing this by the radiometrically determined age.[1]

Experiments such as those just described have shown that the rate of groundwater movement is

[1]A discussion of radiometric dating appears in Chapter 19.

highly variable. Although a typical rate for many aquifers is thought to be about 15 meters per year (about 4 centimeters per day), velocities more than 15 times this figure have been measured in exceptionally permeable materials.

SPRINGS

Springs have aroused the curiosity and wonder of people for thousands of years. The fact that springs were, and to some people still are, rather mysterious phenomena is not difficult to understand, for here is water flowing freely from the ground in all kinds of weather in seemingly inexhaustible supply, but with no obvious source. As a result, some rather interesting (although incorrect) explanations for the source of springs were proposed. Some of these have managed to live on to the present. One such explanation is that springs draw their water from the ocean. However, just how the salt is removed and how the water is elevated to the great heights it reaches in mountain springs remain unanswered questions. Another explanation for springs, one supported by Aristotle, suggested that the water originated in cold subterranean caverns where water vapor condensed from the air that had penetrated the earth and produced the needed water supply.

Not until the middle of the 17th century did the French physicist Pierre Perrault invalidate the age-old assumption that precipitation could not adequately account for the amount of water emanating from springs and flowing in rivers. Over a period of years, Perrault computed the quantity of water that fell on the Seine River basin. He then calculated the mean annual runoff by measuring the river's discharge, and after allowing for the loss of water by evaporation, he showed that there was sufficient water remaining to feed the springs. Thanks to Perrault's pioneering efforts and the measurements by many afterward, we now know that the source of springs is water from the zone of saturation and that the ultimate source of this water is precipitation.

When the water table intersects the earth's surface, a natural flow of groundwater results, which we call a **spring** (Figure 10.6). One situation leading to

FIGURE 10.6
Thousand Springs, Gooding County, Idaho. (Photo by C. F. Bowen, U.S. Geological Survey)

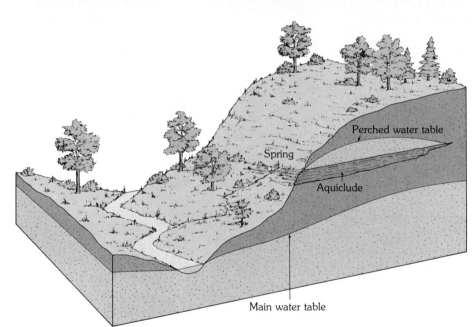

FIGURE 10.7
When an aquiclude is situated above the main water table, a localized zone of saturation may result.

Perched water table

Spring

Aquiclude

Main water table

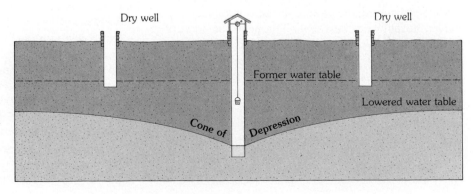

FIGURE 10.8
A cone of depression in the water table often forms around a pumping well. If heavy pumping lowers the water table, the shallow wells may be left dry.

Dry well

Dry well

Former water table

Lowered water table

Cone of Depression

the formation of a spring is illustrated in Figure 10.7. Here an aquiclude is situated above the main water table. As water percolates downward, a portion of it is intercepted by the aquiclude, thereby creating a localized zone of saturation called a **perched water table.** Springs, however, are not confined to places where a perched water table creates a flow at the surface. Indeed, there are a wide variety of spring types because subsurface conditions vary greatly from place to place. Even in areas underlain by impermeable crystalline rocks, permeable zones may exist in the form of fractures or solution channels. If these openings fill with water and intersect the ground surface along a slope, a spring will result.

WELLS

The most common device used by humans for removing groundwater is the **well,** an opening bored into the zone of saturation. Wells serve as reservoirs into which groundwater moves and from which it can be pumped to the surface. Digging for water dates back many centuries and continues to be an important method of obtaining water today. By far the single greatest use of this water in the United States is for irrigation. More than 65 percent of the groundwater used each year is for this purpose. Industrial uses rank a distant second, followed by the amount used in cities and rural areas.

The water table level may fluctuate considerably during the course of a year, dropping during dry seasons and rising following periods of rain. Therefore, to insure a continuous supply of water, a well should penetrate below the water table. Whenever water is withdrawn from a well, the water table in the vicinity of the well is lowered. The extent of this effect, which is termed **drawdown,** decreases with increasing distance from the well. The result is a depression in the water table, roughly conical in shape, known as a **cone of depression** (Figure 10.8). Since the cone of depression increases the hydraulic gradient near the well, groundwater will flow more rapidly toward the opening. For most small domestic wells, the cone of depression is hardly appreciable. However, when wells are used for irrigation or for industrial purposes, the withdrawal of water can be great enough to cre-

ate a very wide and steep cone of depression that may substantially lower the water table in an area and cause nearby shallow wells to become dry (Figure 10.8).

Digging a successful well is a familiar problem for people in areas where groundwater is the primary source of supply. One well may be successful at a depth of 10 meters (33 feet) whereas a neighbor may have to go twice as deep to find an adequate supply. Still others may be forced to go deeper or try a different site altogether. Some of the causes for unsuccessful wells, as well as reasons for differences in the depths of successful wells are illustrated in Figure 10.9 on page 238. In Figure 10.9A, a perched water table is responsible for variation. The successful well tapped the saturated zone above a lens of clay whereas the unsuccessful well missed this zone and did not go deep enough to reach the main water table. Figure 10.9B illustrates the difficulties encountered when wells are drilled into crystalline rocks. Since groundwater is confined largely to fractures, drilling a successful well will often depend on the chance intersection of the well with an adequate network of fractures.

ARTESIAN WELLS

To many people the term *artesian* is applied to any well drilled to great depths. This use of the term is incorrect. Others believe that an artesian well must flow freely at the surface. Although this is a more correct notion than the first, it represents too narrow a definition. The term **artesian** may be applied correctly to any situation in which groundwater under pressure rises above the level of the aquifer. As we shall see, this does not always mean a free-flowing surface discharge.

For an artesian system to exist, two conditions must be met (Figure 10.10A): (1) water must be confined to an aquifer that is inclined so that one end can receive water; and (2) aquicludes, both above and below the aquifer, must be present to prevent the water from escaping. When such a layer is tapped, the pressure created by the weight of the water above will cause the water to rise. If there were no friction,

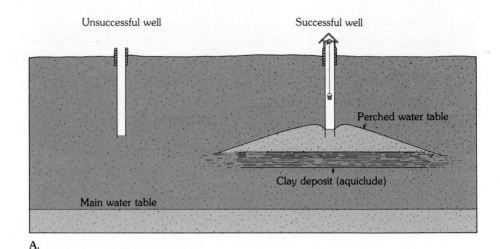

Unsuccessful well   Successful well

Perched water table

Clay deposit (aquiclude)

Main water table

A.

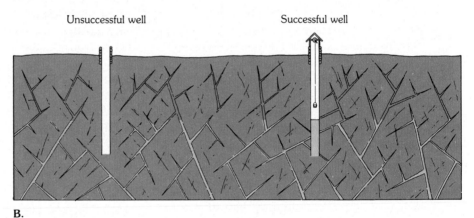

Unsuccessful well   Successful well

B.

FIGURE 10.9
A. The neighbor with the successful well tapped a perched water table. The neighbor with the dry well will have to dig deeper before being successful. **B.** Groundwater in crystalline rock is confined primarily to fractures. A successful well depends upon the chance intersection with an adequate network of fractures.

the water in the well would rise to the level of the water at the top of the aquifer. However, friction reduces the height of this pressure surface. The greater the distance from the recharge area (area where water enters the inclined aquifer), the greater the friction and the less the rise of water. In Figure 10.10A, well 1 is a **nonflowing artesian well,** because at this location the pressure surface is below ground level. When the pressure surface is above the ground and a well is drilled into the aquifer, a **flowing artesian well** is created (well 2, Figures 10.10A and B). It is important to realize that not all artesian systems are wells. Artesian springs also exist. Here groundwater reaches the surface by rising through a natural fracture rather than through an artificially produced hole.

Artesian systems act as conduits, often transmitting water great distances from remote areas of recharge

to points of discharge. A well-known artesian system in South Dakota is a good example of this. In the western part of the state, the edges of a series of sedimentary layers have been bent up to the surface along the flanks of the Black Hills. One of these beds, the permeable Dakota Sandstone, is sandwiched between impermeable strata and gradually dips into the ground toward the east. When the aquifer was first tapped, water poured from the ground surface, creating fountains many meters high (Figure 10.11). In some places, the force of the water was sufficient to power waterwheels. Scenes such as the one pictured in Figure 10.11, however, can no longer occur, because thousands of additional wells now tap the same source. This depleted the reservoir and the water table in the recharge area was lowered. As a consequence, the pressure dropped to the point where many wells stopped flowing altogether and

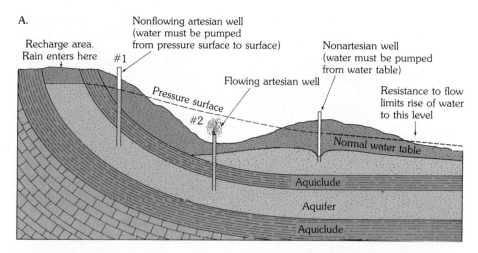

A.

Recharge area.
Rain enters here

#1

Nonflowing artesian well
(water must be pumped
from pressure surface to surface)

Nonartesian well
(water must be pumped
from water table)

Pressure surface

Flowing artesian well

#2

Resistance to flow
limits rise of water
to this level

Normal water table

Aquiclude

Aquifer

Aquiclude

B.

FIGURE 10.10
A. Artesian systems occur when an inclined aquifer is surrounded by impermeable beds. **B.** A flowing artesian well, Gallatin County, Montana. (Photo by J. R. Stacy, U.S. Geological Survey)

had to be pumped. On a different scale, city water systems can be considered examples of artificial artesian systems (Figure 10.12). The water tower, into which water is pumped, would represent the area of recharge, the pipes the confined aquifer, and the faucets in homes the flowing artesian wells.

PROBLEMS ASSOCIATED WITH GROUNDWATER WITHDRAWAL

As with many of our valuable natural resources, groundwater is being exploited at an ever-increasing rate. In some areas, overuse threatens the groundwater supply. In addition, a number of costly problems

related to the withdrawal of groundwater may accompany and compound the difficulties associated with diminishing groundwater resources.

The tendency for many natural systems is to establish or attempt to establish a condition of equilibrium. The groundwater system is no exception. The water table level represents the balance between the rate of infiltration and the rate of discharge and withdrawal. Any imbalance will either raise or lower the water table. Long-term imbalances can lead to a significant drop in the water table if there is either a decrease in groundwater recharge, such as occurs during a prolonged drought, or an increase in the rate of groundwater discharge or withdrawal. As we would expect, water tables have gradually dropped in many areas

FIGURE 10.11
A "gusherlike" flowing artesian well in South Dakota in the early part of the century. Thousands of additional wells now tap the same confined aquifer; thus, the pressure has dropped to the point that many wells stopped flowing altogether and have to be pumped. (Photo by N. H. Darton, U.S. Geological Survey)

where withdrawal has increased steadily. This problem has been especially acute in portions of the arid and semiarid western United States where groundwater is essential to both agriculture and industry. In many places groundwater is literally being "mined," for even if pumping were to cease immediately, the groundwater reservoir would not be replenished for centuries. The following example illustrates this:

> Consider for example, a location in the dry, southwestern United States where annual recharge to an aquifer is on the order of only two-tenths of an inch of water. In such areas it is not uncommon to pump two feet or more of water per year for irrigation or other uses. In this oversimplified example, if the entire aquifer were pumped at that rate, yearly pumpage would be equivalent to 120 years' recharge, and ten years of pumping would remove a 1200-year accumulation of water. New recharge during the pumping period would be negligible. Mechanical problems and economic factors prevent complete dewatering of an aquifer, but the example is valid in principle.[1]

SUBSIDENCE

As we shall see later in this chapter, surface subsidence can result from natural processes related to groundwater. However, the ground may also sink when water is pumped from wells faster than natural recharge processes can replace it. This effect is particularly pronounced in areas underlain by thick layers of unconsolidated sediments. As the water is withdrawn, the water pressure drops and the weight of the overburden is transferred to the sediment. The greater pressure packs the sediment grains tightly together and the ground subsides.

[1]U.S. Geological Survey, *Water of the World,* 1968, p. 16.

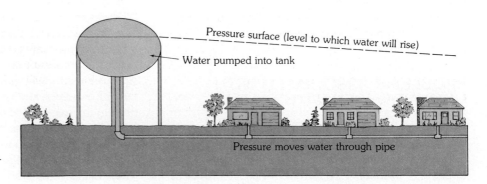

FIGURE 10.12
City water systems can be considered to be artificial artesian systems.

Pressure surface (level to which water will rise)

Water pumped into tank

Pressure moves water through pipe

Many areas may be used to illustrate land subsidence resulting from the excessive pumping of groundwater from relatively loose sediment. A classic example in the United States occurred in the San Joaquin Valley of California. Here, in a region of extensive irrigation, the water table beneath the valley has gradually been drawn down by as much as 30 meters. As a consequence, the land has subsided up to 3 meters in some places. Similar effects have been documented in the Houston and Panhandle areas of Texas, as well as in Las Vegas and New Orleans. Damage to buildings, water and sewer lines, and roads can be extensive and costly. In addition, subsidence in coastal areas may require the construction of levees to keep out the encroaching sea.

Perhaps the most spectacular example of subsidence has occurred in Mexico City, which is built upon a former lake bed. In the first half of this century thousands of wells were sunk into the water-saturated sediments beneath the city. As water was withdrawn, portions of the city subsided by as much as 6 to 7 meters. In some places buildings have sunk to such a point that access to them from the street is at what used to be the second floor level!

SALTWATER CONTAMINATION

In many coastal areas the groundwater resource is being threatened by the encroachment of salt water. In order to understand this problem, we must examine the relationship between fresh groundwater and salt groundwater. Figure 10.13A is a diagrammatic cross section that illustrates this relationship in a coastal area underlain by permeable homogenous materials. Since fresh water is less dense than salt water, it floats on the salt water and forms a large, lens-shaped body that may extend to considerable depths below sea level. In such a situation, if the water table is 1 meter above sea level, the base of the freshwater body will extend to a depth of about 40 meters below sea level. Stated another way, the depth of the fresh water below sea level is about 40 times greater than the elevation of the water table above sea level. Thus, when excessive pumping lowers the water table by a certain amount, the bottom of the freshwater zone will rise by 40 times that amount. Therefore, if groundwater withdrawal continues to exceed recharge, there will come a time when the elevation of the salt water will be sufficiently high to

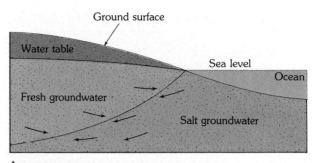

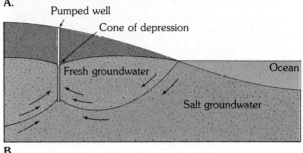

FIGURE 10.13
A. Since fresh water is less dense than salt water, it floats on the salt water and forms a lens-shaped body that may extend to considerable depths below sea level. **B.** When excessive pumping lowers the water table by a certain amount, the base of the freshwater zone will rise by 40 times that amount. The result may be saltwater contamination of wells.

be drawn into wells, thus contaminating the freshwater supply (Figure 10.13B). Deep wells and wells near the shore are usually the first to be affected.

In urbanized coastal areas, the problems created by excessive pumping are compounded by a decrease in the rate of natural recharge. As more and more of the surface is covered by streets, parking lots, and buildings, infiltration into the soil is diminished.

In an attempt to correct the problem of saltwater contamination of groundwater resources, a network of recharge wells may be used. These wells allow waste water to be pumped back into the groundwater system. A second method of correction is accomplished by building large basins. These basins collect surface drainage and allow it to seep into the ground. On Long Island, where the problem of saltwater contamination was recognized more than 40 years ago, both of these methods have been employed with considerable success.

GROUNDWATER POLLUTION

The pollution of groundwater is a serious matter, particularly in areas where aquifers supply a large part of the water supply. A very common type of groundwater pollution is sewage. Its sources include an ever-increasing number of septic tanks, as well as inadequate or broken sewer systems and barnyard wastes.

If water contaminated with bacteria from sewage enters the groundwater system, it may become purified through natural processes. The harmful bacteria may be mechanically filtered out by the sediment through which the water percolates, destroyed by chemical oxidation, and/or assimilated by other organisms. In order for purification to occur, however, the aquifer must be of the correct composition. For example, extremely permeable aquifers such as highly fractured crystalline rock, coarse gravel, or cavernous limestone have such large openings that contaminated groundwater may travel long distances without being cleansed. In this case, the water flows too rapidly and is not in contact with the surrounding

material long enough for purification to occur. This is the problem at well 1 in Figure 10.14. On the other hand, when the aquifer is composed of sand or permeable sandstone, the water can sometimes be purified within distances as short as a few tens of meters. The openings between sand grains are large enough to permit water movement, yet the movement of the water is slow enough to allow ample time for its purification (well 2, Figure 10.14).

Sometimes sinking a well can lead to groundwater pollution problems. If the well pumps a sufficient quantity of water, the cone of depression will locally increase the slope of the water table. In some instances, the original slope may even be reversed. This could lead to the contamination of wells that yielded unpolluted water before heavy pumping began (Figure 10.15). Also recall that the rate of groundwater movement increases as the slope of the water table steepens. This could produce problems because a faster rate of movement allows less time for the water to be purified in the aquifer before it is pumped to the surface.

Sanitary landfills and garbage dumps represent

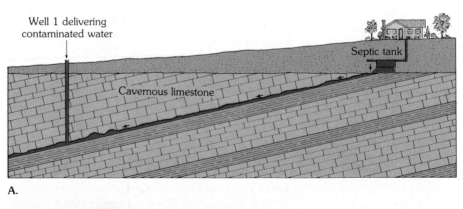

A.

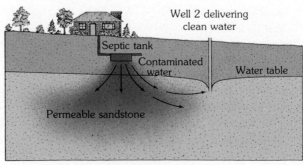

B.

FIGURE 10.14
A. Although the contaminated water has traveled more than 100 meters before reaching well 1, the water moves too rapidly through the cavernous limestone to be purified.
B. Since the aquifer is sandstone, the water is purified in a relatively short distance.

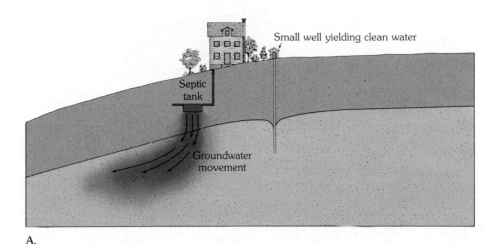

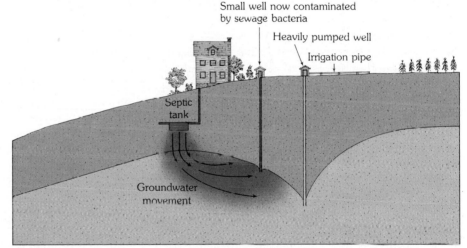

FIGURE 10.15
A. Originally the outflow from the septic tank moved away from the small well. **B.** The heavily pumped well changed the slope of the water table, causing contaminated groundwater to flow toward the small well.

another source of pollutants that may endanger the groundwater supply of an area. As rainwater oozes through the refuse, it may dissolve a variety of organic and inorganic materials, some of which may be harmful. If water containing material leached from the landfill reaches the water table, it will mix with the groundwater and contaminate the supply. Since groundwater movement is usually slow, the polluted water may go undetected for a considerable period of time. When the problem is finally discovered, the volume of contaminated water may be very large. Thus, even if the source of pollution is eliminated immediately (which is most unlikely), the problem will linger for many years until the contaminated water has migrated from the area of use.

HOT SPRINGS, GEYSERS, AND GEOTHERMAL ENERGY

HOT SPRINGS

By definition, the water in **hot springs** is 6–9°C (10–15°F) warmer than the mean annual air temperature for the localities where they occur. In the United States alone, there are over 1000 such springs (Figure 10.16).

Mineral explorations over the world have shown that temperatures in deep mines and oil wells usually rise with an increase in depth below the surface. Temperatures in such situations increase an average of about 2°C per 100 meters (1°F per 100 feet). Therefore when groundwater circulates at great

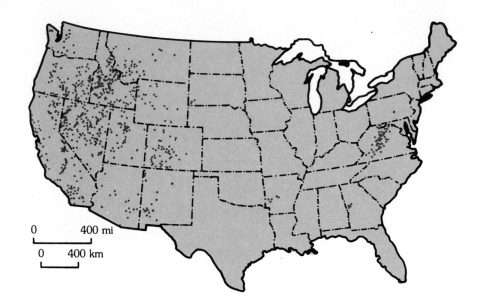

FIGURE 10.16
Distribution of hot springs and geysers in the United States. Note the concentration in the West, where igneous activity has been most recent. (After G. A. Waring, U.S. Geological Survey Professional Paper 492, 1965)

0   400 mi

0   400 km

depths, it becomes heated, and if it rises to the surface, the water emerges as a hot spring. The water of some hot springs in the United States, particularly in the East, is heated in this manner. However, the great majority (over 95 percent) of the hot springs (and geysers) in the United States are found in the West. The reason for such a distribution is that the source of heat for most hot springs is cooling igneous rock, and it is in the West that igneous activity has occurred more recently.

GEYSERS

Geysers are intermittent hot springs or fountains where columns of water are ejected with great force at various intervals, often rising 30–60 meters (100–200 feet). After the jet of water ceases, a column of steam rushes out, usually with a thundering roar. Perhaps the most famous geyser in the world is Old Faithful in Yellowstone National Park, which erupts about once each hour (Figure 10.17). Geysers are also found in other parts of the world, including New Zealand and Iceland, where the term *geyser*, meaning "spouter" or "gusher," originated.

Geysers occur where extensive underground chambers exist within hot igneous rocks. As relatively cool groundwater enters the chambers, it is heated by the surrounding rock. At the bottom of the chambers, the water is under great pressure because of the weight of the overlying water. Consequently this water must reach temperatures well above 100°C

(212°F) before it will boil. For example, water at the bottom of a 300-meter (1000-foot) water-filled chamber must attain a temperature of nearly 230°C to boil. As the temperature rises, water expands and some flows out at the surface. This loss of water reduces the pressure on the remaining water in the chamber. The reduced pressure lowers the boiling point and the water deep within the chamber quickly turns to steam and the geyser erupts (Figure 10.18). Following the eruption, cool groundwater again seeps into the chamber and the cycle begins anew.

Groundwater from hot springs and geysers usually contains more material in solution than groundwater from other sources because hot water is a more effective dissolver than cold. When the water contains much dissolved silica, *geyserite* is deposited around the spring. *Travertine,* a form of calcite, is a characteristic deposit at hot springs in limestone regions (Figure 10.19). Some hot springs contain sulfur, which gives water a poor taste and unpleasant odor. Undoubtedly Rotten Egg Spring, Nevada, is such a situation.

FIGURE 10.17
Old Faithful, one of the world's most famous geysers, emits as much as 45,000 liters (almost 12,000 gallons) of hot water and steam about once each hour. (Photo by James E. Patterson)

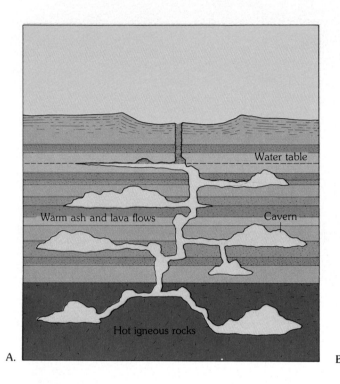

A.

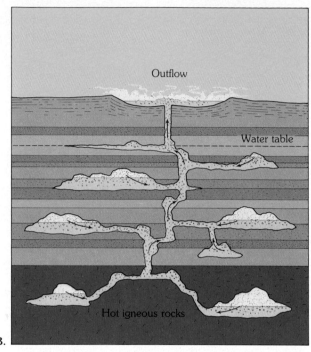

B.

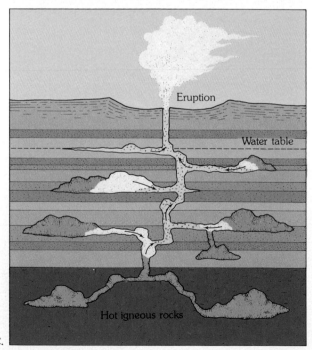

C.

FIGURE 10.18

Idealized diagrams of a geyser. A geyser can form if the heat is not distributed by convection. **A.** In this figure, the water near the bottom is heated to near its boiling point. The boiling point is higher there than at the surface because the weight of the water above increases the pressure. **B.** The water higher in the geyser system is also heated; therefore, it expands and flows out at the top, reducing the pressure on the water at the bottom. **C.** At the reduced pressure on the bottom, boiling occurs. The bottom water flashes into steam, and the expanding steam causes an eruption.

FIGURE 10.19
Mineral deposits surrounding Mammoth Hot Springs, Yellowstone National Park.
(Photo by E. J. Tarbuck)

FIGURE 10.20
Steam wells for geothermal power production near Ejido Hidalgo, Mexico. (Photo by James E. Patterson)

GEOTHERMAL ENERGY

Many natural geyser areas around the world are potential sites for tapping **geothermal energy,** that is, natural steam used for power generation. In New Zealand, Italy, Mexico, the Soviet Union, and the United States, underground supplies of superheated steam are now being used to provide power for generating electricity (Figure 10.20, page 247). In the United States the first commercial geothermal power plant was built in 1960 at "The Geysers," north of San Francisco. The most favorable geological factors for a geothermal reservoir of commercial value include:

1 A potent source of heat, such as a large magma chamber. The chamber should be deep enough to insure adequate pressure and a slow rate of cooling and yet not be so deep that the natural circulation of water is inhibited. Magma chambers of this type are most likely to occur in regions of recent volcanic activity;

2 Large and porous reservoirs with channels connected to the heat source, near which water can circulate and be stored;

3 Capping rocks of low permeability that inhibit the flow of water and heat to the surface. A deep and well-insulated reservoir is likely to contain much more stored energy than an uninsulated, but otherwise similar, reservoir.[1]

It is too early to judge whether natural steam has the potential to satisfy an important part of the world's requirements for electrical power, but with the need to develop new sources of energy, the possibilities are definitely worth exploring.

THE GEOLOGIC WORK OF GROUNDWATER

The primary erosional work carried out by groundwater is that of dissolving rock. Since soluble rocks, especially limestone, underlie millions of square kilometers of the earth's surface, it is here that groundwater carries on its rather unique and important role as an erosional agent. Although nearly insoluble in

[1]Adapted from U.S. Geological Survey, *Natural Steam for Power,* 1968.

pure water, limestone is quite easily dissolved by water containing small quantities of carbonic acid. Most natural water contains this weak acid because rainwater readily dissolves carbon dioxide from the air and from decaying plants. Therefore, when groundwater comes in contact with limestone, the carbonic acid reacts with the calcite in the rocks to form calcium bicarbonate, a soluble material that is then carried away in solution.

CAVERNS

Among the most spectacular results of groundwater's erosional handiwork is the creation of limestone **caverns.** Most are relatively small, yet some have spectacular dimensions. In the United States, Carlsbad Caverns in southeastern New Mexico and Mammoth Cave in Kentucky are famous examples. One chamber in Carlsbad Caverns has an area equivalent to fourteen football fields and enough height to accommodate the U.S. Capitol Building (see chapter-opening photo). At Mammoth Cave, the total length of interconnected caverns approaches 50 kilometers (30 miles).

Most caverns are believed to be created at or below the water table in the zone of saturation. Here the groundwater follows lines of weakness in the rock, such as joints and bedding planes. As time passes, the dissolving process slowly creates cavities and gradually enlarges them into caverns. The material that is dissolved by the groundwater is carried away and discharged into streams.

Certainly the features that arouse the greatest curiosity for most cavern visitors are the stone formations that often exhibit quite bizarre patterns and give some caverns a wonderland appearance. These features are created by the seemingly endless dripping of water over great spans of time. The calcite that is left behind produces the limestone we call travertine. These cave deposits, however, are also commonly called *dripstone,* an obvious reference to their mode of origin.

Although the formation of caverns takes place in the zone of saturation, the deposition of dripstone is not possible until the caverns are above the water table in the zone of aeration. This commonly occurs as nearby streams cut their valleys deeper, causing the water table to drop as the elevation of the river drops. As soon as the chamber is filled with air, the stage is set for the decoration phase of cavern building to begin.

The various dripstone features found in caverns are collectively called **speleothems,** no two of which are exactly alike (Figure 10.21). Perhaps the most familiar speleothems are **stalactites.** These icicle-like pendents hang from the ceiling of the cavern and form where water seeps through cracks above. When the water reaches the air in the cave, a small amount of evaporation takes place. Thus, some of the carbon dioxide in solution escapes and a residue of calcium carbonate remains. Deposition occurs as a ring around the edge of the water drop. As drop after drop follows, each leaves an infinitesimal trace of calcite behind, and a hollow limestone tube is created. Water then moves through the tube, remains suspended momentarily at the end, contributes a tiny ring of calcite, and falls to the cavern floor. The stalactite just described is appropriately called a *soda straw* (Figure 10.22). Often the hollow tube of the soda straw becomes plugged or its supply of water increases. In either case, the water is forced to flow, and hence deposit, along the outside of the tube. As

FIGURE 10.21
Speleothems are of many types, including stalactites, stalagmites, and columns. New Cave, Carlsbad Caverns National Park, New Mexico. (Photos by E. J. Tarbuck)

FIGURE 10.22
A. A "live" solitary soda straw stalactite. (Photo by Clifford Stroud, National Park Service). **B.** A soda straw "forest" in Carlsbad Caverns. (Photo courtesy of the National Park Service)

deposition continues, the stalactite takes on the more common conical shape.

Speleothems that form on the floor of a cavern and reach upward toward the ceiling are called **stalagmites.** The water supplying the calcite for stalagmite growth falls from the ceiling and splatters over the surface. As a result, stalagmites do not have a central tube and are usually more massive in appearance and rounded on their upper ends than stalactites.

KARST TOPOGRAPHY

Many areas of the world have landscapes that to a large extent have been shaped by the dissolving power of groundwater. Such areas are said to exhibit

karst topography. The term is derived from a plateau region located along the northeastern shore of the Adriatic Sea in the border area between Yugoslavia and Italy where such topography is strikingly developed. In the United States, karst landscapes occur in areas of Kentucky, Tennessee, southern Indiana, and central and northern Florida. Generally, arid and semiarid areas do not develop karst topography. When solution features exist in such regions, they are likely to be remnants of a time when more humid conditions prevailed.

Karst areas characteristically exhibit an irregular terrain punctuated with many depressions, called **sinkholes** or **sinks.** In the limestone areas of Kentucky and southern Indiana, there are literally tens of

250

thousands of these depressions varying in depth from just a meter or two to a maximum of more than 50 meters.

Sinkholes commonly form in one of two ways. Some develop gradually over many years without any physical disturbance to the rock. In these situations, the limestone immediately below the soil is dissolved by downward-seeping rainwater that is freshly charged with carbon dioxide. These depressions are usually not deep and are characterized by relatively gentle slopes. By contrast, sinkholes can also form suddenly and without warning when the roof of a cavern collapses under its own weight. Typically, the depressions created in this manner are steep-sided and deep. When they form in populous areas, they may represent a serious geologic hazard. The crater-like sinkhole in Figure 10.23 on page 252 began forming in Winter Park, Florida, on May 8, 1981, just one day before this photograph was taken. Newspaper accounts, such as the one that follows, were front page news and made this sinkhole one of the most publicized ever.

SINKHOLE NIBBLES AWAY AT FLORIDA CITY

WINTER PARK, Fla. (AP)—A giant sinkhole—already several hundred feet wide after swallowing a three-bedroom bungalow, half a swimming pool and six Porsches—nibbled away at a side street yesterday and threatened a main thoroughfare.

"It has slowed down, but it hasn't quit," said Winter Park Fire Capt. Gus LaGarde.

The crater, estimated at between 450 and 600 feet wide and 125 to 170 feet deep, grew by eight to 10 feet yesterday and was filling with water, authorities said.

It developed Friday night and opened rapidly Saturday, when it gulped the wood-frame cottage, cars and part of a foreign car lot and wrecked the city's $150,000 municipal swimming pool.

It was slowly eating its way west yesterday, leaving a group of businesses, their backs lost in Saturday's slide, hanging at the edge of the pit, LaGarde said.

The hole devoured most of a side street yesterday and was about 50 feet from one main thoroughfare and moving closer to several others, he said.

"We're still losing some of the perimeter," he said. "It doesn't appear to get any deeper . . . it continues to eat up the roadway, power poles, anything that gets in the way."

Residents and owners of nearby homes and businesses were warned on Saturday to leave until the sinkhole stopped growing. Some people rented trucks and began moving furniture and other property.[1]

Sinkhole formation is not uncommon in northern and central Florida. In fact, the Winter Park event was just one of three that occurred in the area over a two-week span. In each case the collapse at the surface was probably triggered by a lowering of the water table brought about by a severe drought. As the water table dropped, the roofs of the underground cavities lost support and fell into the voids below.

In addition to a surface pockmarked by sinkholes, karst regions characteristically show a striking lack of surface drainage. Following a rainfall, the runoff is funneled below ground by way of sinks where it then flows through caverns until it finally reaches the water table. When streams do exist at the surface, their paths are usually short. The names of such streams often give a clue to their fate. In the Mammoth Cave area of Kentucky, for example, there is Sinking Creek, Little Sinking Creek, and Sinking Branch. Other sinkholes become plugged with clay and debris to create small lakes or ponds.

---

[1]Courtesy of Associated Press.

A.

FIGURE 10.23
A. An aerial view of a large sinkhole that formed in Winter Park, Florida, in May, 1981. (Photo courtesy of George Remaine, *Orlando Sentinel Star*).
B. A closer view of the Winter Park sinkhole. (Photo courtesy of Barbara Vitaliano, *Orlando Sentinel Star*)

B

252

REVIEW QUESTIONS

1   Compare and contrast the zones of aeration and saturation. Which of these zones contains groundwater?

2   Although we usually think of tables as being flat, the water table generally is not. Explain.

3   What is an effluent stream? How does an influent stream differ?

4   Distinguish between porosity and permeability.

5   What is the difference between an aquiclude and an aquifer?

6   Under what circumstances can a material have a high porosity but not be a good aquifer?

7   As illustrated in Figure 10.5, groundwater moves in looping curves. What factors cause the water to follow such paths?

8   Briefly describe the important contribution to our understanding of groundwater movement made by Henry Darcy.

9   When an aquiclude is situated above the main water table, a localized saturated zone may be created. What term is applied to such a situation?

10   Two neighbors each dig a well. Although both wells penetrate to the same depth, one neighbor is successful and the other is not. Describe a circumstance that might explain what happened.

11   What is meant by the term *artesian?*

12   In order for artesian wells to exist, two conditions must be present. List these conditions.

13   When the Dakota Sandstone was first tapped, water poured freely from many artesian wells. Today these wells must be pumped. Explain.

14   Why is the pumping of water in some areas of the southwestern United States a serious problem?

15   Briefly explain what happened in Mexico City as the result of excessive groundwater withdrawal.

16   In a particular coastal area the water table is 4 meters above sea level. Approximately how far below sea level does the fresh water reach?

17   Why does the rate of natural groundwater recharge decrease as urban areas develop?

18   Which aquifer would be most effective in purifying polluted groundwater: coarse gravel, sand, or cavernous limestone?

19   What is the source of heat for most hot springs and geysers? How is this reflected in the distribution of these features?

20   Name two common speleothems and distinguish between them.

21   Speleothems form in the zone of saturation. True or False? Briefly explain your answer.

22   Areas whose landscapes largely reflect the erosional work of groundwater are said to exhibit what kind of topography?

23   Describe two ways in which sinkholes are created.

KEY TERMS

aquiclude (p. 234)

aquifer (p. 234)

artesian (p. 237)

belt of soil moisture (p. 231)

capillary fringe (p. 232)

cavern (p. 248)

cone of depression (p. 237)

Darcy's law (p. 235)

drawdown (p. 237)

effluent stream (p. 233)

flowing artesian well (p. 238)

geothermal energy (p. 248)

geyser (p. 244)

groundwater (p. 232)

head (p. 235)

hot spring (p. 243)

hydraulic gradient (p. 235)

influent stream (p. 233)

karst topography (p. 250)

nonflowing artesian well (p. 238)

perched water table (p. 237)

permeability (p. 234)

porosity (p. 233)

sinkhole (sink) (p. 250)

speleothem (p. 249)

spring (p. 235)

stalactite (p. 249)

stalagmite (p. 250)

water table (p. 232)

well (p. 237)

zone of aeration (p. 232)

zone of saturation (p. 232)

11

GLACIERS AND GLACIATION

11

A glacier is a thick ice mass that originates on land from the accumulation, compaction, and recrystallization of snow and shows evidence of past or present movement. Although glaciers are found in many parts of the world today, most are located in remote areas. Literally thousands of relatively small glaciers exist in mountainous regions. Such glaciers are generally confined to mountain valleys and are termed **alpine glaciers** (Figure 11.1). On a different scale, **continental glaciers,** or ice sheets, are massive accumulations of ice that are not confined to valleys, but cover a large portion of a landmass. Two massive continental glaciers exist on earth today, one covering most of the island of Greenland and the other covering a large portion of Antarctica (Figure 11.2). Their combined area represents almost 10 percent of the earth's land surface. The Greenland ice sheet covers 80 percent of this large island, occupying about 1.7 million square kilometers and averaging nearly 1500 meters thick. However, when compared to the glacier covering the continent of Antarctica, the Greenland ice sheet seems quite small.

Two alpine glaciers merging to form a medial moraine. St. Elias Mountains, Yukon Territory. (Photo by Warren Hamilton, U.S. Geological Survey)

257

FIGURE 11.1
Aerial view of South Cascade Glacier, an alpine glacier about 3 kilometers long in the North Cascade Range, Washington. The snowline and cracks called crevasses are clearly visible. (Photo by Austin Post, U.S. Geological Survey)

258

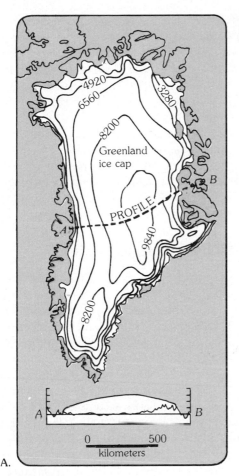

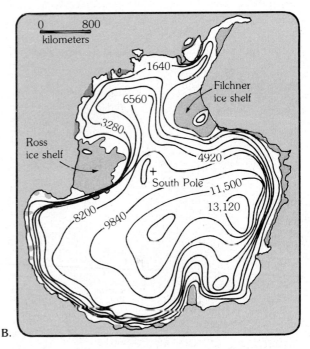

FIGURE 11.2

The ice sheets of **A.** Greenland and **B.** Antarctica are the only continental glaciers on earth today. The contours indicate the depth of ice (in feet). (Courtesy of Ward's Nat ural Science Establishment, Inc., Rochester, N.Y.)

GLACIERS AND THE HYDROLOGIC CYCLE

Earlier we learned that the earth's water is in constant motion. Time and time again the same water is evaporated from the oceans into the atmosphere, precipitated upon the land, and carried by rivers and underground flow back to the sea. However, when precipitation falls at high elevations or high latitudes, the water may not immediately make its way toward the sea. Rather, it may become part of a large mass of moving ice known as a glacier. Although the ice will eventually melt and continue its path to the sea, it can be stored in a glacier for many tens, hundreds, or even thousands of years. For example, data collected from the ice covering Greenland shows that portions of this glacier are more than 25,000 years old.

How much water is stored as glacial ice? Estimates by the U.S. Geological Survey indicate that only slightly more than two percent of the world's water is accounted for by glaciers. But this small figure may be misleading when the actual amounts of water are considered. The total volume of all alpine glaciers is about 210,000 cubic kilometers, comparable to the combined volume of the world's largest saline and freshwater lakes. Furthermore, 80 percent of the world's ice and nearly two-thirds of the earth's fresh water are represented by Antarctica's glacier, which covers an area almost one and one-half times that of the United States. If this ice melted, sea level would rise an estimated 60 to 70 meters, and the ocean would inundate many densely populated coastal areas (Figure 11.3). The hydrologic importance of the

FIGURE 11.3
North America, showing the present-day coastline compared to the coastline that existed during the last ice-age maximum (18,000 years ago; solid outer line) and the coastline that would exist if present ice sheets in Greenland and Antarctica melted (black inner border). (After R. H. Dott, Jr. and R. L. Battan, *Evolution of the Earth*, New York: McGraw-Hill, 1971. Reprinted by permission of the publisher)

continent and its ice can be illustrated in another way. If Antarctica's ice sheet were melted at a suitable rate it could feed (1) the Mississippi River for more than 50,000 years, (2) all the rivers in the United States for about 17,000 years, (3) the Amazon River for approximately 5000 years, or (4) all the rivers of the world for about 750 years.

As the foregoing discussion illustrates, the quantity of ice on earth today is truly immense. However, present glaciers occupy only slightly more than one-third the area they did in the very recent geologic past. Later in this chapter we will examine this period of extensive glaciation often called the Ice Age.

FORMATION OF GLACIAL ICE

Snow is the raw material from which glacial ice originates; therefore, glaciers form in areas where more snow falls in winter than melts during the summer. Before a glacier is created, snow must be converted into ice. Strictly speaking, snowflakes are not frozen water but delicate crystals formed when water vapor changes directly into a solid—a process called *sublimation*. When temperatures remain below freezing following a snowfall, the fluffy accumulation of hexagonal crystals soon changes. As air infiltrates the spaces between the crystals, the extremities of the crystals evaporate and the water vapor condenses near the centers of the crystals. In this manner snowflakes become smaller, thicker, and more spherical, and the large pore spaces disappear. By this process air is forced out and what was once light, fluffy snow is recrystallized into a much denser mass of small grains having the consistency of coarse sand. This granular recrystallized snow is called **firn** and is commonly found making up old snowbanks near the end of winter. As more snow is added, the pressure on the lower layers increases, compacting the ice grains at depth. Once the thickness of ice and snow exceeds 50 meters, the weight is sufficient to fuse firn into a solid mass of interlocking ice crystals. Glacial ice has now been formed.

MOVEMENT OF A GLACIER

The movement of glacial ice is generally referred to as flow. The fact that glacial movement is described in this way would seem to constitute a paradox—ice is solid, yet it is capable of flow. The way in which ice flows is complex and is of two basic types. The first of these, **plastic flow,** involves movement within the ice. Ice behaves as a brittle solid until the pressure or load upon it is equivalent to the weight of about 50 meters of ice. Once that load is surpassed, ice behaves as a plastic material and flow begins. Such flow occurs because of the molecular structure of ice. Glacial ice consists of layers of molecules stacked one upon the other. The bonds between layers are weaker than those within each layer. Therefore, when a stress exceeds the strength of the bonds between the layers, the layers remain intact and slide over one another.

A second, and often equally important, mechanism of glacial movement consists of the entire ice mass slipping along the ground. With the exception of some glaciers located in polar regions where the ice is probably frozen to the solid bedrock floor, most glaciers are thought to move by this sliding process, called **basal slip.** In this process, meltwater is believed to act as a hydraulic jack and perhaps as a lubricant helping the ice over the rock. The source of the liquid water is related in part to the fact that the melting point of ice decreases as pressure increases. Therefore, deep within a glacier the ice may be at the melting point even though its temperature is below 0°C. In addition, other factors may contribute to the presence of meltwater deep within the glacier. Temperatures may be increased by plastic flow (an effect similar to frictional heating), by heat added from the earth below, and by the refreezing of meltwater that has seeped down from above. This last process relies on the property that as water changes state from liquid to solid, heat (termed latent heat of fusion) is released.

Figure 11.4 illustrates the effects of these two basic types of glacial motion. This vertical profile through a glacier also shows that all the ice does not flow forward at the same rate. Just as in streams, frictional drag with the bedrock floor causes the lower portions of the glacier to move more slowly.

In contrast to the lower portion of the glacier, the upper 50 meters or so is not under sufficient pressure to exhibit plastic flow. Consequently, the ice in this uppermost zone is brittle and is appropriately referred

to as the **zone of fracture.** Incapable of flow, the ice in the zone of fracture is carried along "piggyback" style by the ice below. When the glacier moves over irregular terrain, the zone of fracture is subjected to tension, resulting in cracks called **crevasses** (Figure 11.5). These gaping cracks can make travel across glaciers dangerous and may extend to depths of 50 meters. Below this depth, plastic flow seals them off.

RATES OF GLACIAL MOVEMENT

Unlike streamflow, the movement of glaciers is not readily apparent to the casual observer. If we could watch an alpine glacier move, we would see that, like the water in a river, all of the ice in the valley does not move downvalley at an equal rate. Just as friction with the bedrock floor slows the movement of the ice at the bottom of the glacier, the drag created by the valley walls leads to the flow being greatest in the center of the glacier.

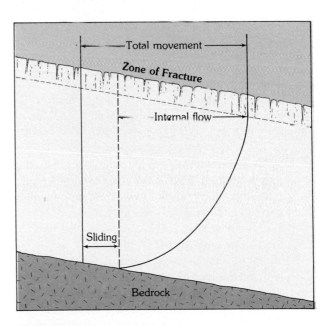

FIGURE 11.4
Glacial movement is divided into two components. Below about 50 meters, ice behaves plastically and flows. In addition, the entire mass of ice may slide along the ground. The ice in the zone of fracture is carried along "piggyback" style. Notice that the rate of movement is slowest at the base of the glacier where frictional drag is greatest.

FIGURE 11.5
Crevasses form in the brittle ice of the zone of fracture. They do not continue down into the zone of flow. (Courtesy of Ward's Natural Science Establishment, Inc., Rochester, N.Y.)

The first measurements of glacial movement were made more than 100 years ago. In this experiment, stakes were carefully placed in a straight line across the top of an alpine glacier. Periodically, the positions of the stakes were recorded, revealing the type of movement just described (Figure 11.6).

How rapidly does glacial ice move? Average velocities vary considerably from one glacier to another. Some move so slowly that trees and other vegetation may become well established in the debris that has accumulated on the glacier's surface, whereas others may move at rates of up to several meters per day. The advance of some glaciers is characterized by periods of extremely rapid movement followed by periods during which movement is practically nonexistent. Such rapid movements are called **surges** (Figure 11.7). For example, Hassanabad Glacier in the Karakoram, a mountain range in Kashmir and northwestern India, advanced 10 kilometers in less than 3

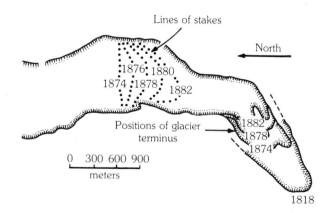

FIGURE 11.6
Map of Rhone Glacier, Switzerland. The ice along the sides of the glacier moves slowest. Also notice that even though the ice front was retreating, the ice within the glacier was advancing. (Courtesy of Robert Scholten)

262

A.

FIGURE 11.7
The surge of Variegated Glacier, an alpine glacier near Yakutat, Alaska, is captured in these two aerial photographs taken one year apart. **A.** August 29, 1964 and **B.** August 22, 1965. During a surge, ice velocities are 20 to 50 times greater than during a quiescent phase. (Photos by Austin Post, U.S. Geological Survey)

B.

months—a rate of almost 130 meters per day. The precise cause or causes of these sporadic, short-lived advances is not well understood. One idea is that the base of the glacier may have been frozen to the bedrock and then sudden melting released the ice. Another hypothesis suggests that a block of stagnant ice at the terminus of a glacier in an alpine valley may act as a dam until the buildup of pressure by the flowing ice behind it forces it to give way.

ECONOMY OF A GLACIER

Glaciers are constantly gaining and losing ice. Snow accumulation and ice formation occur in a region termed the **zone of accumulation,** the outer limits of which are defined by the **snowline.** The elevation of the snowline varies greatly. In polar regions, it may be sea level, whereas in tropical areas, the snowline exists only high in mountain areas, often at altitudes exceeding 4500 meters. Above the snowline, the addition of snow thickens the glacier and promotes movement. Below the snowline, snow from the previous winter as well as some of the glacial ice melts. In addition, large pieces of ice may break off the front of the glacier in a process called **calving.** Calving creates icebergs in places where the glacier has reached the sea or a lake.

Whether the margin of a glacier is advancing, retreating, or remaining stationary depends upon the economy, or budget, of the glacier. That is, it depends upon the balance or lack of balance between accumulation on the one hand and wastage (also termed **ablation**) on the other. If ice accumulation exceeds ablation, the glacial front advances until the two factors balance. At this point, the terminus of the glacier is stationary. Should a warming trend occur that causes ablation to exceed accumulation, the ice front will retreat. As the terminus of the glacier retreats, the extent of the zone of ablation diminishes. Therefore, in time a new balance will be reached between accumulation and ablation, and the ice front will again become stationary.

Whether the margins of a glacier are advancing, retreating, or stationary, the ice within the glacier continues to flow forward. In the case of a receding glacier, the ice simply does not flow forward rapidly enough to offset ablation. This point is illustrated rather well in Figure 11.6. While the line of stakes

within Rhone Glacier continued to move downvalley, the terminus of the glacier slowly retreated upvalley.

GLACIAL EROSION

Glaciers are capable of great amounts of erosion. For anyone who has observed the terminus of an alpine glacier, the evidence of its erosive force is clear. The observer can witness firsthand the release of rock material of various sizes from the ice as it melts. All signs lead to the conclusion that the ice has scraped, scoured, and torn rock from the floor and walls of the valley and carried it downslope. Indeed, as a medium of sediment transport, ice has no equal. Once rock debris is acquired by the glacier, the enormous competency of ice will not allow the debris to settle out like the load carried by a stream or by the wind. Consequently, glaciers can carry huge blocks that no other erosional agent could possibly budge (Figure 11.8). Although today's glaciers are of limited importance as erosional agents, many landscapes that were

FIGURE 11.8

A giant, glacially-transported boulder in Yellowstone National Park. (Photo courtesy of the National Park Service)

modified by the widespread glaciers of the most recent ice age still reflect to a high degree the work of ice.

Glaciers erode the land primarily in two ways. First, as a glacier flows over a fractured bedrock surface, it loosens and lifts blocks of rock and incorporates them into the ice. This process, known as **plucking,** occurs when meltwater penetrates the cracks and joints of bedrock beneath a glacier and freezes. As the water expands, it exerts tremendous leverage that pries the rock loose. In this manner sediment of all sizes, ranging from particles as fine as flour to blocks as big as houses, becomes part of the glacier's load.

The second major erosional process is **abrasion.** As the ice and its load of rock fragments slide over bedrock, they function as a kind of "sandpaper" to smooth and polish the surface below. The pulverized rock produced by the glacial "grist mill" is appropriately called **rock flour.** So much rock flour may be produced that meltwater streams flowing out of a glacier often have the grayish appearance of skimmed milk and offer visible evidence of the grinding power of ice. When the ice at the bottom of a glacier contains large fragments of rock, long scratches and grooves called **glacial striations** may be cut into the bedrock (Figure 11.9). These linear grooves provide

clues as to the direction of ice flow. By mapping the striations over large areas, patterns of glacial flow can often be reconstructed. On the other hand, not all abrasive action produces striations. When the sediment consists primarily of fine silt-sized particles, the rock surfaces over which the glacier moves may become highly polished. The broad expanses of smoothly polished granite in Yosemite National Park, California, provide an excellent example.

The erosional effects of alpine and continental glaciers are quite different. A visitor to an alpine glaciated region is likely to see a sharp and very angular topography. The reason is that as alpine glaciers move downvalley, they tend to accentuate the irregularities of the mountain landscape by creating steeper canyon walls and making bold peaks even more jagged. By contrast, continental ice sheets generally override the terrain and hence tend to subdue rather than accentuate the irregularities they encounter.

Finally, it should be pointed out that, as is the case with other agents of erosion, the rate of glacial erosion is highly variable. This differential erosion by ice is largely controlled by four factors: (1) rate of glacial movement; (2) thickness of the ice; (3) shape, abundance, and hardness of the rock fragments contained in the ice at the base of the glacier; and (4) the erodibility of the surface beneath the glacier. Variations in any or all of these factors from time to time and/or from place to place mean that the features, effects, and degree of landscape modification in glaciated regions can vary greatly.

LANDFORMS CREATED BY GLACIAL EROSION

Although the erosional potential of continental glaciers is enormous, landforms carved by these huge ice masses usually do not inspire the same degree of wonderment and awe as do the erosional features created by alpine glaciers. In regions where the erosional effects of continental ice sheets are significant, glacially scoured surfaces and subdued terrain are the rule. By contrast, erosion by alpine glaciers accentuates an already mountainous topography. Much of the rugged mountain scenery so celebrated for its majestic beauty is the product of glacial erosion.

FIGURE 11.9
Results of glacial abrasion. Glacial striations in bedrock, Worcester County, Massachusetts. (Photo by W. C. Alden, U.S. Geological Survey)

GLACIATED VALLEYS

A hike up a glaciated valley would reveal a number of striking ice-created features. The valley itself is often a dramatic sight. Rather than creating their own valleys, glaciers take the path of least resistance by following the courses of pre-existing stream valleys. Prior to glaciation mountain valleys are characteristically narrow and V-shaped because streams are well above base level and are therefore downcutting. However, during glaciation these narrow valleys undergo a transformation as the glacier widens and deepens them, creating a U-shaped **glacial trough** (Figure 11.10). In addition to producing a broader and deeper valley, the glacier also straightens the valley. As ice flows around sharp curves, its great erosional force removes the spurs of land that extend into the valley. The results of this activity are triangular-shaped cliffs called **truncated spurs.**

Since the intensity of glacial erosion depends in part upon the thickness of the ice, main (trunk) glaciers cut their valleys deeper than do their smaller tributary glaciers. Thus, after the glaciers have receded, the valleys of feeder glaciers stand above the main glacial trough and are termed **hanging valleys.** Rivers flowing through hanging valleys may produce spectacular waterfalls, such as those in Yosemite National Park (Figure 11.11).

As hikers walk up a glacial trough, they may pass a

FIGURE 11.10
Prior to glaciation, a mountain valley is typically narrow and V-shaped. During glaciation, an alpine glacier widens, deepens, and straightens the valley, creating a U-shaped glacial trough. (Photo courtesy of John Montagne.)

FIGURE 11.11
Bridalveil Falls in Yosemite National Park cascades from a hanging valley into the glacial trough below. (Courtesy of Ward's Natural Science Establishment, Inc., Rochester, N.Y.)

FIGURE 11.12
The bowl-shaped depressions at the heads of these small glaciated valleys are termed cirques. (Photo by Bradford Washburn)

series of bedrock depressions on the valley floor that were probably formed by plucking. If these depressions are filled with water, they are called **pater noster lakes,** a reference to the fact that, from high above or on a map, they resemble a string of beads.

At the head of a glacial valley is a very characteristic and often imposing feature associated with an alpine glacier—a **cirque.** As Figure 11.12 illustrates, these bowl-shaped depressions have precipitous walls on three sides but are open on the downvalley side. The cirque represents the focal point of the gla-

cier's growth, that is, the area of snow accumulation and ice formation. Although the origin of cirques is still not totally clear, they are believed to begin as irregularities in the mountainside that are subsequently enlarged by frost wedging and plucking along the sides and bottom of the glacier. After the glacier has melted away, the cirque basin is usually occupied by a small lake called a **tarn** (Figure 11.13).

Sometimes when two glaciers exist on opposite sides of a divide, one flowing away from the other, the common headwall (back wall) between the cir-

FIGURE 11.13
A tarn is a lake occupying a cirque. This example, Iceberg Lake, is in Glacier National Park, Montana. (Photo by Austin Post, U.S. Geological Survey)

ques is largely eliminated as plucking and frost action enlarge each cirque. When this occurs, the two glacial troughs intersect to create a gap or pass from one valley into the other. Such a feature is termed a **col.** Some important and well-known mountain passes that are cols include St. Gotthard Pass in the Swiss Alps, Tioga Pass in the Sierra Nevada, and Berthoud Pass in the Rocky Mountains.

Before leaving the topic of glacial troughs and their associated features, one more rather well-known fea-

ture should be discussed—fiords. **Fiords** are deep, often spectacular, steep-sided inlets of the sea that are present at high latitudes where mountains are adjacent to the ocean (Figure 11.14). Norway, British Columbia, Greenland, New Zealand, Chile, and Alaska all have coastlines characterized by fiords. They represent drowned glacial troughs that were submerged as the ice left the valley and sea level rose following the Ice Age. The depths of fiords are often dramatic, in some instances exceeding 1000 meters.

269

FIGURE 11.14
Fiords are drowned glacial troughs. In late Pleistocene time, the now-drowned valleys
were full of ice. Glacier Bay, Alaska. (Photo by Warren Hamilton, U.S. Geological
Survey)

However, the great depths of these flooded troughs is
only partly explained by the post-Ice Age rise in sea
level. Unlike the situation governing the downward
erosional work of rivers, sea level does not act as base
level for glaciers. As a consequence, glaciers are
capable of eroding their beds far below the surface of
the sea. For example, a 300-meter thick alpine glacier
can carve its valley floor more than 250 meters below
sea level before downward erosion ceases and the ice
begins to float.

ARÊTES AND HORNS

A visit to the Alps, the Northern Rockies, or many
other scenic mountain landscapes carved by alpine

FIGURE 11.15
The Matterhorn, a glacially eroded peak in the Swiss Alps. (Courtesy of Ward's Natural Science Establishment, Inc., Rochester, N.Y.)

glaciers would reveal not only glacial troughs, cirques, pater noster lakes, and the other related features just discussed. In addition, a visitor would likely see sinuous, sharp-edged ridges called **arêtes** and sharp, pyramid-like peaks called **horns** projecting above the surroundings. Both features can originate from the same basic process, the enlargement of cirques produced by plucking and frost action. In the case of the spires of rock called horns, a group of cirques around a single high mountain are responsible. As the cirques enlarge and converge, an isolated horn is produced. Certainly the most famous example is the classic Matterhorn in the Swiss Alps (Figure 11.15). Arêtes can be formed in a similar manner except that the cirques are not clustered around a point but rather exist on opposite sides of a divide. As the cirques grow, the divide separating them is reduced to a very narrow knife-like partition. An

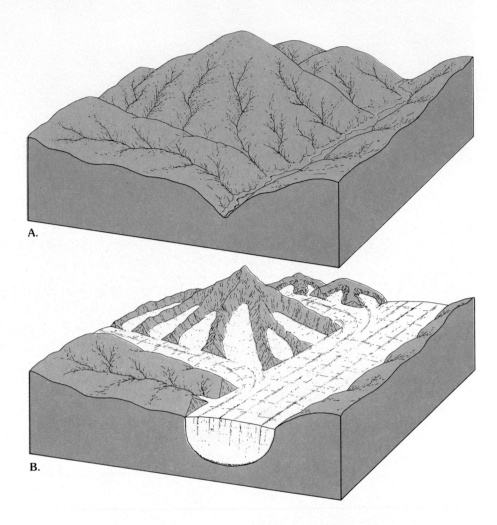

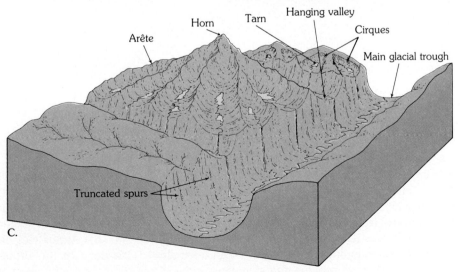

FIGURE 11.16
Landforms created by alpine glaciers. **A.** A mountain mass not affected by glacial erosion. **B.** The mountain mass during the period of maximum glacial activity. **C.** The mountain mass shortly after the glaciers have melted from its valleys. [After William Morris Davis, "The Sculpture of Mountains by Glaciers," *Scottish Geographical Magazine* 22 (1906)]

Arête

Horn

Tarn

Hanging valley

Cirques

Main glacial trough

Truncated spurs

A.

B.

C.

FIGURE 11.17
Roches moutonnée, Black Bay, Lake Athabaska, Saskatchewan. The gentle slope was abraded and the steep side was plucked. The ice moved from right to left. (Photo courtesy of Geological Survey of Canada, catalog number 28521)

arête, however, may also be created in another way. When two glaciers occupy parallel valleys, an arête can form when the divide separating the moving tongues of ice is progressively narrowed as the glaciers scour and widen their valleys.

The landforms created by glacial erosion in an alpine environment are summarized in Figure 11.16.

ROCHES MOUTONNÉES

In many glaciated landscapes, but most frequently where continental glaciers modified the terrain, the ice carves small streamlined hills from protruding bedrock knobs. These asymmetrical knobs of bedrock, called **roches moutonnées,** are formed when glacial abrasion smoothes the gentle slope facing the oncoming ice sheet and plucking steepens the opposite side as the ice rides over the knob (Figure 11.17). When roches moutonnées are present, they can be used to indicate the direction of glacial flow, because the gentler slope is on the side from which the ice advanced.

GLACIAL DEPOSITS

As we have seen, glaciers are capable of acquiring and transporting a huge load of debris as they slowly and relentlessly advance across the land. Of course, these materials must be deposited when the ice eventually melts. Thus in regions of deposition, glacial sediment can play a truly significant role in shaping the physical landscape. For example, in many areas once covered by the continental ice sheets of the recent Ice Age, the bedrock is rarely exposed because glacial deposits that are tens or even hundreds of meters thick completely mantle the terrain. The general effect of these deposits is to reduce the local relief and thus level the topography. Indeed, many of the rural country scenes that are familiar to many of us— rocky pastures in New England, wheat fields in the Dakotas, rolling farmland in the Midwest—result directly from glacial deposition.

Long before the theory of an extensive Ice Age was ever proposed, much of the soil and rock debris covering portions of Europe was recognized as coming from somewhere else. At the time, these "foreign" materials were believed to have been "drifted" into their present positions by floating ice during an ancient flood. As a consequence, the term *drift* was applied to this sediment. Although rooted in an incorrect concept, this term was so well established by the time the true glacial origin of the debris became widely recognized that it remained in the basic glacial vocabulary. Today the word **drift** is an all-embracing term for sediments of glacial origin, no matter how, where, or in what shape they were deposited.

One of the features that distinguishes drift from sediments laid down by other erosional agents is that glacial deposits consist primarily of mechanically weathered rock debris that underwent little or no chemical weathering prior to deposition. Thus, minerals that are notably prone to chemical decomposition, such as hornblende and the plagioclase feldspars, are often conspicuous components in glacial sediments.

Glacial drift is divided by geologists into two distinct types: (1) materials deposited directly by the glacier, which are known as **till;** and (2) sediments laid down by glacial meltwater, called **stratified drift.**

FIGURE 11.18
Glacial till is an unsorted mixture of many different sediment sizes. (Photo by W. C. Alden, U.S. Geological Survey)

LANDFORMS MADE OF TILL

Till is deposited as glacial ice melts and drops its load of rock fragments. Unlike moving water and wind, ice cannot sort the sediment it carries; therefore, deposits of till are characteristically unsorted mixtures of many particle sizes (Figure 11.18). A close examination of this sediment would reveal that many of the pieces are striated and polished as the result of being dragged along by the glacier. Such pieces help distinguish till from other deposits that may also consist of a mixture of different sediment sizes, such as the debris from a mudflow or a rockslide.

When there are boulders found in the till or lying free on the surface, they are called **glacial erratics,** if they were derived from a source outside of the area where they are found. By examining the erratics as well as the mineral composition of the remaining till, geologists are sometimes able to find clues about the path that the glacier took. In portions of New England, as well as other areas, erratics may be seen

FIGURE 11.19
Land cleared of glacial erratics which were then piled into walls about a field near Whitewater, Wisconsin. (Photo by W. C. Alden, U.S. Geological Survey)

dotting pastures and farm fields. In fact, in some places these large rock fragments were cleared from the fields and piled into fences (Figure 11.19). Keeping the fields clear, however, was and is an ongoing chore because each spring the fields have to be cleared of newly exposed erratics that wintertime frost heaving has lifted to the surface.

END AND GROUND MORAINES

Probably the most common term for landforms made of glacial deposits is *moraine*. Originally this term was used by French peasants when referring to the ridges and embankments of debris found near the margins of glaciers in the French Alps. Today, however, moraine has a broader meaning, because it is applied to a number of landforms, all of which are composed primarily of till (Figure 11.20).

An **end moraine** is a ridge of till that forms at the terminus of both alpine and continental glaciers. These relatively common landforms are deposited when a state of equilibrium is attained between ablation and ice accumulation. That is, the end moraine forms when the ice is melting and evaporating near the end of the glacier at a rate equal to the forward advance of the glacier from its region of nourishment. Although the terminus of the glacier is now stationary, the ice continues to flow forward, delivering a continuous supply of sediment in the same manner a conveyor belt delivers goods to the end of a production line. As the ice melts, the till is dropped and the

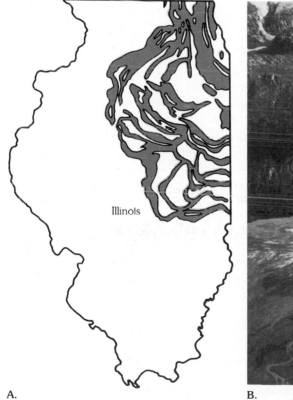

A.

B.

FIGURE 11.20
A. End moraines deposited by the most recent ice sheets in Illinois. (After Illinois State Geological Survey). **B.** End moraines of the Northern Iliamma Glacier, Alaska. The present ice margin can be seen in the background. (Photo by Bradford Washburn)

end moraine grows. Therefore, the longer the ice front remains stable, the larger the ridge of till will become.

Eventually the time comes when ablation exceeds nourishment. At this point the front of the glacier will begin to recede in the direction from which it originally advanced. However, as the ice front retreats, the conveyor belt action of the glacier continues to provide fresh supplies of till to the terminus. In this manner a large quantity of till is deposited as the ice melts away, creating a rock-strewn, undulating plain. This gently rolling layer of till laid down as the ice front receded is termed **ground moraine.** Ground moraine has a leveling effect, filling in low spots and clogging old stream channels, often leading to a derangement of the existing drainage system. In areas where this layer of till is still relatively fresh, such as the northern Great Lakes region, poorly drained swampy land is quite common.

Periodically, a glacier will retreat to a point where ablation and nourishment once again balance. When this happens the ice front stabilizes and a new end moraine is created.

The pattern of end moraine formation and ground moraine deposition may be repeated many times before the glacier has completely vanished. Such a pattern is illustrated by Figure 11.20. It should be pointed out that the outermost end moraine marks the limit of the glacial advance. Because of its special status, this end moraine is also called the **terminal moraine.** On the other hand, the end moraines that were created as the ice front occasionally stabilized during retreat are termed **recessional moraines.** Note that both terminal and recessional moraines are essentially alike; the only difference between them is their relative positions.

End moraines deposited by the last major stage of Ice Age glaciation are prominent features in many parts of the Midwest and Northeast. In Wisconsin, the wooded, hilly terrain of the Kettle Moraine near Milwaukee is a particularly picturesque example. Certainly a well-known example in the Northeast is Long Island. This linear strip of glacial sediment that extends northeastward from New York City is part of an end moraine complex that stretches from eastern Pennsylvania to Cape Cod, Massachusetts. The end moraines that make up Long Island represent materials that were deposited by a continental glacier in the relatively shallow waters off the coast and built up many meters above sea level. Long Island Sound, the narrow body of water separating the island and the mainland, was not built up as much by glacial deposition and was therefore subsequently flooded by the rising sea following the Ice Age.

LATERAL AND MEDIAL MORAINES

Alpine glaciers produce two types of moraines that occur exclusively in mountain valleys. The first of these is called a **lateral moraine** (Figure 11.21). As we learned earlier, when an alpine glacier moves downvalley, the ice erodes the sides of the valley with great efficiency. In addition, large quantities of debris are added to the glacier's surface as rubble falls or slides from higher up on the valley walls and collects on the edges of the moving ice. When the ice eventually melts, this accumulation of debris is dropped next to the valley walls. These ridges of till paralleling the sides of the valley constitute the lateral moraines.

The second type of moraine that is unique to alpine glaciers is the **medial moraine** (see chapter-opening photo). Medial moraines are created when two alpine glaciers coalesce to form a single ice stream. The till that was once carried along the edges of each glacier joins to form a single dark stripe of debris within the newly enlarged glacier. The creation of these dark stripes within the ice stream is one obvious proof that glacial ice moves, because the medial moraine could not form if the ice did not flow downvalley. It is quite common to see several medial moraines within a single large alpine glacier, because a streak will form whenever a tributary glacier joins the main valley.

DRUMLINS

Moraines are not the only landforms deposited by glaciers. In some areas that were once covered by continental ice sheets, a special variety of glacial landscape exists—one characterized by smooth, elongate, parallel hills called **drumlins** (Figure 11.22). Certainly one of the best-known drumlins is Bunker Hill in Boston, the site of the famous Revolutionary War battle in 1775. An examination of Bunker Hill or other less famous drumlins would reveal that they are

FIGURE 11.21
A well-developed lateral moraine deposited by the shrinking Athabaska Glacier in the Canadian Rockies. (Photo by James E. Patterson)

FIGURE 11.22
Drumlins, such as this one in upstate New York, are depositional features associated with continental glaciers. (Courtesy of Ward's Natural Science Establishment, Inc., Rochester, N.Y.)

streamlined asymmetrical hills composed largely of till. They range in height from about 15 to 50 meters and may be up to one kilometer long. The steep side of the hill faces the direction from which the ice advanced, while the gentler, longer slope points in the direction the ice moved. Drumlins are not found as isolated landforms, but rather occur in clusters called *drumlin fields* (Figure 11.23). One such cluster, east of Rochester, New York, is estimated to contain about 10,000 drumlins. Although drumlin formation is not fully understood, their streamlined shape would indicate that they were molded in the zone of plastic flow within an active glacier. It is believed that many drumlins originate when glaciers advance over previously deposited drift and reshape the material.

LANDFORMS MADE OF STRATIFIED DRIFT

As the name implies, stratified drift is sorted according to the size and weight of the particles. Since ice is not capable of such sorting activity, these materials are not deposited directly by the glacier as till is, but rather reflect the sorting action of the glacial meltwater that was responsible for dropping them. Accumulations of stratified drift often consist largely of sand and gravel, that is, bed load material, because the finer rock flour remains suspended and therefore is commonly carried far from the glacier by the meltwater streams. An indication that stratified drift consists primarily of sand and gravel is reflected in the fact that in many areas these deposits are actively mined as a source of aggregate for road work and other construction projects.

OUTWASH PLAINS AND VALLEY TRAINS

At the same time that an end moraine is forming, water from the melting glacier cascades over the till, sweeping some of it out in front of the growing ridge of unsorted debris. Meltwater generally emerges from the ice in rapidly moving streams that are often choked with suspended material and carry a substantial bed load as well. As the water leaves the glacier, it moves onto the relatively flat surface beyond and rapidly loses velocity. As a consequence, much of its bed

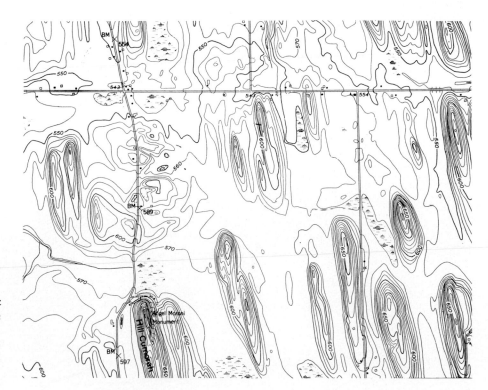

FIGURE 11.23
Portion of a drumlin field shown on the Palmyra, New York, 7.5 minute topographic map. North is at the top. The drumlins are steepest on the north side, indicating that the ice advanced from this direction.

load is dropped and the meltwater begins weaving a complex pattern of braided channels (see Figure 9.15). In this way a broad, ramplike surface composed of stratified drift is built adjacent to the downstream edge of most end moraines. When the feature is formed in association with an ice sheet, it is termed an **outwash plain,** and when largely confined to a mountain valley, it is usually referred to as a **valley train.**

Often outwash plains are pockmarked with basins or depressions known as **kettles** (Figure 11.24). Kettles also occur in deposits of till. Kettles are formed when a block of stagnant ice becomes wholly or partly buried in drift and ultimately melts, leaving a pit in the glacial sediment. Although most kettles do not exceed two kilometers in diameter, some with diameters exceeding 10 kilometers occur in Minnesota. Likewise, the typical depth of most kettles is less than 10 meters, although the vertical dimensions of some approach 50 meters. In many cases, water eventually fills the depression and forms a pond or lake.

ICE-CONTACT DEPOSITS

When the wasting terminus of a glacier shrinks to a critical point, flow virtually stops and the ice becomes stagnant. With time, meltwater flowing over, within, and at the base of the motionless ice deposits stratified drift. Then, as the supporting ice melts away, the stratified sediment is left behind in the form of hills, terraces, and ridges. Such accumulations are collec-

tively termed **ice-contact deposits** and are classified according to their shape.

When the ice-contact stratified drift is in the form of a mound or steep-sided hill, it is called a **kame** (Figure 11.25). Some kames represent bodies of sediment deposited by meltwater in openings within or depressions on top of the ice. Others originate as deltas or fans built outward from the ice by meltwater streams. Later, when the stagnant ice melts away, these various accumulations of sediment collapse to form isolated, irregular mounds.

When glacial ice occupies a valley, terraces known as **kame terraces** may be built along the sides of the valley. These features are commonly narrow masses of stratified drift laid down between the glacier and the side of the valley by streams that drop debris along the margins of the shrinking ice mass.

Finally, on some glacial landscapes long, narrow, sinuous ridges composed largely of sand and gravel are present. Some ridges are more than 100 meters high with lengths in excess of 100 kilometers. The dimensions of many others, however, are far less spectacular. Known as **eskers,** these ridges were deposited by meltwater rivers flowing in confined channels within, on top of, and beneath a mass of motionless, stagnant glacial ice (Figure 11.26). While many sediment sizes were carried by the torrents of meltwater in the ice-banked channels, only the coarser material could settle out of the turbulent stream.

FIGURE 11.24
Kettle pond in gravel near the terminus of Baird Glacier in southeastern Alaska. (Photo by A. F. Buddington, U.S. Geological Survey)

FIGURE 11.25
White Kame in Kettle Moraine State Forest, Wisconsin. (Photo by G. J. Knudson, Wisconsin Department of Natural Resources)

FIGURE 11.26
The retreat of the glacier reveals an esker, the sinuous ridge of sand and gravel in the
center of the photograph. (Photo by Bradford Washburn)

THE GLACIAL THEORY AND THE ICE AGE

At various points in the preceding pages mention was
made of the Ice Age, a time when ice sheets and
alpine glaciers were far more extensive than they are
today. As was noted earlier, there was a time when
the most popular explanation for what we now know
to be glacial deposits was that the materials had been
drifted in by means of icebergs or perhaps simply

swept across the landscape by a catastrophic flood. What convinced geologists that an extensive ice age was responsible for these deposits as well as for many other features?

In 1821 a Swiss engineer, Ignaz Venetz, presented a paper suggesting that glacial features occurred at considerable distances from the existing margins of glaciers in the Alps. The inference therefore was that the glaciers had once occupied positions farther downvalley. In 1836, Louis Agassiz, another Swiss scientist, who doubted the theory of widespread glacial activity put forth by Venetz and later by Jean de Charpentier, set out to prove that the idea was not valid. Instead, his fieldwork in the Alps convinced him of the merits of his colleagues' hypothesis. A year later Agassiz authored the theory of a great ice age that had had extensive and far-reaching effects—an idea that was to give Agassiz widespread fame.

The proof of the glacial theory proposed by Agassiz and others constitutes a classic example of applying the principle of uniformitarianism. Realizing that certain features are produced by no other known process but glacial action, they were able to begin reconstructing the extent of now-vanished ice sheets based on the presence of features and deposits found far beyond the margins of present-day glaciers. In this manner the development and verification of the glacial theory continued during the nineteenth century, and through the efforts of many scientists, a knowledge of the nature and extent of former ice sheets became clear.

By the turn of the twentieth century, geologists had largely determined the areal extent of Ice Age glaciation. Further, during the course of their investigations they had discovered that many glaciated regions had not one layer of drift but several. Moreover, close examination of these older deposits showed well-developed zones of chemical weathering and soil formation as well as the remains of plants that require warm temperatures. The evidence was clear; there had not been just one glacial advance but several, each separated by long periods when climates were as warm or warmer than at present. The Ice Age then had not simply been a time when the ice advanced over the land, lingered for a while, and then receded. Rather, the period was a very complex event characterized by a number of advances and withdrawals of glacial ice. In North America four major stages of glaciation have been identified (Figure 11.27). Each was named for the midwestern state where the deposits of that ice sheet are well exposed and/or were first studied. These are, in order of occurrence, the Nebraskan, Kansan, Illinoian, and Wisconsinan. Additional evidence from Europe,

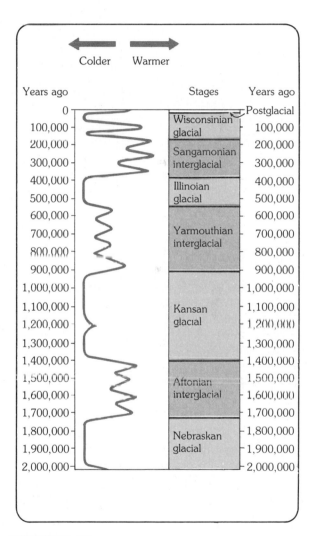

FIGURE 11.27

The Pleistocene epoch was marked by fluctuating climatic conditions that led to alternating glacial and interglacial periods. Four separate stages of glaciation are recognized for North America. (After D. B. Ericson and G. Wollin, "Pleistocene Climates and Chronology in Deep-Sea Sediments," *Science* 162 (1968) : 1233. Copyright © 1968 by the American Association for the Advancement of Science)

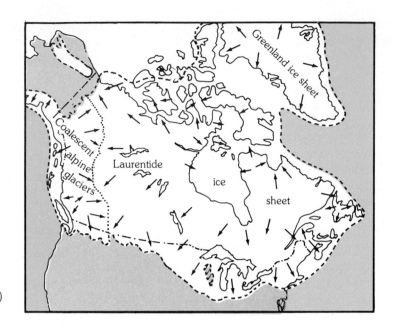

FIGURE 11.28
At their maximum extent, Pleistocene glaciers covered about 10 million square kilometers (4 million square miles) of North America.

Alaska, and elsewhere further indicates that other glacial advances probably preceded the Nebraskan. During the glacial age, ice left its imprint on almost 30 percent of the earth's land area, including about 10 million square kilometers of North America, 5 million square kilometers of Europe, and 4 million square kilometers of Siberia (Figure 11.28). The amount of glacial ice in the Northern Hemisphere was roughly twice that of the Southern Hemisphere. The primary reason is that the southern polar ice could not spread far beyond the margins of Antarctica. By contrast, North America and Eurasia provided great expanses of land for the spread of ice sheets.

Today we know that the Ice Age began between two and three million years ago. This means that most of the major glacial stages occurred during a division of the geologic calendar called the **Pleistocene epoch**. Although the Pleistocene is commonly used as a synonym for the Ice Age, note that this epoch does not encompass all of the last glacial period. The Antarctic ice sheet, for example, is believed to have formed about 10 million years ago.

SOME INDIRECT EFFECTS OF ICE AGE GLACIERS

In addition to the massive erosional and depositional work carried on by Pleistocene glaciers, the ice sheets had other, sometimes profound, effects upon the landscape. For example, as the ice advanced and retreated, animals and plants were forced to migrate. This led to stresses that some organisms could not tolerate. Hence, such creatures as the giant ground sloth, the fearsome-looking sabre-tooth tiger, and prehistoric elephants called mastodons and mammoths became extinct. Furthermore, many present-day stream courses bear little resemblance to their preglacial routes. The Missouri River once flowed northward toward Hudson Bay, while the Mississippi River followed a path through central Illinois and the head of the Ohio River reached only as far as Indiana. Other rivers that today carry only a trickle of water but nevertheless occupy broad channels are testimony to the fact that they once carried torrents of glacial meltwater.

In areas that were centers of ice accumulation, such as Scandinavia and the Canadian Shield, the land has been slowly rising over the past several thousand years. As Figure 11.29 illustrates, uplifting of almost 300 meters has occurred in the Hudson Bay region. This, too, is the result of the continental glaciers. But how can glacial ice cause crustal movement? We now believe the land is rising because the added weight of the three-kilometer-thick mass of ice caused downwarping of the earth's crust. Following the removal of this immense load, the crust has been

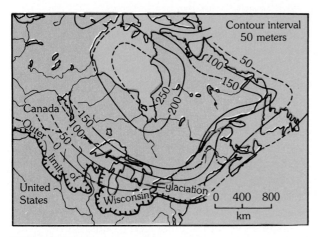

FIGURE 11.29

In northern Canada where there was the greatest accumulation of glacial ice, the weight caused downwarping of the crust. When the ice melted the crust began to rebound, leaving old shorelines situated high above sea level. (After P. B. King, "Tectonics of Quaternary Time in Middle North America," fig. 4, p. 836 in H. E. Wright and David G. Frey, eds. *The Quaternary of the United States,* 1965. Reprinted by permission of Princeton University Press)

adjusting by gradually rebounding upward ever since.[1]

Certainly one of the most interesting and perhaps dramatic effects of the ice age was the fall and rise of sea level that accompanied the advance and retreat of the glaciers. Earlier in this chapter it was pointed out that sea level would rise by an estimated 60 or 70 meters if the water locked up in the Antarctic ice sheet were to melt completely (see Figure 11.3). Such an occurrence would flood many densely populated coastal areas. Although the total volume of glacial ice today is great, exceeding 25 million cubic kilometers, during the ice age the volume of glacial ice amounted to about 70 million cubic kilometers, or 45 million cubic kilometers more than at present. Since we know that the snow from which glaciers are made ultimately comes from the evaporation of ocean water, the growth of ice sheets must have caused a worldwide drop in sea level. Indeed, estimates suggest that sea level was as much as 130 meters lower than today. Thus, land that is presently flooded by

[1]For a more complete discussion of this concept, termed *isostatic adjustment,* see Chapter 18.

the oceans was dry. The Atlantic Coast of the United States lay more than 100 kilometers to the east of New York City; France and Britain were joined where the famous English Channel is today; Alaska and Siberia were connected across the Bering Strait; and Southeast Asia was tied by dry land to the islands of Indonesia.

While the formation and growth of ice sheets was an obvious response to significant changes in climate, the existence of the glaciers themselves triggered important climatic changes in the regions beyond their margins. In arid and semiarid areas on all of the continents, temperatures and thus evaporation rates were lower, and at the same time moderate precipitation totals were experienced. This cooler, wetter climate resulted in the formation of many lakes called **pluvial lakes,** from the Latin term *pluvia* meaning *rain.* In North America, the greatest concentration of pluvial lakes occurred in the vast Basin and Range region of Nevada and Utah. By far the largest of the lakes in this region was Lake Bonneville. With maximum depths exceeding 300 meters and an area of 50,000 square kilometers, Lake Bonneville was nearly the same size as present-day Lake Michigan. As the ice sheets waned, the climate again grew more arid, and the lake levels lowered in response. Although most of the lakes completely disappeared, a few small remnants of Lake Bonneville remain, the Great Salt Lake being the largest and best known.

CAUSES OF GLACIATION

A great deal is known about glaciers and glaciation. Much has been learned about glacier formation and movement, the extent of glaciers past and present, and the features created by glaciers, both erosional and depositional. However, a widely accepted theory for the causes of glacial ages has not yet been established. Although nearly 150 years have elapsed since Louis Agassiz proposed his theory of a great ice age, no complete agreement exists as to the causes of such events.

While widespread glaciation has been a very rare occurrence in the earth's history, the ice age that encompassed the Pleistocene epoch is not the only glacial period for which a record exists. Other earlier glaciations are indicated by deposits called **tillite,** a rock formed when glacial till is lithified. Such deposits, found in strata of several different ages, usually contain striated rock fragments and some overlie grooved and polished bedrock surfaces or are associated with sandstones and conglomerates that show features of outwash deposits. Two Precambrian glacial episodes have been identified in the geologic record, the first approximately two billion years ago and the second about 600 million years ago. Further, a well-documented record of an earlier glacial age is found in late Paleozoic rocks that are about 250 million years old and which exist on several landmasses.

Any theory that attempts to explain the causes of glacial ages must successfully answer two basic questions. The first question is: What causes the onset of glacial conditions? In order for continental ice sheets to have formed, average temperatures must have been somewhat lower than at present and perhaps substantially lower than throughout much of geologic time. Thus, a successful theory would have to account for the gradual cooling that finally leads to glacial conditions. The second question that requires an answer is: What caused the alternation of glacial and interglacial stages that have been documented for the Pleistocene epoch? While the first question deals with long-term trends in temperature that occur on a scale of millions of years, this second question relates to much shorter-term changes.

Although the literature of science contains a vast array of theories relating to the possible causes of glacial periods, we will discuss only a few major ideas in an effort to summarize current thought.

Probably the most attractive theory for explaining the fact that extensive glaciations have occurred only a few times in the geologic past comes from the theory of plate tectonics.[1] Not only does this theory provide geologists with explanations about many previously misunderstood processes and features, it also provides a possible explanation for some hitherto unexplainable climatic changes, including the onset of glacial conditions. Since glaciers can form only on the continents, we know that landmasses must exist somewhere in the higher latitudes before an ice age can commence. Many believe that ice ages have only

occurred when the earth's shifting crustal plates have carried the continents from tropical latitudes to more poleward positions.

Glacial features in present-day Africa, Australia, South America, and India indicate that these regions experienced an ice age near the end of the Paleozoic era, about 250 million years ago. For many years this puzzled scientists. Was the climate in these relatively tropical latitudes once like it is today in Greenland and Antarctica? Until the plate tectonics theory was formulated, there had been no reasonable explanation. Today scientists realize that the areas containing these ancient glacial features were joined together as a single supercontinent located at latitudes far to the south of their present positions. Later, this landmass broke apart and its pieces, each moving on a different plate, drifted toward their present locations (Figure 11.30). It is now believed that during the geologic past continental drift accounted for many dramatic climatic changes as landmasses shifted in relation to one another and moved to different latitudinal positions. Changes in oceanic circulation also must have occurred, altering the transport of heat and moisture and consequently the climate as well. Since the rate of plate movement is very slow—on the order of a few centimeters per year—appreciable changes in the positions of the continents occur only over great spans of geologic time. Thus, climatic changes brought about by continental drift are extremely gradual and happen on a scale of millions of years.

Since climatic changes brought about by moving plates are extremely gradual, the plate tectonics theory cannot be used to explain the alternation between glacial and interglacial climates that occurred during the Pleistocene epoch. Therefore we must look to some other triggering mechanism that may cause climatic change on a scale of thousands rather than millions of years. Today many scientists believe or strongly suspect that the climatic oscillations that characterized the Pleistocene may be linked to variations in the earth's orbit. This hypothesis was first developed and strongly advocated by the Yugoslavian scientist Milutin Milankovitch and is based on the premise that variations in incoming solar radiation are a principal factor in controlling the climate of the earth. Milankovitch formulated a comprehensive mathematical model based on the following elements:

---

[1] A brief overview of the theory appears in Chapter 1 and a more extensive discussion may be found in Chapter 16.

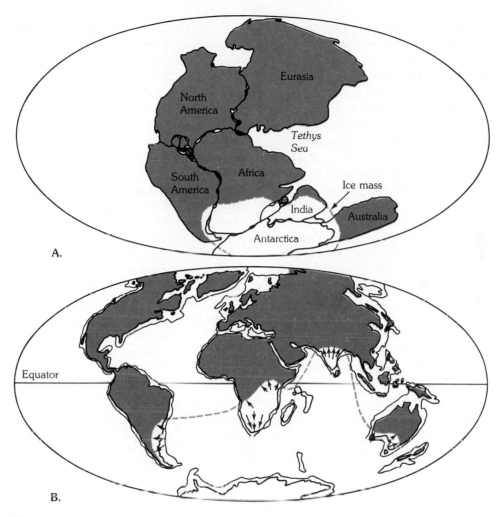

FIGURE 11.30
A. The supercontinent Pangaea showing the area covered by glacial ice 300 million years ago. **B.** The continents as they are today. The shading outlines areas where evidence of the ancient ice sheets exists. The dashed line joining the glaciated regions indicates how large the ice sheet would have had to be if the continents had been in their present positions at the time of glaciation. (A. and B. after R. F. Flint and B. J. Skinner, *Physical Geology,* 2nd ed., p. 418, New York: John Wiley & Sons, 1977)

1 Variations in the shape *(eccentricity)* of the earth's orbit about the sun;

2 Changes in *obliquity;* that is, changes in the angle that the axis makes with the plane of the earth's orbit; and

3 The wobbling of the earth's axis, called *precession.*

Using these factors, Milankovitch calculated variations in the receipt of solar energy and the corresponding surface temperature of earth back into time in an attempt to correlate these changes with the climatic fluctuations of the Pleistocene. In explaining climatic changes that result from these three variables, it should be noted that they cause little or no variation in the total amount of solar energy reaching the ground. Instead, their impact is felt because they change the degree of contrast between the seasons. Somewhat milder winters in the middle to high latitudes means greater snowfall totals, while cooler summers would bring a reduction in snowmelt.

Over the years the astronomical theory of Milankovitch has been widely accepted, then largely rejected, and now, in light of recent investigations, is once again very popular. Among recent studies that have added credibility and support to the astronomical theory is one in which deep-sea sediments containing certain climatically sensitive microorganisms were analyzed to establish a chronology of temperature changes going back nearly one-half million years.[1] This time scale of climatic change was then compared to astronomical calculations of eccentricity, obliquity, and precession to determine if a correlation did indeed exist. Although the study was very involved and mathematically complex, the conclusions were straightforward. The authors found that major variations in climate over the past several hundred thousand years were closely associated with changes in the geometry of the earth's orbit; that is, cycles of climatic change were shown to correspond closely with the periods of obliquity, precession, and orbital eccentricity. More specifically, they stated: "It is concluded that changes in the earth's orbital geometry are the fundamental cause of the succession of Quaternary ice ages."[1]

Let us briefly summarize the theories that were just described. The theory of plate tectonics provides us with an explanation for the widely spaced and nonperiodic onset of glacial conditions at various times in the geologic past, while the astronomical theory proposed by Milankovitch and recently supported by the work of J. D. Hays and his colleagues furnishes an explanation for the alternating glacial and interglacial episodes of the Pleistocene.

In conclusion, it should be emphasized at this point that the ideas just discussed do not represent the only possible explanations for glacial ages. Although interesting and attractive, these theories are certainly not without critics nor are they the only theories currently under study in an attempt to piece together the puzzle of the causes of ice ages. Other factors may and, in fact, probably do enter into the picture.

---

[1] J. D. Hays, John Imbrie, and N. J. Shackelton, "Variations in the Earth's Orbit: Pacemaker of the Ice Ages," *Science*, 194 (4270): 1121–32.

[1] J. D. Hays, et al, p. 1131. The term *Quaternary* refers to the period on the geologic calendar that encompasses the last few million years.

REVIEW QUESTIONS

1 What is a glacier? Under what circumstances does glacial ice form?

2 Where are glaciers found today? What percentage of the earth's land area do they cover? How does this compare to the area covered by glaciers during the Pleistocene?

3 Describe glacial flow. In an alpine glacier does all of the ice move at the same rate? Explain.

4 Why do crevasses form in the upper portion of a glacier but not below 50 meters?

5 Under what circumstances will the front of a glacier advance? Retreat? Remain stationary?

6 Describe the processes of glacial erosion.

7 How does a glaciated mountain valley differ from a mountain valley that was not glaciated?

8 List and describe the erosional features you might expect to see in an area where alpine glaciers exist or have recently existed.

9 What is glacial drift? What is the difference between till and stratified drift? What general effect do glacial deposits have on the landscape?

10 List the five basic moraine types. What do all moraines have in common? What is the significance of terminal and recessional moraines?

11 How do kettles form?

12 What direction was the ice sheet moving that affected the area shown in Figure 11.23? Explain how you were able to determine this.

13 What are ice-contact deposits? Distinguish between kames and eskers.

14 The development of the glacial theory is a good example of applying the principle of uniformitarianism. Explain briefly.

15 In North America four major stages of glaciation have been recognized. List them in the order they occurred.

16 During the Pleistocene epoch the amount of glacial ice in the Northern Hemisphere was about twice as great as in the Southern Hemisphere. Briefly explain why this was the case.

17 List three indirect effects of Ice Age glaciers.

18 How might plate tectonics help explain the cause of ice ages? Can plate tectonics explain the alternation between glacial and interglacial climates during the Pleistocene?

KEY TERMS

ablation (p. 264)

abrasion (p. 265)

alpine glacier (p. 257)

arête (p. 271)

basal slip (p. 261)

calving (p. 264)

cirque (p. 268)

col (p. 269)

continental glacier (p. 257)

crevasse (p. 261)

drift (p. 274)

drumlin (p. 276)

end moraine (p. 275)

esker (p. 279)

fiord (p. 269)

firn (p. 260)

glacial erratic (p. 274)

glacial striations (p. 265)

glacial trough (p. 266)

glacier (p. 257)

ground moraine (p. 276)

hanging valley (p. 266)

horn (p. 271)

ice-contact deposit (p. 279)

kame (p. 279)

kame terrace (p. 279)

kettle (p. 279)

lateral moraine (p. 276)

medial moraine (p. 276)

outwash plain (p. 279)

pater noster lakes (p. 268)

plastic flow (p. 261)

Pleistocene epoch (p. 282)

plucking (p. 265)

pluvial lake (p. 283)

recessional moraine (p. 276)

roche moutonnée (p. 273)

rock flour (p. 265)

snowline (p. 264)

stratified drift (p. 274)

surge (p. 262)

tarn (p. 268)

terminal moraine (p. 276)

till (p. 274)

tillite (p. 283)

truncated spur (p. 266)

valley train (p. 279)

zone of accumulation (p. 264)

zone of fracture (p. 261)

12

DESERTS
AND
WINDS

DESERTS

The word *desert* literally means *deserted* or *unoccupied*. Many desert areas are not truly deserted and indeed, in some cases, many people live there. Nevertheless, the world's dry regions are probably the least familiar land areas on earth outside of the polar realm. For example, one popular image of deserts is that they consist of mile after mile of drifting sand dunes. It is true that sand accumulations do exist in some areas and may be striking features, but they represent only a small percentage of the total desert area. In the Sahara, the world's largest desert, sand accumulations cover only 10 percent of the surface, while in the sandiest of all deserts, the Arabian, about 30 percent is sand covered. A more typical surface consists of barren rock or expanses of stoney ground.

Another commonly held but incorrect perception of dry lands is that they are practically lifeless. Although reduced in amount and different in character, plant life is usually present (Figure 12.1). Whereas desert plants differ widely from place to place, they all have one characteristic in common—they have developed adaptations that make them highly tolerant of drought. Such plants, called **xerophytes,** may have waxy leaves, stems, or branches that reduce water loss. Others have small leaves or none at all.

The San Luis Valley in southern Colorado. At the valley's eastern edge, at the base of the Sangre de Cristo Mountains, lie the Great Sand Dunes, some of the highest dunes in the world. (Photo by Stephen A. Trimble)

291

FIGURE 12.1
A scene in the Sonoran Desert near Tucson, Arizona. Dry environments are often far from being lifeless. (Photo by R. Scott Dunham)

292

Further, the roots of some species often extend to great depths in order to tap the moisture found there, while others produce a shallow but widespread root system that enables them to quickly absorb great amounts of moisture from the infrequent desert showers. Often the stems of desert plants are thickened by a spongy tissue that can store enough water to sustain the plant until the next rainfall. Thus, although they are often widely dispersed and provide little ground cover, plants of many kinds flourish in the desert.

DISTRIBUTION OF DRY LANDS

The desert (arid) and steppe (semiarid) regions of the world encompass about 48 million square kilometers, nearly one-third of the earth's land surface. No other climatic group covers so large a land area. The world map showing the distribution of desert and steppe regions reveals that dry lands are concentrated in the subtropics and in the middle latitudes (Figure 12.2).

The heart of the low-latitude dry climates lies in the vicinities of the Tropics of Cancer and Capricorn. A glance at Figure 12.2 shows a virtually unbroken desert environment stretching across North Africa to northwestern India. In addition to this single great expanse, lesser areas of subtropical desert include northern Mexico and the southwestern United States, parts of southern Africa, the west coast of South America, and a large portion of Australia. The existence of this dry subtropical realm is primarily the result of the prevailing global pattern of air pressure.

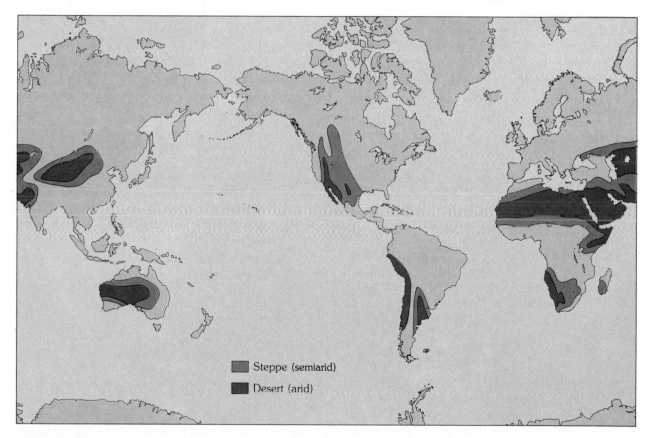

FIGURE 12.2
Arid and semiarid climates cover about 30 percent of the earth's land surface. No other climatic group covers so large an area.

That is to say, coinciding with the low-latitude dry regions are zones of high atmospheric pressure. These semipermanent pressure systems are characterized by dry subsiding air. Such conditions generally preclude cloud formation and precipitation.

Unlike their low-latitude counterparts, middle-latitude deserts and steppes are not controlled by the subsiding air masses associated with high pressure. Instead, these dry lands exist principally because of their position in the deep interiors of large land masses far removed from the oceans. In addition, the presence of high mountains across the paths of prevailing winds further acts to separate these areas from water-bearing, maritime air masses. In North America the Coast Ranges, Sierra Nevada, and Cascades are the foremost barriers, while in Asia, the great Himalayan chain prevents the summertime monsoon flow of moist Indian Ocean air from reaching into the interior. Because the Southern Hemisphere lacks extensive land areas in the middle latitudes, only a small area of desert and steppe are found in this latitude range existing primarily near the southern tip of South America in the rainshadow of the towering Andes.

WHAT IS MEANT BY "DRY"?

What is meant by the term *dry?* Sometimes it is arbitrarily defined by a single rainfall figure; for example, 25 centimeters per year of precipitation. However, the concept of dryness is a relative one that refers to any situation in which a water deficiency exists. Hence climatologists define **dry climate** as one in which yearly precipitation is not as great as the potential loss of water by evaporation. Dryness then is not only related to annual rainfall totals, but it is also a function of evaporation which, in turn, is closely dependent upon temperature. As temperatures climb, potential evaporation also increases. Twenty-five centimeters of rain may support only a sparce vegetative cover in Nevada while the same amount of precipitation falling in northern Scandinavia is sufficient to support forests.

From a climatic standpoint, perhaps the single most characteristic feature of deserts, aside from the fact that annual precipitation totals are small, is that the amount of rain received each year is very unreliable. Generally the smaller the average annual rainfall, the greater its variability. As a result, yearly aver-

ages are often misleading. There are usually more years when rainfall totals are below the average than above.

GEOLOGIC PROCESSES IN ARID CLIMATES

The angular hills, the sheer canyon walls, and the pebble or sand covered surface of the desert contrast sharply with the rounded hills and curving slopes of more humid places. Indeed, to a visitor from a humid region, a desert landscape may seem to have been shaped by forces different than those operating in well-watered areas. However, while the contrasts may be striking, they are not a reflection of different processes but merely the differing effects of the same processes operating under contrasting climatic conditions.

In humid regions relatively fine textured soils support an almost continuous cover of vegetation that mantles the surface. Here the slopes and rock edges are rounded. Such a landscape reflects the strong influence of chemical weathering in a humid climate. By contrast, much of the weathered debris in deserts consists of unaltered rock and mineral fragments— the result of mechanical weathering processes. In dry lands rock weathering of any type is greatly reduced because of the lack of moisture and the scarcity of organic acids from decaying plants. Chemical weathering, however, is not completely lacking in deserts. Over long spans of time clays and thin soils do form and many iron-bearing silicate minerals oxidize, producing the rust-colored stain found tinting some desert landscapes.

Most of the time, desert stream courses (called **washes** in the western United States) are dry (Figure 12.3A). This fact is often quite obvious even to the casual observer who, while traveling, notices the number of bridges with no streams beneath them or the number of dips in the road where dry washes cross. However, when the rare heavy showers do come, so much rain falls in such a short time that all of it cannot soak in. Since the vegetative cover is sparse, runoff is largely unhindered and consequently rapid, often creating flash floods along valley floors (Figure 12.3B). Such floods, however, are quite unlike floods in humid regions. A flood on a river like the Mississippi may take several days to reach its crest and then

A.

B.

FIGURE 12.3
Tanque Verde wash, Tucson, Arizona. **A.** During a dry period and **B.** following a rain.
(Photos by Tad Nichols)

to subside again, while desert floods arrive suddenly and likewise subside in a short time. Because much of the surface material is not anchored by vegetation, the amount of erosional work that occurs during one of these short-lived events is impressive.

Unlike the drainage in humid regions, stream courses in arid regions are seldom well integrated. That is, desert streams lack an extensive system of tributaries. In fact, a basic characteristic of deserts is that most of the streams that originate in them are small and die out before reaching the sea. Because the water table is usually far below the surface, few desert streams can draw upon it. Without a steady supply of water, evaporation soon depletes the stream and the remaining water sinks into the ground. The few permanent streams that do cross arid regions, such as the Colorado and Nile rivers, originate outside the desert, often in well-watered mountains. Here the water supply must be great to compensate for the losses occurring as the stream crosses the desert. For example, after the Nile leaves the rainy regions that are its source, it traverses almost 2000 kilometers of the Sahara without the contribution of a single tributary.

It should be emphasized that running water, although an infrequent occurrence, nevertheless does most of the erosional work in deserts. This is contrary to a commonly held belief that wind is the most important erosional agent sculpturing desert landscapes. Although wind erosion is more significant in dry areas than elsewhere, most desert landforms are nevertheless carved by running water. As we shall see shortly, the main role of wind is in the transportation and deposition of sediment, creating and shaping the ridges and mounds we call dunes.

TRANSPORTATION OF SEDIMENT BY WIND

Moving air, like moving water, is turbulent and able to pick up loose debris and transport it to other locations. Just as in a stream, the velocity of wind increases with height above the surface. Also like a stream, wind transports fine particles in suspension while heavier ones are carried as bed load. However, the transport of sediment by wind differs from that of running water in two significant ways. First, wind has

a low density compared to water; thus it is not capable of picking up and transporting coarse materials. Second, because wind is not confined to channels, it can spread sediment over large areas, as well as high into the atmosphere.

BED LOAD

The **bed load** carried by wind consists of sand grains. Observations in the field and experiments using wind tunnels indicate that windblown sand moves by skipping and bouncing along the surface—a process termed **saltation**. The term is not a reference to salt, but instead derives from the Latin word meaning "to jump." The movement of sand grains begins when wind reaches a velocity sufficient to overcome the inertia of the resting particles. At first, the sand rolls along the surface. Upon striking another grain, one or both of the grains may jump into the air. Once in the air, the sand is carried forward by the wind until gravity pulls the grain back toward the surface. When the sand hits the surface, it either bounces back into the air or dislodges other grains which then jump upward. In this manner a chain reaction is established, filling the air near the ground with saltating sand grains in a short period of time (Figure 12.4).

Bouncing sand grains never travel far from the surface. Even when winds are very strong, the height of the saltating sand seldom exceeds one meter and under less extreme conditions is usually confined to heights no greater than one-half meter (Figure 12.5). Some sand grains are too large to be thrown into the air by impact from other particles. When this is the case, the energy provided by the impact of the smaller saltating grains drives the larger grains forward. Estimates indicate that between 20 and 25 percent of the sand transported in a sandstorm is moved in this way.

SUSPENDED LOAD

Unlike sand, dust can be swept high into the atmosphere by the wind. Since dust is often composed of rather flat particles that have large surface areas when compared to the weight of the particle, it is relatively easy for turbulent air to counterbalance the pull of gravity and keep these fine particles suspended for extended periods. Although both silt and clay can be carried in suspension, silt commonly makes up the bulk of the **suspended load** because the reduced

FIGURE 12.4
A cloud of saltating sand grains moving up the gentle slope of a dune. (Photo by Stephen A. Trimble)

FIGURE 12.5
A granite outcrop undercut by the abrasive action of wind-blown sand. Since sand grains are never lifted far above the surface, only the lower part of the outcrop was affected. (Photo by K. Segerstrom, U.S. Geological Survey)

297

FIGURE 12.6
Dust storms like this one in Colorado were relatively common during the 1930s in the Dust Bowl region of the Great Plains. (Courtesy of the U.S. Department of Agriculture)

level of chemical weathering in deserts produces only small amounts of clay.

Fine particles are easily carried by the wind, but they are not easily acquired by the turbulent air. The reason is that the wind velocity is practically zero within a very thin layer close to the ground. Thus, the wind cannot lift the sediment by itself. Instead, the dust must be ejected or spattered into the moving currents of air by bouncing sand grains or other disturbances. This idea is illustrated nicely by a dry country road on a windy day. Left undisturbed, little dust is raised by the wind. However, as a car or truck moves over the road, the previously smooth layer of silt is disturbed, creating a thick cloud of dust.

Although the suspended load is usually deposited relatively near its source, high winds are capable of carrying large quantities of dust great distances (Figure 12.6). In the 1930s, silt picked up in Kansas was transported to New England and beyond into the North Atlantic. Similarly, dust blown from the Sahara has been traced as far as the West Indies.

FIGURE 12.7
A blowout, Sioux County, Nebraska. The remnant behind which the horse is standing indicates the level of the land prior to the formation of the blowout. (Photo by N. H. Darton, U.S. Geological Survey)

WIND EROSION

Compared to running water and moving ice, wind is a relatively insignificant erosional agent. Recall that even in deserts, few major erosional landforms are created by the wind. Although wind erosion is not restricted to arid and semiarid regions, it does its most effective work in these areas. In humid places moisture binds particles together and vegetation anchors the soil so that wind erosion is negligible. For wind to be effective, dryness and scanty vegetation are important prerequisites. When such circumstances exist, wind may pick up, transport, and deposit great quantities of fine sediment. During the 1930s parts of

FIGURE 12.8
Desert pavement composed of
angular rock fragments. (Photo
by Stephen A. Trimble)

the Great Plains experienced great dust storms. The plowing under of the natural vegetative cover for farming, followed by severe drought, made the land ripe for wind erosion, and led to the area being labeled the Dust Bowl.

One way that winds erode is by **deflation,** the lifting and removal of loose material. Although the effects of deflation are sometimes difficult to notice because the entire surface is being lowered at the same time, they can be significant. In portions of the 1930s Dust Bowl, vast areas of land were lowered by as much as one meter in only a few years.

The most noticeable results of deflation in some places are shallow depressions which are quite appropriately called **blowouts** (Figure 12.7). In the Great Plains region, from Texas north to Montana, thousands of blowouts are visible on the landscape. They range in size from small dimples less than one meter deep and three meters wide to depressions that approach 50 meters in depth and several kilometers across. The factor that controls the depths of these basins (that is, acts as base level) is the local water table. When blowouts are lowered to the water table, damp ground and vegetation prevent further deflation.

In portions of many deserts the surface is characterized by a layer of coarse pebbles and gravels that are too large to be moved by the wind. Such a layer, called **desert pavement,** is created as the wind lowers the surface by removing fine material until eventually only a continuous cover of coarse sediment remains (Figure 12.8). Once desert pavement becomes established, a process that may take hundreds of years, the surface is effectively protected from further deflation.

Like glaciers and streams, wind erodes by **abrasion.** In dry regions as well as along some beaches, windblown sand cuts and polishes exposed rock surfaces. However, abrasion is often credited for accomplishments beyond its actual capabilities. Such features as balanced rocks that stand high atop narrow pedestals, and intricate detailing on tall pinnacles are not the results of abrasion. Since sand seldom travels more than a meter above the surface, the wind's sandblasting effect is obviously quite limited in vertical extent. Abrasion by windblown sand, however, does create interestingly shaped stones called **ventifacts** (Figure 12.9). The side of the stone exposed to the prevailing wind is abraded, leaving it polished, pitted, and with sharp edges. If the wind is not

FIGURE 12.9
Ventifacts. (Photo by M. R.
Campbell, U.S. Geological
Survey)

consistently from one direction, or if the pebble becomes reoriented, it may have several faceted surfaces.

WIND DEPOSITS

Although wind is relatively unimportant as a producer of erosional landforms, wind deposits are significant features in some regions. Accumulations of windblown sediment are particularly conspicuous landscape elements in the world's dry lands and along many sandy coasts. Wind deposits are of two distinctive types: (1) mounds and ridges of sand from the wind's bed load and (2) extensive blankets of silt that once were carried in suspension.

SAND DEPOSITS

As is the case with running water, wind drops its load of sediment when wind velocity falls and the energy available for transport diminishes. Thus sand begins to accumulate wherever an obstruction across the path of the wind slows the movement of the air. Unlike many deposits of silt, which form blanket-like layers over large areas, winds commonly deposit

sand in mounds or ridges called **dunes** (Figure 12.10).

As moving air encounters an object, such as a clump of vegetation or a rock, the wind sweeps around and over it leaving a shadow of slower moving air behind the obstacle as well as a smaller zone of quieter air just in front of the obstacle. Some of the saltating sand grains moving with the wind come to rest in these wind shadows. As the accumulation of sand continues, it becomes a more imposing barrier to the wind and thus a more efficient trap for even more sand. If there is a sufficient supply of sand and the wind blows steadily for a long enough time, the mound of sand grows into a dune.

A profile of a dune shows an asymmetrical shape with the leeward slope being steep and the windward slope more gently inclined. Sand moves up the gentle slope on the windward side by saltation. Just beyond the crest of the dune, where the wind velocity is reduced, the sand accumulates. As more sand collects, the slope steepens and eventually some of it slides or slumps under the pull of gravity. In this way the leeward slope of the dune, called the **slip face,**

FIGURE 12.10
Crest of high dunes, Great Sand Dunes National Monument, Colorado. (Photo by
Stephen A. Trimble)

maintains an angle of about 34 degrees, the angle of
repose for loose dry sand.[1] Continued sand accumulation coupled with periodic slides down the slip face
result in the slow migration of the dune in the direction of air movement.

As sand is deposited on the slip face, layers form
which are inclined in the direction the wind is blowing. These sloping layers are called **cross beds** (Figure 12.11). When the dunes are eventually buried
under other layers of sediment and become part of
the sedimentary rock record, their asymmetrical
shape is destroyed, but the cross beds remain. By
studying the orientation of these beds, geologists can

[1]Recall from Chapter 8 that the angle of repose is the steepest
angle at which material remains stable.

FIGURE 12.11
Cross-bedding in the Navajo Sandstone, Zion National Park, Utah. (Photograph used by permission of Dennis Tasa)

determine the average direction of ancient winds. This information, together with other data, is then used in reconstructing climates of the geologic past. This knowledge of past climatic conditions, in turn, aids in determining earlier positions of the earth's moving lithospheric plates.

TYPES OF SAND DUNES

Although often complex, dunes are not just random heaps of sediment. Rather they are accumulations that usually assume patterns that are surprisingly consistent. Addressing this point, a leading early investigator of dunes, the British engineer R. A. Bagnold, observed: "Instead of finding chaos and disorder, the observer never fails to be amazed at a simplicity of form, an exactitude of repetition, and a geometric order unknown in nature on a scale larger than of crystalline structure. In places, vast accumulations of sand weighing millions of tons, move inexorably, in regular formation, over the surface of the country retaining their shape. . . ."

Barchan Dunes Solitary sand dunes shaped like crescents and with their tips pointing downwind are called **barchan dunes** (Figure 12.12). These dunes form where supplies of sand are limited and the surface is relatively flat, hard, and lacking vegetation. They migrate slowly with the wind at a rate of up to

FIGURE 12.12
Barchan dunes. The gentle slope is on the side from which the prevailing winds blow. (Aerial view by John S. Shelton. Side view by G. K. Gilbert, U.S. Geological Survey)

15 meters per year. Their size is usually modest with the largest barchans reaching heights of about 30 meters while the maximum spread between their horns approaches 300 meters. When the wind direction is nearly constant, the crescent form of these dunes is nearly symmetrical. However, when the wind direction is not perfectly fixed, one tip becomes longer than the other.

Transverse Dunes In regions where vegetation is sparse or absent and sand is very plentiful, the dunes form a series of long ridges that are separated by troughs and oriented at right angles to the prevailing wind. Because of this orientation, they are termed **transverse dunes** (Figure 12.13 on pages 304–5). Typically, many coastal dunes are of this type. In

addition, they are common in many arid regions where the extensive surface of wavy sand is sometimes called a sand sea.

Longitudinal Dunes Also called **seif dunes, longitudinal dunes** are long ridges of sand that form more or less parallel to the prevailing wind and where sand supplies are limited. Apparently the prevailing wind direction must vary somewhat, but still remain in the same quadrant of the compass. Although the smaller types are only three or four meters high and several tens of meters long, in some large deserts longitudinal

FIGURE 12.13 →
Transverse dunes in Great Sand Dunes National Monument, Colorado. (Photo by Stephen A. Trimble)

FIGURE 12.14
A vertical loess bluff near the Mississippi River in southern Illinois. (Photo by James E. Patterson)

dunes can reach great size. For example, in portions of North Africa, Arabia, and central Australia, these dunes may approach a height of 100 meters and extend for distances of more than 100 kilometers (62 miles).

Parabolic Dunes   Unlike the other dunes that have been described thus far, **parabolic dunes** form where vegetation partially covers the sand. The shape of these dunes resembles the shape of barchans except that their tips point into the wind rather than downwind. Parabolic dunes often form along coasts where there are strong onshore winds and abundant sand. If the sand's sparse vegetative cover is disturbed at some spot, deflation creates a blowout. Sand is then transported out of the depression and

deposited as a curved rim which grows higher as deflation enlarges the blowout.

LOESS

In some parts of the world the surface topography is mantled with deposits of windblown silt. Over periods of perhaps thousands of years dust storms deposited this material which is called **loess.** As can be seen in Figure 12.14, when loess is breached by streams or road cuts it tends to maintain vertical cliffs and lacks any visible layers. The distribution of loess indicates that there are two primary sources for this sediment: deserts and glacial outwash deposits. The thickest and most extensive deposits of loess in the world occur in western and northern China, where accumulations of 30 meters are not uncommon and thicknesses of more than 100 meters have been measured. It is this fine, buff-colored sediment that gives the Yellow River (Hwang Ho) and the adjacent Yellow Sea their names. The source of China's 800,000 square kilometers of loess are the extensive desert basins of central Asia.

In the United States, deposits of loess are significant in many areas, including South Dakota, Nebraska, Iowa, Missouri, and Illinois as well as portions of the Columbia Plateau in the Pacific Northwest. The correlation between the distribution of loess and important farming regions in the Midwest and eastern Washington is not just a coincidence, because soils derived from this wind-deposited sediment are among the most fertile in the world. Unlike the deposits in China, the loess in the United States, as well as in Europe, is an indirect product of glaciation, for its source was deposits of outwash. During the retreat of the glacial ice, many river valleys were choked with sediment that was provided by meltwater. Strong westerly winds sweeping across the barren floodplains picked up the finer sediment and dropped it as a blanket on the east side of the valleys. Such an origin is confirmed by the fact that loess deposits are thickest and coarsest on the lee side of such major glacier drainage outlets as the Mississippi and Illinois rivers and rapidly thin with increasing distance from the valleys. Furthermore, the angular mechanically weathered particles composing the loess are essentially the same as the rock flour produced by the grinding action of glaciers.

THE EVOLUTION OF A DESERT LANDSCAPE

Since arid regions typically lack permanent streams, they are characterized as having **interior drainage,** that is, a discontinuous pattern of intermittent streams that do not flow out of the desert to the ocean. In the United States, the dry Basin and Range region provides an excellent example. The region includes southern Oregon, all of Nevada, western Utah, southeastern California, as well as southern Arizona and New Mexico. The name Basin and Range is an apt description for this almost 800,000 square kilometer region, since it is characterized by more than 200 relatively small mountain ranges which rise 900–

1500 meters above the basins that separate them. In this region, as in others like it around the world, erosion is carried out for the most part without reference to the ocean (ultimate base level) because drainage is in the form of local interior systems. Even in areas where permanent streams flow to the ocean, few tributaries exist, and thus only a relatively narrow strip of land adjacent to the stream has sea level as the ultimate level of land reduction.

The block models shown in Figure 12.15 depict the stages of landscape evolution in a mountainous desert such as the Basin and Range region. During and following the uplift of the mountains, running water begins carving the elevated mass and depositing large quantities of debris in the basin. During this early stage relief is greatest, for as erosion lowers the mountains and sediment fills the basins, elevation differences diminish.

When the occasional torrents of water produced by sporadic rains move down the mountain canyons, they are heavily loaded with sediment. Emerging from the confines of the canyon, the runoff spreads over the gentler slopes at the base of the mountains and quickly loses velocity. Consequently most of its load is dumped within a short distance. The result is a cone of debris at the mouth of a canyon known as an **alluvial fan** (Figure 12.16). Since the coarsest material is dropped first, the head of the fan is steepest, having a slope of perhaps 10 to 15 degrees. Moving down the fan, the size of the sediment and the steepness of the slope decrease and merge imperceptibly with the basin floor. An examination of the fan's surface would likely reveal a braided channel pattern because of the water shifting its course as successive channels became choked with sediment. Over the years, a fan enlarges, eventually coalescing with fans from adjacent canyons to produce an apron of sediment called a **bajada** along the mountain front.

On the rare occasions when rainfall is abundant, streams may flow across the bajada to the center of the basin, converting the basin floor into a shallow **playa lake.** Playa lakes are temporary features that last only a few days or at best a few weeks before evaporation and infiltration remove the water. The dry, flat lake bed that remains is termed a **playa.** Playas are typically composed of fine silts and clays, and occasionally encrusted with salts precipitated during

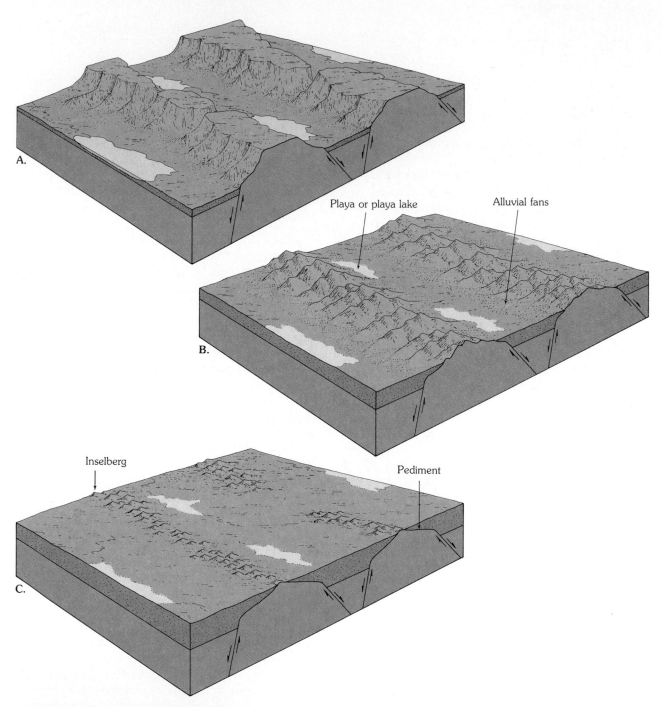

FIGURE 12.15
Stages of landscape evolution in a mountainous desert. As erosion of the mountains and deposition in the basins continue, relief diminishes.
A. Early stage. **B.** Middle stage. **C.** Late stage.

A.

FIGURE 12.16

A. Aerial view of alluvial fans in Death Valley, California. The size of the fan depends upon the size of the drainage basin. As these fans grow, they will eventually coalesce to form a bajada. (Photo courtesy of Fairchild Aerial Photograph Collection, Whittier College). B. A portion of the Furnace Creek, California, topographic map showing excellent alluvial fan development. (Courtesy of the U.S. Geological Survey)

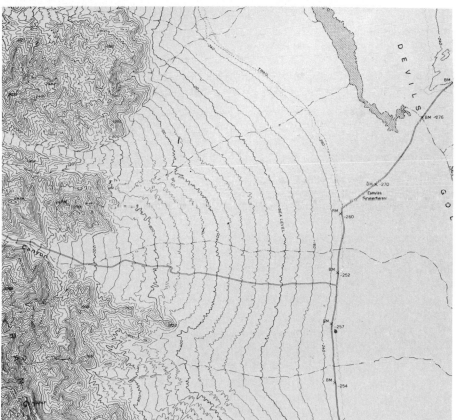

B.

FIGURE 12.17
A playa is a dry, flat lake bed. Following rains or periods of snowmelt in the adjacent
mountains, a shallow, temporary playa lake forms in the basin. As water evaporates,
salts are deposited. (Photo by C. B. Hunt, U.S. Geological Survey)

evaporation (Figure 12.17). These precipitated salts
may be unusual. A case in point is the sodium borate
(better known as borax) mined from ancient playa
lake deposits in Death Valley, California.

With time the mountain front is worn back and a
sloping bedrock platform, called a **pediment,** is created adjacent to the steep mountain front. A pediment is an erosional surface, usually covered by a
thin veneer of debris, that is formed by the action of
running water. Just how the water carves the pediment, however, is unclear and still a matter for
debate.

With the ongoing dissection of the mountain mass
into an intricate series of valleys and sharp divides as
well as the accompanying sedimentation, the local
relief continues to diminish. After more time passes,
the steady retreat of the mountain front enlarges the

pediment. Eventually this pediment growth results in
nearly the entire mountain mass being consumed.
Thus, by the late stages of erosion, the mountain
areas are reduced to a few large bedrock knobs projecting above the surrounding pediment and sediment-filled basin. These isolated erosional remnants
on an old-age desert landscape are called **inselbergs,**
a German word meaning "island mountains."

Each of the stages of landscape evolution in an
arid climate depicted in Figure 12.15 can be observed in the Basin and Range region. Recently
uplifted mountains in an early stage of erosion are
found in southern Oregon and northern Nevada.
Death Valley, California, and southern Nevada fit
into the more advanced middle stage, while the late
stage, with its inselbergs and extensive pediments,
can be seen in southern Arizona.

REVIEW QUESTIONS

1   "Most deserts consist of mile after mile of drifting sand dunes." True or False? Provide some examples to support your answer.

2   How extensive are the desert and steppe regions of the earth?

3   What is the primary cause of subtropical deserts? Of middle-latitude deserts?

4   In which hemisphere (northern or southern) are middle-latitude deserts most common?

5   Why is rock weathering reduced in deserts?

6   What is the most important erosional agent in deserts?

7   What is a wash?

8   Describe the way in which wind transports sand. During very strong winds, how high above the surface can sand be carried?

9   Why is wind erosion relatively more important in arid regions than in humid areas?

10   What factor limits the depths of blowouts?

11   How do sand dunes migrate?

12   Four major dune types are recognized. Indicate which type of dune is associated with each of the statements below.
   (a)   Sometimes called seif dunes.
   (b)   Dunes whose tips point into the wind.
   (c)   Long sand ridges oriented at right angles to the wind.
   (d)   Often form along coasts where strong winds create a blowout.
   (e)   Solitary dunes whose tips point downwind.
   (f)   Long sand ridges that are oriented more or less parallel to the prevailing wind.

13   Although sand dunes are the best-known wind deposits, accumulations of loess are very significant in some parts of the world. What is loess? Where are such deposits found? What are the origins of this sediment?

14   Why is sea level (ultimate base level) not a significant factor influencing erosion in desert regions?

15   Describe the features and characteristics associated with each of the stages in the evolution of a mountainous desert. Where in the United States can these stages be observed?

KEY TERMS

abrasion (p. 299)

alluvial fan (p. 307)

bajada (p. 307)

barchan dune (p. 302)

bed load (p. 296)

blowout (p. 299)

cross beds (p. 301)

deflation (p. 299)

desert pavement (p. 299)

dry climate (p. 294)

dune (p. 300)

inselberg (p. 310)

interior drainage (p. 307)

loess (p. 306)

longitudinal (seif) dune (p. 303)

parabolic dune (p. 306)

pediment (p. 310)

playa (p. 307)

playa lake (p. 307)

saltation (p. 296)

slip face (p. 300)

suspended load (p. 296)

transverse dune (p. 303)

ventifact (p. 299)

wash (p. 294)

xerophyte (p. 291)

13

SHORELINES

13

The waters of the ocean are constantly in motion. The restless nature of the water is more noticeable along the shore—the dynamic interface between land and sea. Here we can observe the rhythmic rise and fall of the tides and see the waves constantly rolling in and breaking. Sometimes the waves are low and gentle. At other times, they pound the shore with an awesome fury.

Although it may not be readily apparent to the occasional visitor, the shoreline is constantly being shaped and modified by the moving ocean waters. However, the nature of present-day shorelines is not just the result of the relentless attack of the land by the sea. Indeed, the shore is a complex zone whose unique character is the result of many geologic processes. For example, practically all coastal areas were affected by the worldwide rise in sea level that accompanied the melting of glaciers at the close of the Pleistocene epoch. As the sea edged landward, the shoreline became superimposed upon landscapes that had been shaped by such processes as stream erosion, glaciation, volcanic activity, and the forces of mountain building.

WAVES

Wind-generated waves provide most of the energy that shapes and modifies shorelines. Where the land and sea meet, waves that may have traveled unimpeded for hundreds or thousands of kilometers

Wave erosion is straightening this once-irregular shoreline. (Photo by Kenneth Hasson)

315

suddenly encounter a barrier that will not allow them to advance farther. Stated another way, the shore is the location where a practically irresistible force confronts an almost immovable object. The conflict that results is never-ending and sometimes dramatic.

The undulations of the water surface, called waves, derive their energy and motion from the wind. If a breeze of less than 3 kilometers (2 miles) per hour starts to blow across still water, small wavelets appear almost instantly. When the breeze dies, the ripples disappear as suddenly as they formed. However if the wind exceeds 3 kilometers per hour, more stable waves gradually form and progress with the wind.

All waves are described in terms of the characteristics illustrated in Figure 13.1. The tops of the waves are the *crests,* which are separated by *troughs.* The vertical distance between trough and crest is called the **wave height,** and the horizontal distance separating successive crests is the **wave length.** The **wave period** is the time interval between the passage of successive crests at a stationary point. The height, length, and period that are eventually achieved by a wave depend upon three factors: (1) the wind speed; (2) the length of time the wind has blown; and (3) the **fetch,** or distance that the wind has traveled across open water. As the quantity of energy transferred from the wind to the water increases, the heights of the waves increase as well. In the open ocean, wave heights of 1 to 4 meters are common, although storms may produce much higher waves. Because winds are often gusty and turbulent, the waves covering the surface of the ocean are often quite irregular in height and length.

When wind stops or changes direction or if waves leave the stormy area where they were created, they continue on without relation to local winds. The waves also undergo a gradual change to *swells,* which are lower and longer, and may carry a storm's energy to distant shores. Because many independent wave systems exist at the same time, the sea surface acquires a complex, irregular pattern. Hence the sea waves we watch from the shore are usually a mixture of swells from faraway storms and waves created by local winds.

An important point to remember is that in the open sea the motion of the wave is different from the motion of the water particles within it. It is the wave form that moves forward, not the water itself. Each water particle moves in a circular path during the passage of a wave (Figure 13.2). As a wave passes, a water particle returns almost to its original position. This is demonstrated by observing the behavior of a floating cork as a wave passes. The cork merely seems to bob up and down and sway slightly to and fro without advancing appreciably from its original position.[1] Because of this, waves in the open sea are called **waves of oscillation.** The energy contributed by the wind to the water is transmitted not only along the surface of the sea but downward as well. Friction causes a progressive loss of energy with an increase in depth, until at a depth equal to about one-half the wave length, the movement of water particles becomes negligible. This is shown by the rapidly diminishing diameters of water-particle orbits in Figure 13.2.

As long as a wave is in deep water it is unaffected by water depth. However when a wave approaches the shore the water becomes shallower and influences wave behavior. The wave begins to "feel bottom" at a water depth equal to about one-half its wave length. Since some energy is used in moving small particles of sediment back and forth, the wave slows. As the wave continues to advance toward the shore the slightly faster seaward waves catch up, decreasing the wave length. As the speed and length of the wave diminish, the wave steadily grows higher. Finally a critical point is reached when the steep wave front is unable to support the wave, and it collapses, or *breaks* (Figure 13.3). What had been a wave of oscillation now becomes a **wave of translation** in which the water advances up the shore. The turbulent water created by breaking waves is called **surf.**

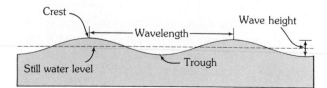

FIGURE 13.1
Characteristics of a wave.

---

[1]The wind does drag the water slightly forward, causing the surface circulation of the oceans.

Following the uprush of water onto the beach, a seaward backwash occurs. The water from expended waves most commonly moves seaward in a broad sheet that produces the undertow so often felt by swimmers. Sometimes the backwash occurs in narrow, localized channels and if strong enough, it may pull swimmers into deep water.

WAVE EROSION

During periods of calm weather wave action is minimal. However, just as streams do most of their work during floods, so too waves do most of their work during storms. The impact of high, storm-induced waves against the shore can at times be awesome in its violence. Each breaking wave may hurl thousands of tons of water against the land, sometimes causing the earth to literally tremble. The pressures exerted by Atlantic waves, for example, average nearly 10,000 kilograms per square meter (more than 2000 pounds per square foot) in winter. During storms the force is even greater. During one such storm a 1350-ton portion of a steel and concrete breakwater was ripped from the rest of the structure and moved to a useless position toward the shore at Wick Bay, Scot-

land. Five years later the 2600-ton unit that replaced the first met a similar fate. There are many such stories that demonstrate the great force of breaking waves. It is no wonder then that cracks and crevices are quickly opened in cliffs, seawalls, breakwaters, and anything else that is subjected to these enormous shocks. Water is forced into every opening, causing air in the cracks to become highly compressed by the thrust of crashing waves. When the wave subsides, the air expands rapidly, dislodging rock fragments and enlarging and extending pre-existing fractures. One especially dramatic example of the force of waves may be seen along some cliffed coasts where *blowholes* exist (Figure 13.4). During storms, water and air are forced into the seaward opening and emerge at the surface as a gusherlike column of water. The noise accompanying these upward bursts of spray resembles that of an erupting geyser.

In addition to the erosion caused by wave impact and pressure, **abrasion,** the sawing and grinding action of the water armed with rock fragments, is also important. In fact, abrasion is probably more intense in the surf zone than in any other environment. Smooth, rounded stones and pebbles along the shore are obvious reminders of the grinding action of rock against rock in the surf zone. Further, such

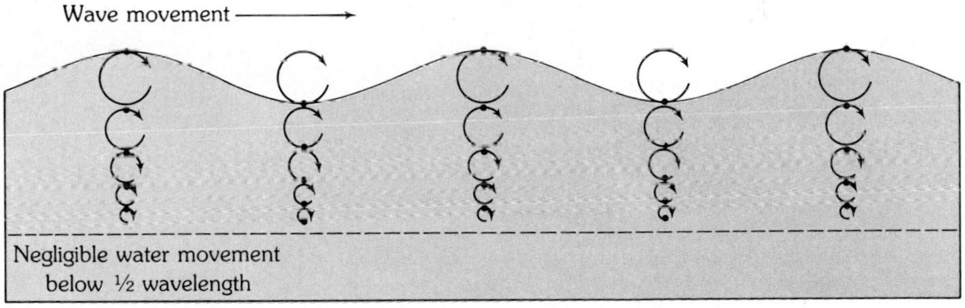

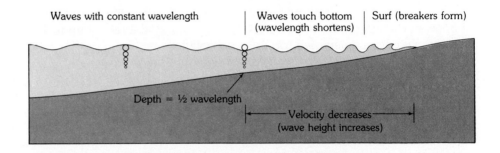

FIGURE 13.2
Movement of water particles with the passage of a wave.

Wave movement ———→

Negligible water movement below ½ wavelength

FIGURE 13.3
Changes that occur when a wave moves onto shore.

Waves with constant wavelength | Waves touch bottom (wavelength shortens) | Surf (breakers form)

Depth = ½ wavelength

Velocity decreases (wave height increases)

FIGURE 13.4
The pounding waves coupled with the bombardment of a cliff by rock fragments.sometimes produce a sea cave. As the sea cave extends into the cliff, it narrows and may erode to the surface near the shore. Such an opening is called a blowhole. When waves crash into the seaward opening of the cave, a column of spray emerges from the blowhole with the noise of an erupting geyser. (Photo by James E. Patterson)

FIGURE 13.5
Cliff undercut by wave erosion north of Guaymas, Sonora, Mexico. (Courtesy of Ward's Natural Science Establishment, Inc., Rochester, N.Y.)

fragments are used as "tools" by the waves as they cut horizontally into the land (Figure 13.5).

Along shorelines composed of unconsolidated material rather than hard rock, the rate of erosion by breaking waves can be extraordinary (Figure 13.6). In parts of Britain, where waves have the easy task of eroding glacial deposits of sand, gravel, and clay, the coast has been worn back 3 to 5 kilometers since Roman times, sweeping away many villages and ancient landmarks. A similar retreat may be seen along the cliffs of Cape Cod, which in places are retreating at a rate of up to 1 meter per year.

WAVE REFRACTION

Most waves approach a shoreline at an angle. However, when they reach the shallow water of a smoothly sloping bottom they are bent and tend to become parallel to the shore. Such bending of the waves is

FIGURE 13.6
This abandoned cottage was once relatively safe from wave attack. As the shoreline retreated, the cottage was exposed to the full force of the sea. (Photo by Kenneth Hasson)

called **refraction** (Figure 13.7). The part of the wave nearest the shore touches bottom and slows down first, while the end that is still in deep water continues forward at its regular speed. The net result is a wave front that may approach nearly parallel to the shore regardless of the original direction of the wave.

Due to refraction, wave impact is concentrated against the sides and ends of headlands projecting into the water, while wave attack is weakened in bays. This differential wave attack along irregular coastlines is illustrated in Figure 13.8. Since the waves reach the shallow water in front of the headland sooner than they do in adjacent bays, they are bent more nearly parallel to the protruding land and strike it from all three sides. Over a long period this process will straighten irregular coastlines.

FIGURE 13.7
Wave bending around the end of a beach at Stinson Beach, California. (Photo by
James E. Patterson)

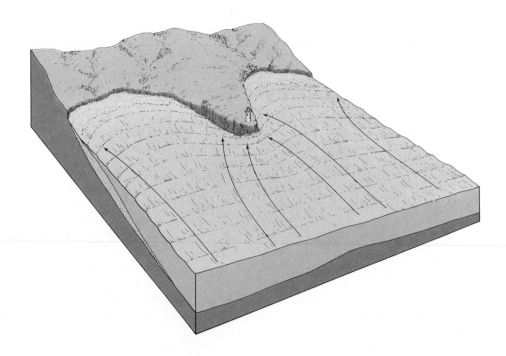

FIGURE 13.8
Wave refraction along an irregular shoreline.

BEACH DRIFT AND LONGSHORE CURRENTS

Although waves are refracted, most still reach the shore at some angle, however slight. Consequently, the uprush of water from each breaking wave is oblique. However, the backflow is in the direction of the slope of the beach. The effect of this pattern of water movement is to transport particles of sediment in a zigzag pattern along the beach (Figure 13.9). This movement is called **beach drift,** and it can transport sand and pebbles hundreds or even thousands of meters each day.

Oblique waves also produce currents within the surf zone that flow parallel to the shore. Since the water here is turbulent, these **longshore currents** easily move the fine suspended sand as well as roll larger sand and gravel along the bottom. When the sediment transported by longshore currents is added to the quantity moved by beach drift, the total amount can be very large. At Sandy Hook, New Jersey, for example, the quantity of sand transported along the shore over a 48-year period averaged almost 750,000 tons per year. For a 10-year period at Oxnard, California, more than 1.5 million tons of sediment moved along the shore each year.

There should be little wonder that beaches have been characterized as "rivers of sand." At any point along a beach there is likely to be more sediment that was derived elsewhere than material eroded from the shore area immediately behind it. It is also worth noting that much of the sediment composing beaches is not wave-eroded debris. Rather, in many areas sediment-laden rivers that discharge into the ocean are the major source of material. Hence, if it were not for beach drift and longshore currents, many beaches would be nearly sandless.

HUMAN INTERFERENCE WITH SHORELINE PROCESSES

The natural movement of sand by longshore currents and beach drift sometimes creates problems for those who live along the shore. Sand is either being eroded from a place where people want it to remain or it is being deposited where it is not wanted. In many coastal areas remedies for these problems include the building of such artifical structures as jetties, groins, and breakwaters. In many cases, however, these structures interfere with normal beach processes and interrupt the movement of sand. Such interference

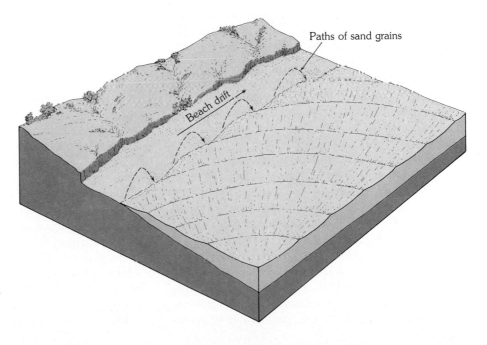

FIGURE 13.9
Beach drift, caused by the uprush of water from oblique waves.

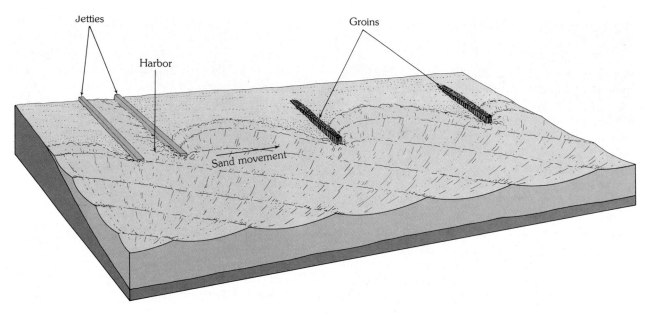

FIGURE 13.10
Jetties and groins trap sand that would otherwise be moved down the beach by the action of waves.

can create many new problems and result in unwanted changes that are very expensive to correct.

Jetties are usually built in pairs and extend into the ocean at the entrances to rivers and harbors. By confining the flow of water to a narrow zone, the ebb and flow caused by the rise and fall of the tides keep the sand in motion and prevent deposition in the channel. However, as illustrated in Figure 13.10, the jetty may act as a dam against which the longshore current and beach drift deposit sand. At the same time, wave activity removes sand on the other side. Since the other side is not receiving any new sand, there is soon no beach at all.

To maintain or widen beaches that lost sand from the action of longshore currents and beach drift, **groins** are often constructed. Groins are short walls built at right angles to the shore so as to trap moving sand (Figure 13.10). These structures often do their job so effectively that the longshore current beyond the groin is sand deficient. As a result, the current removes sand from the beach on the leeward side of the groin. To offset this effect, property owners downcurrent from the structure may erect a groin on their property. In this manner, the number of groins

multiplies. An example of such proliferation is the shoreline of New Jersey, where more than three hundred such structures have been built. Since it has been shown that groins often do not provide a satisfactory solution, they are no longer the preferred method of keeping beach erosion in check. In some areas, a system of *beach nourishment* is used. This simply involves the periodic addition of sand to the beach system. The source of the sand may be the bottom of a nearby lagoon or inland dunes. In some cases, sand is trucked in and added to the beach. In other instances, the sand is added at an upstream location to be distributed down the coast by wave activity. It should be noted, however, that beach nourishment can be an expensive solution. When 24 kilometers of Miami Beach were recently replenished, the cost was $64 million. Furthermore, in some instances, beach nourishment can lead to unwanted environmental effects. For example, beach replenishment at Waikiki Beach, Hawaii, involved replacing coarse calcareous sand with softer, muddier calcareous sand. Destruction of the soft beach sand by breaking waves increased the water's turbidity and killed offshore coral reefs. At Miami Beach, where

A.

B.

FIGURE 13.11
A. The shoreline at Santa Monica pier as it appeared in 1931. **B.** The same area in 1949. The construction of the breakwater disrupted longshore transport and caused the seaward growth of the beach. (Photos courtesy of Fairchild Aerial Photography Collection, Whittier College)

quartz sand was replaced by calcareous sand, the increased turbidity damaged local coral communities.

In order to create a quiet water area near shore to protect boats from the force of large breaking waves, a **breakwater** may be constructed parallel to the shoreline. However, when this is done, the reduced wave activity along the shore behind the structure may allow sand to accumulate. If this happens, the marina will eventually fill with sand while the downstream beach erodes and retreats. At Santa Monica, California, where the building of a breakwater created such a problem, the city had to install a dredge to remove sand from the protected quiet water zone and deposit it down the beach where longshore currents and beach drift could recirculate the sand (Figure 13.11).

As the foregoing examples illustrate, whenever people interfere with natural shoreline processes, the beach system responds. Clearly, any human action that does not consider the potential effects on down-shore areas only results in more problems. Deposition at one site leads to erosion elsewhere and the configuration of the beach is changed.

Sometimes human activities far from the coast can also have a dramatic impact on a shoreline. Recall that the supply of sand for many beaches comes from rivers flowing into the sea. Waves then transfer this sediment along the shoreline by beach drift and longshore currents. However, when a dam is built on a river, its load of sediment, which once supplied the beach, is intercepted and deposited in a reservoir. Although little new sand is being added, the work of waves along the shore continues. The result is increased beach erosion. Without a supply of sand, the beach will eventually disappear.

SHORELINE FEATURES

Along the rugged and irregular New England coast or along the steep shorelines of the West Coast, the effects of wave erosion are often easy to see.

FIGURE 13.12
Elevated wave-cut platform along the California coast near San Francisco. A new platform is being created at the base of the cliff. (Photo by John S. Shelton)

FIGURE 13.13
Sea arch and sea stack along the California coast west of Santa Cruz. (Photo by John S. Shelton)

Wave-cut cliffs, as their name implies, originate by the cutting action of the surf against the base of coastal land. As erosion progresses, rocks overhanging the notch at the base of the cliff crumble into the surf and the cliff retreats (Figure 13.5). A relatively flat, benchlike surface, the **wave-cut platform,** is left behind by the receding cliff (Figure 13.12). The platform broadens as wave attack continues. Some of the debris produced by the breaking waves remains along the water's edge as part of the beach, while the remainder is transported farther seaward.

Headlands that extend into the sea are vigorously attacked by waves because of refraction. The surf erodes the rock selectively, wearing away the softer or more highly fractured rock at the fastest rate. At first, sea caves may form. When two caves on opposite sides of a headland unite, a **sea arch** results. Finally the arch falls in, leaving an isolated remnant, or **sea stack,** on the wave-cut platform (Figure 13.13). Eventually it too will be consumed by the action of the waves.

Where beach drift and longshore currents are active, several features related to the movement of sed-iment along the shore may develop. **Spits** are elongated ridges of sand that project from the land into the mouth of an adjacent bay (Figure 13.14). Often the end in the water hooks landward in response to wave-generated currents. The term **baymouth bar** is applied to a sand bar that completely crosses a bay, sealing it off from the open ocean (Figure 13.15A). Such a feature tends to form across bays where currents are weak, allowing a spit to extend to the other side. A **tombolo,** a ridge of sand that connects an island to the mainland or to another island (Figure 13.15B), forms in much the same manner as a spit.

The gently sloping coastline found along the Gulf Coast and much of the eastern shore of the United States south of New York City is frequently characterized by **barrier islands,** which are low offshore ridges of sand that parallel the coast (Figure 13.16). The lagoons separating these narrow islands from the shore represent zones of relatively quiet water that allow small craft traveling between New York and northern Florida to avoid the rough waters of the North Atlantic.

 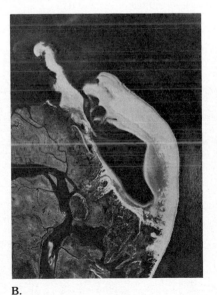 

A.   B.   C.

FIGURE 13.14
This sequence of aerial photographs shows the growth of a spit at Little Egg Inlet, New Jersey, over a 23-year period. **A.** In 1940, the spit was just beginning to develop. **B.** By 1957, the spit was about 400 meters wide and 1600 meters long. **C.** By 1963, the curving spit had grown 300 meters longer. (Photos courtesy of the U.S. Geological Survey)

A.

B.

FIGURE 13.15
These features were created because sediment was moved by beach drift and long-shore currents. **A.** Baymouth bar. (Photo by E. C. Stebinger, U.S. Geological Survey). **B.** Tombolo. (Photo by E. S. Bastin, U.S. Geological Survey)

FIGURE 13.16
Barrier islands, Cape Hatteras, North Carolina. (Photo courtesy of NASA)

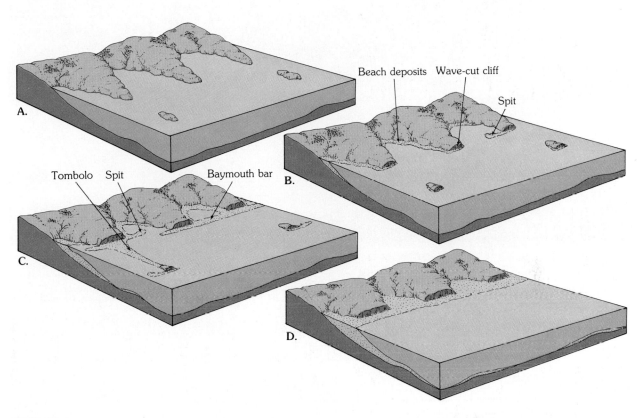

FIGURE 13.17
Development of an initially irregular coastline.

How barrier islands originate is still not certain. They possibly form in three or more ways. Some are thought to have originated as spits that were subsequently severed from the mainland by wave erosion or by the general rise in sea level following the last episode of glaciation. Some barrier islands may have been created when turbulent waters in the line of breakers heaped up sand that was scoured from the bottom. Since these sand barriers rise above normal sea level, the piling up of sand likely resulted from the work of storm waves at high tide. Finally, some studies suggest that barrier islands may be former sand dune ridges that originated along the shore during the last glacial period, when sea level was lower. When the ice sheets melted, sea level rose and flooded the area behind the beach-dune complex.

There is little question that a shoreline soon undergoes modification regardless of its initial configuration. At first most coastlines are irregular, although the degree of and reason for the irregularity may vary considerably from place to place. Along a coastline that is characterized by varied geology, the pounding surf may at first increase its irregularity because the waves will erode the weaker rocks more easily than the stronger ones. However it is commonly agreed that if a shoreline remains stable, marine erosion and deposition will eventually produce a more regular coast. Figure 13.17 illustrates the evolution of an initially irregular coast. As waves erode the headlands, creating cliffs and a wave-cut platform, sediment is carried along the shore. Some material is deposited in the bays, while other debris is formed into spits and baymouth bars. At the same time rivers fill the bays with sediment. Ultimately a smooth coast results.

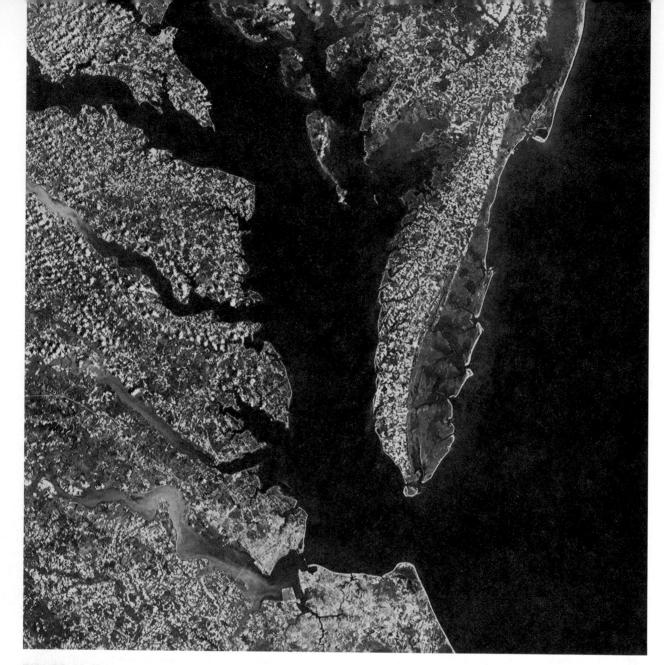

FIGURE 13.18
Satellite image of a portion of the East Coast showing Chesapeake Bay, an estuary created when a river mouth was submerged by a rise in sea level. (Courtesy of NASA)

EMERGENT AND SUBMERGENT COASTS

The great variety of present-day shorelines suggests that they are complex areas. Indeed, to understand the nature of any particular coastal area, many factors must be considered, including rock types, size and direction of waves, frequency of storms, tidal range,

and submarine profile. Moreover, recent tectonic events and changes in sea level must also be taken into account. These many variables make shoreline classification difficult.

One way that many geologists classify coasts is based upon changes that have occurred with respect to sea level. This commonly used but incomplete classification divides coasts into the two categories of emergent and submergent. **Emergent coasts** develop either because an area has been uplifted or as a result of a drop in sea level. Conversely, **submergent coasts** are created when sea level rises or the land adjacent to the sea subsides.

In some areas the coast is clearly emergent because rising land or a falling water level exposes wave-cut cliffs and platforms above sea level. Excellent examples include portions of coastal California where uplift has occurred in the recent geological past. The elevated wave-cut platform shown in Figure 13.12 illustrates this situation. In the case of the Palos Verdes Hills, south of Los Angeles, seven different terrace levels exist, indicating seven episodes of uplift. The ever-persistent sea is now cutting a new platform at the base of the cliff. If uplift follows, it too will become an elevated marine terrace.

Other examples of emergent coasts include regions that were once buried beneath great ice sheets. When glaciers were present, their weight depressed the crust; when the ice melted, the crust began to gradually spring back. Consequently, prehistoric shoreline features may now be found high above sea level. The Hudson Bay region of Canada is such an area, portions of which are still rising at a rate of more than one centimeter per year.

In contrast to the preceding examples, other coastal areas show definite signs of submergence. The shoreline of a coast that has been submerged in the relatively recent past is often highly irregular because the sea typically floods the lower reaches of river valleys. The ridges separating the valleys, however, remain above sea level and project into the sea as headlands. These drowned river mouths, which are often called **estuaries,** characterize many coasts today. Along the Atlantic Coast, the Chesapeake and Delaware bays are examples of large estuaries created by submergence (Figure 13.18). The picturesque coast of Maine, particularly in the vicinity of Acadia National Park, is another excellent example of an area that was flooded by the post-glacial rise in sea level and transformed into a highly irregular coastline.

It should be kept in mind that most coasts have complicated geologic histories. With respect to sea level, many have at various times emerged and then submerged. Each time they may retain some of the features created during the previous situation.

TIDES

Tides are periodic changes in the elevation of the ocean surface at a specific location. Their rhythmic rise and fall along coastlines have been known since antiquity, and other than waves, they are the easiest ocean movements to observe (Figure 13.19). Although known for centuries, tides were not explained satisfactorily until Sir Isaac Newton applied the law of gravitation to them. Newton showed that there is a mutual attractive force between two bodies, and that since oceans are free to move, they are deformed by this force. Hence tides result from the gravitational attraction exerted upon the earth by the moon, and to a lesser extent by the sun.

To illustrate how tides are produced, we will assume that the earth is a rotating sphere covered to a uniform depth with water (Figure 13.20). It is easy to see how the moon's gravitational force can cause the water to bulge on the side of the earth nearest the moon. In addition, however, an equally large tidal bulge is produced on the side of the earth directly opposite the moon. Both tidal bulges are caused, as Newton discovered, by the pull of gravity, a force that is inversely proportional to the square of the distance between the two objects. In this case the objects involved are the moon and the earth. Because the force of gravity decreases as distance increases, the moon's gravitational pull on the earth is slightly greater on the near side of the earth than on the far side. The result of this differential pulling is to stretch (elongate) the earth. Although the solid earth is stretched by the moon's gravitational pull, the amount of elongation is very slight. However, the world's oceans, which are mobile, are deformed quite dramatically by this effect to produce the two opposing tidal bulges.

Since the position of the moon changes slightly in a single day, it is the tidal bulges that remain in place while the earth rotates beneath them. Therefore the earth will carry an observer at any given place alternately into areas of deeper and shallower water. When he is carried into the regions of deeper water, the tide rises, and as he is carried away, the tide falls. Therefore, during one day the observer would experience two high tides and two low tides. In addition to the earth rotating, the tidal bulges also move as the moon revolves about the earth every 28 days. As a

A.

B.

FIGURE 13.19
A. High tide and **B.** low tide in the Bay of Fundy at Parrsboro, Nova Scotia. (Photos by G. Blouin, Information Canada Photothèque)

result, the tides, like the time of moonrise, occur about 50 minutes later each day. After 28 days one cycle is complete and a new one begins.

There may be an inequality between the high tides during a given day. Depending upon the position of the moon, the tidal bulges may be inclined to the equator as in Figure 13.20B. This figure illustrates that an observer in the Northern Hemisphere experiences a high tide on the side of the earth under the moon that is considerably higher than the high tide half a day later. On the other hand, a Southern Hemisphere observer would experience the opposite effect.

The sun also influences the tides, but because it is so far away, the effect is considerably less than that of the moon. In fact, the tide-generating potential of the sun is slightly less than half that of the moon. Near the times of new and full moons, the sun and moon are

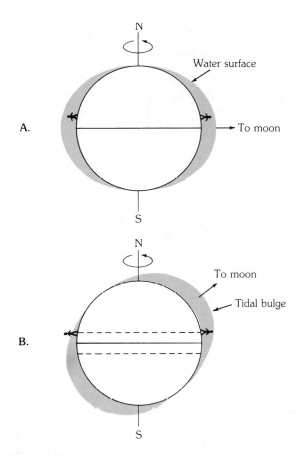

FIGURE 13.20
Tidal bulges on an ocean covered earth with the moon **A.** in the plane of the earth's equator and **B.** above the plane of the equator. In the latter situation the observer experiences unequal high tides.

aligned and their forces are added together (Figure 13.21A). Accordingly, the two tide-producing bodies cause higher tidal bulges (high tides) and lower tidal troughs (low tides). These are called the **spring tides.** Spring tides create the largest daily tidal range, that is, the largest variation between high and low tides. Conversely, at about the time of the first and third quarters of the moon, the gravitational forces of the moon and sun on the earth are at right angles, and each partially offsets the influence of the other (Figure 13.21B). As a result, the daily tidal range is less. These are the **neap tides.**

Although the discussion thus far explains the basic causes and patterns of tides, these theoretical considerations cannot be used to predict either the height or the time of actual tides at a particular place. The shape of coastlines and the configuration of ocean basins greatly influence the tide. Consequently, tides at various places respond differently to the tide-producing forces. This being the case, the nature of the tide at any location can be determined most accurately by actual observation. The predictions in tidal tables and the tidal data on nautical charts are based upon such observations.

Tidal current is the term used to denote the horizontal flow of water accompanying the rise and fall of the tide. As the tide rises, water flows in toward the shore as a **flood tide,** submerging the low-lying coastal zone. When the tide falls, a reverse flow, the **ebb tide,** again exposes the drowned portion of the shore. The areas affected by these alternating tidal currents are **tidal flats.** Depending upon the nature of the coastal zone, tidal flats vary in size from narrow strips lying seaward of the beach to extensive zones that may extend for several kilometers.

Although tidal currents are not important in the open sea, tides may create rapid currents in bays, river estuaries, straits, and other narrow places. Off the coast of Brittany, for example, tidal currents which accompany a high tide of 12 meters (40 feet) may attain a speed of 20 kilometers (12 miles) per hour. While tidal currents are not generally believed to be major agents of erosion and sediment transport, notable exceptions occur where tides move through narrow inlets. Here they constantly scour the small entrances to many good harbors that would otherwise be blocked.

TIDES AND THE EARTH'S ROTATION

The tidal drag of the moon is steadily slowing the earth's rotation. The rate of slowing, however, is not great. It has been estimated that the weak but steady braking action of the tide is slowing the earth's rotation at a rate between one second per day per 120,000 years and one second per day per 100,000 years. Although this may seem inconsequential, over millions of years this small effect will become very large. Eventually, thousands of millions of years into the future, rotation will cease and the earth will no longer have alternating days and nights.

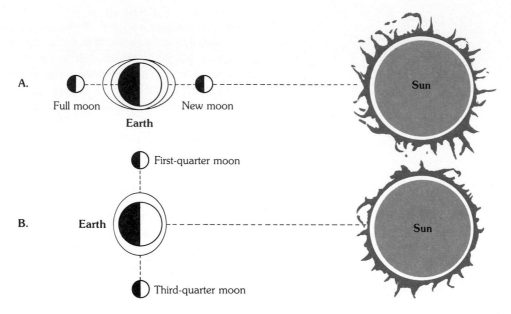

FIGURE 13.21
Relationship of the moon and sun to the earth during **A.** spring tides and **B.** neap tides.

FIGURE 13.22
A 375-million-year-old fossil coral from the Sulphur Springs Range in Nevada. Growth lines on corals such as this one enabled scientists to determine the number of days in an ancient year. (Photo by C. W. Merrian, U.S. Geological Survey)

If the earth is slowing, the length of each day must have been shorter and the number of days per year must have been greater in the geologic past. By studying fossil corals and clam shells, geologists have shown that this is indeed the case. By counting the number of daily growth rings of some well-preserved fossil specimens, the number of days per year can be ascertained (Figure 13.22). Studies using this ingenious technique indicate that 350 million years ago a year had between 400 and 410 days, while 280 million years ago there were 390 days in a year. These figures closely agree with current estimates of the earth's slowing rotation.

TIDAL POWER

With increased public interest in the rising costs and eventual depletion of petroleum, greater attention is being focused upon alternate energy sources. Although several methods of generating electrical energy from the oceans have been proposed, the ocean's energy potential remains largely untapped.

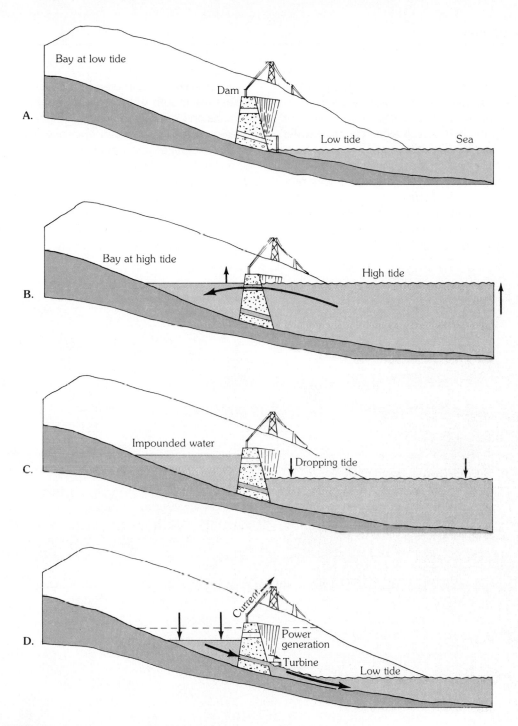

FIGURE 13.23
Simplified diagram showing the principle of the tidal dam. (After John J. Fagan, *The Earth Environment*, Englewood Cliffs, N.J.: Prentice-Hall, 1974. Reprinted by permission)

The development of tidal power is the principal example of energy production from the ocean.

Tides have been used as a source of power for centuries. Beginning in the twelfth century, water wheels driven by the tides were used to power gristmills and sawmills. During the seventeenth and eighteenth centuries, much of Boston's flour was produced at a tidal mill. Today, far greater energy demands must be met and more sophisticated ways of using the force created by the perpetual rise and fall of the ocean must be employed.

Tidal power is harnessed by constructing a dam across the mouth of a bay or an estuary in a coastal area having a large tidal range (Figure 13.23). The narrow opening between the bay and the open ocean magnifies the variations in water level that occur as the tides rise and fall. The strong in-and-out flow that results at such a site is then used to drive turbines and electrical generators.

Tidal energy utilization is exemplified by the tidal power plant at the mouth of the Rance River in France (Figure 13.24). By far the largest yet con-

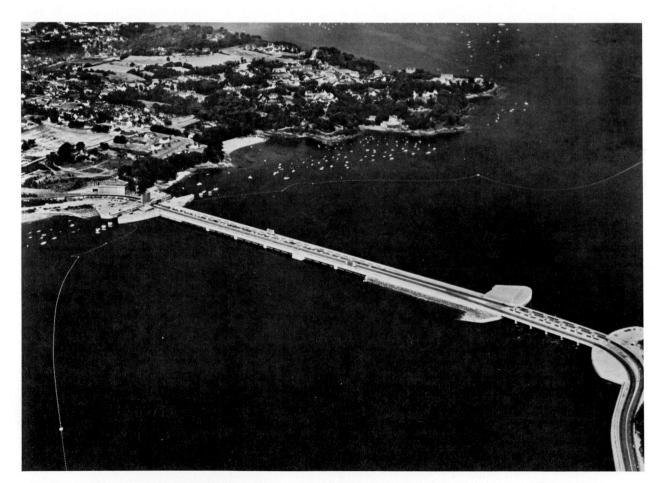

FIGURE 13.24
The world's first tidal power station to produce electricity was built across the Rance River estuary in 1966. The tide which rushes up this estuary on the northern coast of Brittany is one of the highest in the world, reaching 13.5 meters. (Courtesy of French Embassy Press and Information Division)

structed, this plant went into operation in 1966 and produces enough power to satisfy the needs of Brittany and also contribute to the demands of other regions. Much smaller experimental facilities near Murmansk in the Soviet Union and near Taliang in China are also being used to generate electricity. The United States has not yet tapped its tidal power potential, although a site at Passamaquoddy Bay in Maine, where the tidal range approaches 15 meters (50 feet), has been under review for more than 50 years.

Along most of the world's coasts it is not possible to harness tidal energy. If the tidal range is less than 8 meters (25 feet) or narrow, enclosed bays are absent, tidal power development is uneconomical. For this reason, the tides will never provide a very high proportion of our ever-increasing electrical energy requirements. Nevertheless, the development of tidal power may be worth pursuing as Paul R. Ryan points out:

> **Although total tidal power potential represents only a relatively small proportion of world energy requirements, its realization would nevertheless save a significant amount of fossil fuels. Tidal projects worldwide have been estimated to have a potential energy output of 635,000 gigawatts, the equivalent of more than a billion barrels of oil, a year.[1]**

In addition to the fact that electricity produced by the tides consumes no exhaustible fuels (and hence creates no noxious wastes), such facilities disturb the landscape much less than the large reservoirs that are created when rivers are dammed.

---

[1]"Harnessing Power from the Tides: State of the Art," *Oceanus* 22 (1980):64.

REVIEW QUESTIONS

1   List three factors that determine the height, length, and period of a wave.

2   Describe the motion of a water particle as a wave passes (see Figure 13.2).

3   Explain what happens when a wave breaks.

4   Describe two ways in which waves cause erosion.

5   What is wave refraction? What is the effect of this process along irregular coastlines (see Figure 13.8)?

6   Why are beaches often called "rivers of sand"?

7   For what purpose is a groin built? Why might the building of one groin lead to the building of others?

8   Describe the formation of the following features: wave-cut cliff, wave-cut platform, sea stack, spit, baymouth bar, and tombolo.

9   Discuss three possible ways in which barrier islands originate.

10   What observable features would lead you to classify a coastal area as emergent?

11   Are estuaries associated with submergent or emergent coasts? Why?

12   Discuss the origin of ocean tides.

13   Explain why an observer can experience two unequal high tides during one day (see Figure 13.20).

14   How does the sun influence tides?

15   What is meant by flood tide? Ebb tide?

16   How have tides affected the earth's rotation? How did geologists substantiate this idea?

17   What advantages does tidal power production offer? Is it likely that tides will ever provide a significant proportion of the world's electrical energy requirements?

KEY TERMS

abrasion (p. 317)

barrier island (p. 325)

baymouth bar (p. 325)

beach drift (p. 321)

breakwater (p. 323)

ebb tide (p. 331)

emergent coast (p. 329)

estuary (p. 329)

fetch (p. 316)

flood tide (p. 331)

groin (p. 322)

jetty (p. 322)

longshore current (p. 321)

neap tide (p. 331)

refraction (p. 319)

sea arch (p. 325)

sea stack (p. 325)

spit (p. 325)

spring tide (p. 331)

submergent coast (p. 329)

surf (p. 316)

tidal current (p. 331)

tidal flats (p. 331)

tide (p. 329)

tombolo (p. 325)

wave-cut cliff (p. 325)

wave-cut platform (p. 325)

wave height (p. 316)

wave length (p. 316)

wave of oscillation (p. 316)

wave of translation (p. 316)

wave period (p. 316)

EARTHQUAKES

14

QUAKE KILLS HUNDREDS

MANAGUA, Nicaragua (AP)—A disastrous earthquake rolled through this Central American city of 300,000 early yesterday, leaving a heavy toll of death and destruction. Unofficial estimates of the dead ranged as high as 18,000 but that figure appeared to be exaggerated.

"It's like standing on jelly down here," radioed U.S. communications satellite technician Ray Hashberger from a station two miles outside the city.

There were confirmed reports of at least 200 killed and thousands injured and homeless. Bodies and debris littered the streets, buildings were afire, and it appeared the death toll might come to as many as 2,000 dead.

Many of those who were not injured sat on the curbstones in a daze, surrounded by what few possessions they could save from the rubble. Many others fled the city.

Half the downtown section of the city lay in ruins as night fell. The quake devastated 36 blocks in the central area.

Fires burned out of control through the afternoon. The quake, which measured between 6 and 7 on the Richter scale of magnitude, struck at 12:40 am yesterday following a series of lesser jolts.

All normal communications, water and electrical services were out. The U.S. Embassy was among the buildings destroyed.

Jack Burton, an information officer in the U.S. Embassy, was in Lima, Peru, for that city's major earthquake May 31, 1970.

"The Lima quake was more gradual though it was about of the same intensity," he said. "But this quake came on like gangbusters. It knocked us on our knees. We had no warning at all."

One survivor said, "It felt like the end of the world." Thousands roamed the streets as if dazed, and other thousands fled to the countryside as smoke billowed from the rubble.

By last night, the widespread fires were said to be under control.

Smaller tremors continued to hit the city throughout the day and into the evening, loosening debris from already wrecked buildings.[1]

This was not the first earthquake to devastate the city of Managua, and it probably will not be the last. It is estimated that nearly one million earthquakes occur each year. Fortunately few are as devastating as the 1972 Managua earthquake. Generally only a few destructive earthquakes occur worldwide each year, but when they do, they are among the most destructive natural forces on earth. The shaking of the ground coupled with the liquefaction of some soils wreak havoc on manmade structures (Figure 14.1). In addition, when a quake occurs in a populated area, power and gas lines are often ruptured, causing numerous fires. In the 1906 San Francisco earthquake, most of the damage was caused by fires which ran unchecked when broken water mains left firefighters with only trickles of water (Figure 14.2).

WHAT IS AN EARTHQUAKE?

An **earthquake** is the vibration of the earth produced by the rapid release of energy. This energy radiates in all directions from its source, or **focus,** in the form of waves analogous to those produced when a bell is struck, vibrating the air around it. During an earthquake, and for many hours following, the earth could be described as "ringing like a bell." Even though the energy dissipates rapidly with increasing distance from the focus, instruments located throughout the world record the event.

The tremendous energy released by atomic explosions or by volcanic eruptions can produce an earthquake, but these events are usually weak and infrequent. What mechanism does produce a destructive earthquake? Ample evidence exists that the earth is

Trace of the San Andreas fault in the Carrizo Plain region of southern California. (Photo by John S. Shelton)

[1]Courtesy of The Associated Press.

FIGURE 14.1
These leaning apartment houses rest on unconsolidated soil, which imitated quicksand during the 1964 earthquake in Niigata, Japan. Although some of the buildings were hardly damaged, their new orientation left something to be desired. (Courtesy of NOAA)

FIGURE 14.2
San Francisco in flames after the 1906 earthquake. (Reproduced from the collection of the Library of Congress)

340

not a static planet. Numerous ancient wave-cut benches can be found many meters above the level of the highest tides, which indicates crustal uplifting of equal magnitude. Other regions exhibit evidence of extensive subsidence. In addition to these vertical displacements, offsets in fence lines, roads, and other structures indicate that horizontal movement is also prevalent (Figure 14.3). These movements are frequently associated with large fractures in the earth called **faults.** Most of the motion along faults can be satisfactorily explained by the plate tectonics theory. The plate model proposes that large slabs of the earth are continually in motion. These mobile plates interact with neighboring plates, straining and deforming the rocks at their edges. It is along these plate boundaries that most earthquakes occur.

The actual mechanism of earthquake generation eluded geologists until H. F. Reid conducted a study following the great 1906 San Francisco earthquake. This earthquake was accompanied by displacements of several meters along the northern portion of the San Andreas fault, a 950-kilometer (600-mile)-long fracture which runs northward through southern California. This large fault zone separates two great sections of the earth, the North American plate and the Pacific plate. Field investigations determined that during this single earthquake the Pacific plate slid as much as 6 meters (20 feet) in a northward direction past the adjacent North American plate.

Using land surveys conducted several years apart, Reid discovered that during the 50 years prior to the 1906 earthquake the land at distant points on both sides of the fault showed a relative displacement of slightly more than 3 meters (10 feet). The mechanism for earthquake formation which Reid deduced from this information is illustrated in Figure 14.4. Tectonic forces ever so slowly deform the crustal rocks on both sides of the fault as illustrated by the bent features. Under these conditions, rocks are bending and storing elastic energy, much like a wooden stick would if bent. Eventually, the forces holding the rocks together are overcome. As slippage occurs at the weakest point (the focus), displacement will exert strain farther along the fault where additional slippage will occur until most of the built-up strain is released. This slippage allows the deformed rock to "snap back." The vibrations we know as an earthquake occur as the rock elastically returns to its original shape. The "springing back" of the rock was termed **elastic rebound** by Reid, since the rock behaves elastically, much like a stretched rubber band does when it is released.

FIGURE 14.3
This fence was offset 2.5 meters (8.5 feet) during the 1906 San Francisco earthquake. (Photo by G. K. Gilbert, U.S. Geological Survey)

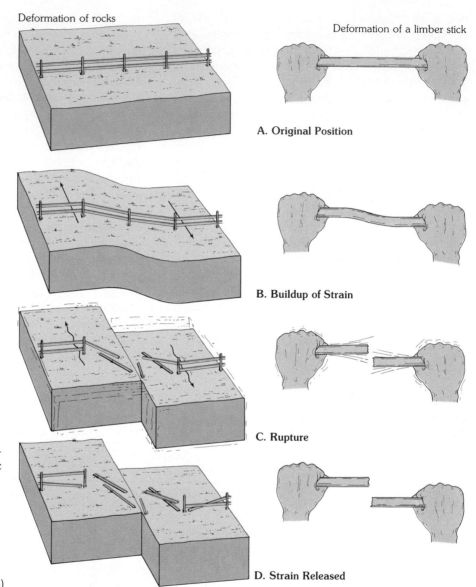

Deformation of rocks

Deformation of a limber stick

A. Original Position

B. Buildup of Strain

C. Rupture

D. Strain Released

FIGURE 14.4
Elastic rebound. As rock is deformed it bends, storing elastic energy. Once the rock is strained beyond its breaking point it ruptures, releasing the stored up energy in the form of earthquake waves. (Modified after Foster, *Physical Geology,* 4th edition, Columbus, Ohio: Charles E. Merrill, 1983)

The intense vibrations of the 1906 San Francisco earthquake lasted about 40 seconds. Although most of the displacement along the fault occurred in this rather short period, additional movements and adjustments in the rocks occurred for several days following the main quake. The adjustments which follow a major earthquake often generate smaller earthquakes called **aftershocks.** Although these aftershocks are usually much weaker than the main earthquake, they can sometimes cause significant destruction to already badly weakened structures. In addition, several small earthquakes called **fore-**shocks often precede a major earthquake by several days or in some cases by as much as several years. Monitoring of these foreshocks has been used as a means of predicting a forthcoming major earthquake. We will consider the topic of earthquake prediction in a later section of this chapter.

The tectonic forces which created the strain that was eventually released during the 1906 San Francisco earthquake are still active. Currently, laser beams[1] are used for establishing the relative motion

[1]Laser beams are used in very precise surveying instruments because of their incredibly accurate straight line qualities.

FIGURE 14.5
Scarp resulting from vertical movement along a fault zone during the 1964 Alaskan earthquake. (Courtesy of the U.S. Geological Survey)

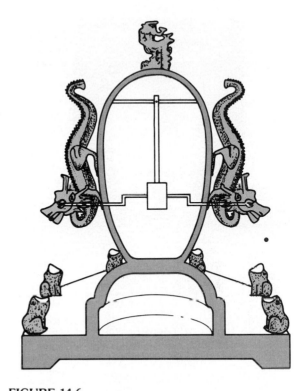

FIGURE 14.6
Ancient Chinese seismograph. During an earth tremor, the dragons located in the direction of the main vibrations would drop a ball into the mouths of the frogs below.

between the opposite sides of this fault. These measurements have revealed a displacement of 2 centimeters per year. Although this rate of movement seems slow, it is indeed appreciable on a geologic time scale. In 30 million years such a rate of displacement is sufficient to slide the western portion of California northward so that Los Angeles, on the northward-moving Pacific plate, would be adjacent to San Francisco on the North American plate. Once the strain along any segment of this active fault again reaches sufficient levels, another slippage with an accompanying earthquake can be expected. It is estimated that great earthquakes should occur about every 50 to 200 years along plate boundaries such as the San Andreas fault. This repetitive process is often described as *stick-slip motion,* since elastic energy is stored over a period of time and then released through slippage.

Not all motion along the San Andreas fault is of the stick-slip type. Along certain portions of this fault the motion is a slow *creep.* Thus, while some sections of the fault are continually creeping, other "locked" segments are building up strain that could result in a major earthquake.

In addition, not all movement along faults is horizontal. Vertical displacement along faults, in which one side is lifted higher in relation to the other, is also common. Figure 14.5 shows a scarp (cliff) produced by just such vertical displacement. In the same man-

ner, the 1964 Good Friday earthquake in Alaska produced a 15 meter vertical offset at one location. Further, many earthquakes occur at such great depths that no displacement is evident at the surface.

SEISMOLOGY

The study of earthquake waves, **seismology,** dates back to attempts made by the Chinese almost 2000 years ago to determine the direction from which these waves originated. The seismic instrument used by the Chinese was a large hollow jar that probably contained a mass suspended from the top (Figure 14.6). This suspended mass (similar to a clock pendulum) was connected in some fashion to the jaws of several dragon figurines that encircled the container. The jaws of each dragon held a metal ball. When earthquake waves reached the instrument, the relative motion between the suspended mass and the jar would dislodge some of the metal balls into the

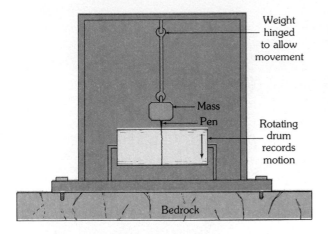

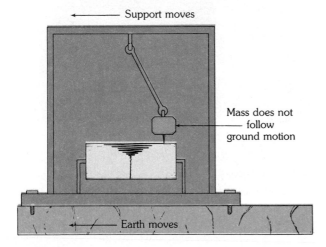

FIGURE 14.7
Principle of the seismograph. The inertia of the suspended mass tends to keep it motionless, while the recording drum, which is anchored to bedrock, vibrates in response to seismic waves. Thus, the stationary mass provides a reference point from which to measure the amount of displacement occurring as the seismic wave passes through the ground below.

waiting mouths of frogs directly below. The Chinese were probably aware that the first strong ground motion from an earthquake is directional, and when it is strong enough, all poorly supported items will topple over in the same direction. Apparently the Chinese used this fact plus the position of the dislodged ball to detect the direction to an earthquake's source. However, the complex motion of seismic waves makes it unlikely that the actual direction to an earth-

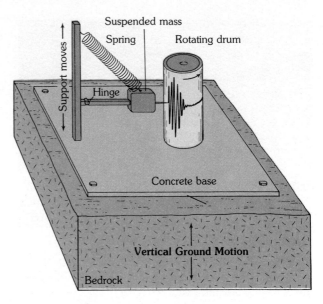

FIGURE 14.8
Seismograph designed to record vertical ground motion.

quake was very often determined.

In principle at least, modern **seismographs,** instruments which record seismic waves, are not unlike the device used by the early Chinese. Seismographs have a mass which is freely suspended from a support that is attached to the ground (Figure 14.7). When the vibration from a distant earthquake reaches the instrument, the **inertia**[1] of the mass keeps it relatively stationary, while the earth and support move. The movement of the earth in relation to the stationary mass is recorded on a rotating drum or magnetic tape.

Earthquakes cause both vertical and horizontal ground motion; therefore, more than one type of seismograph is needed. The instrument shown in Figure 14.7 is designed so that the mass is permitted to swing from side-to-side and thus it detects horizontal ground motion. Usually two horizontal seismographs are employed, one oriented north-south and the other placed with an east-west orientation. Vertical ground motion can be detected if the mass is suspended from a spring as shown in Figure 14.8.

[1]Inertia: Simply stated, objects at rest tend to stay at rest and objects in motion tend to remain in motion unless either is acted upon by an outside force. You probably have experienced this phenomenon when you tried to quickly stop your automobile and your body continued to move forward.

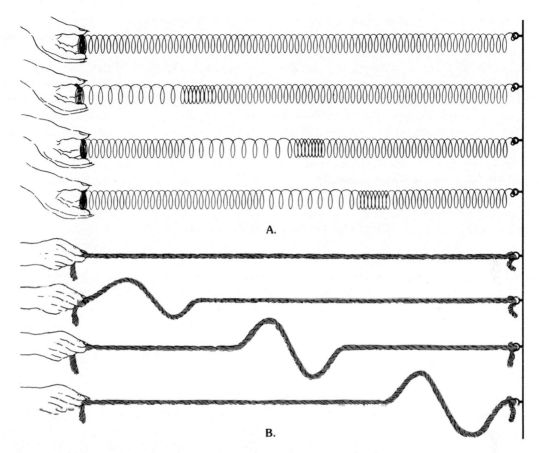

A.

B.

FIGURE 14.9
Types of body waves and their characteristic motion. **A.** P waves cause the particles
in the material to vibrate back and forth in the same direction as the waves move.
B. S waves cause particles to oscillate at right angles to the direction of wave motion.

You can easily demonstrate the principle of a seismograph by attaching a heavy mass to a string. Hold
the other end of the string so that the mass is just off
the floor, then move your hand in rapid side-to-side
strokes. These hand movements represent the vibrations of an earthquake. Notice that the mass remains
relatively motionless. Any movement that occurs will
be small and represents the natural oscillation of the
pendulum (this is similar to the oscillation of a clock
pendulum). Modern seismographs use a mechanism
to remove the effect of the natural oscillation of the
suspended mass.

In order to detect very weak earthquakes, or a
great earthquake that occurred in another part of the
world, seismic instruments are designed to magnify

the actual ground motion. There is, however, a practical limit to the amount of magnification that is possible. Seismographs detect substantial amounts of so-
called background noises caused by winds, pounding
surf on a distant shore, and nearby traffic. Increasing
the sensitivity of seismographs only serves to increase
the background noise which drowns out very weak
earthquake vibrations. Conversely, other instruments
are designed to withstand the violent shaking that
occurs very near the earthquake source.

The records obtained from seismographs, called
seismograms, provide a great deal of information
concerning the behavior of seismic waves. Simply
stated, seismic waves are elastic energy which radiates out in all directions from the focus. The

propagation (transmission) of this energy can be compared to the shaking of gelatin in a bowl which results as some is spooned out. Whereas the gelatin will have one mode of vibration, seismograms reveal that two main groups of seismic waves are generated by the slippage of a rock mass. One of these wave types travels along the outer layer of the earth. These are called **surface waves.** Others travel through the earth's interior and are called **body waves.** Body waves are further divided into two types called **primary** or **P waves** and **secondary** or **S waves.**

The basis on which body waves are divided is their mode of propagation through intervening material. P waves push (compress) and pull (dilate) rocks in the direction the wave is traveling. This wave motion is analogous to that generated by human vocal cords as they move air to and fro to transmit sound. S waves, on the other hand, "shake" the particles at right angles to their direction of travel. This can be illus-

trated by tying one end of a rope to a fence post and shaking the other end while holding the rope tense.

Notice in Figure 14.9 that the propagation of P waves involves changing the volume and shape of the intervening material, while S waves change only the shape. Because solids, liquids, and gases all resist being compressed and will elastically spring back once the force is removed, P waves can travel through all types of matter. On the other hand, because S waves change only the shape of the media through which they travel and because fluids (gases and liquids) do not resist changes in shape, fluids will not transmit S waves.

The motion of surface waves is somewhat more complex (Figure 14.10). As surface waves travel along the ground, they cause it and anything resting upon it to move in a manner similar to the way ocean swells toss a ship about. In addition to the up-and-

FIGURE 14.10
Surface waves consist of two types of wave motion. One motion produces a complex up-and-down motion similar to ocean swells, while the other surface wave whips the ground from side-to-side without any vertical motion.

Direction of wave movement

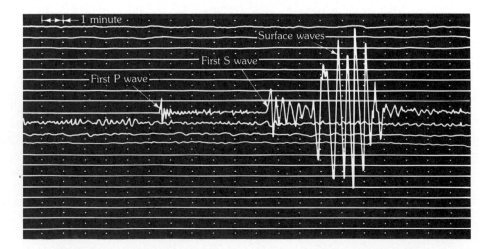

FIGURE 14.11
Typical seismic record. Note the time interval between the arrival of each wave type.

down motion generated, surface waves also have a side-to-side motion which is similar to an S wave oriented in a horizontal plane. This latter motion causes most of the structural damage to buildings and their foundations.

By observing a "typical" seismic record as shown in Figure 14.11, some of the differences among these seismic waves becomes apparent. P waves arrive at the recording station before S waves, which themselves arrive before the surface waves. This is a consequence of their relative velocities. For purposes of illustration, the velocity of P waves through granite within the crust is about 6 km/sec, whereas S waves under the same conditions will travel at 3.5 km/sec. Differences in density and elastic properties of the transmitting material greatly influence the velocities of these waves. In water, for example, the velocity of P waves is 1.5 km/sec, while the S wave velocity is 0 km/sec. However, in any solid material P waves travel about 1.7 times faster than S waves and surface waves can be expected to travel 0.9 times the velocity of the S waves traveling in the layer directly below.

In addition to velocity differences, also notice in Figure 14.11 that the height, or more correctly, the amplitude, of these wave types varies. The S waves have a slightly greater amplitude than the P waves while the surface waves, which cause the greatest destruction, exhibit an even greater amplitude. Because surface waves are confined to a narrow region near the surface and are not spread throughout the earth as P and S waves are, they retain their maximum amplitude longer. Surface waves also have longer periods (time interval between crests), therefore they are often referred to as **long waves,** or **L waves.**

As we shall see, seismic waves are useful in determining the location and magnitude of earthquakes. In addition, seismic waves provide a tool for probing the earth's interior.

LOCATING THE SOURCE OF AN EARTHQUAKE

Recall that the focus is the place where the earthquake originates, usually below ground (Figure 14.12). The **epicenter** is the location on the surface

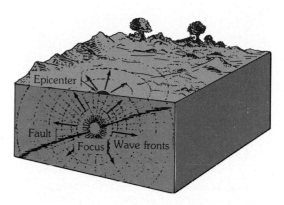

FIGURE 14.12
The focus of most earthquakes is located at depth. The surface location directly above the focus is called the epicenter.

directly above the focus. For shallow earthquakes the difference in velocities of P and S waves provides a method for determining the distance to an earthquake. The principle used is analogous to a race between two autos, one faster than the other. The greater the distance of the race, the greater will be the difference in the arrival times at the finish line. Therefore, the greater the interval between the arrival of the first P wave and the first S wave, the greater the distance to the earthquake.

A system for locating earthquake epicenters was developed through the use of seismograms from earthquakes whose epicenters could be easily pinpointed from physical evidence. From these seismograms, travel-time graphs as shown in Figure 14.13 were constructed. The first travel-time graphs were refined when seismograms from nuclear explosions became available, since the location and time of detonation were precisely known.

Using the sample seismogram in Figure 14.11 and the travel-time curve in Figure 14.13, the distance separating the recording station and the earthquake can be determined. This is accomplished by establishing the time interval between the arrival of the first P wave and the first S wave and then finding the place on the travel-time graphs which exhibits an equivalent time spread between the P and S wave curves. From this information we can determine that this earthquake occurred 3300 kilometers from the recording instrument. Although the distance to an earthquake is established in this manner, its location

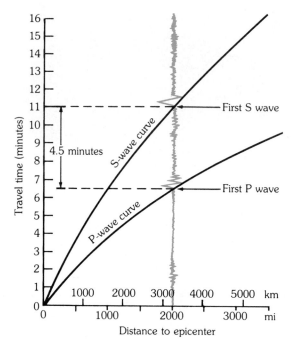

FIGURE 14.13
A travel-time graph is used to determine the distance to an earthquake's epicenter. The difference in arrival times of the first P and S waves in this illustration is 4.5 minutes. Thus the epicenter is roughly 3200 kilometers (2000 miles) away.

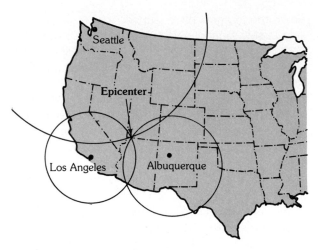

FIGURE 14.14
Earthquake epicenter is located using the distances obtained from three seismic stations.

could be in any direction from the observer. As shown in Figure 14.14, the precise location can be found only when the distance is known from three or more seismic stations. By drawing circles representing the epicenter distance for each of these observatories, an accurate location is established.

The study of earthquakes was greatly bolstered during the 1960s through efforts to discriminate between underground nuclear explosions and natural earthquakes. The United States established a worldwide network of over 100 seismic stations coordinated through Golden, Colorado. The largest of these, located near Billings, Montana, consists of an array of 525 instruments grouped in 21 clusters covering a region 200 km in diameter. Using data from this array, high-speed computers are able to locate an epicenter by a trial and error technique.

EARTHQUAKE BELTS

About 95 percent of the energy released by earthquakes is concentrated in a few relatively narrow zones that wind around the globe (Figure 14.15). The greatest energy is released along a path located near the outer edge of the Pacific Ocean known as the *circum-Pacific belt*. Included in this zone are regions of great seismic activity such as Japan, the Philippines, Chile, and numerous volcanic island chains, as exemplified by the Aleutian Islands. Another major concentration of strong seismic activity runs through the mountainous regions that flank the Mediterranean Sea and continues through Iran and on past the Himalayan complex. Figure 14.15 indicates that yet another continuous belt extends for thousands of kilometers through the world's oceans. This zone coincides with the oceanic ridge system, which is an area of frequent but low-intensity seismic activity.

The areas of the United States included in the circum-Pacific belt lie adjacent to the San Andreas fault and along the western coastal regions of Alaska, including the Aleutian Islands. In addition to these high-risk areas, other sections of the United States are regarded as regions of relatively high earthquake probability during the next 100 years (Figure 14.16).

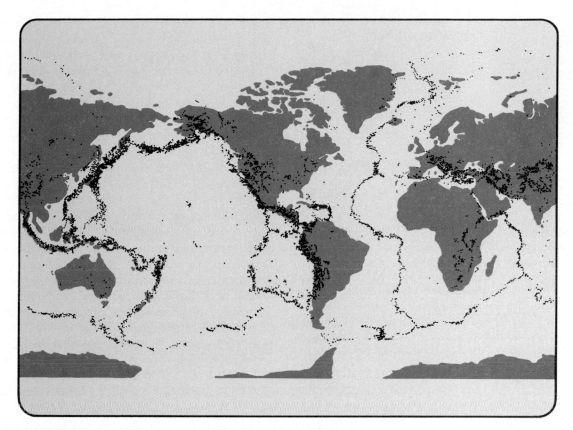

FIGURE 14.15
World distribution of earthquakes for a nine-year period. (Data from NOAA)

One of these regions extends from southern Illinois southward along the Mississippi River. This region is the site of three strong shocks which devastated the town of New Madrid, Missouri, in 1811–12. The drainage of the Mississippi was also altered by these quakes, creating two new lakes within abandoned river channels. Although the total amount of destruction caused by the New Madrid earthquakes was slight, it should be remembered that in the early 1800s this was a sparsely populated frontier region. A similar earthquake in this region today would be truly catastrophic. The cause for earthquakes in a relatively stable region like this is not easily explained. New Madrid is far removed from the tectonic regions near plate boundaries where stress is built between moving sections of lithosphere.

EARTHQUAKE DEPTHS

Evidence from seismic records reveals that earthquakes originate at depths ranging from 5 to nearly 700 kilometers. In a somewhat arbitrary fashion, earthquake foci have been classified by their depth of occurrence. Those with points of origin within 60 kilometers of the surface are referred to as **shallow,** while those generated between 60 and 300 km are considered **intermediate,** and those with a focus greater than 300 km are classified as **deep.** About 90 percent of all earthquakes occur at depths of less than 100 km and all very strong earthquakes appear to originate at shallow depths. For example, the 1906 San Francisco earthquake involved movement within the upper 15 km of the earth's crust, whereas the

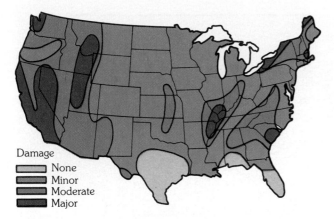

Damage
- None
- Minor
- Moderate
- Major

FIGURE 14.16
Earthquake risk map for the conterminous United States shows the relative risk of seismic destruction based on past occurrences. Although southeastern Missouri and California have each experienced a major earthquake and are therefore assigned the same risk, this does not mean that these areas have an equal probability of being struck by another major earthquake in the immediate future. (From Environmental Science Service Administration)

1964 Alaskan earthquake had a focus depth of 33 km. Seismic data reveal that while shallow-focus earthquakes have been recorded with Richter magnitudes of 8.6, the strongest intermediate depth quakes have had values below 7.5, and deep-focus earthquakes have not exceeded 6.9 in magnitude.

When earthquake data were plotted according to geographic location and depth, several interesting observations were noted. Rather than a random mixture of shallow and deep earthquakes, some very definite distribution patterns emerged. Earthquakes generated along the oceanic ridge system always have a shallow focus and none is very strong. Further, it was noted that almost all deep-focus earthquakes occurred in the circum-Pacific belt, particularly in regions situated landward of deep-ocean trenches. In a study conducted in the southwestern Pacific near the Tonga trench by H. Benioff, it was discovered that foci depths increased with distance from the trench as shown in Figure 14.17. These seismic regions, called **Benioff zones** after the man who extensively studied them, are oriented about 45 de-

FIGURE 14.17
Distribution of earthquake foci in 1965 in the vicinity of Tonga Island. (Data from B. Isacks, J. Oliver, and L. R. Sykes)

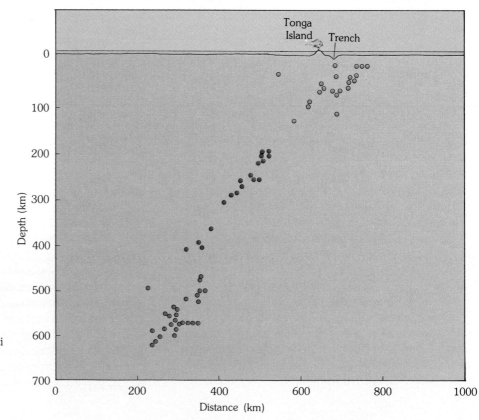

grees to the surface. Why should earthquakes be oriented along a narrow zone which plunges almost 700 kilometers into the earth's interior? We will consider this question further in Chapter 16.

EARTHQUAKE INTENSITY AND MAGNITUDE

Early attempts to establish the intensity of an earthquake relied heavily on descriptions of the event. There was an obvious problem related to this method—people's accounts varied widely, making an accurate classification of a quake's intensity difficult. Then in 1902 a fairly reliable scale based on the amount of damage caused to various types of structures was developed by Giuseppe Mercalli. A modified form of the **Mercalli intensity scale** is presently used by the U.S. Coast and Geodetic Survey (Table 14.1). Nevertheless, the destruction wrought by earthquakes is not always an adequate means for comparison. Many factors, including distance from the epicenter, nature of the surface materials, and building design, cause variations in the amount of

damage sustained. In addition, the epicenters of many earthquakes do not coincide with populated regions of the globe. Consequently, methods were devised which quantitatively established the amount of energy released during an earthquake, a measurement referred to as **magnitude.**

Ideally, the magnitude of an earthquake can be determined from the amount of material which slides along the fault and the distance it is displaced. Even in an ideal setting such as that of the 1906 San Francisco earthquake, where the fault trace is visible and displacement can be measured from physical evidence, this method can only provide a crude estimate of the forces involved. In most earthquakes, the fault does not penetrate the surface, therefore the amount of displacement cannot be measured directly. In 1935, Charles Richter of the California Institute of Technology attempted to rank the earthquakes of southern California into groups of large, medium, and small magnitude. The system he developed determines earthquake magnitudes from the motions measured by seismic instruments.

TABLE 14.1
Modified Mercalli intensity scale.

I	Not felt except by a very few under specially favorable circumstances.
II	Felt only by a few persons at rest, especially on upper floors of buildings.
III	Felt quite noticeably indoors, especially on upper floors of buildings, but many people do not recognize as an earthquake.
IV	During the day felt indoors by many, outdoors by few. Sensation like heavy truck striking building.
V	Felt by nearly everyone, many awakened. Disturbances of trees, poles, and other tall objects sometimes noticed.
VI	Felt by all; many frightened and run outdoors. Some heavy furniture moved; few instances of fallen plaster or damaged chimneys. Damage slight.
VII	Everybody runs outdoors. Damage negligible in buildings of good design and construction; slight to moderate in well-built ordinary structures; considerable in poorly built or badly designed structures.
VIII	Damage slight in specially designed structures; considerable in ordinary substantial buildings with partial collapse; great in poorly built structures. (Fall of chimneys, factory stacks, columns, monuments, walls.)
IX	Damage considerable in specially designed structures. Buildings shifted off foundations. Ground cracked conspicuously.
X	Some well-built wooden structures destroyed. Most masonry and frame structures destroyed with foundations. Ground badly cracked.
XI	Few, if any (masonry) structures remain standing. Bridges destroyed. Broad fissures in ground.
XII	Damage total. Waves seen on ground surfaces. Objects thrown upward into air.

SOURCE: U.S. Coast and Geodetic Survey.

TABLE 14.2
Earthquake magnitudes and
expected world incidence.

Richter Magnitudes	Earthquake Effects	Estimated Number per Year
< 3.5	Generally not felt, but recorded	900,000
3.5–5.4	Often felt, but only minor damage detected	30,000
5.5–6.0	Slight damage to structures	500
6.1–6.9	Can be destructive in populous regions	100
7.0–7.9	Major earthquakes. Inflict serious damage	20
≥ 8.0	Great earthquakes. Produce total destruction to nearby communities	One every 5–10 years

SOURCE: *Earthquake Information Bulletin* and others.

Today a refined **Richter scale** is used worldwide to describe earthquake magnitude. Using Richter's scale, the magnitude is determined by measuring the amplitude of the largest wave recorded on the seismogram. Although seismographs greatly magnify the ground motion, large-magnitude earthquakes will cause the recording pen to be displaced farther than small-magnitude earthquakes. In order for seismic stations worldwide to obtain the same magnitude for a given earthquake, adjustments must be made for the weakening of the seismic waves as they move from the focus, and for the sensitivity of the recording instrument. Richter established 100 kilometers as the standard distance and the Wood-Anderson instrument as the standard recording device.

The largest earthquakes ever recorded have Richter magnitudes near 8.6. These great shocks released approximately 10^{26} ergs of energy—roughly equivalent to the detonation of 1 billion tons of TNT. Apparently, earthquakes having a Richter magnitude greater than 9 do not occur. Conversely, earthquakes with a Richter magnitude of less than 3.5 are usually not felt by humans. With the advent of more sensitive instruments, tremors of a magnitude of −2 have been recorded. Table 14.2 shows how earthquake magnitudes and their effects are related.

As we have seen, earthquakes vary enormously in strength; consequently, the wave amplitudes generated vary by thousands of times as well. To accommodate this wide variation, Richter used a logarithmic scale to express magnitude. On this scale a tenfold increase in wave amplitude corresponds to an increase of one on the magnitude scale. Thus, the amplitude of the largest surface wave for a 5-magnitude earthquake is 10 times greater than the wave amplitude produced by an earthquake having a magnitude of 4. Further, each unit of magnitude increase on the Richter scale equates to roughly a 30-fold increase in the energy released. Thus, an earthquake with a magnitude of 6.5 releases 30 times more energy than one with a magnitude of 5.5, and roughly 900 times more energy than a 4.5-magnitude quake. A major earthquake with a magnitude of 8.5 releases millions of times more energy than the smallest earthquakes felt by humans. This should dispel the notion that a moderate earthquake decreases the chances for the occurrence of a major quake in the same region. Thousands of moderate tremors would be needed to release the amount of energy equal to one "great" earthquake.

Some of the world's major earthquakes and their corresponding Richter magnitudes are listed in Table 14.3. Great earthquakes such as these can be expected in a tectonically active region every 50 to 200 years. The region of the San Andreas fault which participated in the San Francisco earthquake in 1906 has not generated a large earthquake in over 75 years—an alarming statistic to the residents of this region.

Recent studies have shown that the Richter scale does not adequately differentiate those earthquakes

TABLE 14.3
Some notable worldwide and U.S. earthquakes.

Year	Location	Deaths (est.)	Magnitude	Comments
1290	Chihli (Hopei), China	100,000		
1556	Shensi, China	830,000		Possibly the greatest natural disaster
1737	Calcutta, India	300,000		
1755	Lisbon, Portugal	70,000		Tsunami damage extensive
[1]1811–12	New Madrid, Missouri	Few		Three major earthquakes
[1]1886	Charleston, South Carolina	60		
[1]1906	San Francisco, California	700	8.25	Fires caused extensive damage
1908	Messina, Italy	120,000	7.5	
1920	Kansu, China	180,000	8.5	
1923	Tokyo, Japan	150,000	8.2	Fire caused extensive destruction
1960	Southern Chile	5700	8.5–8.7	Possibly the largest magnitude earthquake ever recorded
[1]1964	Alaska	131	8.4–8.6	
1970	Peru	66,000	7.8	Great rockslide
[1]1971	San Fernando, California	65	6.5	Damage exceeded $500,000,000
1975	Liaoning Province, China	Few	7.5	First major earthquake to be predicted
1976	Tangshan, China	650,000	7.6	Not predicted

[1]U.S. earthquakes
SOURCE: U.S. National Oceanic and Atmospheric Administration.

which have very high magnitudes. Since all of the very strongest quakes have nearly equal wave amplitudes, the Richter scale becomes saturated at this level. Consequently, methods to extend the Richter scale so that it can accurately measure the magnitudes of the largest earthquakes have been proposed. One method analyzes very long-period seismic waves for this purpose. On this extended scale, the 1906 San Francisco earthquake with a surface wave magnitude of 8.3 would be demoted to 7.9, whereas the 1964 Alaskan earthquake with an 8.4–8.6 magnitude would be increased to 9.2. Using this system, the strongest earthquake on record is the 1960 Chilean earthquake with a magnitude of 9.5.

EARTHQUAKE DESTRUCTION

The most violent earthquake to jar North America this century—the Good Friday Alaskan Earthquake—occurred at 5:36 P.M. on March 27, 1964. Felt throughout that state, the earthquake had a magnitude of 8.4–8.6 on the Richter scale and reportedly lasted 3 to 4 minutes. In that short period of time 114 persons were killed, thousands were left homeless, and the economy of the state was badly disrupted (Figure 14.18). Had the schools and business districts been open, the toll surely would have been higher. The total financial loss in Alaska has been estimated at 300 million dollars, roughly twenty times the original 15 million dollar purchase price of the state. The location of the epicenter and the towns which were hardest hit by the quake are shown in Figure 14.18. Within 24 hours of the initial shock, 28 aftershocks were recorded, 10 of which exceeded a magnitude of 6 on the Richter scale.

Many factors determine the amount of destruction which will accompany an earthquake. The most obvious of these are the magnitude of the earthquake and the proximity of the quake to a populated area. Fortunately, most earthquakes are small and occur in remote regions of the earth. However, about twenty major earthquakes are reported annually, one or two of which are catastrophic.

During an earthquake, the region within 20 to 50 kilometers (12.5 to 30 miles) of the fault will experience roughly the same degree of ground shaking, but beyond this limit the vibration deteriorates rapidly.

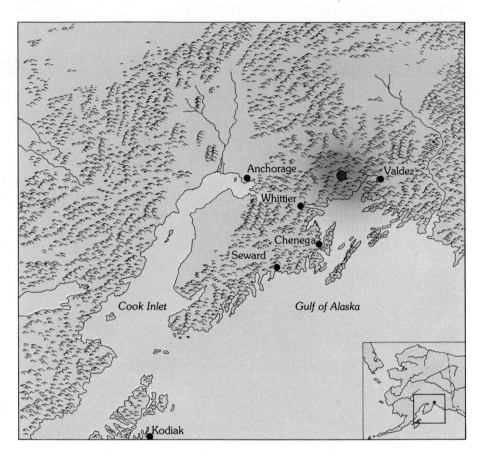

FIGURE 14.18
Region most affected by the Good Friday earthquake of 1964. Note the location of the epicenter (red dot). (After U.S. Geological Survey)

Occasionally, during exceptionally great earthquakes such as the New Madrid earthquake of 1811, the area of influence can be much larger. The epicenter of this earthquake was located directly south of Cairo, Illinois, and the vibrations were felt from the Gulf of Mexico to Canada, and from the Rockies to the Atlantic seaboard.

DESTRUCTION CAUSED BY SEISMIC VIBRATIONS

The 1964 Alaskan earthquake provided geologists with new insights into the role of ground shaking as a destructive force. As the energy released by an earthquake travels along the earth's surface, it causes the ground to vibrate in a complex manner by moving up and down as well as from side to side. The amount of structural damage attributable to the vibrations depends on several factors, including (1) the intensity and duration of the vibrations; (2) the nature of the

material upon which the structure rests; and (3) the design of the structure.

All of the multistory structures in Anchorage were damaged by the vibrations; the more flexible wood frame residential buildings fared best. However, many homes were destroyed when the ground failed. A striking example of how construction variations affect earthquake damage is shown in Figure 14.19. We can see in this photo that the steel-frame building on the left withstood the vibrations, whereas the relatively rigid concrete structure was badly damaged.

The greatest loss of life from an earthquake can be partially attributable to the type of structures that were inhabited. In 1556 in the Shensi region of China, an estimated 830,000 persons perished when an early morning earthquake struck the region. Many of these people lived in dwellings carved out of a compacted windblown sediment called loess (see Chapter 12). The walls of these structures failed,

FIGURE 14.19
Damage caused to the five-story J. C. Penney Co. building, Anchorage, Alaska. Very
little structural damage was incurred by the adjacent building. (Courtesy of NOAA)

allowing the roofs to bury the inhabitants. The dense
population and present-day construction practices in
this region are such that a repetition of the 1556
event is possible.

Most of the large structures in Anchorage were
damaged even though they were built to conform to
the earthquake provisions of the Uniform Building
Code of California. Perhaps some of that destruction
can be attributed to the unusually long duration of
this earthquake, which was estimated at 3 to 4
minutes. Most earthquakes consist of tremors lasting

from 20 seconds to one minute. The San Francisco
earthquake of 1906 was felt for about 40 seconds.

Although the region within 20 to 50 kilometers of
the epicenter will experience about the same degree
of ground shaking, the destruction varies considerably within this area. This difference is mainly attributable to the nature of the ground on which the structures are built. Soft sediments, for example, generally
amplify the vibration more than solid bedrock. Thus,
the buildings located in Anchorage, which were situated on unconsolidated sediments, experienced

heavy structural damage. By contrast, most of the town of Whittier, although located much nearer to the epicenter than Anchorage, rests on a firm foundation of granite and hence suffered much less damage from the seismic vibrations. However, Whittier was damaged by a seismic sea wave.

This great earthquake had its effects felt thousands of kilometers from the epicenter as it set the earth vibrating like a large tuning fork. Amplified ground waves generated **seiches,** the rhythmic sloshing of water in lakes, reservoirs, and smaller enclosed basins, in areas as far away as the Gulf of Mexico. Off the coast of Texas, 2-meter waves damaged small craft, while smaller waves were noticed in swimming pools in both Texas and Louisiana. Seiches can be particularly dangerous when they occur in reservoirs retained by earthen dams. These waves have been known to slosh over reservoir walls and weaken the structure, thereby endangering the lives of those downstream. Another response to these large-scale

vibrations is the fluctuation of water levels in wells. During this earthquake, numerous wells along a belt from South Dakota to Georgia were affected. The largest reported fluctuation in water level was 3.5 meters.

TSUNAMI

Most of the deaths associated with the 1964 Alaskan quake were caused by **seismic sea waves,** or **tsunami.**[1] These destructive waves have popularly been called "tidal waves." However, this name is not accurate since these waves are not generated by the tidal effect of the moon or sun.

Most tsunamis result from the vertical displacement of the ocean floor during an earthquake as illustrated in Figure 14.20. Once formed, a tsunami

---

[1]Seismic sea waves were given the name *tsunami* by the Japanese who have suffered a great deal from them. The term *tsunami* is now used worldwide.

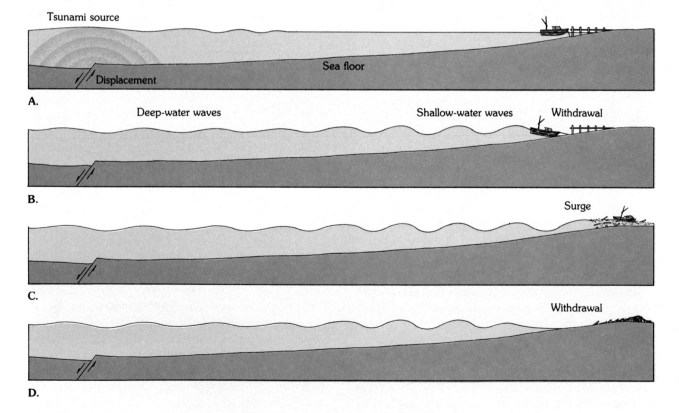

FIGURE 14.20
Schematic drawing of a tsunami generated by displacement of the ocean floor. The size and spacing of the swells are not to scale.

resembles the ripples formed when a pebble is dropped into a pond. In contrast to ripples, tsunamis advance at speeds between 500 and 800 kilometers per hour. Despite this striking characteristic, a tsunami in the open ocean can pass undetected because its height is usually less than one meter and the distance between wave crests ranges from 100 to 700 kilometers. However, upon entering shallower coastal waters, these destructive waves are slowed down and the water begins to pile up to heights that occasionally exceed 30 meters. As a tsunami approaches the shore, it appears as a rapid rise in sea level with a turbulent and chaotic surface.

Usually the first warning of a tsunami is a rather rapid withdrawal of water away from beaches (Figure 14.20). Residents of coastal areas have learned to heed this warning and move to higher ground. About 5 to 30 minutes later the retreat of water is followed by a surge capable of extending hundreds of meters inland. In a successive fashion, each surge is followed by rapid oceanward retreat of the water. These waves, separated by intervals of between 10 and 60 minutes, are able to traverse large stretches of the ocean before their energy is totally dissipated. The tsunami generated by the 1960 Chilean earthquake, in addition to completely destroying villages along an 800-kilometer stretch of coastal South America, traveled 17,000 kilometers across the Pacific to Japan. Here, about 22 hours after the quake, considerable destruction was inflicted upon southern coastal villages of Honshu, the major island of Japan. For several days afterwards, tidal gauges located in Hilo, Hawaii, were able to detect these diminishing waves as they bounced about the Pacific.

The tsunami generated in the 1964 Alaskan earthquake inflicted heavy damage to the communities in the vicinity of the Gulf of Alaska, completely destroying the town of Chenega. Kodiak was also heavily damaged and most of its fishing fleet destroyed when a seismic sea wave carried many vessels into the business district (Figure 14.21). The deaths of 107 persons have been attributed to this tsunami. By contrast, only 9 persons died in Anchorage as a direct result of the vibrations. Tsunami damage following the Alaskan earthquake extended along much of the west coast of North America, and in spite of a 1-hour

FIGURE 14.21
A tsunami washed this fishing fleet into the heart of the village of Kodiak, Alaska. (Photo by W. R. Hansen, U.S. Geological Survey)

warning, 12 persons perished in Crescent City, California, where all of the deaths and most of the destruction was caused by the fifth wave. The first wave crested about 4 meters (13 feet) above low tide and was followed by three progressively smaller waves. Believing that the tsunami had ceased, people returned to the shore, only to be met by the fifth and most devasting wave, which, superimposed upon high tide, crested about 6 meters higher than the level of low tide.

Although most tsunamis are generated by earthquakes, a volcanic eruption in the ocean can generate this destructive phenomenon as well. For example, the 1883 volcanic explosion of Krakatoa generated a tsunami which drowned an estimated 36,000 coastal residents of Java and Sumatra.

In 1946, a large tsunami struck the Hawaiian Islands without warning. A wave more than 15 meters high left several coastal villages in shambles. This destruction motivated the United States Coast and Geodetic Survey to establish a tsunami warning system for the coastal areas of the Pacific. From seismic observatories throughout the area, warnings of large earthquakes are reported to the Tsunami Warning Center in Honolulu. Using tidal gauges, a determination is made as to whether a tsunami has been formed. Within an hour a warning is issued. Although tsunamis travel very rapidly, there is sufficient time to evacuate all but the region nearest the epicenter (Figure 14.22). For example, a tsunami generated near the Aleutian Islands would take 5 hours to reach Hawaii and one generated near the coast of Chile would travel 15 hours before reaching Hawaii. Fortunately, most earthquakes do not generate tsunamis. On the average, only about 1.5 destructive tsunamis are generated worldwide each year. Of these, only one every ten years can be considered catastrophic.

FIRE

Fire was only a minor consequence in the 1964 Alaskan earthquake, but often it is the most destructive

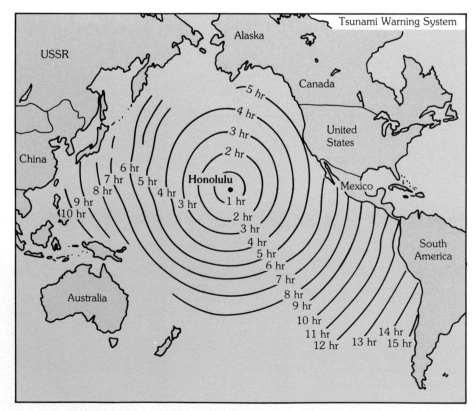

FIGURE 14.22
Tsunami travel times to Honolulu, Hawaii, from locations throughout the Pacific. (From NOAA)

result. The 1906 earthquake centered near the city of San Francisco reminds us of the formidable threat of fire. The central city contained mostly large, older wooden structures and brick buildings. Although extensive damage was done to many of the unreinforced brick buildings, the greatest destruction was caused by innumerable fires which started when gas and electrical lines were severed. The fires raged out of control for three days and devastated over 500 blocks of the city (see Figure 14.2). The problem was compounded by the initial ground shaking which broke the city's water lines into hundreds of unconnected pieces.

The fire was finally contained when buildings were dynamited along a wide boulevard to provide a fire break. Although only a few deaths were attributed to the fires, that is not always the case. An earthquake which rocked Japan in 1923 triggered an estimated 250 fires, which devastated the city of Yokohama and destroyed more than half the homes in Tokyo. Over 100,000 deaths were attributed to the fires which were driven by unusually high winds.

LANDSLIDES AND GROUND SUBSIDENCE

In the 1964 Alaskan earthquake, it was not ground vibrations directly, but landslides and ground subsidence triggered by the vibrations that probably caused the greatest damage to structures. At Valdez and Seward the violent shaking caused deltaic materials to liquify; the subsequent slumping carried both waterfronts away. Because of the threat of recurrence, the entire town of Valdez was relocated about 7 kilometers away in a region of stable ground. The destruction at Valdez was compounded by the tragic loss of 31 lives. While waiting for an incoming vessel, the 31 persons and the dock slid into the sea.

Most of the damage in the city of Anchorage was also attributed to landslides caused by the shaking and lurching ground. Many homes were destroyed in Turnagain Heights when a layer of clay lost its strength and carried over 200 acres of land toward the ocean (Figure 14.23). The destruction was so complete that this area was bulldozed over and made into a park, which was appropriately named "Earthquake Park." Downtown Anchorage was equally disrupted as blocks of earth broke loose and sections of

the main business district dropped by as much as 3 meters (10 feet).

EARTHQUAKE PREDICTION AND CONTROL

The vibrations that shook the San Fernando, California area on the morning of February 9, 1971, inflicted 64 deaths and almost a billion dollars in damages (Figure 14.24)—all of this from an earthquake that lasted 60 seconds and had a moderate rating of 6.6 on the Richter scale. Fortunately, because of the early hour, freeways, businesses, and schools were sparsely occupied, reducing the possible toll. Also, had the Lower Van Norman Lake Dam, badly damaged during the earthquake, actually broken, 80,000 additional lives might have been lost which would have made it the most catastrophic event ever in the United States. This quake in populous southern California reemphasized the need for reliable methods of earthquake prediction and control.

Japan's location in an earthquake-prone region has resulted in great interest in earthquake prediction there. The Japanese have established a complex seismic network extending 200 kilometers (125 miles) out into the ocean. Here on the ocean floor where background noise is slight, the Japanese plan to monitor microearthquakes (foreshocks), which precede the main earthquake. It is hoped that by monitoring these seismic activities some pattern will emerge which can be used to accurately predict forthcoming tremors.

In California, uplift or subsidence of the land and changes in movement of a fault zone from a slow creep to a locked position have been found to precede moderate earthquakes. It therefore seems reasonable that the prediction of earthquakes may be feasible by continually monitoring ground tilt, fault movement, and seismic activity. Some monitoring networks are already operating in the earthquake-prone regions of the United States; others have been proposed.

Although no reliable method of short-range prediction has yet been devised, a few successful predictions have been made. In 1966, an earthquake in Tashkent, U.S.S.R., was predicted by monitoring the

FIGURE 14.23
View of the Turnagain Heights slide in Anchorage, shortly after the earthquake. (Courtesy of the U.S. Geological Survey)

radon level in wells. Radon is an inert gas generated by the radioactive decay of radium, a small amount of which is found in certain rocks. Normally this gas is locked within rock, but during the buildup of stress, the newly formed cracks allow for its release. In February, 1975, an earthquake in northeast China was predicted only hours before it occurred. By warning an estimated 3 million people to remain outdoors on a cold evening, tens of thousands of lives were believed to be spared. Western observers confirmed Chinese reports that almost 90 percent of the struc-

tures in the city of Haicheng were heavily damaged. The rather large foreshocks which preceded this earthquake aided the prediction and also prompted the people to heed the warning.

Unfortunately, the Chinese were able to predict, but not pinpoint, the exact date of the great Tangshan earthquake of 1976. Their long-range warning of an upcoming earthquake was not precise enough to save as many as 650,000 persons estimated to have lost their lives and another 780,000 who were injured. The Chinese have also had false alarms. In a

FIGURE 14.24
Collapsed overpass to Golden State Freeway, San Fernando earthquake, 1971.
(Photo by R. W. Wallace, U.S. Geological Survey)

province near Hong Kong, people evacuated their dwellings for over a month, but no earthquake followed. The debate that would have to precede an evacuation order for a large city in the United States, such as Los Angeles, would be considerable. The cost of the evacuation, loss of work time, and innumerable other problems associated with an evacuation would have to be weighed against the earth-

quake's probability. Earthquake prediction must become a more proven science before such warnings will be heeded.

The actual control of earthquakes is another matter altogether. The discovery that humans have inadvertantly triggered earthquakes has given earth scientists some encouragement. The most convincing evidence that people can initiate earthquakes came

361

between 1962 and 1966, when studies of the seismic activity at the Rocky Mountain Arsenal near Denver were conducted. For a period of 80 years prior to 1962, the U.S. Coast and Geodetic Survey reported no significant earthquake activity in the Denver region. In 1962 the arsenal began disposing wastes from its chemical warfare production into a well over 3600 meters deep. During the period of fluid waste injection, from April, 1962 to September, 1965, about 700 microearthquakes were reported, 75 intense enough to be felt. The injection of water under pressure is believed to have "lubricated" the fault, which had been building up strain over the years. This lubricating effect is not one of making rocks along the fault zone slippery. Rather the water exerts an outward force which is directed perpendicular to the fault plane. The induced outward force opposes the natural inward force caused by the weight of rock piled above. When the injection was halted for about a year, a marked drop in seismic activity was also detected. When pumping resumed, the frequency of tremors increased markedly.

Other earthquakes caused by human activity have occurred in regions adjacent to large reservoirs such as Lake Mead on the Arizona-Nevada border. Ever since Lake Mead was filled in 1936, hundreds of small tremors have been recorded. They are thought to have been caused by the added weight of the lake, and perhaps aided by the "lubricating" effect of water seeping into the rock below. Another large reservoir in India is believed responsible for triggering a disastrous earthquake in which 200 persons were killed. Underground nuclear explosions have also been responsible for initiating numerous small aftershocks, although none has been as great as the explosion itself.

The hope of many scientists is that we may someday be able to reduce the threat of earthquakes by triggering numerous small earthquakes using fluid injections or nuclear explosions. Such methods would slowly and continually release the elastic strain that might otherwise build up and be released as a high-magnitude earthquake. Recall, however, that many thousands of minor tremors are required to equal the energy released by one strong earthquake. This fact coupled with the inaccessibility of many fault zones makes the possibility of earthquake control not quite as feasible as it first appears. A favorable condition for earthquake control in California is the shallow depth of earthquake foci, making drilling operations possible. Many tests will have to be made in remote regions before we dare risk such a venture along a fault system like the San Andreas, at least in those portions located in populous regions.

REVIEW QUESTIONS

1 What is an earthquake? Under what circumstances do earthquakes occur?

2 How are faults, foci, and epicenters associated?

3 Earthquakes occur only in the rigid lithosphere, not in the plastic asthenosphere. Using the elastic rebound idea, explain this phenomenon.

4 Faults which are experiencing no active creep may be considered "safe." Rebut or defend this statement.

5 Describe the principle of a seismograph.

6 Contrast the motion produced by P waves with the movements created by S waves.

7 P waves move through solids, liquids, and gases whereas S waves move only through solids. Explain.

8 Using Figure 14.13, determine the distance between an earthquake and a seismic station if the first S wave arrives 3 minutes after the first P wave.

9 Although the Mercalli intensity scale was more reliable than earlier attempts at describing earthquake intensity, it too has drawbacks. Briefly describe these limitations.

10 How does earthquake magnitude differ from earthquake intensity?

11 For each increase of one on the Richter scale, wave amplitude increases __ times.

12 An earthquake measuring 7 on the Richter scale releases about __ times more energy than an earthquake with a magnitude of 6.

13 List three factors that affect the amount of destruction caused by seismic vibrations.

14 In addition to the destruction created directly by seismic vibrations, list three other types of destruction associated with earthquakes.

15 How might earthquakes be controlled in the future?

KEY TERMS

aftershock (p. 342)

Benioff zone (p. 350)

body wave (p. 346)

deep focus (p. 349)

earthquake (p. 339)

elastic rebound (p. 341)

epicenter (p. 347)

fault (p. 341)

focus (p. 339)

foreshock (p. 342)

inertia (p. 344)

intermediate focus (p. 349)

long (L) waves (p. 347)

magnitude (p. 351)

Mercalli intensity scale (p. 351)

primary (P) waves (p. 346)

Richter scale (p. 352)

secondary (S) waves (p. 346)

seiche (p. 356)

seismic sea wave (p. 356)

seismogram (p. 345)

seismograph (p. 344)

seismology (p. 343)

shallow focus (p. 349)

surface wave (p. 346)

tsunami (p. 356)

THE
EARTH'S
INTERIOR

15

Although the earth's interior lies just below us, its accessibility to direct observation is very limited. For example, when we compare our efforts at probing the earth's interior through drilling to our accomplishments in planetary exploration, this fact becomes very apparent. On the scale of our solar system, our recent space probes to the planet Saturn are equivalent to a hole drilled about 1600 kilometers into the earth, far deeper than the 9-kilometer depth that has actually been penetrated. Even though volcanic activity can be considered as a window into the earth's interior because materials are brought up from below, this activity allows only a glimpse of the outer 200 kilometers of our planet—only a small fraction of the earth's 6370-kilometer radius.

Fortunately geologists have learned a great deal about the earth's composition through space exploration, by high-pressure laboratory experimentation, and from samples of the solar system (meteorites) that frequently collide with the earth. More importantly, many clues to the physical conditions inside our planet have been obtained through the study of seismic waves generated by earthquakes and nuclear explosions. As seismic waves pass through the earth, they carry information about the materials through which they were transmitted to the surface. Hence, when carefully analyzed, seismic records provide an x-raylike picture of the earth's interior.

PROBING THE EARTH'S INTERIOR

Much of our knowledge of the earth's interior comes from the study of P (compressional) and S (shear) waves that penetrate the earth and emerge at some

Seismic exploration is used to probe the earth's subsurface zones. (Courtesy of Gulf Oil)

367

distant point. Simply stated, the technique involves accurately measuring the time required for seismic waves to travel from an earthquake or nuclear explosion to a seismographic station. Since the time required for P and S waves to travel through the earth depends upon the properties of the rock materials encountered, seismologists search for variations in travel times that cannot be accounted for simply by differences in the distances traveled. These variations correspond to changes in rock properties.

A simple analogy might help clarify this method of probing the earth's interior. For illustration, suppose you travel along the same route each morning to a specific destination. On every day of the week your trip takes the same time, except for Fridays when it always takes several minutes longer. The task is for someone to establish the travel time for your daily trips (without following you) and determine why the Friday trip takes longer. Do you make stops along the way or somehow get delayed on Fridays and not other days? Seismologists must explain similar variations in travel time. Since the earth is not a homogeneous body, anomalous travel times occur because waves passing through the earth encounter differences in materials at various depths.

One major problem is that to obtain accurate travel times, the exact location and time of the seismic event must be established. For earthquakes, this information can only be obtained from the seismic waves themselves, which makes measurement somewhat uncertain. Nuclear test explosions, on the other hand, have the advantage over earthquakes in that the precise time and location are known. However, despite the limitations of studying seismic waves generated by earthquakes, seismologists during the first half of this century were able to detect the major layers of the earth. Yet, not until the early 1960s, when nuclear testing was in its heyday and arrays consisting of hundreds of sensitive seismographs were deployed, were the fine structures of the earth first established with certainty.

THE NATURE OF SEISMIC WAVES

To examine the earth's composition and structure, some of the basic properties of wave transmission, or propagation, must first be studied. As stated in the preceding chapter, seismic waves travel out from

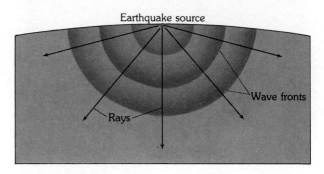

FIGURE 15.1
Seismic waves travel in all directions from an earthquake source (focus) as wave fronts. The direction of this motion can also be shown as rays, lines drawn perpendicular to the wave fronts.

their source in all directions as a wave front. For purposes of description, the common practice is to consider the path taken by these waves as rays, or lines drawn perpendicular to the wave front as shown in Figure 15.1. Significant characteristics of seismic waves include:

1   The velocity of seismic waves depends on the density and elasticity of the intervening material. Seismic waves travel most rapidly in rigid materials which elastically spring back to their original shape when the strain is removed. For instance, a crystalline rock would transmit seismic waves more rapidly than a layer of unconsolidated material.

2   Within a given layer the speed of seismic waves generally increases with depth because pressure increases and squeezes the rock into a more compact elastic material.

3   Compressional waves (P waves), which oscillate back and forth in the same direction as their direction of motion, travel through solids as well as liquids because when compressed these materials behave elastically; that is, they spring back to their original shape when the strain is removed (Figure 15.2A). Shear waves (S waves), which oscillate at right angles to their direction of motion, cannot travel through liquids because, unlike solids, liquids have no shear strength (Figure 15.2B). That

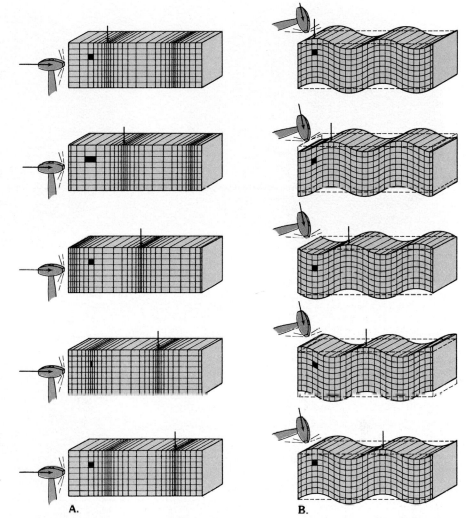

FIGURE 15.2
The transmission of P and S waves through a solid. **A.** The passage of P waves causes the intervening material to suffer alternate compressions and expansions. **B.** The passage of S waves causes a change in shape without changing the volume of the material. Because liquids behave elastically when compressed (they spring back when the stress is re moved), they will transmit P waves. However, since liquids do not resist changes in shape, S waves cannot be transmitted through liquids. (From O. M. Phillips, *The Heart of the Earth,* San Francisco: Freeman, Cooper and Co., 1968)

A.

B.

is, when liquids are subjected to forces that act to change their shape, they simply flow.

4 In all materials, P waves travel faster than S waves.

5 When seismic waves pass from one material to another, the wave energy is refracted (bent).[1] In addition, some of the energy is reflected from the **discontinuity** (boundary between the two dissimilar materials). This is similar to what happens to light when it passes from air into water.

---
[1]Refraction occurs provided that the ray is not traveling perpendicular to the boundary.

Thus, depending on the nature of the layers through which they pass, seismic waves are speeded up or slowed down, bent, and in some cases, stopped altogether. These observable changes in seismic wave motions enable seismologists to probe the earth's interior.

If the earth were a perfectly homogeneous body, seismic waves would spread through the earth in all directions as shown in Figure 15.3. Seismic waves traveling through such an ideal planet would travel in a straight line at a constant speed. However, this is not the case for the earth. It so happens that the seismic waves that reach seismographic stations located

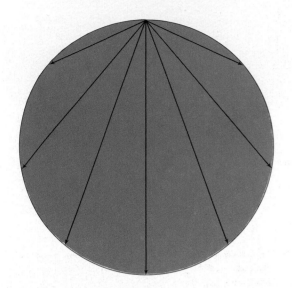

FIGURE 15.3
Seismic waves travel through a planet with uniform
properties along linear paths and at constant velocities.

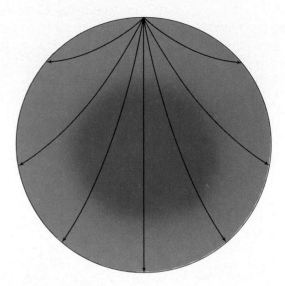

FIGURE 15.4
Wave paths through a planet where velocity increases
with depth.

farther from an earthquake travel at faster speeds
than those that are recorded at more nearby observatories. As stated earlier, this general increase in
speed with depth is a consequence of increased pressure which enhances the elastic properties of deeply
buried rock. As a result, the paths of seismic rays
through the earth are refracted (bent) as shown in
Figure 15.4.

As more sensitive seismographs were developed,
it became apparent that in addition to gradual velocity changes, rather abrupt changes also occur at particular depths. Since these discontinuities were detected worldwide, seismologists concluded that the
earth must be composed of distinct layers or shells of
varying composition or structure (Figure 15.5). Compositional layering is believed to have resulted from
density sorting that took place during an early molten
period in the earth's history. During this period heavy
substances sank while lighter components floated
upward. Structural layering, on the other hand, represents material of the same composition that has
undergone a phase change. Phase changes occur
when rock has melted or almost entirely melted, or
when the atoms in minerals have rearranged themselves into tighter crystalline structures in response to
the enormous pressures existing at great depths.

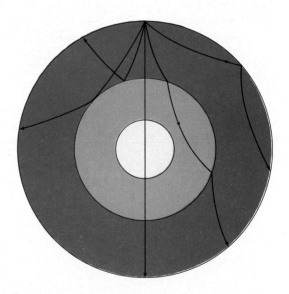

FIGURE 15.5
A few of many possible paths seismic rays take through
the earth.

Seismological data gathered from numerous seismographic stations have been continuously compiled
and analyzed for many years. From this information,
seismologists have, over the past 75 years, developed a detailed picture of the earth's interior (Figure

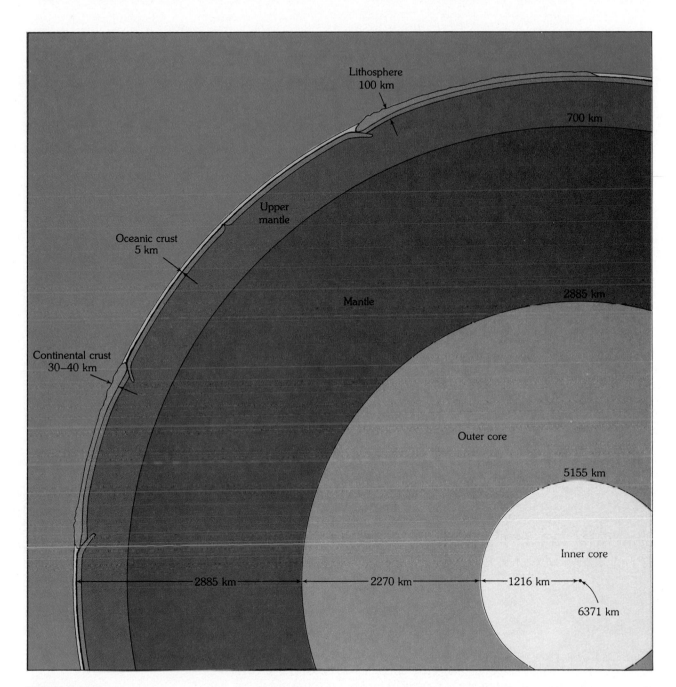

FIGURE 15.6
Cross-sectional view of the earth showing internal structure.

15.6). This model is continually rechecked and fine-tuned as more data become available and as new seismic techniques are employed. Further, laboratory studies which experimentally determine the proper-ties of various earth materials under the extreme environments found deep in the earth add to this body of knowledge.

Based upon this seismological data the earth has

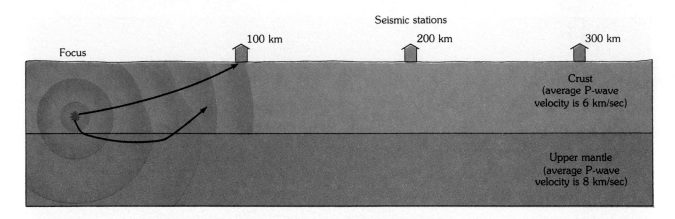

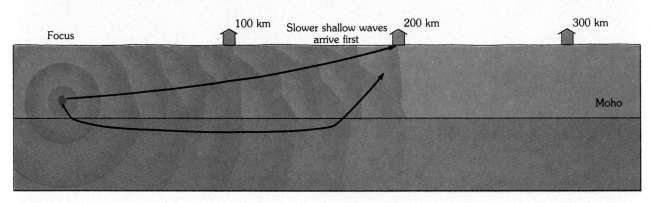

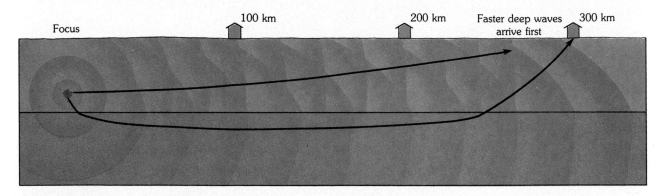

FIGURE 15.7
Idealized paths of seismic waves traveling from an earthquake focus to three seismographic stations. The two nearest recording stations receive the slower waves first, because the waves traveled a shorter distance. However, beyond 200 kilometers, the first waves received are the higher-velocity waves that passed through the mantle.

been divided into four major layers: (1) the **crust,** a very thin outer layer; (2) the **mantle,** a rocky layer located below the crust and having a maximum thick-ness of 2885 kilometers; (3) the **outer core,** a layer about 2270 kilometers thick which exhibits characteristics of a mobile liquid; and (4) the **inner core,** a solid

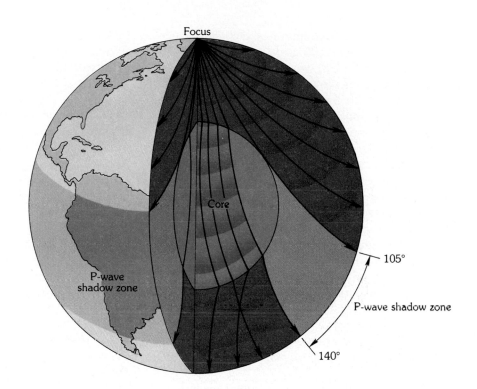

FIGURE 15.8
The abrupt change in physical properties at the mantle-core boundary causes the wave paths to bend sharply. This abrupt change in wave direction results in a shadow zone for P waves between about 105 and 140 degrees.

metallic sphere, about 1216 kilometers in radius. As we shall see, within these layers some notable features have also been discovered.

DISCOVERING THE EARTH'S STRUCTURE

In 1909, a pioneering Yugoslavian seismologist, Andrija Mohorovičić, presented the first convincing evidence for layering within the earth.[1] The boundary he discovered separates crustal rocks from rocks of different composition in the underlying mantle and was named the **Mohorovičić discontinuity** in his honor. For reasons that are obvious, the name for this boundary was quickly shortened to **Moho.**

By carefully examining the seismograms of shallow earthquakes, Mohorovičić found that seismographic stations located more than 200 kilometers from an earthquake obtained appreciably faster average travel velocities for P waves than stations located nearer the quake. In particular, P waves that reached the closer station first had velocities that averaged about 6 kilometers per second. By contrast, the seismic energy recorded at more distant stations traveled

at speeds which approached 8 kilometers per second. This abrupt jump in velocity did not fit the general pattern that had been observed previously. From this data, Mohorovičić concluded that below 30 kilometers there exists a layer with markedly different physical properties than that of the earth's outer shell.

Figure 15.7 illustrates how Mohorovičić came to this important conclusion. Notice that the first ray to reach the seismographic station located 100 kilometers from the epicenter traveled the shortest route directly through the crust. However, at the seismographic station that was 300 kilometers from the epicenter, the first P wave entered and traveled in the mantle, a zone of higher velocity. Thus, although this wave traveled a greater distance, it reached the recording instrument sooner than any of the direct rays, because a large portion of its journey was through a region having a different composition. This principle is analogous to a driver taking a by-pass route around a large city during rush hour. Although this alternate route is longer, it may be faster.

A few years later another major boundary was discovered by the German seismologist Beno Gutenberg. This discovery was based primarily on the

---

[1]Discontinuities in the earth had been predicted by earlier researchers, but their arguments for a central core were inconclusive.

observation that P waves diminish and eventually die out completely about 105 degrees from an earthquake. Then, about 140 degrees away, the P waves reappear, but about two minutes later than would be expected based on the distance traveled. This belt where direct seismic waves are absent is about 35 degrees wide and has been named the **shadow zone**[1] (Figure 15.8). Gutenberg realized that the shadow zone could be explained if the earth contained a core composed of material unlike the overlying mantle and had a radius of 3420 kilometers. The core must somehow hinder the transmission of P waves in a manner similar to the light rays blocked by an opaque object which casts a shadow. However, rather than actually stopping the P waves, the shadow zone is produced by the bending of P waves which enter the core as shown in Figure 15.8.

It was further learned that S waves could not propagate through the core; therefore, geologists concluded that at least a portion of this region is liquid (Figure 15.9). This conclusion was further supported

by the observation that P-wave velocities suddenly decrease about 40 percent as they enter the core. Since melting would reduce the elasticity of rock, all evidence points to the existence of a liquid layer below the rocky mantle.

In 1936, the last major subdivision of the earth's interior was predicted by the discovery of seismic waves believed to be reflected from a boundary within the core. Hence, a core within a core was discovered. The actual size of the inner core was not accurately calculated until the early 1960s when underground nuclear tests were conducted in Nevada. Because the precise location and time of the explosions were known, echoes from seismic waves which bounced off the inner core provided an accurate means of determining its size (Figure 15.10). From these data and subsequent studies, the inner core was found to have a radius of about 1216 kilometers. Further, P waves passing through the inner core have appreciably faster travel times than those penetrating the outer core exclusively. The apparent increase in the elasticity of the inner core material is considered evidence for the solid nature of this innermost region.

---

[1]As more sensitive instruments were developed, weak and delayed P waves that enter this zone via reflection were detected.

FIGURE 15.9
View of the earth showing the paths of P and S waves. Any location more than 105 degrees from the epicenter will not receive direct S waves, since the outer core will not transmit them. A small shadow zone exists (from 105 to 140 degrees) for P waves. The P-wave shadow zone is caused by the bending of these waves as they pass from more rigid mantle material to less rigid core material. The seismic waves that pass through the center of the earth increase in velocity, revealing the existence of the solid inner core.

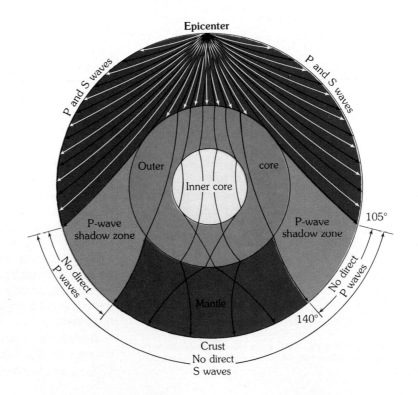

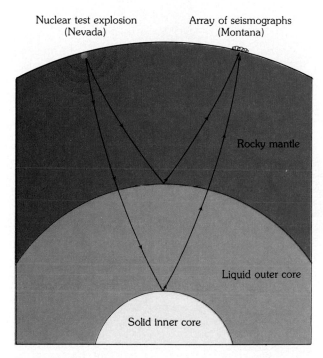

Nuclear test explosion
(Nevada)

Array of seismographs
(Montana)

Rocky mantle

Liquid outer core

Solid inner core

FIGURE 15.10
Travel times of seismic waves generated from nuclear test explosions were used to accurately measure the depth of the inner core. An array of seismographs located in Montana detected the "echoes" that bounced back from the boundary of the inner core.

Over the past 25 years, advances in seismology and rock mechanics have allowed for much refinement of the gross view of the earth's interior that has been presented to this point. Some of these refinements as well as other properties of these major divisions, including their densities and compositions, will be considered next.

THE CRUST

The crust of the earth is on the average less than 20 kilometers thick, making it the thinnest layer so far encountered. However, along this eggshell-thin layer great variations in rock composition and thickness exist. Whereas the crustal rocks of the continental masses are roughly 35 kilometers thick, the oceanic crust is much thinner, averaging only 5 kilometers. In a few exceptionally prominent mountainous regions, the crust obtains its greatest thickness, exceeding 60 kilometers. By contrast, in the stable continental interior, its thickness is closer to 30 kilometers.

The discovery that crustal rocks of the deep-ocean basins are compositionally different from those of the continental masses was first established through studies of seismic velocities. P-wave travel times indicate that velocities of 6 kilometers per second are typical for continental rocks, whereas velocities of 7 kilometers per second are recorded for the oceanic crust. Laboratory experiments were designed to determine which earth materials could produce travel times most like those recorded for these rocky layers. From these experiments, as well as from direct observations, the average composition of continental rocks has been compared with that of the igneous rock granite. Like granite, the continental crust is believed to be enriched in the elements potassium, sodium, and silicon and to have an average density about 2.8 times that of water. Although numerous granitic intrusions and equivalent metamorphic rocks such as gneiss can be found, large outpourings of basalt and volcanic chains composed of andesitic rocks are also abundant. Consequently, indications are that the continental crust's average composition is more similar to rocks of intermediate composition, such as andesite and diorite, than that of "true" granite.

Until recently, geologists could only speculate on the composition of the deep-oceanic crust, which lies beneath 4 kilometers of seawater as well as hundreds of meters of sediments. With the development of the deep-sea drilling ship *Glomar Challenger,* the recovery of core samples from the ocean floor became possible (Figure 15.11). As predicted, the samples obtained were predominately basaltic; indeed, they were different from the rocks which compose the continents. Recall that volcanic eruptions of basaltic material are known to have generated the islands located within the deep-ocean basins.

THE MANTLE

Over 80 percent of the earth's volume is contained within the mantle, a 2885-kilometer thick shell of rock extending from the base of the crust (Moho) to the liquid outer core. Our knowledge of the mantle's composition comes from experimental data, as well as from the examination of material intruded into the crust from below. In particular, the rocks composing kimberlite pipes, in which diamonds are often found,

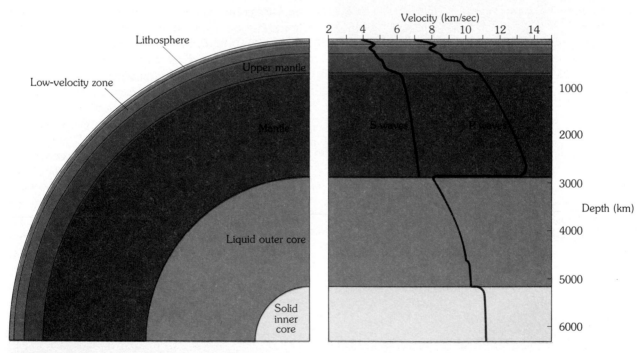

FIGURE 15.12

Variations in P and S wave velocities with depth. Abrupt changes in average wave velocities delineate the major features of the earth's interior. At a depth of about 100 kilometers, the sharp decrease in wave velocity corresponds to the top of the low-velocity zone. Two other bends in the velocity curves occur in the upper mantle at depths
of about 400 and 700 kilometers. These variations are thought to be caused by minerals that have undergone phase changes, rather than resulting from compositional differences. The abrupt decrease in P-wave velocity and the absence of S waves at 2885
kilometers marks the core-mantle boundary. The liquid inner core will not transmit S
waves and within this layer the propagation of P waves is slowed. As the P waves enter the solid inner core, their velocity once again increases. (Data from Bruce A. Bolt)

are thought to have originated at depths approaching 200 kilometers, well within the mantle. These kimberlite deposits are composed of peridotite, a rock that contains iron and magnesium-rich silicate minerals, mainly olivine and pyroxene, plus lesser amounts of garnet. Further, because S waves are readily propagated through the mantle, we conclude that it behaves as an elastic solid. Thus, the mantle is described as a solid rocky layer, the upper portion of which has the same composition as the rock peridotite.

As might be expected, this simple picture of the mantle is far from complete. Any working model of the mantle must explain the temperature distribution calculated for this layer. Whereas the crust has a large increase in temperature with depth, this same trend does not continue downward into the mantle. Rather, the temperature increase with depth in the mantle is apparently much more gradual. This means that the mantle has an effective method of transmitting heat outward. If heat were transmitted through the mantle by conduction, as occurs in the crust, the lower mantle would, out of necessity, be hundreds of times hotter than the outer mantle since the conduction of heat through rock is very slow. Consequently, most geologists conclude that some form of mass transport (convection) of hot rock must exist within the mantle. This being the case, the rock of the mantle, where temperatures and pressures are extreme, must be capable of flow.

If this is true, how does the rocky mantle transmit S waves, which can only travel through solids, and at the same time flow like a fluid? This apparent contradiction can be resolved if the material behaves like a solid under certain conditions and like a fluid the remainder of the time. Geologists generally describe material of this type as exhibiting *plastic* behavior. This means that when the material encounters short-lived stresses, such as those produced by seismic waves, the material behaves like an elastic solid. However, in response to long-term stresses, this same rocky material will flow. This also explains why S waves can penetrate the mantle, yet at the same time, this layer is not able to store elastic energy like a brittle solid and is thus incapable of generating earthquakes. This apparently unusual phenomenon is not restricted to mantle rocks. The manmade substances Silly Putty and some taffy candies also exhibit plastic behavior. When struck with a hammer these materials shatter like a brittle solid. However, when slowly pulled apart they flow plastically. From this example do not get the idea that the mantle is composed of soft puttylike material. Rather it is composed of red-hot, solid rock, which under extreme pressures unknown on the surface of the earth, exhibits the ability to flow.

More recent efforts to probe the upper mantle have confirmed earlier speculations that finer divisions also exist. One of the most significant of these subdivisions is a region located between the depths of 100 and 250 kilometers called the **low-velocity zone.** When penetrating this zone, P and S waves show a marked decrease in velocity (Figure 15.12). The most probable explanation for the observed slowing of seismic energy is the existence of molten rock. This molten material is believed to exist in discrete pockets as mixtures of melt and crystals which make up less than 10 percent of the zone. Although found below the oceanic crust and portions of the continents, this low-velocity zone does not encircle the earth. It is notably absent, for example, below the older shield areas of the continents.

The discovery of the low-velocity zone supports a proposal made earlier that a zone of weak rock exists below 100 kilometers (Figure 15.13). This region of weak material is called the **asthenosphere.** Unlike the low-velocity zone, which is absent under portions of the continents, the asthenosphere is thought to be a global feature. Further, this weak zone may in some regions extend downward as far as 700 kilometers, but only the upper zone consists of partially molten rock. As can be seen in Figure 15.14, this weak layer exists because the rock at this level is nearer its melting point than the rock above or below it. Thus, like red-hot iron, the rock within this zone is easily deformed.

Situated above the asthenosphere is the cool brittle layer about 100 kilometers thick called the **lithosphere** (Figure 15.12). Actually the lithosphere includes the entire crust as well as the uppermost mantle and is defined as that layer of the earth cool enough to behave like a brittle solid.

Movements within the weak asthenosphere are believed to induce motion in the rigid lithosphere above. The discovery of this weak layer was an important contribution to the theory of plate tectonics, which proposes that lithospheric plates move about the face of the earth. It is also from the upper

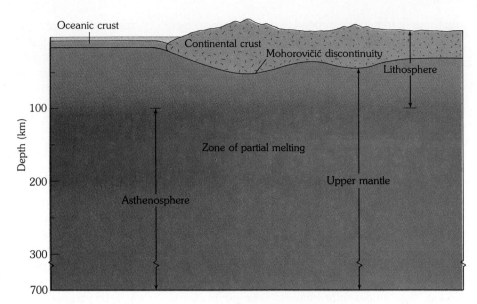

FIGURE 15.13
Respective locations of the asthenosphere and lithosphere.

asthenosphere that some of the molten rock associated with volcanic activity originates.

At the depth of about 400 kilometers a relatively abrupt increase in seismic velocity has been detected (Figure 15.12). While the velocity increase at the crust-mantle boundary is thought to represent a change in composition, the increase at the 400-kilometer level is believed to be the result of a phase change. A phase change occurs when a mineral adjusts its structure in response to changes in temperature and/or pressure. Laboratory studies show that the mineral olivine, $(Mg,Fe)_2SiO_4$, which is one of the main constituents in the rock peridotite, will collapse to a more compact, high-pressure mineral at the pressures experienced at this depth. This structural change could explain the increased seismic velocities observed.

Another boundary is believed to have been detected from seismic velocity variations at a depth of 700 kilometers (Figure 15.12). At this depth the minerals in peridotite are believed to break down into simple metallic oxides. Thus, from 700 to 2885 kilometers the mantle is believed to consist primarily of iron oxide (FeO), magnesium oxide (MgO), and silicon dioxide (SiO_2), rather than silicate minerals having the same gross composition.

THE CORE

As with the other layers discussed thus far, the fact that the earth contained a central core was estab-

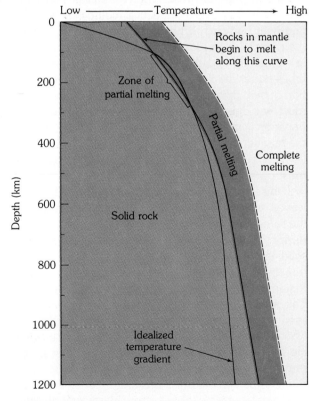

FIGURE 15.14
The relationship between the proposed temperature gradient and melting-point curve of mantle material. In the zone between 100 and 250 kilometers the mantle material slightly exceeds its melting point, which accounts for the existence of the low-velocity zone.

lished from seismological data. Having a radius of 3486 kilometers, this dense sphere inside the earth is larger than the planet Mars. Extending from the inner edge of the mantle to the center of the earth, the core constitutes about one-sixth of the earth's volume and nearly one-third of its total mass. Pressures at the center are millions of times greater than the air pressure at the earth's surface and temperatures are estimated to be between 3000 and 5000°C. As more precise seismic data became available, the core was found to consist of a liquid outer layer about 2270 kilometers thick, and a solid inner portion with a radius of 1216 kilometers.

One of the most interesting characteristics of the core is its great density. At the core-mantle boundary the density is nearly ten times greater than water and at the center the density is 13.5 times greater than water. Even under the extreme pressures at those depths, the common silicate minerals found in the crust with densities 2.6 to 3.5 times that of water could not be compacted enough to account for the great densities calculated for the core. Consequently, attempts were undertaken to determine the earth material which could account for this property.

Surprisingly enough, meteorites provided an important clue to the earth's internal composition. Since meteorites are part of the solar system, they are assumed to be representative samples of the material from which the earth originally accreted. Their composition ranges from metallic types made primarily of iron, to stoney meteorites composed of rocky substances that closely resemble the rock peridotite. Because the earth's crust contains a much smaller percentage of iron than is dictated by the relative abundance of iron in the debris of the solar system, geologists concluded that the interior of the earth must be enriched in this heavy material. Further, iron is the only abundant substance found in the solar system that exhibits the proper density.

Although the core is predominantly iron, it cannot be pure iron. Experiments indicate that the density of pure iron under the extreme pressures of the core is about 10 percent higher than the density that was actually established. This being the case, the suggestion has been made that the core must also contain some lighter elements which alloy with iron and lower its density. This idea is supported by the fact that the best estimates of core temperatures are below the melting environment for pure iron. Thus, if the outer

core were pure iron it would have long ago crystallized, a condition that contradicts seismological data. A liquid outer core can be explained by the addition of lighter elements which, when mixed with iron, lower its melting point. The elements most likely to alloy with iron and account for the core's observed density and the liquid state of the outer core are sulfur and oxygen. However, other substances, including silicon and carbon, are undoubtedly present in minor amounts.

Although the existence of a metallic central core is well established, efforts to explain the core's origin are more speculative. The most widely accepted scenario suggests that the core formed early in the earth's history from what was originally a relatively homogeneous body. During the period of accretion the entire earth was heated by the energy released by the infalling material. Sometime late in this period of growth, the earth's internal temperature was sufficiently high to mobilize the accumulated material. Blobs of heavy iron-rich material collected and sank toward the center. Simultaneously, lighter substances may have floated upward to generate the mantle, and possibly portions of the crust as well. In a short time, geologically speaking, the earth took on a layered configuration perhaps not much different than what we find today.

How can we explain the existence of a liquid outer core when the inner core, which must be hotter, is solid? Most probably in its formative stage the entire core was liquid. Further, this liquid iron alloy was in a state of vigorous mixing. However, during the last 3.5 billion years, the material of the core has been slowly segregating. As the core cooled, a portion of the iron components gradually migrated downward while some of the lighter components floated upward toward the outer edge of the core. The sinking iron-rich components, depleted of the lighter elements which act to depress the melting point, began to solidify.[1] The downward migration and crystallization of the heavier material releases gravitational energy and the heat of fusion which drives the currents in the remaining liquid shell above.

Our picture of the core with its solid inner sphere surrounded by a mobile liquid shell is further supported by the existence of the earth's magnetic field,

---

[1]A similar phenomenon occurs in the Arctic Ocean where salt is driven from seawater during the formation of ice. Here the seawater is colder and denser than the salt-free ice directly above.

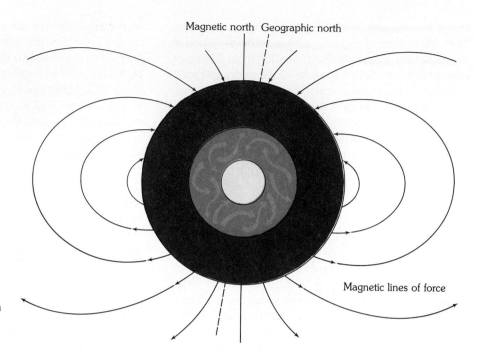

Magnetic north Geographic north

Magnetic lines of force

FIGURE 15.15
The earth's magnetic field is thought to be generated by the vigorous mixing of molten iron alloy in the liquid outer core.

which acts as if a large bar magnet were situated deep within the earth. However, we know that the source of the magnetic field cannot be permanently magnetized material, because the earth's interior is too hot for any material to retain its magnetism. The most widely accepted mechanism explaining the earth's magnetic field requires that the core be made of a material which conducts electricity, such as iron, and one that is mobile enough to circulate (Figure 15.15). Both of these conditions are met by the model of the earth's core that was established on the basis of seismological data.

REVIEW QUESTIONS

1 List the major differences between P and S waves.

2 How does the boundary between the crust and mantle (Moho) differ from the boundaries that occur at depths of about 400 and 700 kilometers?

3 Describe the lithosphere. In what important way is it different from the asthenosphere?

4 Describe the chemical (mineral) makeup of the four principle layers of the earth.

5 Why was it difficult for seismologists to obtain precise travel-time data before the turn of the century?

6 Describe the method first used to accurately measure the size of the inner core.

7 How were the first samples (in place) of the deep-ocean floor obtained?

8 What evidence did Gutenberg use for the existence of the earth's central core?

9 Suppose the shadow zone for P waves was located between 120 and 160 degrees, rather than between 105 and 140 degrees. What would this indicate about the size of the core?

10 Explain why the asthenosphere is able to flow like a fluid, yet has the ability to transmit S waves which cannot travel through fluids.

11 Why are meteorites considered important clues to the composition of the earth's interior?

12 What evidence is provided by seismology to indicate that the outer core is liquid? What other evidence exists for a molten outer core?

13 Why is it possible for the outer core to be molten when the inner core (which has a higher temperature) is in the solid state?

14 In the preceding chapter the statement was made that earthquakes occur only in brittle rock capable of storing elastic energy. Below 100 kilometers the rock is weak and flows freely when stress is applied. How then can we account for the existence of deep-focus earthquakes?

KEY TERMS

asthenosphere (p. 377)
crust (p. 372)
discontinuity (p. 369)
inner core (p. 372)

lithosphere (p. 377)
low-velocity zone (p. 377)
mantle (p. 372)

Mohorovičić discontinuity, or Moho (p. 373)
outer core (p. 372)
shadow zone (p. 374)

PLATE TECTONICS

16

Early in this century geologic thought about the age of the ocean basins was dominated by a belief in their antiquity. Moreover, most geologists accepted the geographic permanency of the oceans and continents. Mountains were believed to result from the contraction of the earth caused by gradual cooling from a once molten state. As the interior cooled and contracted, the earth's solid outer skin was deformed by folding to fit the shrinking planet. Mountains were therefore regarded as analogous to the wrinkles on a dried-out piece of fruit. This model of the earth's tectonic[1] processes, however inadequate, was firmly entrenched in the geologic thought of the time. Even changes in sea level, evident from the record of marine fossils found deep in the continental interiors, were explained using the model of a gradually contracting earth. As the earth's solid outer shell was deformed, some regions subsided and were inundated by the sea, while other areas emerged as dry land.

---

[1]Tectonics refers to the deformation of earth's crust and results in the formation of structural features such as mountains.

The Sinai Peninsula is bordered by rifts believed to be caused by sea-floor spreading. (Courtesy of NASA)

385

During the last few decades spectacular developments have taken place in the earth sciences. Due to the vast accumulation of new data, our ideas about the structure and workings of the earth have changed dramatically. Earth scientists now realize that the positions of landmasses are not fixed. Rather, the continents gradually migrate across the globe. The splitting of continental blocks has resulted in the formation of new ocean basins, while older segments of the sea floor are continually being recycled in areas where we find deep-ocean trenches. Further, because of this movement, once disjointed segments of continental material have collided and formed the earth's great mountain ranges. In short, a revolutionary new model of the earth's tectonic processes has emerged in marked contrast to what was accepted just a few decades ago.

This profound reversal of scientific opinion has been appropriately described as a scientific revolution. Like other scientific revolutions, an appreciable length of time elapsed between the idea's inception and its general acceptance. The revolution began in the early part of the twentieth century as a relatively straightforward proposal that the continents drifted about the face of the earth. After many years of heated debate, the idea of drifting continents was rejected by the vast majority of earth scientists as being improbable. The concept of a mobile earth was particularly distasteful to North American geologists because the majority of evidence came from the Southern Hemisphere. This fact is evidenced by the meager amount of material concerning continental drift in the scientific literature in the United States between 1930 and 1950. However, during the 1950s and 1960s new evidence began to rekindle interest in this abandoned proposal. By 1968 these new developments led to the unfolding of a far more encompassing theory than continental drift—a theory known as plate tectonics.

In this chapter we will examine the events which led to this dramatic reversal of scientific opinion in an attempt to provide some insight into how science works. We will briefly trace the developments that took place from the inception of the concept of continental drift through the general acceptance of the theory of plate tectonics. The evidence gathered to support the concept of a mobile earth will also be provided.

CONTINENTAL DRIFT: AN IDEA BEFORE ITS TIME

The idea that continents, particularly South America and Africa, fit together like pieces of a jigsaw puzzle originated with improved world maps. However, little significance was given this idea until 1915, when Alfred Wegener[1], a German meteorologist and geophysicist, published an expanded version of a 1912 lecture in his book *The Origin of Continents and Oceans*. In this monograph, Wegener set forth the basic outline of his radical hypothesis of **continental drift.** One of his major tenets suggested that a supercontinent he called **Pangaea** (meaning ''all land'') once existed (Figure 16.1). He further hypothesized that about 200 million years ago this supercontinent began breaking into smaller continents, which then ''drifted'' to their present positions. Wegener and others who advocated this position collected substantial evidence to support these claims. The fit of South America and Africa, ancient climatic similarities, fossil evidence, and rock structures all seemed to support the idea that these now separate landmasses were once joined.

FIT OF THE CONTINENTS

Like a few others before him, Wegener first suspected that the continents might have been joined when he noticed the remarkable similarity between the coastlines on opposite sides of the South Atlantic. However, his use of present-day shorelines to make a fit of the continents was challenged immediately by other earth scientists. These opponents correctly argued that shorelines are continually modified by erosional processes and even if continental displacement had taken place, a good fit today would be unlikely. Further, abundant fossil evidence exists that indicates

---

[1]Wegener's ideas were actually preceded by those of an American geologist, F. B. Taylor, who in 1910 published a paper on continental drift. Taylor's paper provided little collaborating evidence for continental drift which may have been the reason that it had a relatively small impact on the geologic community.

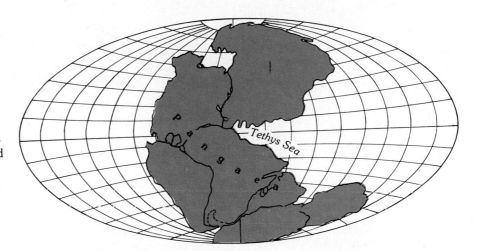

FIGURE 16.1
Reconstruction of Pangaea as it is thought to have appeared 200 million years ago. (After R. S. Dietz and J. C. Holden. *Journal of Geophysical Research* 75 : 4943. Copyright by American Geophysical Union)

most of the world's land areas have experienced periods of either uplift or subsidence in the recent geologic past. This would have markedly altered the position of the global coastlines. Wegener appeared to be aware of these problems, and, in fact, his original jigsaw fit of the continents was only very crude.

A much better approximation of the outer boundary of the continents is the seaward margin of the continental shelf. Today the continental shelf's edge lies several hundred meters below sea level. In the early 1960s, Sir Edward Bullard and two associates produced a map with the aid of computers that attempted to fit the continents at a depth of about 900 meters. The remarkable fit that was obtained is shown in Figure 16.2. Although the continents overlap in a few places, these are regions where streams have deposited large quantities of sediment, thus enlarging the continents. The overall fit obtained by Bullard and his associates was better than even the supporters of the continental drift theory suspected it would be.

FOSSIL EVIDENCE

Although Wegener was intrigued by the remarkable similarities of the shorelines on opposite sides of the Atlantic, he at first thought the idea of a mobile earth improbable. Not until he came across an article citing fossil evidence for the existence of a land bridge connecting South America and Africa did he begin to take his own idea seriously. Through a search of the

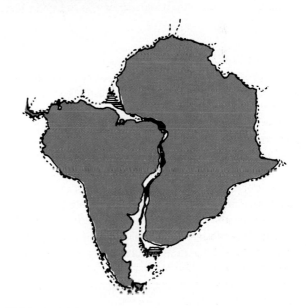

FIGURE 16.2
The best fit of South America and Africa along the continental slope at a depth of 500 fathoms (about 900 meters). (After A. G. Smith. "Continental Drift." In *Understanding the Earth,* edited by I. G. Gass. Courtesy of Artemis Press)

literature Wegener learned that most paleontologists were in agreement that some type of land connection was needed to explain the existence of identical fossils on the widely separated landmasses. This requirement was particularly true for Mesozoic life forms.

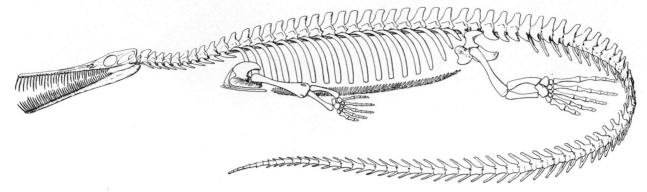

FIGURE 16.3
Drawing of a fossil *Mesosaurus* skeleton. Fossil remains of this and other organisms found on the continents of Africa and South America appear to link these landmasses during the Mesozoic era. (Courtesy of The American Museum of Natural History)

To add credibility to his argument for the existence of Pangaea, Wegener used the already documented evidence that several fossil organisms exist which could not have made the journey across the vast oceans presently separating the continents. In particular, the fossil fern *Glossopteris* was known to have been widely dispersed in the southern continents of Africa, Australia, and South America during the Mesozoic era. Later, fossil remains of *Glossopteris* were discovered in Antarctica as well. In addition, remains of a species of swimming reptile called *Mesosaurus* were found in eastern South America and western Africa (Figure 16.3). Although this reptile probably swam in the shallow waters of these regions, *Mesosaurus* was clearly not capable of making the long journey across the Atlantic Ocean. For Wegener, fossils provided undeniable proof that these landmasses were once joined together as the supercontinent Pangaea.

In his book, Wegener also cited the distribution of present-day organisms as evidence to support the concept of drifting continents. For example, modern organisms with similar ancestries clearly had to evolve in isolation during the last few tens of millions of years. Most obvious of these are the Australian marsupials, which have a direct fossil link to the marsupial opossums found in the Americas.

How could these fossil flora and fauna be so similar in places separated by thousands of kilometers of open ocean? The idea of land bridges was the most widely accepted solution to the problem of migration (Figure 16.4). We know for example that during the recent glacial period the lowering of sea level allowed animals to cross the narrow Bering Straits between Asia and North America. Was it possible then that land bridges once connected Africa and South America? We are now quite certain that land bridges of this magnitude did not exist, for their remnants should still lie below sea level, but are nowhere to be found.

ROCK TYPE AND STRUCTURAL SIMILARITIES

Anyone who has worked a picture puzzle knows that in addition to the pieces fitting together, the picture must be continuous as well. The picture that must match in the "Continental Drift Puzzle" is represented by the rock types and mountain belts found on the continents. If the continents were once together, the rocks found in a particular region on one continent should closely match in age and type with those found in adjacent positions on the matching continent. For example, a good correlation between rocks found in northwestern Africa was made with rocks in eastern Brazil. Recent re-examination of this early evidence has supported Wegener's claim. In both regions, 550 million-year-old rocks lie adjacent to rocks dated at more than 2 billion years in such a manner that the line separating them is continuous when the two continents are brought together (Figure 16.5).

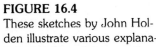

FIGURE 16.4
These sketches by John Holden illustrate various explanations for the occurrence of similar species on landmasses that are presently separated by vast oceans. (Reprinted with permission of John Holden)

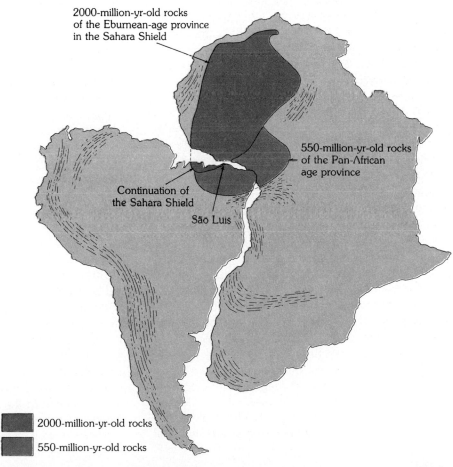

FIGURE 16.5
The jigsaw fit of Africa and South America illustrates that rocks of equivalent age match when the continents are reassembled. (After P. M. Hurley)

2000-million-yr-old rocks of the Eburnean-age province in the Sahara Shield

550-million-yr-old rocks of the Pan-African age province

Continuation of the Sahara Shield

São Luis

2000-million-yr-old rocks

550-million-yr-old rocks

Further evidence to support the concept of continental drift comes from several mountainous belts which appear to terminate at one coastline only to reappear again on a landmass across the ocean. For instance, the mountain belt that includes the Appalachians trends northeastward through the eastern United States and disappears off the coast of Newfoundland. Mountains of comparable age and structure are found in Greenland and Northern Europe. When these landmasses are reassembled as in Figure 16.1, the mountain chains form a nearly continuous belt. Numerous other rock structures exist that appear to have formed at the same time and were subsequently split apart.

Wegener was very satisfied that the similarities in rock structure on both sides of the Atlantic linked these landmasses. In fact, he was too zealous with this evidence and incorrectly suggested that glacial moraines in North America matched up with those of Northern Europe. In his own words, "It is just as if we were to refit the torn pieces of a newspaper by matching their edges and then check whether the lines of print run smoothly across. If they do, there is nothing left to conclude except that the pieces were in fact joined this way."

PALEOCLIMATIC EVIDENCE

Since Alfred Wegener was a climatologist by training, he was keenly interested in obtaining paleoclimatic (ancient climatic) data in support of continental drift. His efforts in this area were rewarded when he found evidence for apparently dramatic climatic changes. For instance, glacial deposits indicated that near the end of the Paleozoic era (between 220 and 300 million years ago), ice sheets covered extensive areas of the Southern Hemisphere. Layers of glacial till were found at the same stratigraphic position in southern Africa and South America, as well as in India and Australia. Below these beds of glacial debris lay striated and grooved bedrock. In some locations the striations and grooves indicated the ice had moved from the sea onto land (Figure 16.6). Much of the land area containing evidence of this late Paleozoic glaciation presently lies within 30 degrees of the equator in a subtropical or tropical climate.

Could the earth have gone through a period sufficiently cold to have generated extensive continental glaciers in what is presently a tropical region? Wegener rejected this explanation because during the late Paleozoic, large tropical swamps existed in the Northern Hemisphere. These swamps with their lush vegetation eventually became the major coal fields of the eastern United States, Europe, and Siberia. As Wegener proposed, a better explanation is provided if the landmasses are fitted together as a supercontinent and then moved nearer to the South Pole. This would account for the conditions necessary to generate extensive expanses of glacial ice over much of the Southern Hemisphere. At the same time this shift would place the northern landmasses nearer the tropics and account for their vast coal deposits.

How does a glacier develop in hot, arid Australia? How do land animals migrate across wide expanses of open water? As compelling as this evidence may have been, fifty years passed before most of the scientific community would accept it and the logical conclusions to which it led.

THE GREAT DEBATE

Wegener's proposal did not attract much open criticism until 1924 when his book was translated into English. From this time on, until his death in 1930, his drift hypothesis encountered a great deal of hostile criticism. To quote the respected American geologist R. T. Chamberlin, "Wegener's hypothesis in general is of the foot-loose type, in that it takes considerable liberty with our globe, and is less bound by restrictions or tied down by awkward, ugly facts than most of its rival theories. Its appeal seems to lie in the fact that it plays a game in which there are few restrictive rules and no sharply drawn code of conduct." W. B. Scott, former president of the American Philosophical Society, expressed the prevalent American view of continental drift in fewer words when he described the theory as "utter damned rot!"

One of the main objections to Wegener's hypothesis stemmed from his inability to provide a mechanism for continental drift. Wegener proposed two possible energy sources for drift. One of these, the tidal influence of the moon, was presumed by Wegener to be strong enough to give the continents a westward motion. However, the prominent physicist

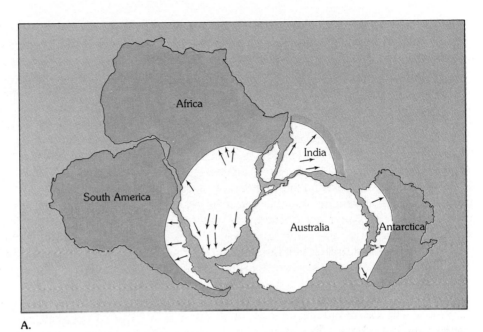

A.

FIGURE 16.6
A. Direction of ice movement in the southern supercontinent called Gondwanaland by the founders of the continental drift concept. **B.** Glacial striations in the bedrock of Hallet Cove, South Australia, indicate direction of ice movement. (Photo by W. B. Hamilton, U.S. Geological Survey)

B.

Harold Jeffreys quickly countered with the argument that tidal friction of the magnitude needed to displace the continents would bring the earth's rotation to a halt in a matter of a few years. Further, Wegener proposed that the larger and sturdier continents broke through the oceanic crust, much like ice breakers cut through ice. However, no evidence existed to suggest that the ocean floor was weak enough to permit passage of the continents without themselves being appreciably deformed in the process. By 1929 criticisms of Wegener's ideas were pouring in from all areas of the scientific community. Despite these afronts, Wegener wrote the fourth and final edition of his book, maintaining his basic hypothesis and adding supporting evidence.

In 1930, Wegener made his third and final trip to the Greenland ice sheet. Although the primary focus of this expedition was to study the harsh winter weather on the ice-covered island, he planned to test his continental drift hypothesis as well. Wegener felt that by precisely establishing the locations of specific points and then measuring their changes over a period of years, he could demonstrate the westward drift of Greenland with respect to Europe. In November, 1930, while returning from Eismitte (an experimental station located in the center of Greenland), Wegener perished along with a companion. His intriguing idea, however, did not die with him.

Why was Wegener not able to overturn the established view of his day? Although his hypothesis was correct in principle, it also contained many incorrect details. For example, the continents do not break through the ocean floor and tidal energy is not the driving mechanism for continental displacement. In order for any scientific viewpoint to gain universal acceptance, supporting evidence from all realms of science must be found. This same idea was stated very well by Wegener himself in response to his critics, when he said, "Scientists still do not appear to understand sufficiently that all earth sciences must contribute evidence toward unveiling the state of our planet in earlier times, and the truth of the matter can only be reached by combining all this evidence." Wegener's great contribution to our understanding of the earth notwithstanding, *all* of the evidence did not support the continental drift hypothesis as he had proposed it. Therefore, Wegener himself answered the very question he must have asked many times— "Why do they reject my proposal?"

Although most of Wegener's contemporaries opposed his views, even to the point of openly ridiculing them, a few considered his ideas plausible. Among the most notable of this latter group was the eminent South African geologist Alexander du Toit and the well-known Scottish geologist Arthur Holmes. In 1937, du Toit published *Our Wandering Continents,* in which he eliminated some of Wegener's errors and added a great deal of new evidence in support of this revolutionary idea. Arthur Holmes contributed to the cause by proposing a plausible driving mechanism for continental drift. In Holmes' book *Physical Geology,* he suggested that convection currents operating within the mantle were responsible for propelling the continents across the globe. Although even to this day geologists are not in agreement on the nature of the driving mechanism for continental drift, the concept proposed by Holmes is still one of the most appealing.

For these few geologists who continued the search, the concept of continents in motion evidently provided enough excitement to hold their interest. Others undoubtedly viewed continental drift as a solution to previously unexplainable observations.

CONTINENTAL DRIFT AND PALEOMAGNETISM

Very little new light was shed on the continental drift hypothesis between the time of Wegener's death in 1930 and the early 1950s. Little was known about the land beneath the sea, which makes up over 70 percent of the earth's surface and was in fact the key to unraveling the secrets of our planet. Perhaps the initial impetus for the renewed interest in continental drift came from rock magnetism, a comparatively new field of study.

Early workers studying rock magnetism set out to investigate ancient changes in the earth's magnetic field in hopes of better understanding the nature of the present-day magnetic field. Anyone who has used a compass to find direction knows that the earth's magnetic field has a north pole and a south pole. These magnetic poles align closely, but not exactly, with the respective geographic poles. In many respects the earth's magnetic field is very much like that produced by a simple bar magnet. Invisible lines of force pass through the earth and extend from one pole to the other (Figure 16.7). A compass needle, itself a small magnet free to move about, becomes aligned with these lines of force and thus points toward the magnetic poles.

The technique used to study ancient magnetic fields relies on the fact that certain rocks contain minerals which serve as fossil compasses. These iron-rich

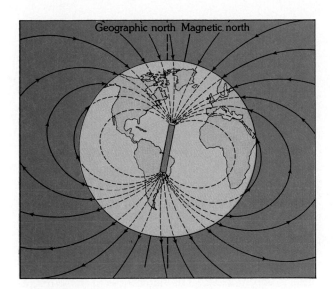

FIGURE 16.7
The earth's magnetic field consists of lines of force much like a giant bar magnet would produce if placed at the center of the earth.

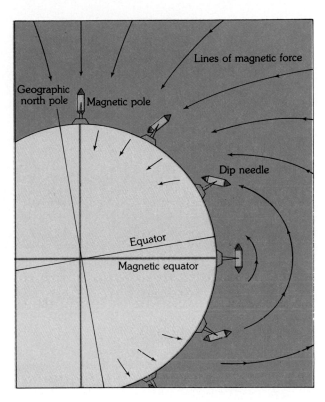

FIGURE 16.8
The earth's magnetic field causes a dip needle (compass oriented in a vertical plane) to align with the lines of magnetic force. The dip angle decreases uniformly from 90 degrees at the magnetic poles to 0 degrees at the magnetic equator. Consequently, the distance to the magnetic poles can be determined from the dip angle.

minerals such as magnetite are abundant, for example, in lava flows of basaltic composition. When heated above a certain temperature called the **Curie point**, these magnetic minerals lose their magnetism. However, when these iron-rich grains cool below their Curie point (about 580° C) they become magnetized in the direction parallel to the existing magnetic field. Once the minerals solidify, the magnetism they possess will remain "frozen" in this position. In this regard, they behave much like a compass needle inasmuch as they "point" toward the existing magnetic poles. If the rock is moved or the magnetic pole changes position, the rock magnetism will, in most instances, retain its original alignment. Rocks formed thousands or millions of years ago thus "remember" the location of the magnetic poles at the time of their formation and are said to possess fossil magnetism, or **paleomagnetism.**

One important aspect of rock magnetism is that the magnetized minerals not only indicate the direction to the poles (like a compass), they also provide a means of determining the latitude of their origin. To envision how latitude can be established from paleomagnetism, imagine a compass needle mounted in a vertical plane rather than horizontally like an ordinary compass. As shown in Figure 16.8, when this modified compass (dip needle) is located over the north magnetic pole it will point straight down. However, as this dip needle is moved closer to the equator, the angle of dip is reduced until it becomes horizontal at the equator. Thus, from the dip needle's angle of inclination, one can determine the latitude. In a similar manner, the orientation of the paleomagnetism in rocks indicates the latitude of the rock at the time it became magnetized.

A study conducted in Europe in the 1950s of several lava flows of different ages led to an interesting discovery. The magnetic alignment in the iron-rich minerals in lava flows of different ages was found to vary widely. A plot of the position of the magnetic

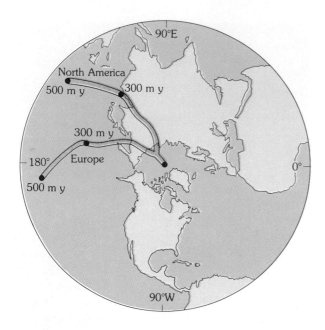

FIGURE 16.9
Simplified apparent polar wandering paths for North America and Europe. If these landmasses are brought together to close the North Atlantic, the paths roughly coincide.

north pole through time revealed that during the past 500 million years the position of the pole had gradually wandered from a location near Hawaii northward through eastern Siberia and finally to its present location (Figure 16.9). This was clear evidence that either the magnetic poles had migrated through time, an idea known as **polar wandering,** or the continents had drifted.

Although the magnetic poles are known to move, studies of the magnetic field indicate that the average positions of the magnetic poles correspond closely to the positions of the geographic poles. This is consistent with our knowledge of the earth's magnetic field, which is generated in part by the rotation of the earth about its axis. If the geographic poles do not wander appreciably, which we believe is true, neither can the magnetic poles. Therefore, a more acceptable explanation for the apparent polar wandering was provided by the continental drift hypothesis. If the magnetic poles remain stationary, their apparent movement can be produced by moving the continents.

The later idea was further supported by comparing the latitude of Europe as determined from rock magnetism with evidence from paleoclimatic studies. In particular, during the period when coal-producing swamps covered much of Europe, paleomagnetic evidence places Europe near the equator—a fact consistent with the tropical environment indicated by the coal deposits.[1]

Further evidence for continental drift came a few years later when a polar wandering curve was constructed for North America (Figure 16.9). To nearly everyone's surprise the curves for North America and Europe had similar paths, except that they were separated by about 30 degrees longitude. When these rocks solidified, could there have been two magnetic north poles which migrated parallel to each other? This is very unlikely. The differences in these migration paths, however, can be reconciled if the two presently separated continents are placed next to one another, as we now believe they were prior to opening of the Atlantic Ocean.

Although this new data rekindled interest in continental drift, it by no means caused a major swing in opinion. For one thing, the techniques used in extracting paleomagnetic data were relatively new and untested. Furthermore, rock magnetism tends to weaken with time, and rocks can also obtain a secondary magnetization. Despite these problems and other conflicting evidence, some researchers were convinced that continental drift had indeed occurred. A new era had begun.

A SCIENTIFIC REVOLUTION BEGINS

During the 1950s and 1960s great technological strides permitted extensive and detailed mapping of the ocean floor. From this work came the discovery of a global oceanic ridge system. Examination of the Mid-Atlantic Ridge reveals a trend which parallels the continental margins on both sides of the Atlantic (see Figure 1.11). Also of importance was the discovery of a central rift valley extending for the length of the Mid-Atlantic Ridge, an indication that great tensional

---

[1]Fossil trees found in this coal lack tree rings (growth rings), a characteristic of tropical vegetation.

forces were at work. In addition, high heat flow and some volcanism were found to characterize the oceanic ridge system.

In other parts of the ocean additional discoveries were being made. Earthquake studies conducted in the vicinity of the deep-ocean trenches suggested activity was occurring at great depths beneath the ocean. Flat-topped seamounts hundreds of meters below sea level showed signs of formerly being islands. Of equal importance, dredging of the oceanic crust was unable to bring up rock that was older than Mesozoic age. Could the ocean floors actually be geologically young features?

SEA-FLOOR SPREADING

In the early 1960s, all of these newly discovered facts were put together by Harry Hess of Princeton University into a hypothesis later to be termed **sea-floor spreading.** Hess was so lacking in confirmed data that he presented his paper as an "essay in geopoetry." Unlike its forerunner, continental drift, which essentially neglects the ocean basins, sea-floor spreading is centered on the activity beyond our direct view.

In Hess' now classic paper, he proposed that the ocean ridges are located above upwelling portions of large convection cells in the mantle (Figure 16.10). As rising material from the mantle spreads laterally, sea floor is carried in a conveyor belt fashion away from the ridge crest. Further, tensional tears at the ridge crest produced by the diverging lateral currents provide pathways for magma to intrude and generate new oceanic crust. Thus, as the sea floor moves away from the ridge crest, newly formed crust replaces it. Hess further proposed that the downward limbs of these convection cells are located beneath the deep-ocean trenches. Here, according to Hess, the older portions of the sea floor are gradually consumed as they descend into the mantle. As one researcher summarized, "No wonder the ocean floor was young—it was constantly being renewed!"

With the sea-floor spreading hypothesis in place, Harry Hess had initiated another phase of this scien-

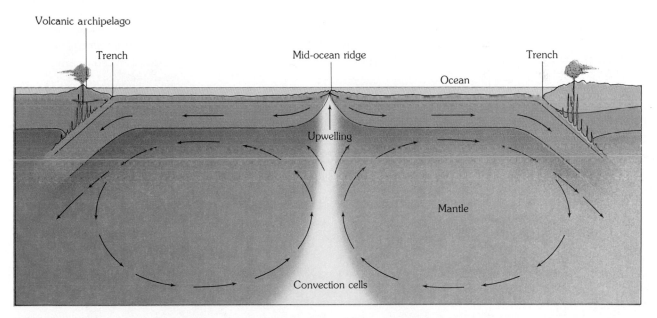

FIGURE 16.10
Sea-floor spreading. Harry Hess proposed that upwelling of mantle material along the mid-ocean ridge system created new sea floor. The convective motion of mantle material carries the sea floor in a conveyor belt fashion to the deep-ocean trenches, where the sea floor descends into the mantle.

tific revolution. The conclusive evidence to support his ideas came a few years later from the work of a young English graduate student, Fred Vine, and his supervisor, D. H. Matthews. The greatness of Vine's and Matthews' work was that they were able to connect two previously unrelated ideas: Hess' sea-floor spreading hypothesis and the newly discovered geomagnetic reversals.

GEOMAGNETIC REVERSALS

About the time that Hess formulated his ideas, geophysicists had begun to accept the fact that the earth's

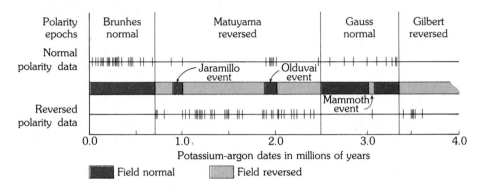

FIGURE 16.11
Time scale of main magnetic reversals in the more recent past. (Data from Allan Cox and G. B. Dalrymple)

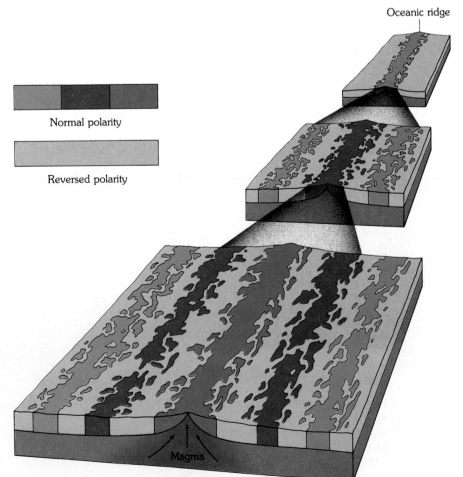

FIGURE 16.12
New sea floor records the polarity of the magnetic field at the time it formed. Hence it behaves much like a tape recorder, as it continually keeps count of changes in the earth's magnetic field.

magnetic field periodically reverses polarity; that is, the north magnetic pole becomes the south magnetic pole and vice versa. The cause of these reversals is apparently linked to the fact that the earth's magnetic field changes in intensity. Recent calculations, for example, indicate that the magnetic field has weakened about 5 percent over the past century. If this trend continues for the next thousand years or so, we might expect the earth's magnetic field to become very weak or nonexistent. During periods when the earth's magnetic field is very weak, some external influence, such as sunspot activity, could possibily contribute to a reversal of polarity. After a reversal has taken place, the field would rebuild itself with opposite polarity. A rock solidifying during one of the periods of reverse polarity will be magnetized with the polarity opposite of those rocks being formed today. When rocks exhibit the same magnetism as the present magnetic field, they are said to possess **normal polarity,** while those rocks exhibiting the opposite magnetism are said to have **reverse polarity.** Using the potassium-argon method of radiometric dating, the polarity of the earth's magnetic field has been reconstructed for a period of several million years (Figure 16.11).

A significant relationship between magnetic reversals and the sea-floor spreading hypothesis was developed from data obtained when very sensitive instruments called **magnetometers** were towed by research vessels across a segment of the ocean floor located off the west coast of the United States. Here workers from the Scripps Institution of Oceanography discovered alternating stripes of high- and low-intensity magnetism which trended in roughly a north-south direction. This relatively simple pattern of magnetic variation defied explanation until 1963, when Fred Vine and D. H. Matthews[1] tied the discovery of the high- and low-intensity stripes to Hess' concept of sea-floor spreading. Vine and Matthews suggested that the stripes of high-intensity magnetism are regions where the paleomagnetism of the ocean crust is of the normal type. Consequently, these positively magnetized rocks enhance the existing magnetic field. Conversely, the low-intensity stripes represent regions where the ocean crust is polarized in the reverse direction and, therefore, weaken the existing magnetic field. But how do parallel stripes of

---

[1] This idea was also put forth a few months earlier by L. W. Morely, but his paper was rejected for publication because of its highly speculative nature.

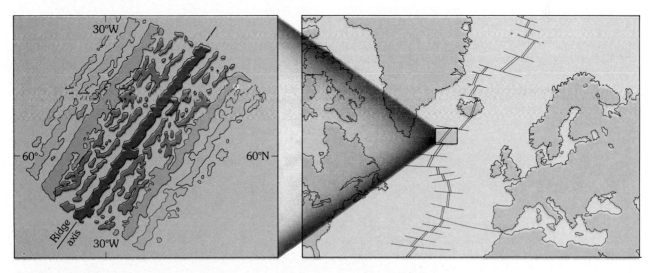

FIGURE 16.13
The symmetrical pattern of magnetic variation found across the Reykjanes Ridge just southwest of Iceland. The colored stripes are zones where the basaltic lavas have normal magnetization. The areas between the colored stripes represent zones of reverse magnetization. (After J. R. Heirtzler, S. Le Pichon, and J. G. Baron, *Deep-Sea Res.* 13 (1966): 247. Reprinted with permission from Pergamon Press, Ltd.)

normally and reversely magnetized rock become distributed across the ocean floor?

Vine and Matthews reasoned that as new basalt was added to the ocean floor at the oceanic ridges, it would be magnetized according to the existing magnetic field. Since the new rock is added in approximately equal amounts to both trailing edges of the spreading oceanic floor, we should expect strips of equal size and polarity to parallel both sides of the ocean ridges (Figure 16.12, page 396). This explanation of the alternating strips of normal and reverse polarity, which lay as mirror images across the ocean ridges, was the strongest evidence so far presented in support of the concept of sea-floor spreading.

A short time later, Vine's and Matthews' proposal was supported by a study group from Columbia University's Lamont-Doherty Geological Observatory. Towing magnetometers across a segment of the Reykjanes Ridge lying south of Iceland, researchers found that the magnetic variations there were indeed symmetrical with the ridge crest (Figure 16.13). By 1968, magnetic variations having similar patterns were identified paralleling most oceanic ridges.

Now that the dates of the more recent magnetic reversal have been established, the rate at which spreading occurs at the various ridges can be determined accurately. In the Pacific Ocean, for example, the magnetic stripes are much larger for corresponding time intervals than those of the Atlantic Ocean. Hence, we conclude that a faster spreading rate exists for the spreading center of the Pacific as compared to the Atlantic. When we apply absolute dates to these magnetic events, we find that the spreading rate for the North Atlantic Ridge is only 1 or 2 centimeters per year.[1] The rate is somewhat faster for the South Atlantic. The spreading rates for the East Pacific Rise generally range between 3 and 8 centimeters per year, with a maximum rate of about 10 centimeters per year in one segment. Thus, not only had Vine and Matthews discovered a magnetic tape recorder that detailed changes in the earth's magnetic field, this recorder could also be used to determine the rate of sea-floor spreading.

There is now general agreement that paleomagnetism was the most convincing evidence set forth in support of the concepts of continental drift and sea-floor spreading. By 1968 geologists began reversing their stand on this issue in a manner not unlike a magnetic reversal. The tide of scientific opinion had indeed switched in favor of a mobile earth.

---

[1]Note that each side of the ridge spreads at this rate.

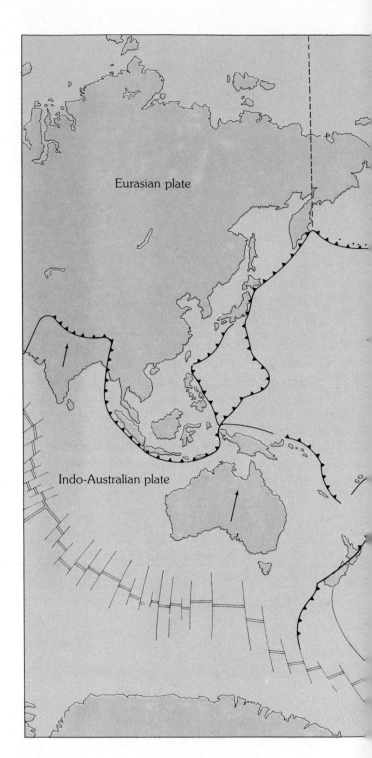

FIGURE 16.14
Mosaic of rigid plates that constitute the earth's outer shell. A. Divergent boundary. B. Convergent boundary, and C. Transform fault boundary. (After W. B. Hamilton, U.S. Geological Survey)

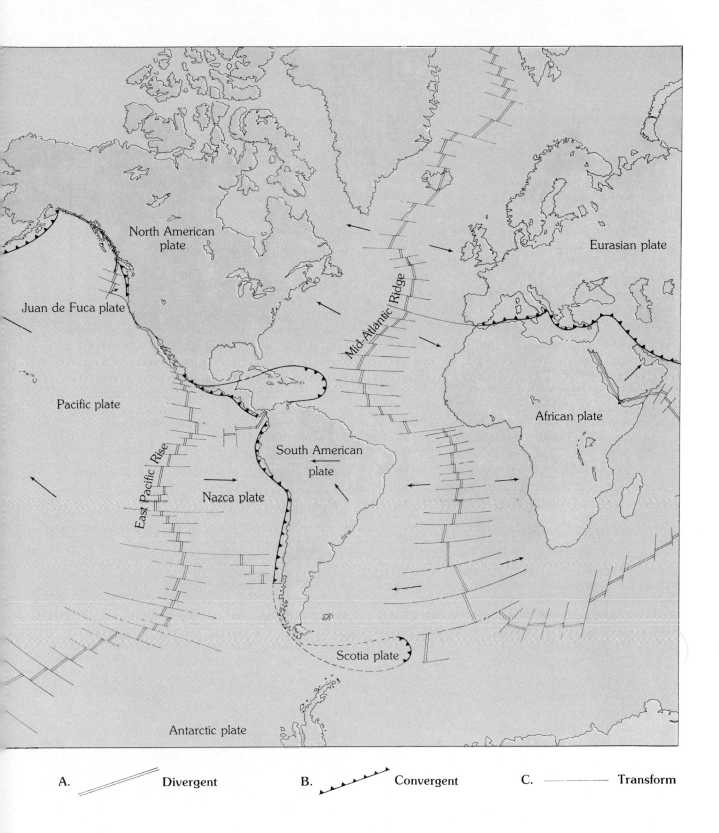

North American plate

Juan de Fuca plate

Pacific plate

Nazca plate

East Pacific Rise

South American plate

Mid Atlantic Ridge

Eurasian plate

African plate

Scotia plate

Antarctic plate

A. ⟋⟋⟋ Divergent   B. ⩗⩗⩗ Convergent   C. ——— Transform

PLATE TECTONICS: A MODERN VERSION OF AN OLD IDEA

By 1968, the concepts of continental drift and sea-floor spreading were united into a much more encompassing theory known as plate tectonics. The implications of plate tectonics are so far-reaching that this theory can be considered the framework from which to view most other geologic processes. Since this concept is relatively new, it most surely will be modified as additional information becomes available; however, the main tenets appear to be sound and are presented here in their current state of refinement.

The theory of plate tectonics states that the outer, rigid lithosphere consists of several individual segments, called **plates.** About twenty plates of various sizes have been identified (Figure 16.14). Of these, the largest is the Pacific plate, which is located mostly within the ocean proper, except for a small sliver of North America that includes southwestern California and the Baja Peninsula. Notice from Figure 16.14 that all of the other large plates contain both continental and oceanic crust—a major departure from the continental drift theory, which proposed that the continents moved through, not with, the ocean floor. Most of the smaller plates, on the other hand, consist exclusively of oceanic material, as for example, the Nazca plate located off the west coast of South America. Although not clearly defined in Figure 16.14, one small plate that roughly coincides with Turkey is located exclusively within a continent.

The lithosphere overlies a zone of much weaker and hotter material known as the asthenosphere. Hence, the lithospheric plates form a rigid outer shell supported from below by the more "plastic" material of the asthenosphere. A relationship appears to exist between the thickness of the lithospheric plates and the nature of the crustal material that caps them. Plates are thinnest in the oceans, where their thickness varies from 80 to 100 kilometers. By contrast, continental blocks are 100 kilometers or more thick and in some regions may approach 400 kilometers thick.

One of the main tenets of the plate tectonics theory is that each plate moves as a dictinct unit in relation to other plates. The mobile behavior of the rock within the asthenosphere is believed to allow this motion in the earth's rigid outer shell. As the plates move, the distance between two cities on the same plate, New York and Denver, for example, remains constant, while the distance between New York and London, which are located on different plates, is constantly changing. Since each plate moves as a distinct unit, all major interactions between plates occur along plate boundaries. Thus, most of the earth's seismic activity, volcanism, and mountain building, occurs along these dynamic margins.

PLATE BOUNDARIES

For some time now, tectonic activity has been known to be restricted to narrow zones, such as the so-called *Ring of Fire* that encircles the Pacific. Thus, the first approximations of plate margins relied on the distribution of earthquake and volcanic activity. Later work indicated the existence of three distinct types of plate boundaries, each differentiated by the movement it exhibits (Figure 16.15). These are:

1 **Divergent boundaries**—where plates move apart, resulting in upwelling of material from the mantle to create new sea floor.

2 **Convergent boundaries**—where plates move together, causing one of the slabs of lithosphere to be consumed into the mantle as it descends beneath an overriding plate.

3 **Transform fault boundaries**—where plates slide past each other without creating or destroying lithosphere.

In the following sections we will briefly summarize the nature of these three types of plate boundaries. Then, in the next two chapters, the role of these plate margins in sea-floor spreading and mountain building will be considered in more detail.

DIVERGENT BOUNDARIES

Divergent boundaries, where plate spreading occurs, are situated at the crests of oceanic ridges. Here, as the plates move away from the ridge axis, the gaps created are immediately filled with molten rock that oozes up from the hot asthenosphere. This material cools slowly to produce new slivers of sea floor. In a continuous manner, successive separations and in-

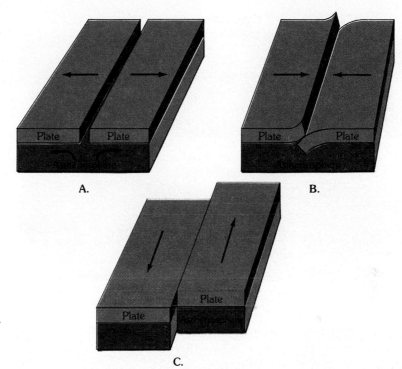

FIGURE 16.15
Schematic of plate boundaries.
A. Divergent boundary.
B. Convergent boundary.
C. Transform fault boundary.

jections of magma add new oceanic crust (lithosphere) between the diverging plates. As noted earlier, this mechanism is called sea-floor spreading and produced the floor of the Atlantic Ocean during the past 160 million years.

Not all spreading centers are as old as the Mid-Atlantic Ridge and not all are found in the middle of large oceans. The Red Sea is believed to be the site of a recently formed divergent boundary. Here the Arabian Peninsula separated from Africa and began to move toward the northeast. Consequently, the Red Sea is providing oceanographers with a view of how the Atlantic Ocean may have looked in its infancy. Another result of sea-floor spreading in the recent geologic past is the Gulf of California.

When a spreading center develops within a continent, the landmass may split into smaller segments as Wegener had proposed for the breakup of Pangaea. The fragmentation of a continent is thought to be initiated by an upward movement of hot rock from below. The effect of this activity is to upwarp the crust directly above the hot rising plume. The crustal stretching associated with the doming generates numerous tensional cracks as shown in Figure 16.16A.

Then, as the hot plume spreads laterally from the region of upwelling, the broken lithosphere is pulled apart. Gradually, the broken slabs slide downward into the gaps created by the diverging plates (Figure 16.16B). The large downfaulted valleys generated by this process are called **rifts,** or **rift valleys.** The Great Rift Valley of East Africa is an excellent example of such a feature (see Figure 17.20). If the spreading process continues in East Africa, the rift valley will lengthen and deepen, eventually extending out into the ocean. At this point the valley will become a narrow linear sea with an outlet to the ocean similar to the Red Sea today (Figure 16.16C). The zone of rifting will remain the site of igneous activity, continually generating new sea floor in an ever-expanding ocean basin (Figure 16.16D).

The Great Rift Valley of East Africa appears to represent the initial stage in the breakup of a continent (see Figure 17.20). The extensive volcanic activity believed to accompany continental rifting is exemplified by large volcanic mountains such as Mount Kilimanjaro and Mount Kenya. If the rift valleys in Africa remain active, East Africa will eventually part from the mainland in much the same way the Arabian

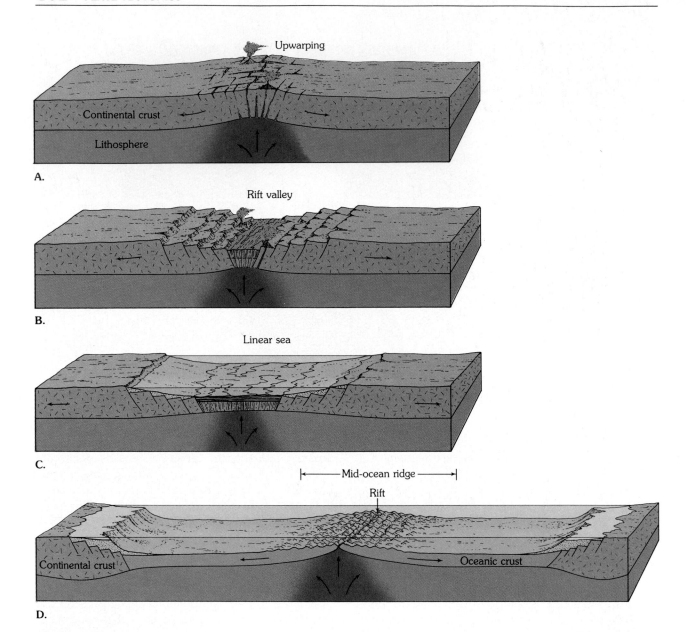

FIGURE 16.16
A. Rising magma upwarps the crust, causing numerous cracks in the rigid lithosphere.
B. As the crust is pulled apart, large slabs of rock sink, generating a rift zone. **C.** Fur
ther spreading generates a narrow sea. **D.** Eventually, an expansive ocean basin and
ridge system are created.

Peninsula did just a few million years ago. However, not all rift valleys develop into full-fledged spreading centers. Running through the central United States is an aborted rift zone extending from Lake Superior to Kansas. This once active rift valley is filled with rock that oozed upward into the crust more than 500 million years ago. Why one rift valley continues to develop while others are abandoned is not yet known.

CONVERGENT BOUNDARIES

At spreading centers new lithosphere is continually being generated; however, since the total surface area of the earth remains constant, lithosphere must also be destroyed. The zone of plate convergence is the site of this destruction. When two plates collide, the leading edge of one is bent downward, allowing it to descend beneath the other. Upon entering the hot asthenosphere, the plunging plate begins to warm and lose its rigidity. Generally the descending plate is relatively cold and approaches 100 kilometers in thickness. Thus, depending upon its angle of descent, it may reach a depth of 700 kilometers before its leading edge is completely assimilated within the material of the upper mantle.

Although all convergent zones are basically similar, the nature of plate collisions is influenced by the type of the crustal material involved. Collisions can occur between two oceanic plates, one oceanic and one continental plate, or two continental plates, as shown in Figure 16.17. Whenever the leading edge of a plate capped with continental crust converges with oceanic crust, the less dense continental material apparently remains "floating," while the more dense oceanic slab sinks into the asthenosphere. The region where an oceanic plate descends into the asthenosphere because of convergence is called a **subduction zone.** As the oceanic plate slides beneath the overriding plate, the oceanic plate bends, thereby producing a **deep-ocean trench** adjacent to the zone of subduction (Figure 16.17A). Trenches formed in this manner may be thousands of kilometers long and 8 to 11 kilometers deep (Figure 16.18).

Oceanic-Continental Convergence   During a collision between an oceanic slab and a continental block, the oceanic crust is bent, permitting its descent into the asthenosphere (Figure 16.17A). As the oceanic slab descends, some of the soft sediments carried upon the sinking plate are scraped off by the overriding continental material. Studies conducted in the coastal regions of western Mexico, where the Cocos plate is being subducted, indicate that roughly half of the sediment carried on the descending plate can be removed in this manner. Therefore, this process contributes to the already substantial accumulation of sediment deposited along the continental margin as the result of erosion on the continent.

Upon entering the hot asthenosphere, the downward moving plate and the remaining water-saturated sediments carried upon it begin to melt. Although the process is poorly understood, the partial melting of this mixture of basaltic rocks and sediments generates magmas having a composition similar to the rock andesite or occasionally granite. The newly formed magmas created in this manner are less dense than the rocks of the mantle. Consequently, when sufficient quantities have accumulated, the molten rock will slowly buoy upward. Most of the rising magma will be emplaced in the overlying continental crust where it will cool and crystallize at a depth of several kilometers. The remaining magma may eventually migrate to the surface, where it can give rise to numerous and occasionally explosive volcanic eruptions. The volcanic portions of the Andes Mountains are believed to have been produced by such activity when the Nazca plate melted as it descended beneath the continent of South America (see Figure 16.14). The frequent earthquakes occurring within the Andes testify to the activity beyond our view.

Mountains such as the Andes that are believed to be produced in part by volcanic activity associated with the subduction of oceanic lithosphere are called **volcanic arcs.** Two volcanic arcs are located in the western United States. One of these, the Cascade Range, is composed of several well-known volcanic mountains, including Mounts Rainier, Shasta, and St. Helens. The second is the Sierra Nevada, in which Yosemite National Park is located. The Sierra Nevada system is the older of the two and has been inactive for several million years as evidenced by the absence of volcanic cones. Here erosion has stripped away most of the obvious traces of volcanic activity and left exposed the large, crystallized magma chambers that once fed lofty volcanoes. As the recent eruptions of Mount St. Helens testify, the Cascade Range is still quite active. The magma here arises from the melting of a small remaining segment of the Juan de Fuca plate. Since the rate of subduction at this plate is slow, the volcanoes of the Cascades are thought to be magma deficient, which partly accounts for their rather sporadic activity.

Oceanic-Oceanic Convergence   When two oceanic slabs converge, one descends beneath the other, initiating volcanic activity in a manner similar to that

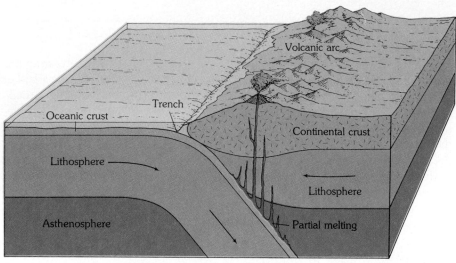

A.

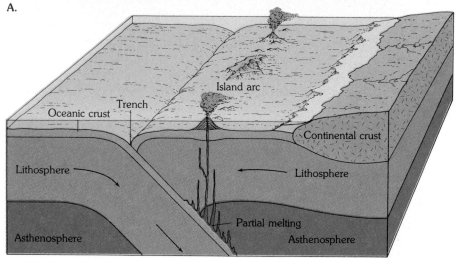

B.

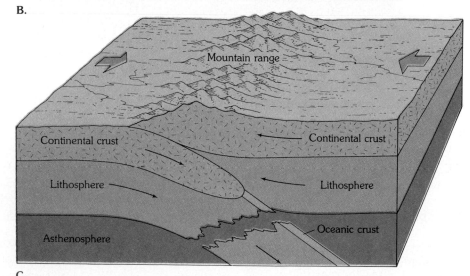

C.

FIGURE 16.17
Zones of plate convergence.
A. Oceanic-continental.
B. Oceanic-oceanic.
C. Continental-continental.

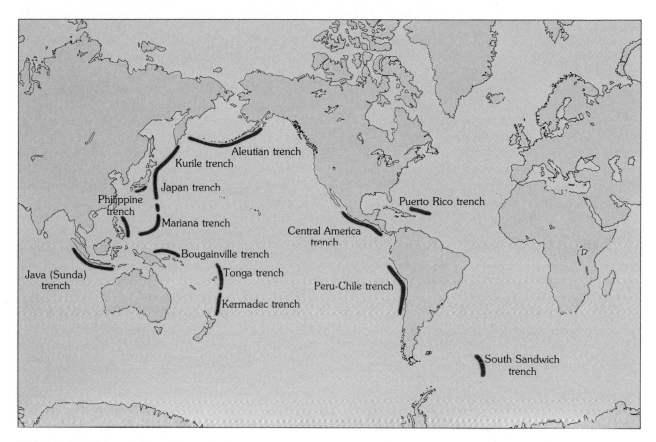

FIGURE 16.18
Distribution of the world's major oceanic trenches.

which occurs at an oceanic-continental convergent boundary. However, in this case, the volcanoes form on the ocean floor rather than on land (Figure 16.17B). If this volcanic activity is sustained, dry land will eventually emerge from the ocean depths. In early stages of development, this newly formed land consists of a chain of small volcanic islands called an **island arc.** The Aleutian, Mariana, and Tonga islands exemplify such features. Island arcs such as these are generally located a few hundred kilometers from an ocean trench where active subduction of the lithosphere is occurring. Adjacent to the island arcs just mentioned are the Aleutian trench, Mariana trench, and the Tonga trench, respectively.

Over an extended period, numerous episodes of volcanic activity build large volcanic piles on the ocean floor. This volcanic activity, plus the buoyancy

of the intrusive igneous rock emplaced within the crust below, gradually increases the size and elevation of the developing arc. This growth, in turn, increases the amount of eroded sediments added to the sea floor. Some of these sediments reach the trench and are deformed and metamorphosed by the compressional forces exerted by the two converging plates. The result of these diverse activities is the development of a mature island arc composed of a complex system of volcanic rocks, folded and metamorphosed sedimentary rocks, and intrusive igneous rocks. Examples of mature island arc systems are the Alaskan Peninsula, the Philippines, and Japan.

Continental-Continental Convergence When two plates carrying continental crust collide, neither plate will subduct beneath the other (Figure 16.17C). This is thought to occur because of the light composition, and thus buoyant nature, of continental rocks. Such a collision is believed to have occurred when the once separated continent of India "rammed" into Asia and produced the Himalayas, perhaps the most spectacular mountain range on earth. During this collision, the continental crust buckled, fractured, and was generally shortened. In addition to the Himalayas, several other complex mountain systems, including the Alps, Appalachians, and Urals are thought to have formed during continental collisions.

Prior to a continental collision, the landmasses involved are separated by the oceanic crust formed during an earlier episode of sea-floor spreading. As the continental blocks converge, the intervening sea floor is subducted beneath one of the plates. The partial melting of the descending oceanic slab and the sediments carried with it generate a volcanic arc. Depending on the location of the subduction zone, the volcanic arc could develop on either of the converging landmasses, or if the subduction zone developed at an appreciable distance into the ocean, an island arc would form. In any case, erosion of the newly formed volcanic arc would add large quantities of sediment to the already sediment-laden continental margins. Eventually, as the intervening sea floor was consumed, these continental masses would collide, thereby squeezing, folding, and generally deforming the sediments as if they were placed in a gigantic vise. The result would be the formation of a new mountain range composed of deformed sedimentary rocks and fragments of the volcanic arc.

After continents collide, the descending oceanic material is believed to break from the continental block and continue moving downward, becoming completely assimilated into the mantle. However, because of its buoyancy, continental lithosphere cannot be carried very far into the mantle. In the case of the Himalayas, the leading edge of the Indian plate was forced partially under Asia, generating an unusually thick accumulation of continental lithosphere. This accumulation accounts, in part, for the high elevation of the Himalayas and the Tibetian Plateau to the north.

TRANSFORM FAULTS

The third type of plate boundary is the transform fault, which is located where plates slide past one another without the production of crust, as occurs along oceanic ridges, or without the destruction of crust, as occurs at oceanic trenches. Transform faults roughly parallel the direction of plate movement and were first identified where they join segments of the oceanic ridge system (Figure 16.19). At first inspection these enormous fractures appear to be simple strike-slip faults along which horizontal motion has offset the oceanic ridge system. However, the relative motion along these fault zones was found to be in the opposite direction required to produce the offsets observed.

The true nature of transform faults was provided in 1965 by J. Tuzo Wilson of the University of Toronto. Wilson suggested that these large fractures connected the global active belts into a continuous network that divides the earth's outer shell into several rigid plates. Thus, Wilson became the first to suggest the earth was made of individual plates, while at the same time identifying the zones along which relative motion between the plates is made possible. In this latter role, transform faults provide the means by which the oceanic crust created at the ridge crests can be transported to its site of destruction, the deep-ocean trenches. Figure 16.20 illustrates this role. Notice that the Juan de Fuca plate moves in a southeasterly direction, eventually being subducted under the west coast of the United States. The southern end of this relatively small plate is bounded by the Mendocino escarpment. This transform fault boundary connects an active spreading center to a subduction zone. Therefore, the fault facilitates the movement of the crustal material created at the ridge crest to its destination beneath the North American continent. Also notice that while the Juan de Fuca plate moves in a southeasterly direction, movement along the San Andreas fault facilitates the northwestwardly drift of the Pacific plate, including a portion of California.

Wilson called these special faults transform faults because the relative motion of the plates can be changed, or transformed, along them. As we saw in the preceding example, divergence occurring at a spreading center can be transformed into convergence at a subduction zone. Since transform faults connect convergent and divergent boundaries in various combinations, other changes in relative plate motion are possible along transform faults.

Most commonly, transform faults join two ridge

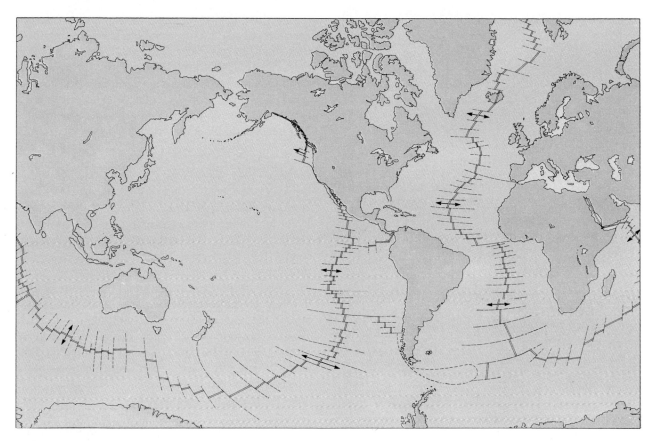

FIGURE 16.19
The relationship between the oceanic ridge system and transform faults. Where transform faults offset ridge segments, they permit the ridge to change direction (curve) as
can be seen in the Atlantic Ocean.

segments and occasionally even two trenches. When transform faults connect two spreading centers as shown in Figure 16.21, the newly created sea floor is moving in opposite directions in the region between the two ridges. Elsewhere along this fracture zone, the relative motion changes so that both plates are moving in the same direction. Thus, the only active part of the fault lies between the two offset ridge segments. This active zone is also a site of frequent, but generally weak, seismic activity.

TESTING THE MODEL

Immediately after the plate tectonics theory was proposed, an avalanche of data began to be compiled from all of the earth sciences to test this revolutionary idea. The model just presented will surely be modified to fit this wealth of data; however, the basic premises appear to be sound and able to withstand the test of time.

Although most geologists have accepted this theory energetically, there remains an ever-diminishing number who reject it in part or in total. Some of the evidence supporting continental drift and sea-floor spreading has already been presented in this chapter. In addition, some of the evidence which was instrumental in solidifying the support for this new concept follows. It should be pointed out that much of this evidence was not new, rather it was a new interpretation of old data that swayed the tide of opinion. Further, some of the data was compiled to refute

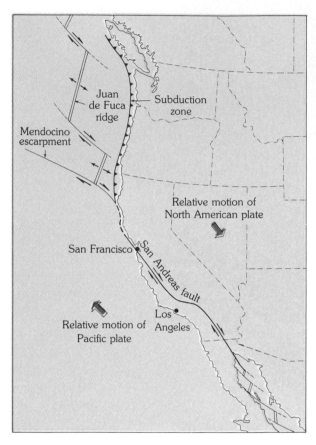

FIGURE 16.20
The role of transform faults in permitting relative motion between adjacent plates. The Mendocino escarpment permits sea floor generated at the Juan de Fuca ridge to move southeastward past the Pacific plate.

rather than support global tectonics. As one researcher stated, "My observations are not compatible with sea-floor spreading, and I shall prepare a critical demonstration that this is so and thus demolish this nutty idea and we can all get back to work." However, he, like many others, found that his results were indeed compatible with this new theory. The revolution had produced a new ruling theory.

PLATE TECTONICS AND EARTHQUAKES

By 1968, the basic outline of global tectonics was firmly established. In this same year, three Lamont-Doherty Observatory seismologists, B. Isacks, J. Oliver, and L. R. Sykes, published papers demonstrating how much more successful the new plate tectonics model was than older models in accounting for the global distribution of earthquakes (Figure 16.22). In particular these seismologists were able to account for the existence of deep-focus earthquakes and their close association with trench-volcanic arc systems. Further, the absence of deep-focus earthquakes along the oceanic ridge system was also shown to be consistent with the new model.

In the preceding chapter we noted the close association between plate boundaries and earthquakes. In trench regions, where slabs of lithosphere plunge into the mantle, this association is most spectacular. When the depths of earthquake foci and their location within the trench systems are plotted, an interesting pattern emerges (Figure 16.23). Notice that most shallow-focus earthquakes occur near the trench ar-

FIGURE 16.21
Transform faults often connect offset segments of an oceanic ridge. Note that the lithosphere is moving in opposite directions in the region between the two ridges, while elsewhere along the fracture zone both plates move in the same direction.

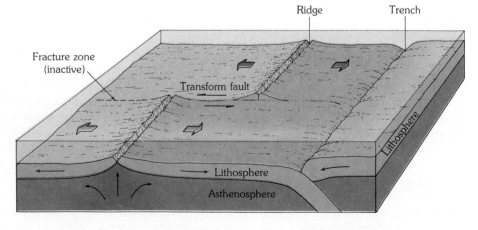

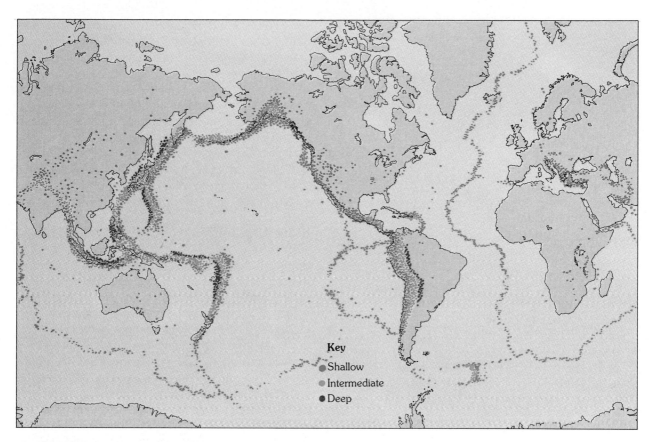

FIGURE 16.22
Distribution of shallow-, intermediate-, and deep-focus earthquakes. (Data from NOAA)

Key
● Shallow
● Intermediate
● Deep

eas, while intermediate- and deep-focus earthquakes are generated away from the trenches and nearer the adjacent volcanic arc. This discovery of a linear distribution of earthquake foci in the trench areas could not be satisfactorily explained by previously existing tectonic models.

Another problem seismologists faced was the inability to identify a mechanism for generating deep-focus earthquakes. Recall that the storage and sudden release of elastic energy was substantiated as the source of energy for shallow-focus earthquakes. However, it was demonstrated early on that at the temperatures and pressures existing at depths of 60 to 100 kilometers, rock would flow under stress rather than behave like a brittle solid.

The Lamont-Doherty seismologists enthusiastically showed that all of these findings were consistent with the plate tectonics theory. In the plate model, the deep-ocean trenches are produced where slabs of oceanic lithosphere are bent and plunge into the mantle below. As the plate descends, it is gradually warmed; however, the zone through which it passes remains relatively cool because new cold material from above is continually supplied. The researchers determined that the center of a 100-kilometer thick slab, which is descending at a rate of several centimeters per year could remain brittle to a depth of 600 to 700 kilometers. Thus, a deep-focus earthquake could be generated in a manner similar to shallow ones, through the release of elastic energy that was stored

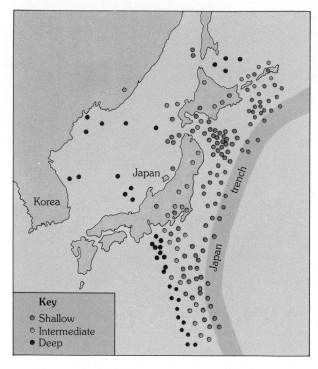

FIGURE 16.23
Distribution of earthquake foci in the vicinity of the Japan trench. (Data from NOAA)

within the descending rigid slab as it met resistance to its downward motion.

As the slab continues to descend into the asthenosphere, deeper and deeper earthquakes are produced. Since the earthquakes occur within the rigid lithosphere rather than within the "plastic" asthenosphere, they provide a method for tracking the plate's descent into the mantle (Figure 16.24). Earthquake data indicate that plates enter the mantle at angles of about 45 degrees. Recall that these zones of inclined seismic activity which extend from the trench into the asthenosphere are called *Benioff zones* after an American seismologist who conducted extensive studies on the distribution of earthquake foci. Extremely few earthquakes have been recorded below 700 kilometers, possibly because the lithosphere has been completely assimilated into the mantle by the time it reaches this depth.

While deep earthquakes occur within the descending plates, numerous large, shallow earthquakes oc-

cur within the overriding plate. These shallow-focus earthquakes are thought to be produced by stresses created as the descending plate scrapes beneath the overriding slab.

The plate model also explains the absence of deep-focus earthquakes along oceanic ridges. According to this model, divergent zones are located where there is a continual upwelling of hot mantle material. Consequently, only the upper layers are brittle enough to generate earthquakes, and hence only shallow-focus earthquakes are observed at the ridges.

EVIDENCE FROM THE DEEP SEA DRILLING PROJECT

Some of the most convincing evidence confirming the sea-floor spreading hypothesis has come from drilling into the sediment on the ocean floor. The source of this important data has been the Deep Sea Drilling Project, a program begun in the late 1960s under the joint sponsorship of several major oceanographic institutions and the National Science Foundation. The primary goal was (and indeed still is) to gather firsthand information about the age and processes of ocean basin formation. Researchers felt that the predictions concerning sea-floor spreading that were based on paleomagnetic data could best be confirmed by the direct sampling of sediments from the floor of the deep-ocean basins. To accomplish this, a new drilling ship, the *Glomar Challenger,* was built (Figure 16.25). The *Glomar Challenger* represented a significant technological breakthrough, because this ship was capable of lowering drill pipe thousands of meters to the ocean floor and then drilling hundreds of meters into the sediments and underlying basaltic crust. What made this possible was the development of a dynamic positioning system that uses sound waves from acoustic beacons that are lowered to the sea floor. Any changes in the ship's position are sensed by a computer that monitors the sound signals from the beacons and automatically controls the ship's side thrusters and main propulsion system. In this manner the position of the *Glomar Challenger* is maintained above the drill hole for extended periods, even in the deepest waters and despite currents, winds, and waves.

Operations began in August, 1968, and shortly thereafter important evidence was gathered in the

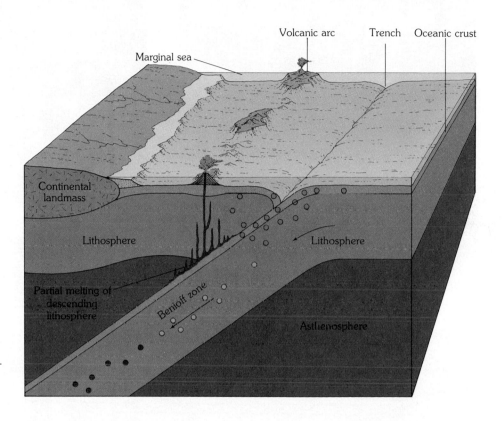

Marginal sea

Volcanic arc   Trench   Oceanic crust

Continental
landmass

Lithosphere

Lithosphere

Partial melting of
descending
lithosphere

Benioff zone

Asthenosphere

FIGURE 16.24
Relationship between the descending plate and depth of earthquake foci.

South Atlantic. At several sites holes were drilled through the entire thickness of sediments to the basaltic rock below. An important objective was to gather samples of sediment from just above the igneous crust as a means of dating the sea floor at each site.[1] Since sedimentation begins immediately after the oceanic crust forms, fossils found in the oldest sediments (that is, those resting directly above the basalt) can be used to date the ocean floor at that site. When the oldest sediment from each drill site was plotted against its distance from the ridge crest, it was revealed that the age of the sediment increased with increasing distance from the ridge. This finding was in agreement with the sea-floor spreading hypothesis which predicted that the youngest oceanic crust is to be found at the ridge crest and that the oldest oceanic crust flanks the continental margins. Further, the rate of sea-floor spreading determined from the ages of sediments was identical to the rate previously estimated from magnetic evidence. Subsequent drilling

in the Pacific Ocean verified these findings. These excellent correlations were a striking confirmation of sea-floor spreading.

The data from the Deep Sea Drilling Project also reinforced the idea that the ocean basins are geologically youthful. To date, no sediment with an age in excess of 160 million years has been found. By comparison, some continental crust has been dated at more than 3.8 billion years.

The thickness of ocean-floor sediments provided additional verification of sea-floor spreading. Drill cores from the *Glomar Challenger* revealed that sediments on the ridge crest are almost entirely absent and that the sediment thickens with increasing distance from the ridge. Since the ridge crest is younger than the areas farther away from it, this pattern of sediment distribution should be expected if the sea-floor spreading hypothesis is correct. Further, measurements in the open ocean have shown that sediment accumulates at a rate of approximately 1 centimeter per 1000 years. Therefore, if the ocean floor were an ancient feature, sediments would be many kilometers thick. However, data from

---

[1]Radiometric dates of the ocean crust itself are unreliable because of the alteration of basalt by seawater.

A.

B.

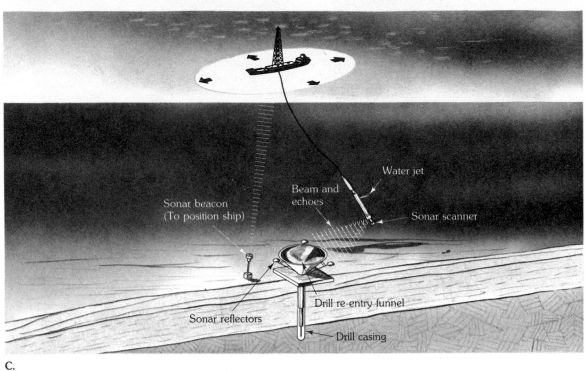

C.

FIGURE 16.25

A. The *Glomar Challenger* is capable of recovering sediment and samples of oceanic crust from the floors of the deepest oceans. The ship is 120 meters (400 feet) long and has a displacement of 10,000 tons. B. Amidships is the towering 42-meter (140-foot) high derrick. This is a view looking straight down from near the top of the derrick, nearly 61 meters (200 feet) above the waterline. C. The *Glomar Challenger* possesses remarkable operating capabilities. The ship can be positioned in water too deep for anchors and remain at a drilling site as long as desired because of a dynamic positioning system that maintains the ship's position within a radius of about 100 meters. When necessary, a previously drilled hole can be re-entered. This may sound simple, but remember that the hole is only 12 centimeters wide and may be thousands of meters below the ship. At the time re-entry is attempted, the drill string is relowered with a sonar scanner that emits sound signals. These signals are echoed back from three reflectors spaced around the funnel. Position information is relayed to the ship and a water jet is used to steer the drill bit directly over the funnel. (Photos courtesy of Victor S. Sotelo, Deep Sea Drilling Project. Part C is based on a National Science Foundation report.)

412

hundreds of drilling sites indicate that the greatest thickness of sediment in the deep-ocean basins is only a few hundred meters. Thus, here is yet another fact that strongly suggests the ocean floor is indeed a young geologic feature.

The Deep Sea Drilling Project has provided an enormous quantity of basic information about the history of the oceans and has confirmed many important aspects of the plate tectonics theory. To summarize, the sea-floor spreading hypothesis was upheld when deep-ocean core samples showed that the age of the oldest sediments and the thickness of sediments increase with increasing distance from the ridge crest. Moreover, data on the thickness and ages of sediments strongly support the idea of geologically youthful ocean basins. In the years to come, the operations of the *Glomar Challenger* or an even more technologically advanced successor will continue the effort to better understand the origin and evolution of the ocean basins.

HOT SPOTS

Mapping of seamounts in the Pacific revealed a chain of volcanic structures extending from the Hawaiian Islands to Midway Island and then continuing north-

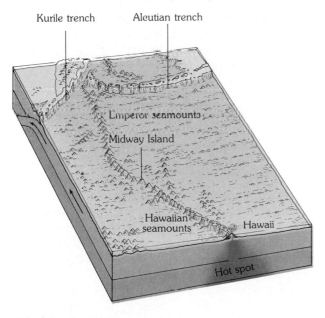

FIGURE 16.26
Chain of islands and seamounts extending from Hawaii to the Aleutian trench.

ward toward the Aleutian trench. Potassium-argon dating of 27 volcanoes in this chain revealed an increase in age with an increase in distance from Hawaii. Suiko Seamount, which is located near the Aleutian trench, is 65 million years old, Midway Island is 27 million years old, and the island of Hawaii rose from the sea less than 1 million years ago (Figure 16.26).

Researchers have proposed that a **hot spot** exists within the mantle and emits magma onto the overlying sea floor. Presumably, as the Pacific plate moved over the hot spot, successive volcanic structures emerged. The age of each volcano indicates the time when it was situated over the relatively stationary hot spot. Kauai is the oldest of the large islands in the Hawaiian chain. Five million years ago, when it was positioned over the hot spot, Kauai was the only Hawaiian Island in existence (Figure 16.27). Visible evidence of the age of Kauai can be seen by examining the extinct volcanoes which have been eroded into jagged peaks and vast canyons. By contrast, the south slopes of the island of Hawaii consist of fresh lava flows and two of Hawaii's volcanoes, Mauna Loa and Kilauea, remain active. Recent evidence indicates that a new volcanic pile is forming on the ocean floor just off the coast of Hawaii. Geologically speaking, it should not be long before another tropical island will be added to the Hawaiian chain.

Although the existence of hot spots is well documented, their exact nature or role in plate tectonics is not altogether clear. Apparently hot spots are unusually warm regions found deep within the earth's mantle. Here the high temperatures produce a rising plume of molten rock which frequently initiates volcanism at the surface. Most evidence indicates that hot spots remain relatively stationary; however, some appear to have exhibited movement. Of the 50 to 120 hot spots believed to exist, about 20 are located near divergent plate boundaries, whereas the others are not associated with plate boundaries. A hot spot beneath Iceland is thought to be responsible for the unusually large accumulation of lava found in that portion of the Mid-Atlantic Ridge. Another hot spot is believed to be located beneath Yellowstone National Park and may be responsible for the large outpourings of lava and volcanic ash which mantle this area. If the Yellowstone region was indeed modified by hot spot volcanism, there is good reason to expect

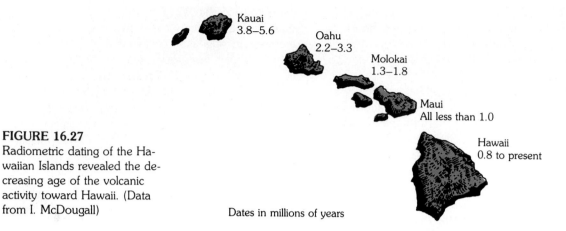

Kauai
3.8–5.6

Oahu
2.2–3.3

Molokai
1.3–1.8

Maui
All less than 1.0

Hawaii
0.8 to present

FIGURE 16.27
Radiometric dating of the Hawaiian Islands revealed the decreasing age of the volcanic activity toward Hawaii. (Data from I. McDougall)

Dates in millions of years

additional activity in the future. Other hot spots are situated beneath the sea floor and have generated numerous seamounts that trend in the direction of plate motion.

THE DRIVING MECHANISM

The plate tectonics theory describes plate motion and the effects of this motion. Therefore, acceptance does not rely on a knowledge of the force or forces moving the plates. This is fortunate, since none of the driving mechanisms yet proposed can account for all of the major facets of plate motion. Nevertheless, the unequal distribution of heat within the earth is accepted by most as the underlying cause of plate movement. The unequal distribution of heat, in turn, is thought by many geologists to generate large convection cells within the mantle (Figure 16.28A, B). The warm, less dense material of the lower mantle rises very slowly in the regions of oceanic ridges. As the material spreads laterally it cools, becomes more dense, and begins to sink back into the mantle, only to be reheated. Note that rocks need not be molten to flow. Just as a hot, solid metal can be pounded into different shapes, so too can rock move when subjected to heat and stress over an extended period. Measurements indicate a higher rate of heat flow at the oceanic ridges than in other oceanic regions—a good indication that some type of thermal convection cells might exist. However, many of the details of their motions remain unclear. How many cells exist? At what depth do they originate? What is the structure of these thermal cells?

Although unequal distribution of heat is generally accepted as the underlying driving force of plates, many geologists are not convinced that large convection cells can exist in the mantle. Thus, many other mechanisms which may play an important role in plate motion have been suggested. One mechanism relies on the fact that the cold oceanic slab has a greater density than the asthenosphere supporting it from below. This being the case, it has been proposed that as a plate begins to descend, the heavy, sinking slab might pull the trailing lithosphere along. This hypothesis is similar to another model which suggests that the elevated position of an ocean ridge could cause the lithosphere to slide under the influence of gravity (Figure 16.28C). These push-pull models are themselves a type of convection current. As the sinking plate enters the mantle, material is forced aside which then migrates toward the ridge systems. The cell is completed as molten rock moves up from the asthenosphere to fill the gap in the diverging oceanic plates.

Some oceans, notably the Atlantic, lack subduction zones; thus, the slab-pull mechanism cannot adequately explain the spreading occurring at these ridges. Other ridge systems are rather subdued, which would reduce the effectiveness of the alternate slab-push model. Perhaps the slab-push and slab-pull phenomena are active in different ridge systems and occasionally even work in tandem.

Another version of the thermal convection model suggests that relatively narrow, hot plumes (hot spots) of rock generate plate motion (Figure 16.28D). These hot plumes are thought to originate near the mantle-core boundary. Upon reaching the lithosphere, these plumes would spread laterally and car-

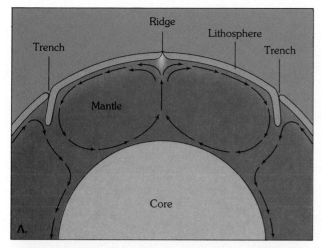

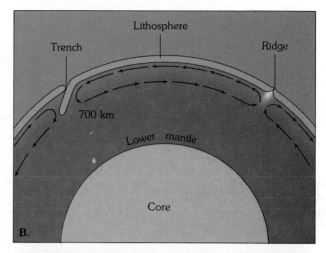

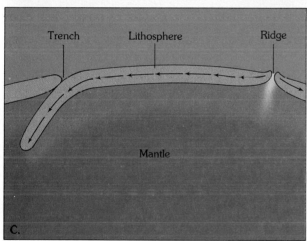

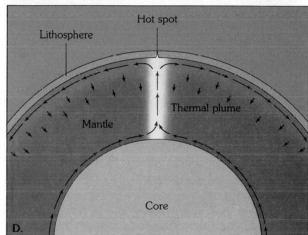

FIGURE 16.28

Proposed models of the driving force for plate tectonics. **A.** Large convection cells in the mantle carry the lithosphere in a conveyor belt fashion. **B.** Convection cells confined to the mobile upper mantle are responsible for plate motion. **C.** Push-pull models are also a type of convection. Here the cold, sinking oceanic slab pulls the trailing sea floor along, while gravity-sliding of material down the elevated ridge crest pushes the slab. **D.** The hot plume model suggests that all upward movement is confined to a few narrow plumes, while downward flow occurs slowly throughout the remaining mantle. (Parts A, B, and D after "The Earth's Mantle," by Peter J. Wyllie, © 1975, by Scientific American, Inc. All rights reserved.)

ry the plates away from the zone of upwelling. The hot plumes usually reveal themselves as volcanic structures growing up from the ocean floor in such places as Iceland. About 20 hot spots have been identified along ridge systems where they may contribute to plate divergence. Recall, however, that some hot plumes, for example, the one which generated the Hawaiian Islands, are not located in ridge areas. Therefore, we must conclude that this model is not without its shortcomings. Perhaps a combination of all these phenomena generates the plate motion observed.

REVIEW QUESTIONS

1   What first led scientists such as Alfred Wegener to suspect that the continents were once joined?

2   What was Pangaea?

3   List the evidence that Wegener and his associates gathered to support the continental drift hypothesis.

4   Early in this century, what was the prevailing view of how land animals migrated across vast expanses of ocean?

5   Briefly explain why the recent acceptance of plate tectonics has been described as a scientific "revolution."

6   How does evidence for a late Paleozoic glaciation in the Southern Hemisphere support the continental drift hypothesis?

7   Explain how paleomagnetism can be used to establish the latitude of a specific place at some distant time.

8   What is meant by sea-floor spreading? Who is credited with formulating the concept of sea-floor spreading?

9   Describe how Fred Vine and D. H. Matthews related the sea-floor spreading hypothesis to magnetic reversals.

10   On what basis were plate boundaries first established?

11   Where is new lithosphere being formed? Destroyed? Why must the production and destruction of the lithosphere be going on at about the same rate?

12   Why is the oceanic portion of a lithospheric plate subducted while the continental portion is not?

13   In what ways may the origin of the Japanese Islands be considered similar to the formation of the Andes Mountains? How do they differ?

14   Differentiate between transform faults and the two other types of plate boundaries.

15   Some people predict that California will sink into the ocean. Is this idea consistent with the concept of plate tectonics?

16   If the "hot spot" concept proves correct, what direction was the Pacific plate moving while the Emperor seamounts were being produced? (See Figure 16.26.) While the Hawaiian seamounts were being produced?

17   With what type of plate boundary are the following places or features associated (be as specific as possible): Himalayas, Aleutian Islands, Red Sea, Andes Mountains, San Andreas fault, Iceland, Japan, Mount St. Helens?

KEY TERMS

continental drift
(p. 386)

convergent boundary
(p. 400)

Curie point (p. 393)

deep-ocean trench
(p. 403)

divergent boundary
(p. 400)

hot spot (p. 413)

island arc (p. 405)

magnetometer (p. 397)

normal polarity
(p. 397)

paleomagnetism
(p. 393)

Pangaea (p. 386)

plate (p. 400)

polar wandering
(p. 394)

reverse polarity
(p. 397)

rift, or rift valley
(p. 401)

sea-floor spreading
(p. 395)

subduction zone
(p. 403)

transform fault
boundary (p. 400)

volcanic arc (p. 403)

17

THE OCEAN FLOOR AND ITS EVOLUTION

17

I f all water were drained from the ocean basins, the exposed surface would not be the quiet, subdued topography as was once thought. Rather, a great diversity of features, including towering mountain chains and deep canyons as well as flat plains would be found. The scenery would be just as varied as that on the continents (Figure 17.1).

The ocean's vast expanse first became apparent through voyages of discovery in the fifteenth and sixteenth centuries. An understanding of the ocean floor's varied topography did not unfold until much later with the historic 3½-year voyage of the H.M.S. *Challenger* (Figure 17.2). From December 1872, to May 1876, the *Challenger* expedition made the first, and still perhaps most comprehensive, study of the global ocean ever attempted by one agency. The 110,000-kilometer (68,000-mile) trip took the ship and its crew of scientists to every ocean except the Arctic. Throughout the voyage they sampled the depth of the water by laboriously lowering a weighted line overboard. Not many years later the knowledge gained by the *Challenger* of the ocean's great depths and varied topography was further expanded with the laying of transatlantic cables. However, as long as ocean depth had to be measured with weighted lines, our knowledge of the sea floor remained slight. Then, in the 1920s

The deep-diving manned submersible *Alvin*. (Photo by Rod Catanach, courtesy of Woods Hole Oceanographic Institution)

a technological breakthrough occurred with the invention of electronic depth-sounding equipment (**echo sounder**).

The echo sounder works by transmitting sound waves toward the ocean bottom (Figure 17.3). A delicate receiver intercepts the echo reflected from the bottom, and a clock precisely measures the time interval to fractions of a second. By knowing the velocity of the sound waves in water (about 1500 meters per second) and the time required for the energy pulse to reach the ocean floor, the depth can be established. The depths determined from continuous monitoring of these echos are normally plotted so that a profile of the ocean floor is produced. Since the invention of the echo sounder, millions of kilometers of continuous sonic-depth determinations have provided a more complete and detailed view of the ocean floor.

Oceanographers studying the topography of the oceans have delineated three major units: the continental margins, the deep-ocean basins, and the mid-ocean ridges. The map in Figure 17.4 outlines these provinces for the North Atlantic, and the profile at the bottom illustrates the varied topography. Such profiles usually have their vertical dimension exaggerated many times—40 times in this case—to make topographic features more conspicuous. Because of this, the slopes shown in the seafloor profile appear to be much steeper than they actually are.

CONTINENTAL MARGINS

The features comprising the **continental margin** include the continental shelf, the continental slope, and the continental rise (Figure 17.5). The first of these parts, the **continental shelf,** is a gently sloping submerged surface extending from the shoreline toward the deep-ocean basin. Since it is underlain by continental-type crust, it is clearly a flooded extension of the continents. The continental shelf varies greatly in width. Almost nonexistent along some continents, the shelf may extend seaward as far as 1500 kilometers along others. On the average, the continental shelf is about 80 kilometers wide and 130 meters deep at the seaward edge. The average inclination of the continental shelf is less than one-tenth of one degree, a drop of only about 2 meters per kilometer. The slope is so slight that it would appear to an observer to be a flat surface.

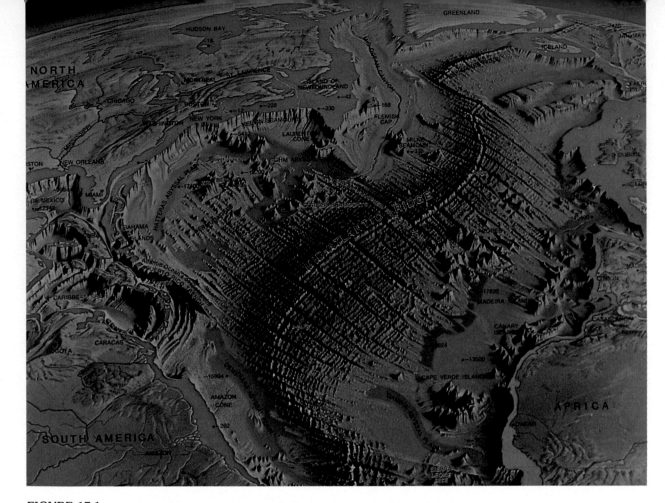

FIGURE 17.1
The ocean floor is characterized by a great diversity of features. This is an artist's view of what would be seen if all of the water were removed from the Atlantic Ocean basin. (From a painting by Heinrich Berann; courtesy of Aluminum Company of America)

FIGURE 17.2
The H.M.S. *Challenger.* (From C. W. Thomson and Sir John Murray, *Report on the Scientific Results of the Voyage of the H.M.S.* Challenger, Vol. 1. Great Britain: Challenger Office, 1895, Plate 1)

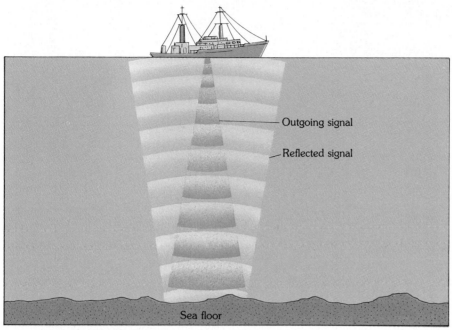

Outgoing signal

Reflected signal

Sea floor

A.

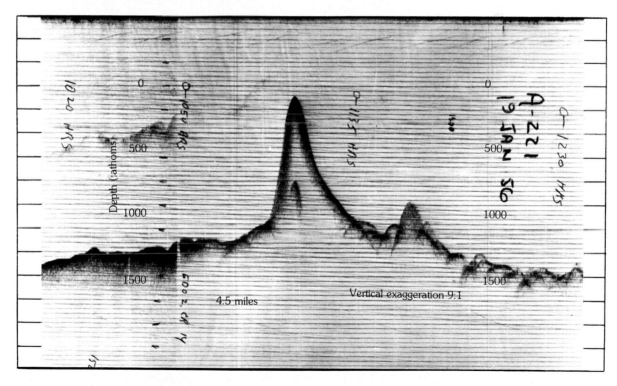

Depth (fathoms)

0

500

1000

1500

0

500

1000

1500

4.5 miles

Vertical exaggeration 9:1

B.

FIGURE 17.3
Profile of the ocean floor. **A.** An echo sounder determines the water depth by measuring the time interval required for a sonic wave to travel from a ship to the sea floor and back. Depth = ½ × sound travel time (speed of sound = 1500 m/sec). **B.** Seafloor profile made by an echo sounder. (Courtesy of Woods Hole Oceanographic Institution)

423

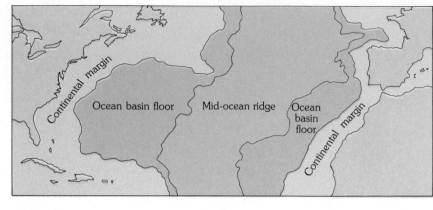

FIGURE 17.4

Major topographic divisions of the North Atlantic and a profile from New England to the coast of North Africa. (After B. C. Heezen, M. Tharp, and M. Ewing, *The Floors of the Oceans,* Geological Society of America Special Paper 65, p. 16)

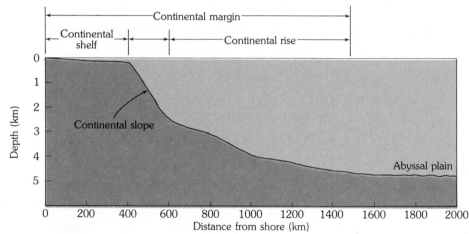

FIGURE 17.5

Schematic profile showing the provinces of the continental margin. (Vertical exaggeration 135 : 1)

Although the continental shelf is relatively feature-less, it is not completely smooth. Some continental shelf areas are mantled by extensive glacial deposits, and are thus quite rugged. The most profound features are long valleys running from the coastline into deeper waters. Many of these valleys are the seaward extensions of river valleys on the adjacent landmass. Such valleys appear to have been excavated during the Pleistocene epoch (Ice Age). During this time great quantities of water were tied up in vast ice sheets on the continents. This caused sea level to drop by 100 meters or more, exposing large areas of the continental shelves (see Figure 11.3). Because of this drop in sea level, rivers extended their courses, and land-dwelling plants and animals inhabited the newly exposed portions of the continents. Dredging off the east coast of North America has produced the remains of numerous land dwellers, including mammoths, mastodons, and horses, adding to the evidence that portions of the continental shelves were once above sea level.

Marking the seaward edge of the continental shelf is the **continental slope.** This feature is characterized by a steep gradient as compared with the shelf, and marks the boundary between continental crust and oceanic crust. Although the inclination of the continental slope varies greatly from place to place, the average is about 5 degrees and in places may exceed 25 degrees.

In regions where trenches do not exist, the steep

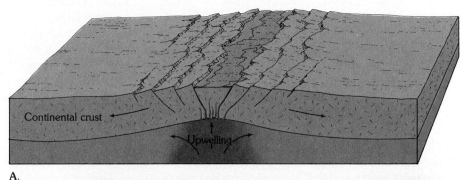

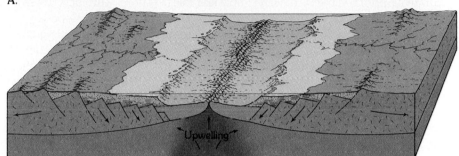

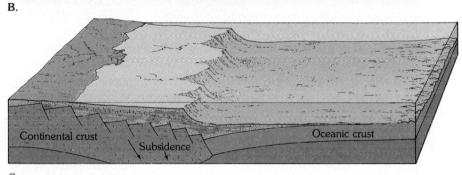

FIGURE 17.6
Stages in the breakup of a continental landmass. **A.** Upwelling of mantle rock causes fragmentation of the crust above. **B.** As the landmass moves away from the zone of upwelling, cooling results in subsidence of the continental margins. **C.** Material derived from the adjacent highlands accumulates into thick wedges of sediment.

continental slope merges into a more gradual incline known as the **continental rise.** Here the gradient lessens to between 4 and 8 meters per kilometer. Whereas the width of the continental slope averages about 20 kilometers, the continental rise may extend for hundreds of kilometers into the deep-ocean basin. This feature consists of a thick accumulation of sediment that moved downslope from the continental shelf to the deep-ocean floor. Although continental rises are relatively featureless, their surfaces are occasionally interrupted by submarine canyons or by submarine volcanoes that have not yet been completely buried by the sediments.

Some continental margins, such as those along the east coast of the United States, are wide and consist of thick accumulations of shallow-water sediments. These sediments are frequently several kilometers thick and are interbedded with limestones that formed during earlier periods of coral reef building, a process that occurs only in shallow water. This evidence led researchers to conclude that these thick accumulations of sediment are produced along a gradually subsiding continental margin.

This explanation fits very well with our knowledge of how new continental margins are produced during the breakup of a continental landmass. Figure 17.6 illustrates the stages in the process. Notice that upwelling of mantle material causes doming, which stretches and fractures the crust. As sea-floor spreading progresses, the stretched and fragmented crust is wedged away from the zone of upwelling where gradual cooling leads to shrinkage and hence subsidence. Sediments carried from adjacent highlands begin to accumulate on the young continental margin. This additional load is believed to contribute to the subsidence.

Passive continental margins of this type are found around the Atlantic Ocean on the trailing edges of the continents. The accretion of material on the passive margins of the continents results in a gradual increase in the size of the landmass. In addition, these vast accumulations of sediments have an important role in mountain building as we shall see in Chapter 18.

Along some mountainous coasts the continental slope descends abruptly into deep-ocean trenches found between the continent and ocean basin. In such cases, the shelf is very narrow or does not exist at all. The side of the trench and the continental slope are essentially the same feature and grade into the adjacent mountains which tower thousands of meters above sea level. These narrow continental margins are primarily located around the Pacific Ocean in areas where the leading edge of a continent is overrunning oceanic lithosphere. Here the stress between the converging plates results in deformation of the continental margin. An example of this activity is found along the west coast of South America. Here the vertical distance from the high peaks of the Andes Mountains to the floor of the deep Peru-Chile trench bordering the continent exceeds 12,000 meters.

SUBMARINE CANYONS AND TURBIDITY CURRENTS

Deep, steep-sided valleys known as **submarine canyons** originate on the continental slope and may extend to depths of 3 kilometers (Figure 17.7). Although some of these canyons appear to be the seaward extensions of river valleys such as the Hudson, many others do not line up in this manner. Furthermore, since these canyons extend to depths far below the maximum lowering of sea level during the Ice Age, we cannot attribute their formation to stream erosion. These features must be created by some process that operates far below the ocean surface. Most available information seems to favor the view that submarine canyons have been excavated by tur-

bidity currents. **Turbidity currents** are downslope movements of dense, sediment-laden water. They are created when sand and mud on the continental shelf and slope are dislodged, perhaps by an earthquake, and are thrown into suspension. Since this mud-choked water is denser than normal sea water, it flows downslope, eroding and accumulating more sediment as it continues to gain speed (Figure 17.8). The erosional work repeatedly carried on by these muddy torrents is thought to be the major force in the excavation of most submarine canyons.

Turbidity currents usually originate along the continental slope and continue across the continental rise, still cutting channels. Eventually they lose momentum and come to rest along the bottom of the ocean basin. As these currents slow, the suspended sediments begin to settle out. First, the coarser sand is dropped, followed by successively finer deposits of silt and then clay. Consequently, these deposits, called **turbidites,** are characterized by a decrease in sediment grain size from bottom to top, a phenomenon known as **graded bedding.**

For many years the existence of turbidity currents in the ocean was a matter of considerable debate among marine geologists. Not until the 1950s did the controversy begin to subside. Two lines of evidence

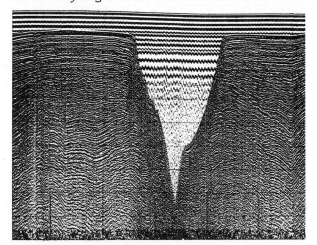

FIGURE 17.7
An echo-sounding profile of the Congo submarine canyon off the west coast of Africa. The bottom of the canyon is about 3 kilometers below the level of the sea floor of the continental shelf. Across the top, the canyon is more than 10 kilometers wide. (Courtesy of K. O. Emery, Woods Hole Oceanographic Institution)

helped establish turbidity currents as important mechanisms of submarine erosion and sediment transportation. The first important evidence came from records of a rather severe earthquake that took place off the coast of Newfoundland in 1929 and resulted in the breakage of 13 transatlantic telephone and telegraph cables. At the time it was presumed that the tremor had caused the multiple breaks. However, when the data were examined, this apparently was not the case. Since the time of each break was known from information provided by automatic re-

corders, a pattern of what had happened could be deduced (Figure 17.9). The breaks high up on the continental slope took place first, almost concurrently with the earthquake. The other breaks occurred in succession over a period of 13 hours. The last one was some 720 kilometers from the source of the quake. The breaks downslope had obviously taken place too long after the tremor to have been caused by the shock of the earthquake. The existence of a turbidity current, triggered by the earthquake, thus appeared as the only plausible alternative. As the

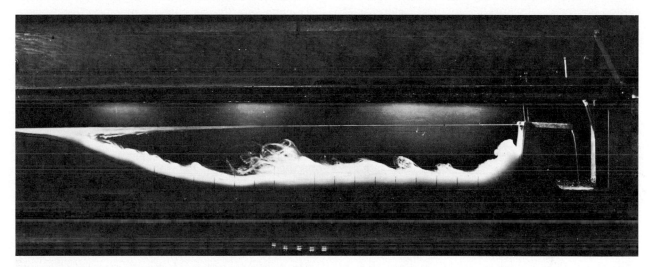

FIGURE 17.8
Turbidity current produced in a water-filled laboratory tank. (Courtesy of H. S. Bell Sedimentation Laboratory, California Institute of Technology)

FIGURE 17.9
Profile of the sea floor showing the events of the November 18, 1929, earthquake off the coast of Newfoundland. The arrows point to cable breaks; the numbers show times of breaks in hours and minutes after the earthquake. Vertical scale is greatly exaggerated. (After B. C. Heezen and M. Ewing, "Turbidity Currents and Submarine Slump and the 1929 Grand Banks Earthquake," *American Journal of Science* 250 : 867)

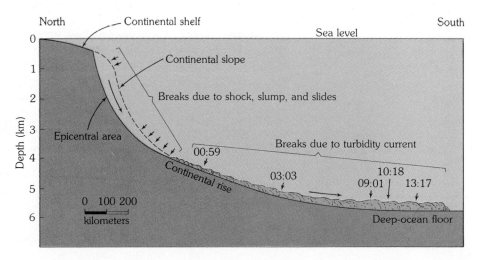

avalanche of sediment-choked water raced downslope, it snapped each of the cables in its path. Investigators calculated that the current reached speeds approaching 80 kilometers per hour on the continental slope and about 24 kilometers per hour on the more gently sloping continental rise.

A second compelling line of evidence relating turbidity currents to submarine erosion and the transportation of sediment came from the examination of deep-sea samples. These cores show that extensive graded beds of sand, silt, and clay exist in the quiet waters of the deep ocean. Some samples also included fragments of plants and animals that live only in the shallower waters of the continental shelves. No mechanism, other than turbidity currents, has been observed that can satisfactorily explain the existence of these deposits.

FEATURES OF THE DEEP-OCEAN BASIN

Between the continental margin and the oceanic ridge system lies the **deep-ocean basin.** The size of this region—almost 30 percent of the earth's surface—is roughly comparable to the percentage of the surface that projects above the sea as land. Here we find remarkably flat regions known as abyssal plains, steep-sided volcanic peaks called seamounts, and deep-ocean trenches, which are extremely deep linear depressions in the ocean floor.

DEEP-OCEAN TRENCHES

Deep-ocean **trenches** are long, relatively narrow features that form the deepest parts of the ocean. Most trenches are located in the Pacific Ocean where some approach or exceed 10,000 meters in depth, and at least a portion of one, the Challenger Deep in the Marianas trench, is more than 11,000 meters below sea level. Table 17.1 presents dimensional characteristics of some of the larger trenches.

Although deep-ocean trenches represent only a very small portion of the area of the ocean floor, they are nevertheless very significant geological features. Trenches are the sites where moving crustal plates are destroyed as they plunge back into the mantle. In addition to the earthquakes created as one plate descends beneath another, volcanic activity is also associated with trench regions. Trenches in the open ocean are paralleled by volcanic island arcs, whereas volcanic mountains such as those making up a portion of the Andes may be found paralleling trenches that are adjacent to continents. The melting of a descending plate produces the molten rock that leads to this volcanic activity.

ABYSSAL PLAINS

Abyssal plains are incredibly flat features; in fact, these regions are likely the most level places on the earth. The abyssal plain found off the coast of Argentina, for example, has less than 3 meters of relief over a distance exceeding 1300 kilometers. The monotonous topography of abyssal plains will occasionally

TABLE 17.1
Dimensions of some deep-ocean trenches.

Trench	Depth (kilometers)	Average Width (kilometers)	Length (kilometers)
Aleutian	7.7	50	3700
Japan	8.4	100	800
Java	7.5	80	4500
Kuril-Kamchatka	10.5	120	2200
Marianas	11.0	70	2550
Middle America	6.7	40	2800
Peru-Chile	8.1	100	5900
Philippine	10.5	60	1400
Puerto Rico	8.4	120	1550
South Sandwich	8.4	90	1450
Tonga	10.8	55	1400

be interrupted by the protruding summit of a buried volcanic structure.

By employing seismic profilers, instruments whose signals penetrate far below the ocean floor, researchers have determined that abyssal plains owe their relatively featureless topography to thick accumulations of sediment that have buried an otherwise rugged ocean floor. The nature of the sediment indicates that these plains consist primarily of sediments transported far out to sea by turbidity currents.

Abyssal plains are found in all of the oceans. However, since the Atlantic Ocean has fewer trenches to act as traps for the sediments carried down the continental slope, it has more extensive abyssal plains that the Pacific.

SEAMOUNTS

Dotting the ocean floors are isolated volcanic peaks called **seamounts** that may rise hundreds of meters above the surrounding topography. Although these steep-sided conical peaks have been discovered in all of the oceans, the greatest number have been identified in the Pacific.

Many of these undersea volcanoes form near oceanic ridges, regions of sea-floor spreading. If the volcano grows rapidly, it may emerge as an island. Examples of volcanic islands in the Atlantic include the Azores, Ascension, Tristan da Cunha, and St. Helena.

While they exist as islands, some of these volcanoes are eroded to near sea level by running water and wave action. Over a span of millions of years the islands gradually sink as the moving plate slowly carries them from the oceanic ridge area. These submerged, flat-topped seamounts are called **guyots**. In other instances, guyots may be remnants of eroded volcanic islands that were formed away from the ridge crest, possibly by hot spot activity. Here subsidence occurs after the volcanic activity ceases and the sea floor cools and contracts.

CORAL REEFS AND ATOLLS

Coral reefs are among the most picturesque features found in the ocean. They are constructed primarily from the calcareous (calcite-rich) skeletal remains and secretions of corals and certain algae. The term *coral reef* is somewhat misleading in that it makes no mention of the skeletons of many small animals and plants found inside the branching framework built by the corals; nor does it reveal that limy secretions of algae help bind the entire structure together.

Coral reefs are confined largely to the warm, clear waters of the Pacific and Indian oceans, although a few occur elsewhere. Reef-building corals grow best in waters with an average annual temperature of about 24°C (75°F). They can survive neither sudden temperature changes nor prolonged exposure to temperatures below 18°C (64°F). In addition, these reef-builders require clear sunlit water. Consequently, the limiting depth of active reef growth is only about 45 meters. Clear blue waters such as those in the Bahamas support active reef building.

In 1831 the naturalist Charles Darwin set out aboard the British ship H.M.S. *Beagle* on its famous five-year expedition that circumnavigated the globe. One outcome of Darwin's studies was the development of a theory on the formation of coral islands, called **atolls**. As Figure 17.10 illustrates, atolls consist of a nearly continuous ring of coral reef surrounding a central lagoon. Darwin's theory explained what seemed to be a paradox; that is, how can corals, which require warm, shallow, sunlit water no deeper than a few tens of meters to live, create structures that reach thousands of meters to the floor of the ocean? Commenting on this in his book *The Voyage of the Beagle*, Darwin states:

> . . . **from the fact of the reef-building corals not living at great depths, it is absolutely certain that throughout these vast areas, wherever there is now an atoll, a foundation must have originally existed within a depth of from 20 to 30 fathoms[1] from the surface.**

The essence of Darwin's theory was that coral reefs form on the flanks of sinking volcanic islands. As the island slowly sinks, the corals continue to build the reef complex upward (Figure 17.11).

> **For as mountain after mountain, and island after island slowly sank beneath the water, fresh bases would be successively afforded for the growth of the corals.**

---

[1]One fathom equals 1.8 meters (6 feet), the approximate distance from fingertip to fingertip of a person with outstretched arms.

FIGURE 17.10
View from space of a group of atolls in the Pacific Ocean. (Courtesy of NASA)

Thus atolls, like guyots, are thought to owe their existence to the gradual sinking of oceanic crust. In succeeding years there were numerous challenges to Darwin's theory. These arguments were not completely put to rest until after World War II when the United States made extensive studies of two atolls (Eniwetok and Bikini) that were going to become sites for testing atomic bombs. Drilling operations at these atolls revealed that volcanic rock did indeed underlie the thick coral reef structure. This finding was a striking confirmation of Darwin's theory.

SEA-FLOOR SEDIMENTS

Except for a few areas, such as near the crests of mid-ocean ridges, the ocean floor is mantled with sediment. Part of this material has been deposited by turbidity currents, and the rest has slowly settled to the bottom from above. The thickness of this carpet of debris varies greatly. In some trenches, which act as traps for sediments originating on the continental margin, accumulations may exceed 9 kilometers. In general, however, sediment accumulations are con-

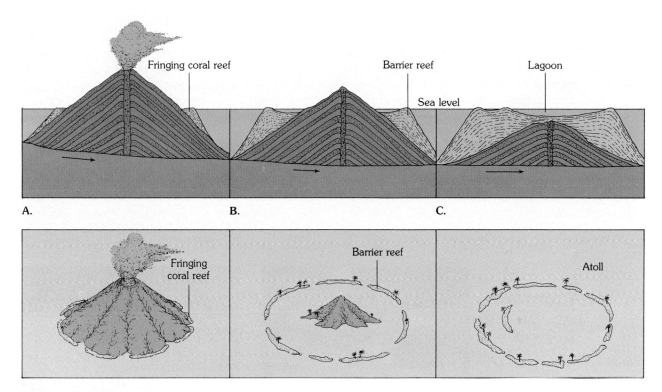

FIGURE 17.11
Cross-sectional view and map views of the formation of a coral atoll.

siderably less. In the Pacific Ocean, uncompacted sediment measures about 600 meters or less, while on the floor of the Atlantic, the thickness varies from 500 to 1000 meters.

Although deposits of sand-sized particles are found on the deep-ocean floor, mud is the most common sediment covering this region. Muds also predominate on the continental shelf and slopes, but the sediments in these areas are coarser overall because of greater quantities of sand. Sampling has revealed that sand deposits are most prevalent as beach deposits along the shore. However, in some cases coarse sediment, which we normally expect to be deposited near the shore, occurs in irregular patches at greater depths near the seaward limits of the continental shelves. While some sand may have been deposited by strong localized currents capable of moving coarse sediment far from shore, much of it appears to be the result of sand deposition on ancient

beaches. Such beaches evidently formed during the Ice Age, when sea level was much lower than today.

Sea-floor sediments can be classified according to their origin into three broad categories: (1) terrigenous ("derived from the land"); (2) biogenous ("derived from organisms"); and (3) hydrogenous ("derived from water"). Although each category is discussed separately, remember that all sea-floor sediments are mixtures. No body of sediment comes from a single source.

Terrigenous sediment consists primarily of mineral grains which were weathered from continental rocks and transported to the ocean. The sand-sized particles settle near shore. However, since the very smallest particles take years to settle to the ocean floor, they may be carried for thousands of kilometers by ocean currents. As a consequence, virtually every part of the ocean receives some terrigenous sediment. However, the rate at which this sediment accumulates on the deep-ocean floor is indeed very slow. From 5000 to 50,000 years are necessary for a 1-centimeter layer to form. Conversely, on the continental margins near the mouths of large rivers, terrigenous sediment accumulates rapidly. In the Gulf of Mexico, for example, the sediment has reached a depth of many kilometers.

Since fine particles remain suspended in the water for a very long time, there is ample opportunity for chemical reactions to occur. Because of this, the colors of the deep-sea sediments are often red or brown. This results when iron on the particle or in the water reacts with dissolved oxygen in the water and produces a coating of iron oxide (rust).

Biogenous sediment consists of shells and skeletons of marine animals and plants (Figure 17.12). This debris is produced mostly by microscopic organisms living in the sunlit waters near the ocean surface. The remains continually "rain" down upon the sea floor.

The most common biogenous sediments are known as *calcareous* ($CaCO_3$) *oozes,* and as their name implies, they have the consistency of thick mud. These sediments are produced by organisms that inhabit warm surface waters. When calcareous

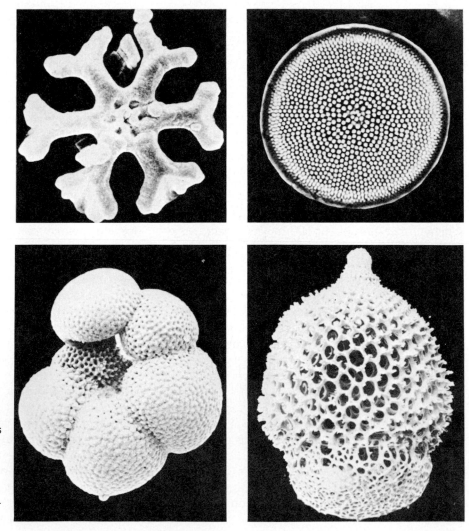

FIGURE 17.12
Enlarged photomicrographs of typical calcareous and siliceous materials from minute animals and plants that compose biogenous sediments. (Courtesy of Deep Sea Drilling Project, Scripps Institution of Oceanography)

hard parts slowly sink through a cool layer of water, they begin to dissolve. This results because cold seawater is rich in carbon dioxide and is thus more acidic than warm water. In seawater deeper than about 4500 meters (15,000 feet), calcareous shells will completely dissolve before they reach the bottom. Consequently, calcareous ooze does not accumulate in the deep-ocean basins.

Other biogenous sediments include *siliceous* (SiO_2) *oozes* and phosphate-rich materials. The former is composed primarily of opaline skeletons of diatoms (single-celled algae) and radiolaria (single-celled animals), while the latter is derived from the bones, teeth, and scales of fish and other marine organisms.

Hydrogenous sediment consists of minerals that crystallize directly from seawater through various chemical reactions. For example, some limestones are formed when calcium carbonate precipitates directly from the water; however, most limestone is composed of biogenous sediment.

One of the principal examples of hydrogenous sediment, and one of the most important sediments on the ocean floor in terms of economic potential are **manganese nodules.** These rounded blackish lumps

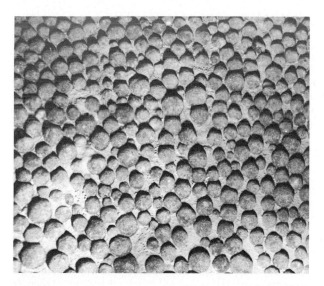

FIGURE 17.13
Manganese nodules photographed at a depth of 2909 fathoms (5323 meters) beneath the *Robert Conrad* south of Tahiti. (Courtesy of Lawrence Sullivan, Lamont-Doherty Geological Observatory)

are composed of a complex mixture of minerals that form very slowly on the floor of the ocean basins (Figure 17.13). In fact, their formation rate represents one of the slowest chemical reactions known. By analyzing the radioactive elements continually incorporated into growing nodules, researchers have determined that the growth rates vary from 0.001 to 0.2 millimeter per 1000 years. Some portions of the sea floor are littered with these deposits whereas others lack them altogether. The presence or absence of nodules has been correlated with the rate of sedimentation. If sediment accumulates too rapidly (at a rate exceeding about 7 millimeters per 1000 years), newly forming nodules are buried and growth ceases. Since nodule growth is exceedingly slow, why are nodules not buried even where sediment accumulates at less than 7 millimeters per 1000 years? Some scientists suggest that benthic animals living in and on the sea floor are responsible for keeping the nodules at the surface. By stirring the sediment, burrowing organisms are thought to produce a slight lifting effect which, in combination with small surface animals that consume newly arrived sediment from nodule surfaces, keeps nodules from being buried.

Although manganese nodules may contain more than 20 percent manganese, the interest in them as a potential resource lies in the fact that other more valuable metals may be enriched in them. In addition to manganese, nodules may contain significant quantities of iron, copper, nickel, and cobalt. All regions containing nodules, however, are not equally good potential sites for mining. Possible mining locations must have abundant nodules (more than 5 kilograms per square meter) and contain the economically optimum mix of cobalt, copper, and nickel. Sites meeting these criteria are relatively limited. Furthermore, before such areas prove to be valuable commercial sources for these metals, the logistics of extracting nodules from the floor of the deep-ocean basins must be worked out.

MID-OCEAN RIDGES

Our knowledge of **mid-ocean ridges,** the sites of seafloor spreading, comes from soundings taken of the ocean floor, core samples obtained from deep-sea drilling, visual inspection using deep-diving submersibles, and even first-hand inspection of slices of ocean floor that have been shoved up onto dry land. Ocean ridge systems are characterized by an elevated position, extensive faulting, and numerous volcanic structures which have developed upon the newly formed crust (Figure 17.1).

Ocean ridges are found in all of the major oceans and represent more than 20 percent of the earth's surface. They are certainly the most prominent topographic features in the oceans, for they form a continuous mountain range which extends for about 65,000 kilometers (40,000 miles) in a manner similar to the seam on a baseball. Although ocean ridges stand high above the adjacent deep-ocean basins, they are much different than the mountains found on the continents. Rather than consisting of thick sequences of folded and faulted sedimentary rocks, oceanic ridges consist of layer upon layer of basaltic rocks that have been faulted and uplifted. The term *ridge* may also be misleading since these features are not narrow, but have widths of from 500 to 5000 kilometers and, in places, may occupy as much as one-half of the total area of the ocean floor.

Despite their large size, most of the accretions of new sea floor occurs along a narrow region centered on the ridge crest. These active **rift zones** are characterized by frequent but generally weak earthquakes and a rate of heat flow that is greater than other crustal segments. Here vertical displacement of large slabs of oceanic crust caused by faulting and the growth of volcanic piles contribute to the characteristically rugged topography of the oceanic ridge system. Further, the rocks along the ridge axis appear very fresh and are nearly void of sediment. Away from the ridge axis the topography becomes more subdued and the thickness of the sediments as well as the depth of the water increase. Gradually the ridge system grades into the flat, sediment-laden abyssal plains of the deep-ocean basin.

During sea-floor spreading new material is added about equally to the two diverging plates; hence, we would expect new ocean floor to grow symmetrically about a centrally located ridge. Indeed, the ridge systems of the Atlantic and Indian oceans are located near the middle of these water bodies and as a consequence are named mid-ocean ridges. However, the East Pacific Rise is situated far from the center of the Pacific Ocean. Despite relatively uniform spreading along the East Pacific Rise, overrunning of lithosphere which formerly floored the Pacific by the adjacent American plates has moved the relative position of the Pacific Rise toward the east.

Partly because of its accessibility to both American and European scientists, the Mid-Atlantic Ridge has been studied more thoroughly than other ridge systems. The Mid-Atlantic Ridge is a gigantic submerged mountain range standing 2500–3000 meters above the adjacent floor of the deep-ocean basins. In a few places, such as Iceland, the ridge has actually grown above sea level. Throughout most of its length, however, this divergent plate boundary lies 2500 meters below sea level. Another prominent feature of the Mid-Atlantic Ridge is a very deep linear valley extending along the ridge axis. In places this rift valley is deeper than the Grand Canyon of the Colorado and two or three times as wide. The name *rift valley* has been applied to this feature because it is so strikingly similar to continental rift valleys such as the Great Rift Valley of East Africa . An examination of Figure 17.1 reveals that this central rift is broken into sections which are offset by transform faults.

THE OCEAN FLOOR AND SEA-FLOOR SPREADING

The concept of sea-floor spreading was formulated in the early 1960s by Harry H. Hess of Princeton University. Later, using deep-driving submersibles, geologists were able to support Hess's thesis that sea-floor spreading occurs along relatively narrow zones located at the crests of ocean ridges (Figure 17.14). Along the East Pacific Rise the active zones of sea-floor formation appear to be only about a kilometer wide, whereas along the Mid-Atlantic Ridge these zones may extend for tens of kilometers. As the plates move apart, magma intrudes into the newly created fracture zone and generates new sections of oceanic crust. This apparently unending process generates new lithosphere that moves from the ridge crest in a conveyor belt fashion.

As various segments of the oceanic ridge system were studied in detail, numerous differences came to light. For example, the East Pacific Rise has a rela-

A.

B.

FIGURE 17.14
A. The deep-diving manned submersible *Alvin* is 7.6 meters long, weighs 16 tons, has a cruising speed of 1 knot, and can reach depths as great as 4000 meters. Typically a pilot and two observers are along during a normal 6- to 10-hour dive. **B.** A photograph taken from the *Alvin* during Project FAMOUS shows lava extrusions in the rift valley of the Mid-Atlantic Ridge. Large toothpastelike extrusions such as this were common features. A mechanical arm is sampling an adjacent blisterlike extrusion. (Courtesy of Woods Hole Oceanographic Institution)

tively fast spreading rate that averages about 6 centimeters per year and reaches a maximum rate of 10 centimeters per year along a section of the ridge located near Easter Island. By contrast, the spreading rate in the North Atlantic is much slower, averaging about 2 centimeters per year. Apparently the rate of spreading strongly influences the appearance of the ridge system. The slow spreading rate along the Mid-Atlantic Ridge is believed to contribute to its very rugged topography and large central rift valley. By contrast, the rapid spreading of the East Pacific Rise is thought to account for its more subdued topography and the lack of a well-defined rift valley through much of its length. Despite these differences, all ridge systems are thought to generate new sea floor in a similar manner.

Although most oceanic crust forms out of view, far below sea level, geologists have been able to exam-ine the structure of the ocean firsthand. In such locations as Newfoundland, Cyprus, and California, slivers of oceanic crust have been elevated high above sea level. From these outcrops it appears that the ocean floor consists of three distinct layers (Figure 17.15A). The upper layer is composed mainly of pillow basalts. The middle layer is made up of numerous interconnected dikes called **sheeted dikes.** Finally, the lower layer is made up of gabbro, the coarse-grained equivalent of basalt, that crystallized at depth. This sequence of rocks is called an **ophiolite complex.** From studies of various ophiolite complexes and related data, geologists have pieced together a scenario for the formation of the ocean floor.

The magma that migrates upward to create new ocean floor originates from partially melted peridotite in the asthenosphere. In the region of the rift zone,

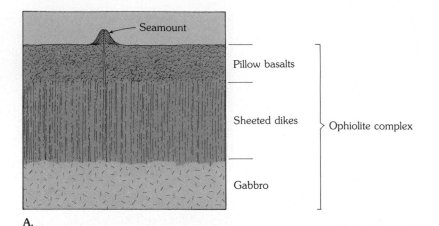

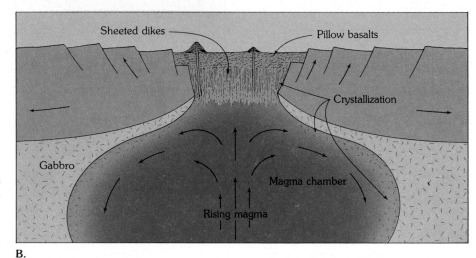

FIGURE 17.15
A. The structure of oceanic crust is thought to be equivalent to the ophiolite complexes that have been discovered elevated above sea level in such places as Cyprus. **B.** The formation of the three units of an ophiolite complex.

this magma source may lie no more than 35 kilometers below the sea floor (Figure 17.15B). Being molten and less dense than the surrounding solid rock, the magma gradually moves upward where it is believed to enter large reservoirs located only a few kilometers below the ridge crest. As the ocean floor is pulled or pushed apart, numerous fractures develop in the crust, permitting this molten rock to migrate upward. During each eruptive phase, the initial flows are thought to be quite fluid and spread over the rift zone in broad, thin sheets. As new lava flows are added to the ocean floor, each is cut by fractures that allow additional lava to migrate upward and form overlying layers. Later in each eruptive cycle, as the magma in the shallow reservoir cools and thickens, shorter flows with a more characteristic pillow form

occur. Recall that pillow lava has the appearance of large, elongate sand bags stacked one atop another (Figure 17.16). Depending upon the rate of flow, the thick pillow lavas may build into volcano-sized mounds. These mounds will eventually be cut off from their supply of magma and be carried away from the ridge crest by sea-floor spreading. The magma which does not flow upward will crystallize at depth to generate thick units of coarse-grained gabbro. This lowest rock unit forms as crystallization takes place along the walls and floor of the magma chamber. In this manner the processes at work at the ridge systems are producing the entire sequence of rocks found in the ophiolite complex.

From the study of ophiolite complexes and mid-ocean ridges, the complex evolutionary history of the

FIGURE 17.16
Pillow lava at Trinity Bay, Newfoundland. (Courtesy of the Geological Survey of Canada, photo no. 152581)

ocean floor is gradually being unraveled. This work has also provided valuable information about the origin, occurrence, and distribution of certain economically important mineral deposits. Several minerals have been mined from ophiolites, including sulfides of copper, zinc, and iron. A well-known example is the ophiolite complex on Cyprus, a significant source of copper for more than 4000 years. Apparently, these deposits represent ores that formed on the sea floor at an ancient oceanic spreading center. Since the mid-1970s active hot springs and metal-rich sulfide deposits have been detected at several localities, including study areas along the East Pacific Rise and the Juan de Fuca Ridge. The deposits are forming where heated seawater rich in dissolved metals and sulfur gushes from the sea floor as particle-filled clouds called *black smokers* (Figure 17.17). Among the most spectacular examples to date are those that were photographed along the East Pacific Rise by scientists aboard the *Alvin*, a deep-diving manned submersible. As shown in Figure 17.18, seawater is believed to infiltrate the hot oceanic crust along the flanks of the ridge. As the water moves through the

newly formed material, it is heated and chemically interacts with the basalt, extracting and transporting sulfur, iron, copper, and other metals. Near the ridge axis, the hot, metal-rich fluid rises along faults. Upon reaching the sea floor, the spewing liquid mixes with the cold seawater and the sulfides precipitate to form massive deposits.

Because newly formed sections of the ocean floor are warm, they are also rather buoyant. This buoyancy is thought to cause large blocks to shear from the sea floor and be elevated. Since the rate of spreading along the Mid-Atlantic Ridge is relatively slow, the displacement of oceanic slabs is more pronounced there than along faster spreading centers such as the East Pacific Rise. Thus, along the Mid-Atlantic Ridge, uplifted sections form nearly vertical walls that border the central rift zone. As sea-floor spreading continues, the earlier formed blocks are wedged away from the ridge axis and replaced by more recently formed segments of ocean crust. This process contributes to the imposing height of the Mid-Atlantic Ridge as well as to its rugged topography.

The primary reason for the elevated position of a

FIGURE 17.17
This is one of two dozen vents called "black smokers" that were found by the *Alvin* at 21°N latitude on the East Pacific Rise in May, 1979. The "smoke" is actually hot, mineral-rich water that has circulated through the ocean crust and picked up iron, copper, and zinc. When the hot solution hits the cold seawater, it precipitates sulfide ores that now coat the vents. (Courtesy of Woods Hole Oceanographic Institution)

ridge system is the fact that newly created oceanic crust is hot, and therefore occupies more volume than cooler rocks of the deep-ocean basin. As the young lithosphere travels away from the spreading center, it gradually cools and contracts. This thermal contraction accounts in part for the greater ocean depths that exist away from the ridge. Almost 100 million years must pass before cooling and contraction cease completely. By this time, rock that was once a part of a majestic oceanic mountain system is located in the deep-ocean basin, where it is mantled by thick accumulations of sediment.

OPENING AND CLOSING OF THE OCEAN BASINS

A great deal of evidence has been gathered to support the fact that Wegener's vast continent of Pan-

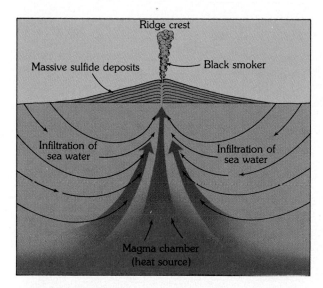

FIGURE 17.18
Massive sulfide deposits can result from the circulation of seawater through the oceanic crust along active spreading centers. As seawater infiltrates the hot basaltic crust, it leaches sulfur, iron, copper, and other metals. The hot, enriched fluid returns to the sea floor near the ridge axis along faults and fractures. Some metal sulfides may be precipitated in these channels as the rising fluid begins to cool. When the hot liquid emerges from the sea floor and mixes with cold seawater, the sulfides precipitate to form massive deposits.

gaea began to break apart about 200 million years ago. An important consequence of this episode of continental rifting was the creation of a "new" ocean basin, the Atlantic. The breakup of Pangaea and the formation of the Atlantic Ocean basin apparently occurred over a span of nearly 150 million years, with the last phase, the separation of Greenland and Eurasia, beginning only about 50 million years ago.

Although continental rifting is well documented, the question that remains open to debate is, What causes a continent to break apart? We have already considered the role of convection currents as a possible driving mechanism for plate motion (see Chapter 16). It seems reasonable to assume that the slow movement of mantle material could initiate continental rifting. However, the shape of the rifted continental margins and the large number of hot spots located along ridge crests led some geologists to a different

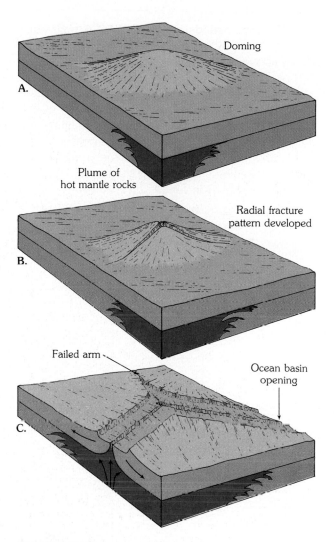

FIGURE 17.19
Diagrammatic representation of the doming and fragmentation of continental crust above an upwelling hot spot. (After Burke and Wilson)

conclusion. They have proposed that hot spots initiate continental fragmentation.

Recall that **hot spots** are plumes of molten rock that are believed to rise from deep within the mantle. Hot spots are generally characterized by large outpourings of basaltic lava for relatively long periods of time. Worldwide, as many as 120 isolated volcanic sites have been attributed to hot spot activity. Since hot spots appear to remain nearly stationary over extended time periods, they form volcanic trails, such as the Hawaiian chain, upon the moving oceanic plates above.

The Canadian geologist J. Tuzo Wilson and his associates have suggested that when a thick segment of continental lithosphere remains stationary over a hot spot for an extended period, the conditions are right for continental rifting. Initially, upwelling of material from below generates a dome, roughly 200 kilometers in diameter, within the overlying continental crust as shown in Figure 17.19A. As the dome enlarges, it fractures with a characteristic three-armed pattern (Figure 17.19B). Rifting continues along two of the arms, resulting in the development of a new ocean basin, while the third arm often fails to develop further (Figure 17.19C). An example of such a three-armed rift system is believed to be represented by the Red Sea, the Gulf of Aden, and the Afar Lowlands (Figure 17.20). Here the arm extending from the Afar Lowlands into the interior of Africa is the failed arm. The two active arms have subsequently generated long, narrow seas.

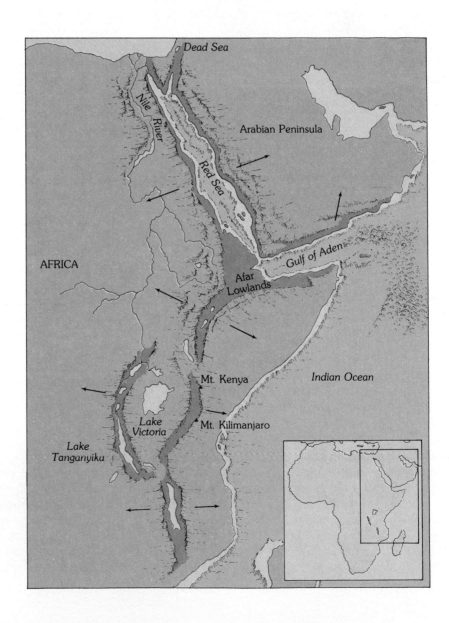

FIGURE 17.20
East African rift valleys and associated features.

Professor Wilson suggested that about 20 hot spots guided the fracturing of Pangaea. Figure 17.21 shows the proposed locations of several hot spots and the associated fractures that could account for the shape of the continental margins that presently border the Atlantic. Notice that in some situations all three arms opened, whereas in others, one of the arms failed. Many of these failed rifts extend into the continental interior, where they are represented by deep, narrow, sediment-filled troughs. Some of the world's major river systems, including the Niger and the Amazon, apparently occupy segments of these failed rifts.

Although Wilson's theory seems to explain the shape of the continents as well as the existence of numerous failed rift valleys, it has not yet gained gen-

eral acceptance. However, if his theory of continental rifting is correct, and hot spots guide the fracturing which precedes sea-floor spreading, then convection currents may only be secondary phenomena. Stated another way, Wilson's theory suggests that the convectional upwelling of magma along the ridge crests may not be the cause of continental rifting, but rather a consequence of it.

In 1966, Wilson also proposed that the Atlantic Ocean opened and then closed to produce the Appalachian mountain belt. Today we find that the Atlantic, Red Sea, and the Gulf of California are opening, while the Mediterranean and Pacific are closing. Further, the sites of former ocean basins are marked by mountain belts such as the Alps, Himalayas, and Urals. In honor of the man who proposed these complex cycles of ocean basin openings and closings, they have come to be called **Wilson cycles** (Figure 17.22).

When an ocean closes and reopens, the zone of fragmentation may not occur at the same location as the suture where the landmasses were joined. During the closing of the proto-Atlantic about 400 million years ago, the suture formed along a mountainous belt extending from Alabama to the British Isles and Norway. However, when the Atlantic began to reopen about 200 million years ago, the split occurred along a somewhat different trend. Thus the size and shape of continental fragments appear to change through time. In the next section we will examine further the concept of "Wilson cycles" as it relates to the formation and breakup of Pangaea.

PANGAEA: BEFORE AND AFTER

Robert Dietz and John Holden have rather precisely projected the gross details of the migrations of individual continents over the past 500 million years. By extrapolating plate motion back in time using such evidence as the orientation of volcanic structures left behind on moving plates, the distribution and movements of transform faults, and paleomagnetism, Dietz and Holden were able to reconstruct Pangaea (see Figure 16.1). The use of radiometric dating helped them establish the time frame for the formation and eventual breakup of Pangaea, and the relatively stationary positions of hot spots through time helped to fix the locations of the continents.

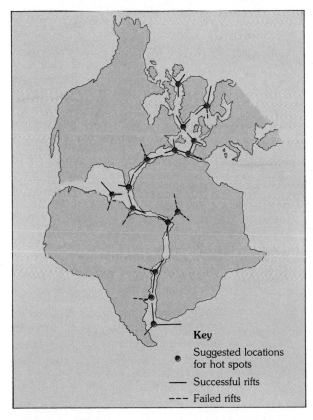

Key

• Suggested locations for hot spots

— Successful rifts

--- Failed rifts

FIGURE 17.21
Possible locations of hot spots and associated three-armed rifts that account for the shapes of the continents surrounding the Atlantic. In most instances two of the arms opened while the third failed. (After Burke and Wilson)

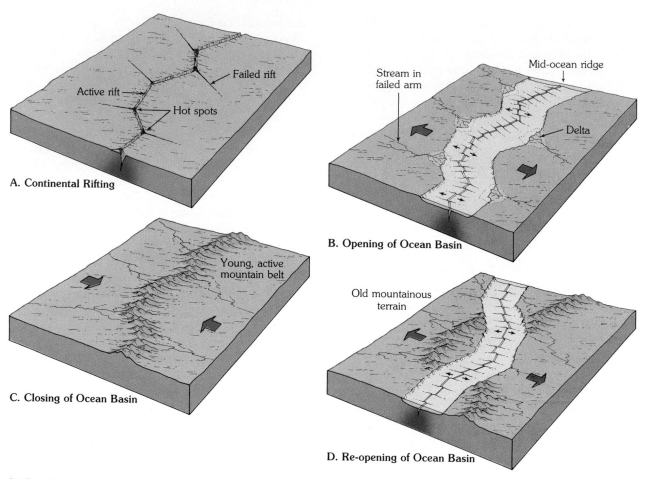

A. Continental Rifting

Failed rift

Active rift

Hot spots

B. Opening of Ocean Basin

Stream in failed arm

Mid-ocean ridge

Delta

C. Closing of Ocean Basin

Young, active mountain belt

D. Re-opening of Ocean Basin

Old mountainous terrain

FIGURE 17.22
The Wilson cycle. Diagrammatic representation of the generation of mountainous terrains through the opening and closing of ocean basins.

BREAKUP OF PANGAEA

The fragmentation of Pangaea began about 200 million years ago. Figure 17.23 illustrates the breakup and subsequent paths taken by the landmasses involved. As we can readily see in Figure 17.23A, two major rifts initiated the breakup. The rift zone between North America and Africa generated numerous outpourings of Triassic-age basalts which are presently visible along the eastern seaboard of the United States. Radiometric dating of these basalts indicates that rifting occurred between 200 and 165 million years ago. This date can be used as the birth date of this section of the North Atlantic. The rift that

formed in the southern landmass of Gondwanaland developed a "Y"-shaped fracture which sent India on a northward journey and simultaneously separated South America-Africa from Australia-Antarctica.

Figure 17.23B illustrates the position of the continents 135 million years ago, about the time Africa and South America began splitting apart to form the South Atlantic. India can be seen halfway into its journey to Asia, while the southern portion of the North Atlantic has widened greatly. By the end of the Cretaceous period, about 65 million years ago, Madagascar had separated from Africa, and the South Atlantic had emerged as a full-fledged ocean

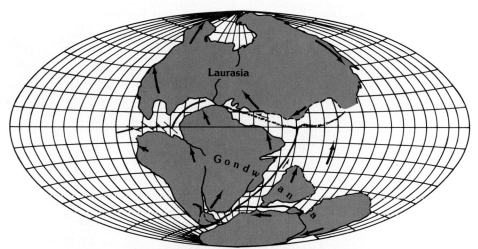

A. 180 Million Years Ago (Triassic Period)

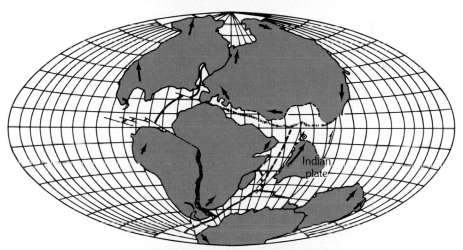

B. 135 Million Years Ago (Jurassic Period)

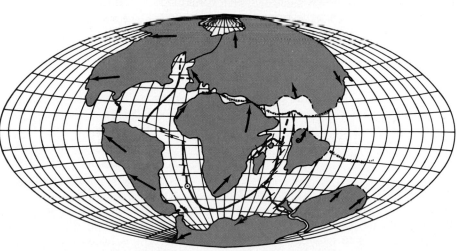

C. 65 Million Years Ago (Cretaceous Period)

FIGURE 17.23A., B., C.
Several views of the breakup of Pangaea over a period of 200 million years according to Dietz and Holden. (Robert S. Dietz and John C. Holden, *Journal of Geophysical Research* 75 : 4939–56, 1970. Copyright by American Geophysical Union)

443

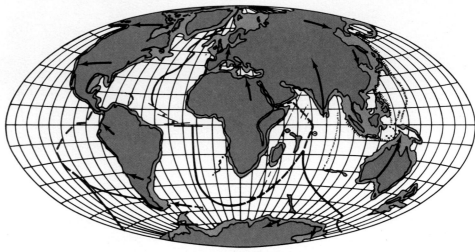

FIGURE 17.23D.

D. Present

(Figure 17.23C). At this juncture, India had drifted over a hot spot that generated numerous fluid basalt flows across a region in western India now called the Deccan Plateau. These lava flows are very similar to those that make up the Columbia Plateau in the Pacific Northwest.

The current map (Figure 17.23D) shows India in contact with Asia, an event that occurred about 45 million years ago and created the highest mountains on earth, the Himalayas, along with the Tibetan Highlands. It is interesting to note that the average height of Tibet is 5000 meters, higher than any spot in the contiguous United States. India's continued northward migration is believed to cause the numerous and often destructive earthquakes which plague that part of the world.

By comparing Figures 17.23C and 17.23D, we can see that the separation of Greenland from Eurasia was a recent event in geologic history. Also notice the recent formation of the Baja Peninsula along with the Gulf of California. This event is thought to have occurred less than 10 million years ago.

BEFORE PANGAEA

Prior to the formation of Pangaea, the landmasses had probably gone through several episodes of fragmentation similar to what we see happening today. Also like today, these ancient continents moved away

from each other only to collide again at some other location. During the period between 500 and 225 million years ago, the fragments of an earlier dispersal began collecting to form the continent of Pangaea. Evidence of these earlier continental collisions include the Ural Mountains of the Soviet Union and the Appalachian Mountains, which flank the east coast of North America.

Available evidence indicates that about 500 million years ago the northern continent of Laurasia was fragmented into three major sections—North America, northern Europe (southern Europe was part of Africa), and Siberia—with each section separated by a sizable ocean. The southern continent of Gondwanaland probably was intact and lay near the South Pole. The first collision is believed to have occurred as North America and Europe closed the pre-North Atlantic. This activity resulted in the formation of the northern Appalachians. Parts of the floor of the former ocean can be seen today high above sea level in Nova Scotia. The sliver of eastern Canada and the United States which lies seaward of the zone of this collision is truly a gift from Europe. It is also thought that before North America and Europe collided, part of Scotland, Ireland, and Norway were attached to the North American plate. While North America and Europe were joining, Siberia was closing the gap between itself and Europe, which lay farther to the west. This closing culminated about 300 to 350 mil-

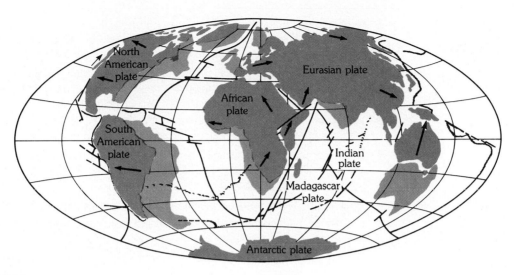

FIGURE 17.24
The world as it may look 50 million years from now. (From "The Breakup of Pangaea," Robert S. Dietz and John C. Holden. Copyright 1970 by Scientific American, Inc. All rights reserved)

lion years ago in the formation of the Ural Mountains. The consolidation of these landmasses completed the northern continent of Laurasia.

During the next 50 million years the northern and southern landmasses converged, producing the supercontinent of Pangaea. At this time (about 250 to 300 million years ago) Africa and North America collided to produce the southern Appalachians.

A LOOK INTO THE FUTURE

After Dietz and Holden drew together the events that over the past 500 million years resulted in the current configuration of the continents, they went one step further and extrapolated plate motion into the future. Figure 17.24 illustrates what they envision the earth's landmasses will look like 50 million years from now. Important changes are seen in Africa, where a new sea emerges as East Africa parts company with the mainland. In North America we see that the Baja Peninsula and the portion of southern California that lies west of the San Andreas fault have slid past the North American plate. If this northward migration takes place as predicted, Los Angeles and San Francisco will pass each other.

In another part of the world, Africa will have moved slowly toward Europe, initiating perhaps the next major mountain building stage on our dynamic planet. Australia is seen on a collision course with New Guinea and Asia, and North and South America are once again separating. These projections into the future, although interesting, must be viewed with caution since many assumptions must be correct for these events to unfold as just described. Nevertheless, similar types of changes in the shapes and positions of the continents will undoubtedly occur for many millions of years to come.

REVIEW QUESTIONS

1 Assuming that the average speed of sound waves in water is 1500 meters per second, determine the water depth if the signal sent out by an echo sounder requires 6 seconds to strike bottom and return to the recorder (see Figure 17.3A).

2 List the three major features that comprise the continental margin. Which of these features is considered a flooded extension of the continent? Which has the steepest slope?

3 How does the continental margin along the west coast of South America differ from the continental margin along the east coast of North America?

4 Defend or rebut the following statement, "Most of the submarine canyons found on the continental slope and rise were formed during the Ice Age when rivers extended their valleys seaward."

5 What are turbidites? What is meant by the term *graded bedding*?

6 Discuss the evidence that helped confirm the existence of turbidity currents in the ocean.

7 Why are abyssal plains more extensive on the floor of the Atlantic than on the floor of the Pacific?

8 What is an atoll? Describe Darwin's theory on the origin of atolls. Was the theory ever confirmed?

9 Differentiate between the three basic types of sea-floor sediment.

10 If you were to examine recently deposited biogenous sediment taken from a depth in excess of 4500 meters (15,000 feet), would it more likely be rich in calcareous materials or siliceous materials? Explain.

11 How are mid-ocean ridges and deep-ocean trenches related to sea-floor spreading?

12 What is the primary reason for the elevated position of the oceanic ridge system?

13 Briefly describe J. Tuzo Wilson's proposal for continental rifting. If Wilson's idea is correct, how does it affect previous notions about convection currents?

14 Describe the continental collisions that created the Appalachian and Ural mountains.

15 Presently Los Angeles is south of San Francisco. However, millions of years from now, their positions may be reversed; that is, Los Angeles may be north of San Francisco. Explain why this could occur.

KEY TERMS

abyssal plain (p. 428)

atoll (p. 429)

biogenous sediment
(p. 432)

continental margin
(p. 421)

continental rise
(p. 425)

continental shelf
(p. 421)

continental slope
(p. 424)

deep-ocean basin
(p. 428)

echo sounder (p. 421)

graded bedding
(p. 426)

guyot (p. 429)

hot spot (p. 439)

hydrogenous sediment
(p. 433)

manganese nodule
(p. 433)

mid-ocean ridge
(p. 433)

ophiolite complex
(p. 435)

rift zone (p. 434)

seamount (p. 429)

sheeted dike (p. 435)

submarine canyon
(p. 426)

terrigenous sediment
(p. 431)

trench (p. 428)

turbidite (p. 426)

turbidity current
(p. 426)

Wilson cycle (p. 441)

18

MOUNTAIN BUILDING AND THE EVOLUTION OF CONTINENTS

18

Mountains are often spectacular features which rise several hundred meters or more above the surrounding terrain. Some occur as single isolated masses; the volcanic cone Kilimanjaro, for example, stands almost 6000 meters (20,000 feet) above sea level overlooking the grasslands of East Africa. Others make up a portion of an extensive mountainous chain, such as the American Cordillera, which runs continuously from the tip of South America through Alaska. Chains such as the Himalayas, are youthful, gigantic mountains that are still rising, while others are very old and nearly worn down, as exemplified by the Appalachian Mountains in the eastern United States.

The name for the processes which collectively produce a mountain system is **orogenesis,** from the Greek *oros* ("mountain") and *genesis* ("to come into being"). Mountain systems show evidence of enormous forces which have folded, faulted, and generally deformed large sections of the earth's crust (Figure 18.1). Although the processes of folding and faulting have contributed to

Imposing east face of the Teton Range. (Courtesy of the National Park Service)

451

the majestic appearance of mountains, much of the credit for their beauty must be given to the work of running water and glacial ice which sculpture these uplifted masses in an unending effort to lower them to sea level.

Geologists have come to realize that in addition to providing spectacular scenery, mountains play a significant role in the evolution of continental crust. Some geologists believe that the continents have gradually grown larger by the addition of linear mountainous terrains to their flanks. These geologists point to the Appalachians of the eastern United States and the Andes of South America as examples of such growth. This hypothesis implies that nearly all continental areas once stood as mountains and were subsequently lowered to their present elevations by erosion. We will return to this idea later in the chapter, but first let us consider the nature of crustal uplifting and examine how rocks are deformed by processes such as folding and faulting.

CRUSTAL UPLIFT

The fossilized shells of marine invertebrates are often found in mountain regions, an indication that the sedimentary rock composing the mountain was once below sea level. This is rather convincing proof that some drastic changes occurred between the time these animals died and when their fossilized remains were discovered. Evidence for crustal uplift such as

FIGURE 18.1
Intensely folded rock strata provide evidence of the forces altering the earth's crust. (Photo by W. B. Hamilton, U.S. Geological Survey)

this is common in the geologic record and is even present in the historical record. For example, Figure 18.2 shows three columns remaining from a Roman temple. The columns have clam borings to a height of about 6 meters (20 feet), indicating that the land upon which the temple was built submerged and was later partially uplifted. These elevated clam borings might also be explained by a recent change in sea level; however, a similar change in sea level is not recorded at any other location for that same time period. Further evidence for crustal uplift can be found along the coastline of the western United States. When a coastal area remains undisturbed for an extended period, wave action cuts a gently sloping

Clam borings

FIGURE 18.2
Remaining columns of the ancient Roman temple of Serapis, Pozzuoli, Italy, in 1836. Clam borings 6 meters above sea level indicate former submergence. (Charles Lyell, *Principles of Geology,* 10th ed., 1867)

bench. In parts of California, ancient wave-cut benches can now be found as terraces, hundreds of meters above sea level (Figure 18.3). Each terrace represents a period when that area was at sea level. Unfortunately, the reasons for uplift are not always as easy to determine as the evidence for the movements.

We know that the force of gravity must play an important role in determining the elevation of the land. In particular, the less dense lithosphere is believed to float on top of the denser and more easily deformed rocks of the asthenosphere. The concept of a floating lithosphere in gravitational balance is called **isostasy.** Perhaps the easiest way to envision isostatic balance is to compare the lithosphere to floating logs. Imagine two logs, one much thicker than the other, floating in water. The larger log will float higher in the water than the smaller log. In the same manner, mountainous regions are believed to represent unusually thick sections of the earth's crust, whereas areas of low elevation do not have such crustal thickness. Mountains, like thick logs, not only stand high above the surface, but also extend farther into the supporting material below (Figure 18.4). This fact has been confirmed by seismic and gravitational data.

Carrying this idea one step further, the lithosphere beneath the oceans must be thinner than that of the continents because its elevation is lower. Although this is true, oceanic rocks also have a greater density than continental rocks, another factor contributing to their lower position.

If the concept of isostasy is correct, we should expect than when weight is added to the crust, the crust will respond by subsiding, and that when weight is removed there will be uplifting. (Visualize what happens to a ship as cargo is being loaded and unloaded.) Evidence for this type of movement exists, strongly supporting the theory of *isostatic adjustment.* When Hoover Dam was built in the 1930s, the impounded waters of Lake Mead and to a lesser degree the millions of tons of sediment collected by it caused regional subsidence and a marked increase in seismic activity. Another classic example is provided by nature. When continental glaciers occupied portions of North America during the Pleistocene epoch,

FIGURE 18.3
Wave-cut terraces on the Palos Verdes Hills south of Los Angeles, California. Once at
sea level, the highest terraces are now about 400 meters above it. (Photo by John S.
Shelton)

the added weight of the 3-kilometer thick mass of ice
caused downwarping of the earth's crust. In the 8000
to 10,000 years since the last ice sheet melted, uplifting of as much as 330 meters has occurred in the
Hudson Bay region, where the thickest ice had accumulated (see Figure 11.29).

As the foregoing examples illustrate, isostatic adjustment can account for considerable crustal movement. Thus, we can now understand why, as erosion
lowers the summits of mountains, the crust will rise in
response to the reduced load. The processes of uplifting and erosion will continue until the deeply buried
portions of the mountains have reached the same

height as the surrounding crust (Figure 18.4). In addition, as the mountains wear down, the weight of the
eroded sediment deposited on the adjacent continental margin will cause it to subside.

To summarize, mountains are unusually thick portions of the earth's crust that remain elevated above
their surroundings because of isostasy. As erosion
removes material, isostatic adjustment gradually
raises the mountains in response. Eventually, the
deepest portions of the mountains are brought up to
the shallower depths of the surrounding crust. The
question still to be answered is, How do these thick
sections of the earth's crust come into existence?

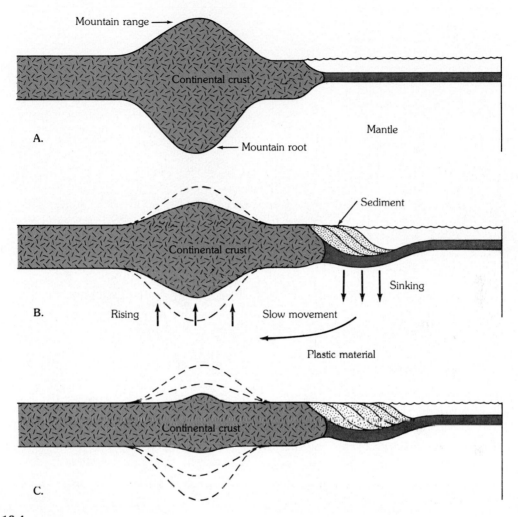

Mountain range ⟶

Continental crust

Mantle

A.

Mountain root

Sediment

Continental crust

B.

Rising

Slow movement

Sinking

Plastic material

Continental crust

C.

FIGURE 18.4

This sequence illustrates how the combined effect of erosion and isostatic adjustment results in a thinning of the crust in mountainous regions.

ROCK DEFORMATION

When rocks are subjected to stresses greater than their own strength, they begin to deform, usually by folding or fracturing. It is easy to visualize how individual rocks break, but how are large rock units bent into intricate folds without being appreciably broken during the process? (Figure 18.5). In an attempt to answer this, structural geologists turned to the laboratory and subjected rocks to stresses while simulat-

ing those conditions believed to exist at various depths within the crust.

Although all rock types deform somewhat differently, the general characteristics of rock deformation were determined from these experiments. Geologists discovered that when stress is applied slowly and under low pressure, rocks first respond by deforming elastically. Changes resulting from **elastic deformation** are reversible; that is, like a rubber band, the rock will return to nearly its original size and shape

FIGURE 18.5
Highly deformed sedimentary strata in the Rocky Mountains of British Columbia near the Sullivan River. (Courtesy of Geological Survey of Canada, photo no. 180345)

when the stress is removed. However, once the elastic limit is surpassed, rocks either rupture or deform plastically. **Plastic deformation** results in permanent changes; that is, the size and shape of a rock unit are altered through folding and flowing. Laboratory experiments confirmed the speculation that at high temperatures and pressures, most rocks deform plastically once their elastic limit is surpassed (Figure 18.6). Rocks tested under surface conditions also deform elastically, but once they exceed their elastic limit, most behave like a brittle solid and rupture. Recall

that the energy for most earthquakes comes from stored elastic energy that is released as rock ruptures and snaps back to its original shape.

One factor that researchers are unable to duplicate in the laboratory is geologic time. We know that if stress is applied quickly, as with a hammer, rocks tend to fracture. On the other hand, these same materials may deform plastically if stress is applied over an extended period. For example, marble benches have been known to sag under their own weight over a period of a hundred years or so. In

456

| Undeformed | Low confining pressure | Intermediate confining pressure | High confining pressure |

FIGURE 18.6
A marble cylinder deformed in the laboratory by applying thousands of pounds of load from above. Each sample was deformed in an environment that duplicated the confining pressure found at various depths. Notice that when the confining pressure was low, the sample deformed by brittle fracture, whereas when the confining pressure was high, the sample deformed plastically. (Courtesy of M. S. Paterson, Australian National University)

nature, small forces applied over long time periods surely play an important role in the deformation of rock strata.

The processes of deformation generate features at many different scales. At one extreme are features on a grand scale—the earth's major mountain systems. At the other extreme, highly localized stresses create minor fractures in the bedrock in most areas. All of these phenomena from the largest folds in the Alps to the smallest fractures in a rock are referred to as *rock structures.* Some of the most prominent rock structures will be discussed in the following sections.

FOLDS

During mountain building, flat-lying sedimentary and volcanic rocks are often bent into a series of broad folds, much like those that would form if you were to hold the ends of a sheet of paper and then push them together. The result of folding is a shortening and

thickening of the crust. Experimental evidence shows that when sedimentary rocks are deep within the crust where confining pressures are great, strata can be deformed into very tight folds stacked one atop the other, without appreciable fracturing. Figure 18.7 illustrates some common folded structures. Linear upfolded forms are commonly called **anticlines,** whereas downfolded structures are typically referred to as **synclines.**[1] Depending upon their orientation, anticlines and synclines are said to be *symmetrical, asymmetrical,* or *overturned,* if one limb has been tilted beyond the vertical (Figure 18.7). An overturned fold can also "lie on its side" so that a plane extending through the axis of the fold would have a

---

[1]By strict definition, an anticline is a structure in which the oldest strata are found in the center. This most typically occurs when strata are upfolded. Further, a syncline is strictly defined as a structure in which the youngest strata are found in the center. This occurs most commonly when strata are downfolded.

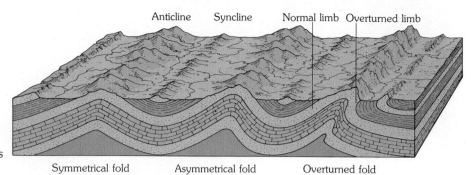

FIGURE 18.7
Block diagram of principal types of folded strata.

Anticline Syncline Normal limb Overturned limb

Symmetrical fold Asymmetrical fold Overturned fold

FIGURE 18.8
Sheep Mountain, a doubly plunging anticline. Note that erosion has cut the flanking sedimentary beds into low ridges that make a "V" pointing in the direction of plunge. (Photo by John S. Shelton)

458

vertical orientation. These *recumbent* folds are common in the Alps where there is evidence that some deformed strata have been shoved as much as 50 kilometers over the adjacent rocks.

Folds do not continue forever; rather, their ends die out much like the wrinkles in cloth. Some folds are said to be *plunging,* since the axis of the fold is plunging into the ground (Figure 18.8). Figure 18.9 shows some examples of plunging folds and the pattern produced when erosion removes the upper layers of these structures and exposes their interiors. Note that the outcrop pattern of an anticline points in the direction it is plunging, while the opposite is true for synclines. A good example of the kind of topography that results when erosional forces attack folded sedimentary strata is found in the Valley and Ridge Province of the Appalachians. Here resistant sandstone beds remain as imposing ridges separated by valleys cut into more easily eroded shale or limestone beds.

Although most folds are caused by compressional stresses that squeeze and crumble strata, some folds are a consequence of vertical displacement. **Monoclines,** broad flexures found on the Colorado Plateau and elsewhere, are such structures. These folds are thought to result from nearly vertical faulting in deep-lying basement rocks as shown in Figure 18.10. Whereas the rigid basement complex responded to vertical stress by fracturing, the relatively flexible sedimentary strata above were deformed by folding.

Broad upwarps in basement rock may also deform the overlying cover of sedimentary strata and gener-

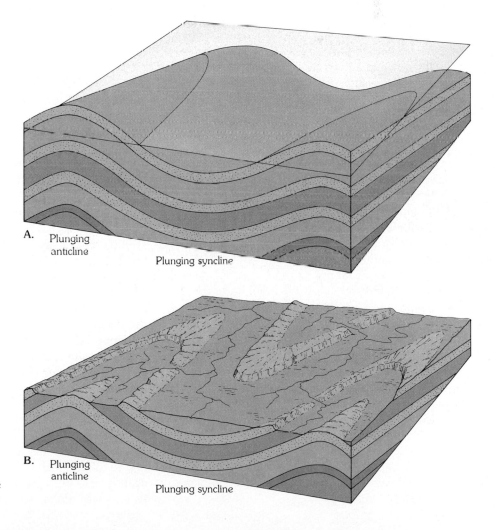

FIGURE 18.9
Plunging folds. **A.** Idealized view. **B.** View after extensive erosion.

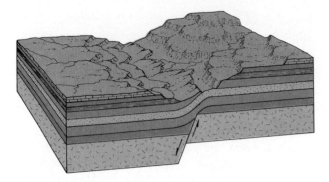

FIGURE 18.10
Monocline consisting of folded sedimentary beds that
were deformed by faulting in the bedrock below.

ate large folds. When this upwarping produces a circular or somewhat elongated structure, the feature is
called a **dome.** Downwarped structures having a similar shape are termed **basins.** The Black Hills of western South Dakota is one such domal structure in
which erosion has stripped away the unwarped sedimentary beds, exposing older igneous and metamorphic rocks in the center. Several basins also exist in
the United States. The basins of Michigan and Illinois
have very gently sloping beds similar to saucers.
Because large basins contain sedimentary beds sloping at such low angles, they are usually identified by
the age of the rocks composing them. The youngest
rocks are found near the center and the oldest rocks
are at the flanks. This is just the opposite order of a
domal structure such as the Black Hills, where the
oldest rocks form the core.

STRIKE AND DIP
When conducting a geological study of a region, the
geologist attempts to identify and describe the dominant structures. In many cases structures are so large
that only a small segment is visible from any one
observation point. In other locations most of the outcrops are hidden by vegetation or by recent sedimentation. Consequently, the reconstruction must be
done from a limited number of outcrops. Despite
these difficulties, a number of mapping techniques
enable geologists to reconstruct the orientation and
shape of the existing structures. In recent years, this
work has been aided by advances in aerial photography and satellite imagery.

Geologic mapping is most easily accomplished in
areas where sedimentary strata are exposed. Since
sediments are usually deposited in horizontal layers,
inclined strata indicate that a period of deformation
occurred following deposition. Two measurements
used to establish the orientation of deformed sedimentary beds are strike and dip. **Strike** is the trend,
or direction, of the strata, whereas **dip** is the angle of
inclination of the bedding surface. Perhaps the easiest way to understand these measurements is to
examine sedimentary rock that is outcropping in an
otherwise flat terrain (Figure 18.11). Strike is defined
as the direction of the line produced by the intersection of the surface represented by the inclined strata
with a horizontal surface, which in this example is the
surface of the land. Dip, on the other hand, is defined
as the angle of maximum inclination of a bed measured in a direction perpendicular to the strike. In the
field, geologists measure the strike (trend) and dip
(inclination) of sedimentary rocks at as many outcrops as practical. These data are then plotted on a
topographic map or an aerial photograph along with
a color-coded description of the rock. From the orientation of the strata, an inferred orientation and
shape of the structure can be established as shown in
Figure 18.12. Using this information, the geologist
can reconstruct the pre-erosional structures, and is
also better able to interpret the region's geologic history.

FAULTS
As indicated in our discussion of earthquakes, faults
are fractures in the earth's crust along which appreciable movement has taken place. Faults are categorized on the basis of the relative movement between
the blocks on both sides of the fault plane. The movement can be horizontal, vertical, or oblique.

Faults having primarily vertical movement are
called **dip-slip faults,** since the displacement is along
the inclination, or dip, of the fault plane. Because
movement along dip-slip faults can be either up or
down the fault plane, two types of dip-slip faults are
recognized. In order to distinguish between the two
types, it has become common practice to call the rock
immediately above the fault surface the *hanging wall*
and to call the rock below the *footwall.* This nomen-
clature arose from prospectors and miners who excavated shafts along fault zones, which are frequently

sites where ores are deposited by hydrothermal solutions. During these operations, the miners would walk on the rocks below the fault trace (the footwall) and hang their lanterns on the rocks above (the hang-

ing wall). Dip-slip faults are classified as **normal faults** when the rock above the fault plane (hanging wall) moves down relative to the rock below (footwall) (Figure 18.13). **Reverse faults** occur when the

A.

B.

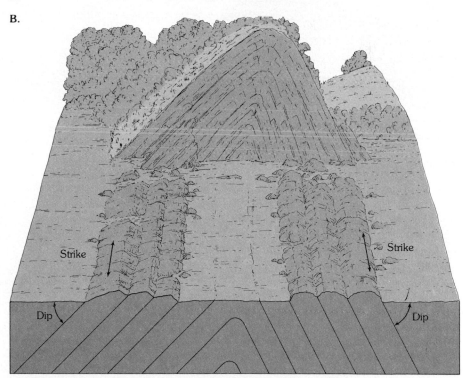

FIGURE 18.11
A. Photo of sedimentary strata sharply folded into an asymmetrical anticline and attacked by erosion. **B.** Drawing of this fold to illustrate the concept of strike and dip. (Reproduced by permission of the Director, Institute of Geological Sciences (NERC); NERC Copyright reserved/Crown Copyright reserved)

Strike

Strike

Dip

Dip

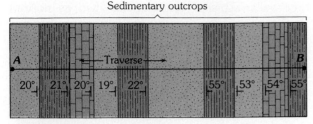

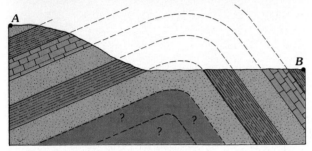

A. Map View

B. Cross-sectional View

FIGURE 18.12
By establishing the strike and dip of outcropping sedimentary beds, (Part A), geologists can infer the orientation of the structure below ground (Part B).

hanging wall moves up relative to the footwall (Figure 18.14). Reverse faults having a very low angle are also referred to as **thrust faults.** In mountainous regions such as the Alps, the American Cordillera, and the Appalachians, thrust faults have displaced rock for as much as 50 kilometers over adjacent strata (Figure 18.15). Thrust faults of this type result from strong compressional stresses. In many instances, thrust faults are thought to form in conjunction with large folds as shown in Figure 18.16.

Faults in which the dominant displacement is along the strike of the fault surface are called **strike-slip faults.** The movement is said to be *right lateral* if the block on the opposite side of the fault moves to your right as you face the fault, and *left lateral* if it moves to your left. Many large strike-slip faults are associated with plate boundaries. When this is the case, the term *transform fault* is applied. Transform faults have nearly vertical dips and serve to connect large structures such as segments of an oceanic ridge. The San Andreas fault in California is a well-known example of a transform fault in which the displace-

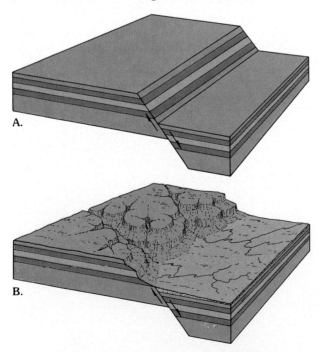

FIGURE 18.13
Block diagrams of a normal fault. **A.** The relative movement of displaced blocks. **B.** How erosion would alter the upfaulted block.

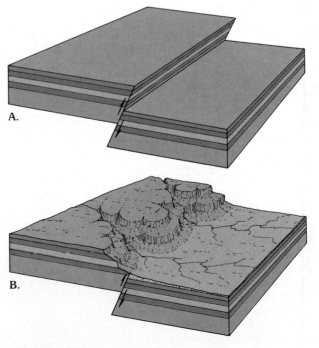

FIGURE 18.14
Block diagrams of a reverse fault. **A.** The relative movement of displaced blocks. **B.** How erosion would alter the upfaulted block.

ment has been on the order of several hundred kilometers. When faults have both vertical and horizontal movement, they are called **oblique-slip faults.**

Up to this point faults have been described as single, planar fractures along which movement has taken place. Most large faults, however, consist of a zone of roughly parallel fractures along which displacement occurs. This is true, for example, along much of the San Andreas fault.

Fault motion provides the geologist with a method of determining the nature of the forces at work within the earth. Normal faults indicate the existence of *tensional stresses* that pull the crust apart. This "pulling

apart" can be accomplished either by uplifting that causes the surface to stretch and break, or by horizontal forces that actually rip the crust apart. Normal faulting is known to occur at spreading centers where plate divergence is prevalent. Here, a central block called a **graben** is bounded by normal faults and drops as the plates separate (Figure 18.17). These grabens produce an elongated valley bounded by upfaulted structures called **horsts.** The Great Rift Valley of East Africa is made up of several large grabens, above which tilted horsts produce a linear mountainous topography. This valley, nearly 6000 kilometers (3600 miles) long, contains the excavation

FIGURE 18.15
The Keystone overthrust. Here dark-colored Cambrian limestone has been thrust over light-colored Jurassic sandstone. (Photo by John S. Shelton)

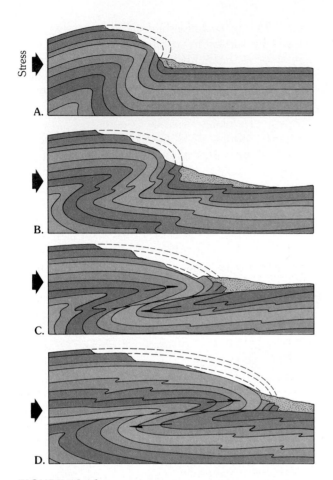

FIGURE 18.16
Stages in the development of an overthrust sheet as a result of folding and thrust faulting. (After A. Heim)

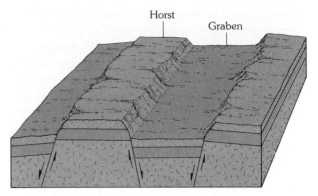

FIGURE 18.17
Diagrammatic sketch of downfaulted block (graben) and upfaulted block (horst).

sites of some of the earliest human fossils. Other rift valleys include the Rhine Valley in Germany and the valley of the Dead Sea in the Middle East.

Since the blocks involved in reverse and thrust faulting are displaced toward one another, geologists conclude that *compressional forces* are at work. The primary regions of this activity are thought to be the convergent zones where plates are colliding. Compressional forces generally produce folds as well as faults and result in a general thickening and shortening of the material involved.

JOINTS

Among the most common features of rocks that are exposed at the earth's surface are fractures called

joints. Unlike faults, joints are fractures along which no appreciable displacement has occurred. Although some joints occur randomly, most occur in roughly parallel groups.

We have already considered two types of joints. Earlier we learned that columnar joints form when igneous rocks cool and develop shrinkage fractures that produce elongated, columnlike structures. Also recall that sheeting produces a pattern of gently curved joints that develop more or less parallel to the surface of large exposed igneous bodies such as batholiths. Here the jointing is thought to result from the gradual expansion that occurs when erosion removes the overlying load.

In contrast to the situations just described, most joints are produced when rocks are deformed, particularly by the tensional and shearing stresses associated with crustal movements. For example, when folding occurs, rocks situated at the axes of the folds are elongated and pulled apart to produce tensional joints. Extensive joint patterns can also develop in response to relatively subtle and often barely perceptible regional unwarping and downwarping of the crust. It should be pointed out that, in many cases, the cause for jointing at a particular locale is not readily apparent.

Many rocks are broken by two or even three sets of intersecting joints that slice the rock into numerous regularly-shaped blocks. These joint sets often exert a strong influence on other geologic processes. For example, chemical weathering tends to be concentrated along joints, and in many areas, groundwater movement and the resulting solution activity in soluble rocks is controlled by the joint pattern. Moreover, a system of joints can influence the direction that stream courses follow. The rectangular drainage pattern described in Chapter 9 is such a case.

MOUNTAIN TYPES

Even though no two mountain ranges are exactly alike, they can be classified according to their most dominant characteristics. Using this approach, four main categories of mountains emerge: (1) folded mountains (complex mountains); (2) volcanic mountains; (3) fault-block mountains; and (4) upwarped mountains (Figure 18.18). Mountain ranges of the same type are commonly found in close proximity forming a mountain system. For example, nearly the entire state of Nevada is composed of numerous elongated fault-block mountains that are separated by structural basins. Further, within any mountainous belt, such as that portion of the American Cordillera in the western United States, mountain ranges representing each of these groups can be found.

In addition to these basic varieties, some regions having mountainous topographies are produced without appreciable crustal deformation. For example, plateaus, which are areas of high-standing rocks that are essentially horizontal, can be deeply dissected into rugged terrains. Although these highlands have the topographical expression of mountains, they lack the structure associated with orogenesis. The opposite situation also exists. For instance, those portions of the Appalachians lying east of the prominent ridges exhibit topography nearly as subdued as the continental interior. Yet, since these areas are composed of upturned and metamorphosed rocks, they are clearly part of the Appalachian Mountains.

In the following sections we will examine three of the four basic mountain types. Volcanic mountains are treated in detail in Chapter 4.

FOLDED MOUNTAINS

Folded mountains comprise the largest and most complex mountain systems. Although folding is often more conspicuous, faulting, metamorphism, and ig-

neous activity are always present in varying degrees (Figure 18.18A). All major mountain belts, including the Alps, Urals, Himalayas, and Appalachians, are of this type. Since folded mountains represent the world's major mountain systems, the process of mountain building is usually described in terms of their formation. Thus a separate section on orogenesis is devoted to the evolution of these often majestic and always complex mountain systems.

FAULT-BLOCK MOUNTAINS

Fault-block mountains are bounded on at least one side by high-angle normal faults (Figure 18.18C). Recall that tensional stresses are responsible for creating normal faults. Some fault-block mountains form in response to broad uplifting, which causes elongation that leads to faulting. Such a situation is exemplified by the fault blocks that rise above the downfaulted rift valleys of East Africa. Other fault-block mountains appear to be formed when single blocks are lifted vertically, high above the adjacent undeformed valleys.

Excellent examples of fault-block mountains are found in the Basin and Range Province, a region that encompasses Nevada and portions of Utah, New Mexico, Arizona, and California. Here the crust has literally been broken into hundreds of pieces, giving rise to nearly parallel mountain ranges, averaging about 80 kilometers in length, which rise precipitously above the adjacent sediment-laden basins. Since the formation of fault-block mountains in the Basin and Range Province was accompanied by volcanic activity, it has been suggested that the period of broad crustal upwarping and tensional fracturing occurred as a result of the buoyancy of the magma emplaced below the surface (Figure 18.19, page 468). As the period of volcanism came to an end, cooling and subsidence tilted the blocks into their present positions. Some geologists disagree with this explanation. They prefer an alternate hypothesis which suggests that changes in a plate boundary along the west coast of North America were responsible. They contend that changes in the movements at the boundary resulted in a switch, from compres-

FIGURE 18.18
Classification of mountains.

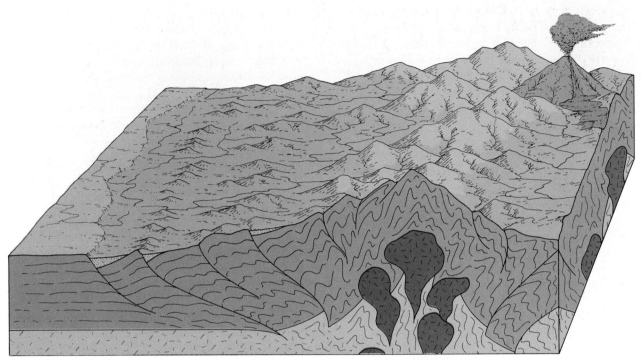

A. Folded (complex)

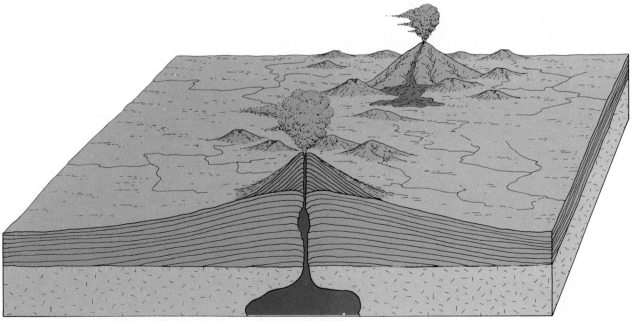

B. Volcanic

466

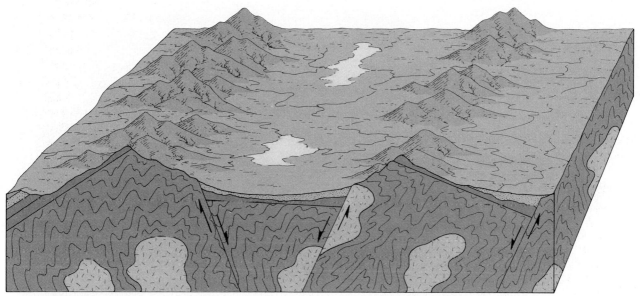

C. Fault-block

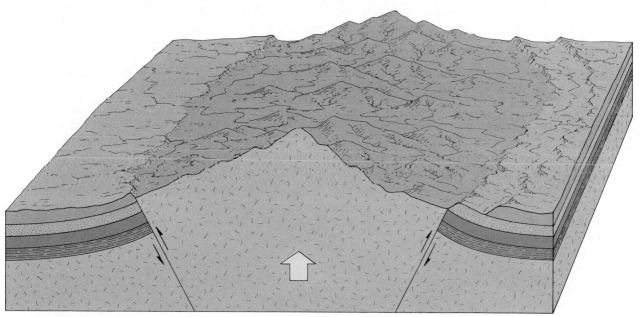

D. Upwarped

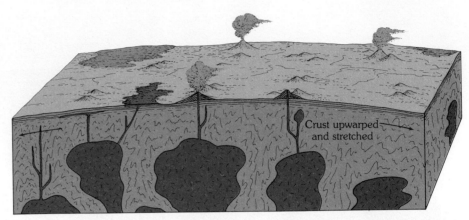

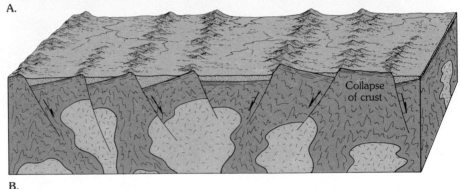

FIGURE 18.19
Formation of the fault-block structures of the Basin and Range Province. **A.** The emplacement of magma into the crust resulted in crustal upwarping and the extrusion of large amounts of volcanic material. **B.** As the magmatic phase came to an end, the cooling crust collapsed along numerous high-angle normal faults.

sional to tensional, in the forces acting upon the region. The tensional forces then extended and faulted the rocks in the region. This deformation, in turn, gave rise to a period of volcanism. Therefore, unlike the preceding explanation, this hypothesis suggests that volcanic activity in the Basin and Range Province did not cause the deformation, but rather was a consequence of the deformation. Regardless of which, if either, explanation is correct, there is general agreement that the fault-block mountains in the region resulted from tensional forces.

Other examples of fault-block mountains in the United States include the Teton Range of Wyoming and the Sierra Nevada of California. Both are faulted along their eastern flanks, which were uplifted as the blocks tilted downward to the west. Looking west from Jackson Hole and Owens Valley respectively, the eastern fronts of these ranges rise over 2 kilometers, making them two of the most precipitous mountain fronts in the United States (see chapter-opening photo).

When a mountain range is said to have been formed by block faulting, this is a reference to the events responsible for the modern topography. It is common for the rocks of the uplifted block to also show evidence of a much earlier period of deformation. The present topography of the Basin and Range, for example, began to form about 15 million years ago. However, during late Mesozoic and early Cenozoic time, these rocks were involved in another mountain building episode. During this earlier phase, igneous activity and thrust faulting deformed the sedimentary rocks in this region. The mountains produced by this activity had been substantially lowered by erosion prior to the more recent period of block faulting.

UPWARPED MOUNTAINS

Upwarped mountains are possibly the mountain type with the greatest diversity. Some, such as the Black Hills in western South Dakota and the Adirondack Mountains in upstate New York, consist of older

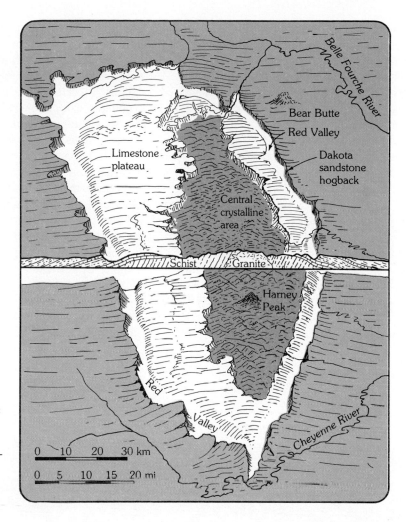

FIGURE 18.20
The Black Hills of South Dakota, an example of an upwarped mountain system in which the resistant igneous and metamorphic central core has been exposed by erosion. (After Arthur N. Strahler, *Introduction to Physical Geography*, 3rd ed. New York: John Wiley & Sons, 1973. Reprinted by permission)

igneous and metamorphic bedrock that was once eroded flat and subsequently mantled with sediment. As the regions were upwarped, erosion removed the veneer of sedimentary strata, leaving a core of igneous and metamorphic rocks standing above the surrounding terrain (Figure 18.20).

Other examples of upwarped mountains are found in the middle and southern Rockies from Montana south through Colorado and into New Mexico. In this group are the Front Range of Colorado, the Sangre de Cristo of New Mexico and Colorado, and the Bighorns of Wyoming. These mountains are structurally much different than the northern Rockies, which include the Canadian Rockies and those portions of the Rockies found in Idaho, western Wyo-

ming, and western Montana. Whereas the latter ranges are composed of thick sequences of sedimentary rocks that were deformed by folding and low-angle thrust faulting, the portion of the southern Rockies bordering the Great Plains was pushed almost vertically upward as part of broad upwarping of the crust, or in some instances, because of displacement along high-angle faults. In general these mountains consist of Precambrian basement rocks covered by relatively thin layers of younger Precambrian or Paleozoic strata. However, since the time of deformation much of this mantle of sedimentary rocks has been eroded from the highest portions of the uplifted blocks, exposing the Precambrian core. In many areas, remnants of these sedimentary layers are visible flanking the crystalline cores of the mountain ranges. They are often easy to identify because the upturned strata form prominent angular ridges called **hogbacks** (Figure 18.21). Examples of exposed Precambrian cores include a number of granitic outcrops that project as steep summits, such as Pikes Peak and Longs Peak in Colorado's Front Range. It has been suggested that the upwarping of the eastern Rockies was in response to compressional stresses generated when the North American plate overrode a portion of the adjacent oceanic plate.

MOUNTAIN BUILDING

Orogenesis has operated during the recent geologic past in several locations around the world. These young mountainous belts include the American Cordillera, which runs along the western margin of the Americas from the Cape of Good Hope to Alaska; the Alpine-Himalaya chain, which extends from the Mediterranean through Iran to northern India and into Indochina; and the mountainous terrains of the western Pacific, which include mature island arcs such as Japan, the Philippines, and Sumatra (Figure 18.22). Most of these young mountain belts have come into existence within the last 100 million years. Some, including the Himalayas, began their growth as recently as 40 million years ago.

In addition to these young complex mountains, several chains of Paleozoic- and Precambrian-aged mountains exist on the earth as well. Although these older structures are deeply eroded and topographically less prominent, they clearly possess the same structural features found in younger mountains. Typical of this older group are the Appalachians in the eastern United States and the Urals in the Soviet Union.

Although complex mountains differ from one another in particular details, all possess the same basic structures. Orogenic belts generally consist of roughly

FIGURE 18.21
A view looking north along the east flank of the Front Range of the Colorado Rockies. The upturned remnants of sedimentary strata (center) once covered the igneous and metamorphic terrain which lies to the west. (Photo by T. S. Lovering, U.S. Geological Survey)

parallel ridges of folded and faulted sedimentary and volcanic rocks, portions of which have been strongly metamorphosed and intruded by somewhat younger igneous bodies. In most cases the sedimentary rocks formed from enormous accumulations of deep-water marine deposits that occasionally exceeded 15,000 meters in thickness, as well as from thinner shallow-water deposits. Moreover, these deformed sedimentary rocks are for the most part older than the mountain building event, indicating an extensive period of quiescent deposition followed by a dramatic episode of deformation.

Careful study of mountainous terrains has revealed that the period of orogenesis is generally quite

FIGURE 18.22
Lofty summit in the Wasatch Range of Utah. This mountain range is just a small portion of the North American Cordillera. (Photo by Stephen A. Trimble)

long, frequently exceeding 100 million years. In addition, reconstruction of these events has shown that deformation generally progressed in a landward direction, so that the deep-water sediments were the first to be deformed. These sediments, which consist primarily of graywackes, volcanic debris, and shales, have been intensely folded, faulted, and metamorphosed as if squeezed by a gigantic vise whose moving jaw migrated from the sea toward the land. After the initial deformation, numerous intrusions of magma further deformed and metamorphosed those deposits, generating the igneous-metamorphic core of the developing mountains. These events caused general thickening of this rock sequence, which began to rise above sea level. As this crystalline core was uplifted and deformed, the thinner shallow-water deposits were shoved toward the continental interior. During this later period of deformation, the shallow-water strata, consisting primarily of relatively clean sandstones, limestones, and shales, were folded and broken along low-angle thrust faults to produce the sedimentary portion of the complex mountains.

THE GEOSYNCLINE CONCEPT

Several theories have been proposed regarding orogenic belt formation. An early proposal suggested that mountains are simply wrinkles in the earth's crust produced as the earth cooled from its original semi-molten state. As the earth cooled and shrank, the crust deformed to conform to a smaller earth, much like the peel of an orange wrinkles as the fruit dries out. However, neither this nor any of the other early proposals were able to withstand careful scrutiny.

Then in the 1850s James Hall, a noted American geologist, made a careful study of the deformed sedimentary strata of that portion of the Appalachians which lies in and around New York state. After discovering such features as shallow-water fossils, ripple marks, and mud cracks, Hall concluded that these enormous accumulations had been deposited in water no deeper than a few hundred meters. But how does a 10,000-meter thick rock unit form in a basin that is only a few hundred meters deep? To account for these enormous accumulations, Hall believed that they must have been deposited in a slowly subsiding trough. According to Hall, this large linear trough, which was later termed a **geosyncline,** subsided

slowly enough to allow sedimentation to keep pace.

About a quarter century later, James Dana, a respected Yale geologist, expanded the concept of geosynclines. Dana proposed that after great thicknesses of sediment had accumulated, horizontal forces directed from the seaward side of the geosyncline began to squeeze the sediments. These compressional forces shortened and thickened the crust, producing a high-standing mountain system while simultaneously pushing much of the sediment deeper into the earth. He thought that these deeply buried sediments melted and generated magma, which then moved upward and intruded the overlying unmelted sediments. Thus, a complex mountain chain consisting of folded and faulted sedimentary and volcanic rocks, surrounding a core of igneous intrusions and metamorphic rocks, was created.

Geologists found that with some modification the geosyncline concept could be applied to other complex mountain systems. From these studies and re-evaluations of the evolutionary history of the Appalachians, it was determined that geosynclines consist of two distinct units (Figure 18.23). The section which James Hall had so ably described consists of shallow-water deposits of relatively clean sandstones, limestones, and shales that accumulated in a trough that came to be called the **miogeosyncline.** Seaward of the miogeosyncline lay the **eugeosyncline** in which mainly deep-water deposits accumulated, including graywackes, lava flows, volcanic debris, and shales.

Although the geosynclinal concept of mountain building has many merits, the underlying cause of orogenesis is not explained. What produced the subsidence in the geosyncline? Why did sediment accumulate, relatively undisturbed for millions of years, and then suddenly go through a period of deformation? Such unanswered questions forced geologists to continue to evaluate the complex problem of mountain building.

With the development of the plate tectonics theory, many of the disturbing questions from the geosynclinal theory were answered. This newest and most widely accepted theory suggests that orogenesis results as large segments of the earth's lithosphere are displaced. According to the theory of plate tectonics, mountain building occurs at convergent plate boundaries. Here the colliding plates provide the compres-

Geosyncline

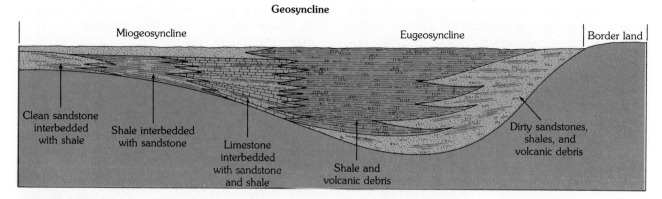

FIGURE 18.23
Idealized cross-section through a geosyncline showing the shallow-water miogeosyncline and the deep-water eugeosyncline.

sional stress to crumble, fault, and metamorphose the thick accumulations of sediments that are deposited along the flanks of landmasses, while melting of the subducted oceanic crust provides a source of magma that intrudes and further deforms these deposits.

Note, however, that mountain building as proposed by the plate tectonics theory should be viewed as an extension of the geosynclinal theory rather than as a concept to replace the older view. Most basic tenets of the geosynclinal theory still remain, because the events that resulted in the formation of complex mountains had been worked out quite well prior to the advent of plate tectonics. Today, however, geologists have a much clearer understanding of the causes of orogenesis.

One major modification of the earlier theory relates to the location and nature of the geosyncline. Rather than being considered a huge linear trough found within the stable continental platform, the geosyncline has been found to consist of sediments that accumulated along the continental margins. Thus, the rock units of the miogeosyncline are now believed to be equivalent to the shallow-water deposits of the continental shelf, whereas the strata of the eugeosyncline are most likely a collection of deep-water sediments and volcanic debris that originated at a volcanic island arc or at a nearby mainland. Although some of these deep-water deposits are thought to be transported by turbidity currents, most eugeosynclinal accumulations are added to the continental margin along subduction zones. Here, sedi-

ments are scraped from a descending plate and plastered against the overriding plate. Since both the miogeosynclinal and the eugeosynclinal deposits exist as more or less wedge-shaped structures rather than as synclinal structures, the terms **miogeoclinal** and **eugeoclinal** are presently used to denote them.

OROGENESIS AT PLATE BOUNDARIES

Modern-day sites for orogenesis are the active volcanic arcs, typified by the islands of Japan and the Andes of South America. Although all volcanic arcs are similar, the Japanese subduction zone occurs where two oceanic plates converge, whereas the Andean arc has formed where oceanic and continental lithosphere collide. Consequently, mountain building at each location follows a somewhat different evolutionary history. Another, and perhaps the most important, cause of orogenesis involves the collision of two continental fragments. Continental collisions not only add to the degree of deformation, but to the difficulty of unraveling the geologic sequence of events. We shall consider each of these sites of mountain building in the following sections.

OROGENESIS AT ISLAND ARCS

Island arcs form where two oceanic plates converge and one is subducted beneath the other. Partial melting of the subducted plate and possibly frictional heating of mantle rocks generate a supply of magma which migrates upward to form the igneous portion of the developing arc system (Figure 18.24A). Over an extended period, numerous episodes of volcanism coupled with the buoyancy of intrusive igneous masses, gradually increase the size and elevation of the developing arc. This growth effectively increases the erosion rate and consequently the amount of sediment added to the adjacent sea floor, as well as to the back-arc basin.

In addition to the sediments derived from land, deep-water sediments are also scraped off the de-

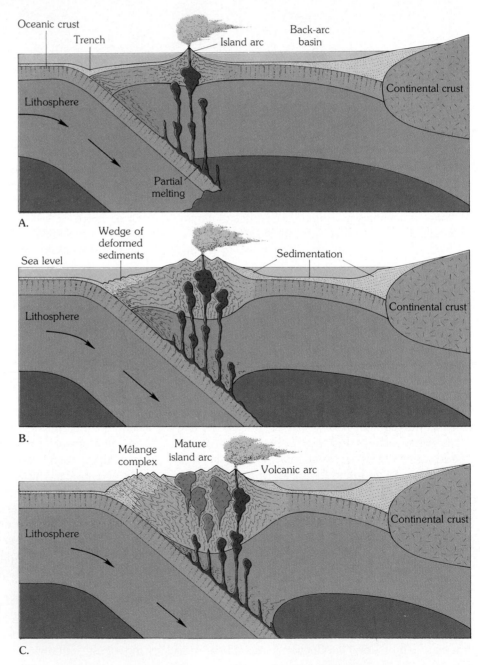

FIGURE 18.24
The development of a mature volcanic island arc at an oceanic-oceanic convergent boundary.

scending oceanic plate and piled onto the landward wall of the trench (Figure 18.24B). The compressional stress exerted by the converging plates causes these sediments, along with the slivers of oceanic crust that have been sheared from the descending plate, to become intricately folded as well as sliced by numerous thrust faults. This activity is thought to generate a thick wedge of highly deformed material lying parallel to, and seaward of, the igneous portion of the arc. This chaotic mixture of sedimentary rock, metamorphic rock, and slivers of sea floor is termed a **mélange.** The metamorphism associated with the formation of a mélange is of the high-pressure, low-temperature type, typified by the blueschist assemblage (see Chapter 7). Continued accretion of this material can result in an accumulation thick enough to stand above sea level.

Landward of the trench, in the volcanic arc, sediments are also being deformed and metamorphosed. Whereas deformation of the mélange is caused primarily by the converging plate, in the area of igneous activity the emplacement of large magma bodies and the associated high-temperature regime enhances the metamorphic processes which are operating. Consequently, the metamorphic rocks here typically contain minerals belonging to the high-temperature regime.

The result of these diverse activities is the development of a mature island arc composed of two roughly parallel orogenic belts (Figure 18.24C). The landward segment is the volcanic arc, composed of large intrusive bodies intermixed with high-grade metamorphic rocks. Located seaward of the volcanic arc is the mélange complex consisting of folded, faulted, and metamorphosed sediments and volcanic debris.

Geologists have only recently come to realize the significance of island arcs in the process of mountain building (Figure 18.25). General agreement exists that the processes operating at modern island arcs represent one of the stages in the formation of continental mountainous belts. In the next section we will

FIGURE 18.25
Three of many volcanoes that comprise the Aleutian arc. This narrow band of volcanism results from the subduction of the Pacific plate. In the distance is the Great Sitkin volcano (1772 meters) which the Aleuts call the "Great Emptier of Bowels," because of its frequent activity. (Photo by Bruce D. Marsh)

consider orogenesis along continental margins in light of what has been learned from the geology of island arcs.

SUBDUCTION-TYPE OROGENESIS ALONG CONTINENTAL MARGINS

Mountain building along continental margins involves the convergence of an oceanic plate and a plate whose leading edge contains continental material. As stated earlier, this type of convergence is believed to generate structures resembling those of a developing volcanic island arc. However, the formation of the volcanic arc-mélange complex is only part of the process involved in creating a mountain system such as the Andes.

The first stage in the development of a complex mountain system is thought to occur prior to the formation of the subduction zone. During this period the continental margin is inactive; that is, it is not a plate boundary but a part of the same plate as the adjoining oceanic crust. The east coast of the United States provides us with a present-day example of an inactive continental margin. Here, as at other inactive continental margins surrounding the Atlantic, deposition of sediment on the continental shelf is producing a wedge of shallow-water (miogeoclinal) sandstones, limestones, and shales (Figure 18.26A). Beyond the continental shelf, turbidity currents (dense slurries of mud and water) are depositing deep-water (eugeoclinal) sediment upon the continental slope and rise.

At some time the continental margin becomes active; a subduction zone forms and the deformation process is initiated (Figure 18.26B). The cause of this event is unknown. Some geologists have suggested that when the accumulation of sediment is sufficiently thick, the depressed oceanic crust shears from the continental mass, thereby initiating subduction. However, this proposal fails to account for the fact that most subduction zones develop seaward of the thickest deposits.

A good place to examine an active continental margin is the west coast of South America. Here the Nazca plate is being subducted beneath the South American plate along the Peru-Chile trench. This subduction zone probably formed in conjunction with the breakup of the supercontinent Pangaea. As South America was torn from Africa and transported westward, the oceanic crust adjacent to the west coast of South America was bent and thrust under the continental plate. However, the oceanic crust does not give way without some effect on the overriding plate. In the case of South America, the Nazca plate apparently deformed the geoclinal sediments that flanked the continental margin, producing the original folded and faulted portion of the complex mountain system we now call the eastern Andes.

Subduction and partial melting of the oceanic plate initiate yet another stage—the development of a volcanic arc (Figure 18.26C). Since the break between the oceanic and continental lithosphere generally forms seaward of the place where the two plates join, the volcanic arc often forms a few hundred kilometers out to sea. The first volcanic mountains of the Andean arc apparently formed seaward of the ancient coastline and probably resembled the present-day volcanic island arcs of the western Pacific. As the descending plate extended itself eastward beneath South America, volcanic activity in the Andes migrated inland of the initial volcanoes, which have for the most part been eroded away. Remnants of the original volcanic arc are the exposed batholiths and metamorphosed terrains composing the crystalline Andes that flank the west coast of South America.

During the development of the volcanic arc, sediment derived from the land as well as that scraped from the subducted plate is plastered against the landward side of the trench. Recall that this chaotic accumulation of metamorphosed rocks and scraps of oceanic crust is called a mélange (Figure 18.26D). One of the best examples of a volcanic arc-mélange complex is found in the western United States and includes the Sierra Nevada and the Coast Range of California. These parallel mountainous belts were produced by the subduction of a portion of the Pacific basin under the western edge of the North American plate. The Sierra Nevada batholith is a remnant of a portion of the volcanic arc that was produced by several surges of magma over a period of tens of millions of years. Subsequent uplifting and erosion have removed most evidence of past volcanic activity and has exposed a core of crystalline rocks. In the trench region, sediments accreted from the subducted plate and those provided by the eroding volcanic arc were intensely folded and faulted into the complex mélange which presently constitutes the Franciscan Formation of the Coast Range of California. Uplifting of the Coast Range took plate quite recently, as evidenced by the unconsolidated sediments still mantling portions of these highlands.

CONTINENTAL COLLISIONS

Up to this point we have discussed the formation of orogenic belts where the leading edge of only one of the two converging plates contained continental crust. However, both of the colliding plates may be carrying continental crust. Because continental lithosphere is evidently too buoyant to undergo any appreciable amount of subduction, a collision between the continental fragments eventually results (Figure 18.27, page 480). An example of such a collision occurred about 45 million years ago when India collided with Asia. India, thought to have been previously a part of Antarctica, was rafted nearly 5000 kilometers due north before the collision occurred. The result was the formation of the spectacular Himalaya Mountains and the Tibetan Highlands. Although most of the oceanic crust which separated these landmasses prior to the collision was subducted, some was caught up in the squeeze along with sediment that lay offshore and can now be found elevated high above sea level. After such a collision, the subducted oceanic plate is believed to decouple from the rigid continental plate and continue its downward path.

The spreading center that separated India from Antarctica and moved it northward is still active; hence, India continues to be thrust into Asia at an estimated rate of a few centimeters per year. However, numerous earthquakes recorded off the southern coast of India indicate that a new subduction zone may be in the making. If formed, it would provide a disposal site for the floor of the Indian Ocean, which is continually being produced at a spreading center located to the southwest. Should this occur, India's northward journey would come to an end and the growth of the Himalayas would cease.

A similar but much older collision is believed to have taken place when the European continent collided with the Asian continent to produce the Ural Mountains, which extend in a north-south direction through the Soviet Union. Prior to the discovery of plate tectonics, geologists had difficulty explaining the existence of mountain ranges such as the Urals which were located deep in the continental interiors. How could thousands of meters of marine sediment be deposited and then become highly deformed while situated in the middle of a large landmass?

Other mountain ranges showing evidence of continental collisions are the Alps and the Appalachians. The Alps are thought to have formed as a result of a collision between Africa and Europe during the closing of the Tethys Sea (see Figure 16.1). During this closure, enormous quantities of sediment were squeezed into huge recumbent folds which were displaced northward as large thrust sheets. Incorporated within these sediments are shreds of oceanic crust trapped along the suture zone where these landmasses met.

Erosion in the Alps has already exposed the relatively large core of high-grade metamorphic rocks generated during the collision. However, large igneous plutons, normally associated with the subduction of an intervening oceanic plate, are lacking in the Alpine belt. The absence of appreciable igneous activity in the Alps has been attributed to the closing of a small ocean basin.

The somewhat older Appalachians are thought to be the result of a collision between North America and northern Africa.[1] Although they have since separated, less than 200 million years ago these two continents were juxtaposed as part of the supercontinent Pangaea. Geologists believe that this joining provided the impetus to generate this once-lofty mountain chain.

Detailed studies of the rocks in the southern Appalachians indicate that the formation of this orogenic belt was more complicated than once thought. Rather than a single continental collision, three distinct episodes of mountain building occurred over a period of nearly 250 million years. According to current data, a large continental mass split to produce proto-North America and proto-Africa plus several smaller crustal fragments about 650 million years ago. (These small fragments, called microcontinents,

---

[1]Recently, geologists have obtained evidence indicating North America may have collided with the west coast of South America rather than Africa to produce the Appalachians.

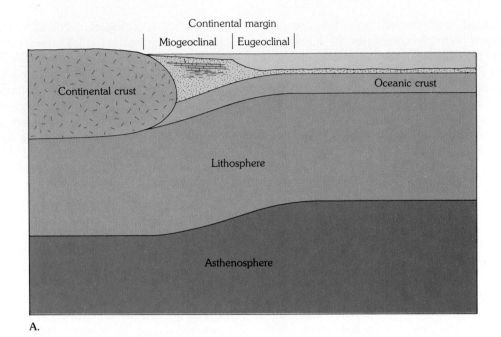

A.

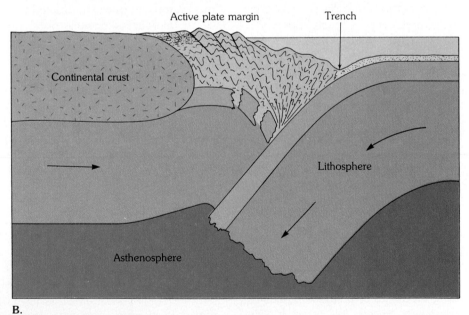

B.

FIGURE 18.26

Subduction-type orogenesis along an active continental margin. **A.** Passive plate margin. **B.** Plate convergence generates a subduction zone. **C.** Partial melting of subducted plate generates the volcanic arc. **D.** Continued growth of the complex mountain system through deformation of the shallow- and deep-water sediments.

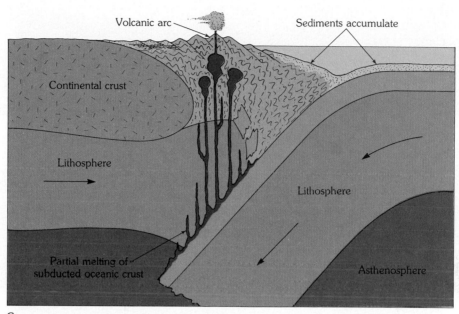

C.

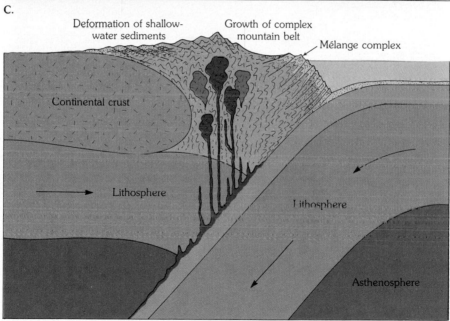

D.

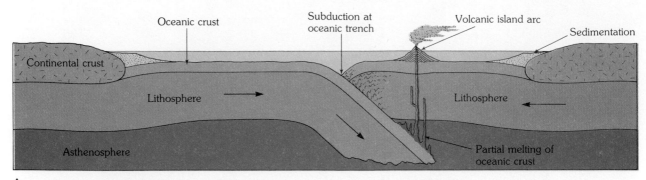

Oceanic crust

Subduction at oceanic trench

Volcanic island arc

Sedimentation

Continental crust

Lithosphere

Lithosphere

Asthenosphere

Partial melting of oceanic crust

A.

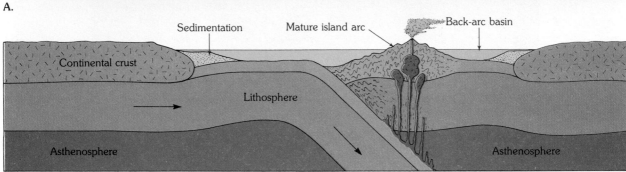

Sedimentation

Mature island arc

Back-arc basin

Continental crust

Lithosphere

Asthenosphere

Asthenosphere

B.

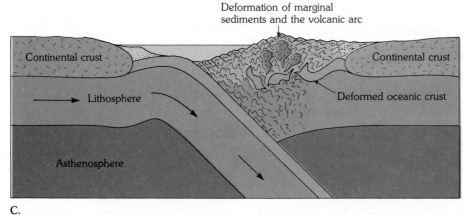

Deformation of marginal sediments and the volcanic arc

Continental crust

Continental crust

Lithosphere

Deformed oceanic crust

Asthenosphere

C.

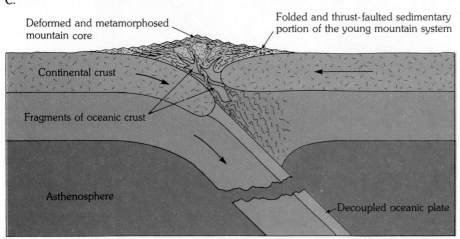

Deformed and metamorphosed mountain core

Folded and thrust-faulted sedimentary portion of the young mountain system

Continental crust

Fragments of oceanic crust

Asthenosphere

Decoupled oceanic plate

D.

FIGURE 18.27

Orogenesis and continental collisions. **A.** Converging plates generate a subduction zone and initiate island arc volcanism. **B.** Sediments scraped from the subducting plate and igneous activity add to the size of the volcanic arc. **C.** Closing of the back-arc basin deforms the entrapped marginal sediments and volcanic arc. **D.** A continental collision closes the ocean basin, resulting in further deformation and metamorphism of the sediments and volcanic rocks.

will be discussed further in the following section.) The orogenesis which generated the Appalachians resulted from the rejoining of proto-North America and proto-Africa and also included the addition of some smaller crustal fragments.

As the two continental blocks began to converge, a subduction zone formed seaward of the ancient coastline of proto-North America. The igneous activity associated with the subducting plate gave rise to a volcanic island arc, perhaps like those which presently rim the western Pacific. Studies in the southern Appalachians indicate that a microcontinent was situated between this island arc and the ancient North American plate. Continued closure of the proto-North Atlantic resulted in a collision of this microcontinent with North America. This orogenic event deformed and metamorphosed this fragment which is recognized today as the crystalline rocks that form the Blue Ridge and Piedmont regions of the Appalachians.

About 380 million years ago, a second orogeny took place as the ocean basin located behind the volcanic arc closed. The metamorphosed remnant of the volcanic arc is the Carolina Slate Belt, which is situated directly east of the Piedmont.

The final orogeny occurred about 300–250 million years ago when Africa collided with North America. This last event is thought to have displaced the earlier accreted terrains farther inland along low-angle thrust faults. At some locations, the total displacement may have exceeded 250 kilometers. This landward displacement also deformed the miogeoclinal sediments which had flanked North America. Today, these folded and faulted sandstones, limestones, and shales compose the essentially unmetamorphosed rocks of the Valley and Ridge Province.

Geologically speaking, shortly after the joining of North America and Africa, the newly formed continent of Pangaea once again began to break into smaller fragments. The decoupling of these continental blocks, as well as prolonged erosion, resulted in the eventual submergence of the eastern flank of the Appalachians. The sedimentation which occurred upon the submerged terrain eventually generated the thick sedimentary wedge currently found along the east coast of the United States. If the Atlantic closes again, these sediments would surely be deformed and metamorphosed into a spectacular mountain

system located seaward of the present Appalachians. In this manner, another sliver of land would be added to the continent.

In summary, the orogenesis of a complex mountain chain, as typified by the Appalachians, is thought to occur as follows:

1 After the breakup of a continental landmass, a thick wedge of sediments is deposited along the inactive continental margins, thereby increasing the size of the newly formed continent.

2 For reasons not yet understood, the ocean basin then begins to close and the continents start to converge.

3 Plate convergence results in the subduction of the intervening oceanic slab, and initiates an extended period of igneous activity. This activity results in the formation of a volcanic arc often located a few hundred kilometers seaward of the ancient coastline.

4 Debris eroded from the volcanic arc and mainland, plus sediment scraped from the descending plate, add to the wedge of sediment along the continental margin.

5 Further convergence causes the narrow sea located behind the volcanic arc to close. This orogenic event deforms and metamorphoses the back-arc sediments and associated volcanic debris (eugeoclinal material) as well as the volcanic arc itself.

6 Eventually, the continents collide. This event and the associated igneous activity further deforms and metamorphoses the entrapped sediments and volcanic arc to produce the crystalline core of the young mountain belt. As this deformed terrain is thrust landward, the shallow-water (miogeoclinal) deposits which once formed the continental shelf are folded and displaced inland along low-angle thrust faults. These folded strata are essentially unmetamorphosed and form the sedimentary portion of the complex mountain system.

7 Finally, a change in the plate boundary ends the growth of the mountainous belt. Only then does the process of erosion become the dominant force in altering the landscape. In addition to the sediment carried seaward, large quantities of

coarse sediment are deposited in intermontane valleys as well as farther inland. Prolonged erosion coupled with isostatic adjustments eventually reduce this mountainous terrain to the average thickness of the continents (see Figure 18.4).

This sequence of events is thought to have been duplicated many times throughout geologic time. However, the rate of deformation and the geologic and climatic settings varied in each instance. Thus, the formation of each mountain chain must be regarded as a unique event.

OROGENESIS AND CONTINENTAL ACCRETION

When originally formulated, the plate tectonics theory suggested two mechanisms for orogenesis. First, continental collisions were proposed to explain the formation of such mountainous terrains as the Alps, Himalayas, and Urals. Second, as typified by the Andes, orogenesis associated with the subduction of oceanic lithosphere was thought to be the underlying tectonic process for many circum-Pacific mountain chains. Recent investigations, however, indicate yet another mechanism of orogenesis. This new proposal suggests that relatively small crustal fragments collide and merge with continental margins and that through this process of collision and accretion, many of the mountainous regions rimming the Pacific have been generated.

What is the nature of the small crustal fragments and where did they come from? Researchers believe that prior to their accretion to a continental block, some of the fragments may have been microcontinents similar in nature to the present-day island of Madagascar. Many others may have been located below sea level and are represented today by submerged platforms rising high above the floor of the western Pacific (see Figure 1.11). Over one hundred of these so-called oceanic plateaus are known to exist. It is believed that these plateaus originated as submerged continental fragments, extinct volcanic arcs, or as submerged volcanic chains associated with hot spot activity.

The widely accepted view today is that as oceanic plates move, they carry the embedded oceanic plateaus or microcontinents to a subduction zone. Here the upper portions of these thickened zones are peeled from the descending plate and thrust in relatively thin sheets upon the adjacent continental block. This newly added terrain increases the width of the continent and may later be overriden and displaced farther inland by colliding with other fragments.

The idea that orogenesis occurs in association with the accretion of small crustal fragments to a continental mass arose principally from studies conducted in the northern portion of the North American Cordillera. Here it was learned that some terrains, principally those in the orogenic belts of Alaska and British Columbia, contain fossil and magnetic evidence to indicate these strata originated nearer the equator. Further, these exotic terrains were found to consist of rock sequences vastly different from those of adjacent terrains (Figure 18.28).

It is now believed that the exotic terrains found in the North American Cordillera were once scattered throughout the eastern Pacific much as we find oceanic plateaus distributed in the western Pacific. Within the last 200 million years, these fragments have migrated toward and collided with the west coast of North America. Apparently, this activity has resulted in the piecemeal addition of fragments to the entire Pacific Coast from the Baja Peninsula to northern Alaska. If this is true, such collisions are responsible for the orogenesis of much of the North American Cordillera. In a like manner, many of the ocean plateaus today will eventually be accreted to active continental margins, thus resulting in the formation of new orogenic belts.

Although it is now widely accepted that most of the orogenic belts rimming the Pacific were generated by microcontinent collisions, many of the details of these events must still be worked out. Subduction with the accompanying removal of thrust sheets evidently plays a key role in this accretionary process. However, the manner in which thin sheets are peeled from the subducting oceanic lithosphere remains uncertain. Further, in many locations where accretion is thought to have occurred, evidence of the volcanic activity normally associated with subduction is lacking. Despite these problems and many other questions, the plate tectonics theory appears to hold the greatest promise for understanding the origin and

FIGURE 18.28
These exposed strata of the so-called Chulitna Terrane located in south-central Alaska consist of a distinctive suite of rocks found nowhere else in North America. The light and dark bands on the left consist of limestones (dark) and basalts (light) which have been folded and overturned such that they presently overlie younger red sandstones and conglomerates of late Triassic age. Presumably these beds were overturned and deformed as they were accreted to Alaska about 90 million years ago. (Photo by David L. Jones, U.S. Geological Survey)

evolution of complex mountains. The geologic history of each mountain system will surely be re-evaluated in terms of this model. This work will undoubtedly shed new light on their evolutionary histories and will also be useful in evaluating the model itself. In this way new insights into our dynamic planet will emerge.

THE ORIGIN AND EVOLUTION OF CONTINENTAL CRUST

In the preceding section, we learned that the theory of plate tectonics provides a model from which to examine the formation of complex mountainous belts. But what roles have plate tectonics and moun-

483

tain building played in the events that led to the origin and evolution of the continents? At this time no single answer to this question has met with overwhelming acceptance. The lack of agreement among geologists can in part be attributed to the complex nature and antiquity of most continental material, which makes deciphering its history very difficult. Whereas oceanic crust has a relatively simple layered structure and a rather uniform composition, the much older continental crust consists of a collage of highly deformed and metamorphosed igneous and sedimentary rocks. Moreover, in many places, such as the Plains states, the basement rocks are mantled by thick accumulations of much younger sedimentary rocks—a fact which inhibits detailed study. Nevertheless, during the last two decades, great strides have been made in unraveling the secrets held by the rocks composing the stable continental interiors.

At one extreme is a proposal that most, if not all, continental crust formed early in the earth's history. According to this view, the continental crust originated during a primeval molten stage and coincided with the segregation of material that produced the earth's core and mantle (see Chapter 20). During this period of chemical segregation, the lighter silica-rich mineral constituents rose to the surface to produce a scum of continental-type rocks. The heavier material, silica-poor but enriched in iron and magnesium, sank to form mantle rock. Shortly after this segregation, the heat liberated by radioactive decay set the crust in motion. Thereafter a process which may have resembled plate tectonics continually reworked and recycled the continental crust. Through such activity, the primitive continental crust was deformed, metamorphosed, and even remelted. The essence of this hypothesis is that the total amount of continental crust has not changed appreciably since its origin; only its distribution and shape have been modified by tectonic activity.

An opposing view, which has gained support in recent years, contends that the continents have grown larger through geologic time by the gradual accretion of material derived from the upper mantle. A main tenet of this hypothesis is that the primitive crust was of an oceanic-type and the continents were small or possibly nonexistent. Then, through the chemical differentiation of mantle material, the continents slowly grew.

This view proposes that the formation of continental material takes place in two distinct phases as shown in Figure 18.29. The first step occurs in the upper mantle directly beneath the oceanic ridges. Here partial melting of the rock peridotite yields basaltic magma which rises to form oceanic crust. The rocks of the ocean floor are higher in silica, potassium, and sodium, and lower in iron and magnesium than the rocks of the upper mantle from which they were derived. As new ocean floor is generated at the ridge crests, older oceanic crust is being destroyed at the oceanic trenches. In trench regions, the subducted oceanic crust is heated sufficiently to cause partial melting. This gives rise to relatively light, silica-rich rocks which are then emplaced in volcanic arcs. This second episode of chemical differentiation produces a magma that is further enriched in silica, potassium, and sodium as compared to its parent material. The subducted oceanic crust, depleted of its lighter constituents, continues to sink and is no longer involved in the process of generating crustal rocks. Thus, the process of mountain building not only restructures continental rocks, but also generates new continental materials.

According to one view, the earliest continental rocks came into existence at a few isolated island arcs. Once formed, these island arcs coalesced to form larger continental masses, while deforming the volcanic and sedimentary rocks which were deposited in the intervening oceans. Eventually this process generated masses of continental crust having the size and thickness of modern continents.

Evidence supporting the view of continental growth comes from research in regions of plate subduction, such as Japan and the western flanks of the Americas. Equally important, however, has been the research conducted in the stable interiors of the continents, particularly in the shield areas. Recall that all continents have these vast, flat lying expanses of highly deformed and metamorphosed igneous and sedimentary rocks. The most common shield terrains consist of immense bodies of granite and granodiorite, which have been strongly metamorphosed into gneisses and later intruded by younger igneous bodies. In addition to these granite-gneiss terrains are the greenstone belts, so named because of the green tinge common to volcanic rocks of basaltic composition that have undergone low-grade regional meta-

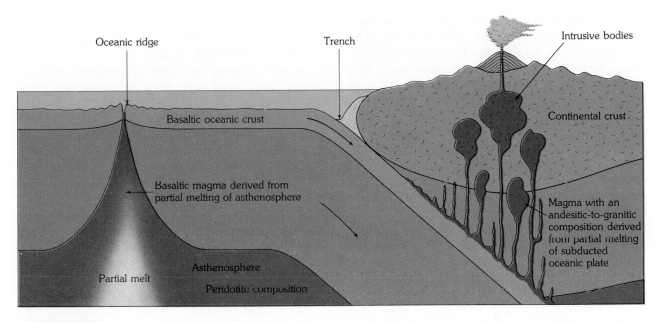

FIGURE 18.29
The two-stage process for transforming material from the asthenosphere into continental crust. Once continental crust is generated, its low density apparently keeps it afloat indefinitely.

morphism. The greenstone belts also contain sedimentary rocks that were strongly deformed and metamorphosed during what appears to be a mountain building episode.

Recent indications are that the rocks of the shield areas are mineralogically and structurally similar to the rocks found at active continental margins where oceanic crust is being consumed. More specifically, the granite-gneiss terrains are chemically similar to the intrusive igneous bodies, such as the Sierra Nevada batholith, which extend along the west coast of the Americas. The ancient rock masses and these younger Mesozoic batholiths are not identical. However, the differences can be explained by the fact that the rocks of the granite-gneiss terrains located in shield areas formed at great depth and were exposed only after long periods of uplift and erosion. On the other hand, the exposed portions of the younger Mesozoic batholiths formed much nearer the surface and thus lack the signs of deep-seated metamorphic alteration. Further, the rocks of the greenstone belts appear to have a chemical composition similar to the lavas and sediments found near modern volcanic

arcs, such as the islands of Japan. The rocks of the greenstone belts are thought to have been deformed along with slivers of the ocean floor and firmly plastered to the continental margin during closings of ancient ocean basins. These rocks would therefore be equivalent to the rocks of a mélange complex, such as the one found in the Coast Range of California.

Radiometric dating of rocks from shield areas, including those in Minnesota and Greenland, has revealed that the oldest terrains formed some 3.8 billion years ago. This date is believed to represent one of the earliest periods of mountain building. At that time, possibly only 10 percent of the present continental crust existed. The next major period of continental evolution may have taken place between 3 and 2.5 billion years ago as indicated by radiometric dates of similar terrains found in the shield areas of Africa and western Australia. It is not known with certainty how many periods of mountain building have occurred since the formation of the earth. The last major period evidently coincided with the closing of the proto-Atlantic and other ancient ocean basins during the formation of the supercontinent Pangaea.

If the continents do in fact grow by accretion of material at their flanks, then the continents have grown larger at the expense of oceanic crust. This view assumes the buoyancy and indestructibility of continental crust. Even the sediment derived from the erosion of continental material that is subducted along with oceanic plates, melts and returns to the continents. Although crustal rock apparently remains afloat indefinitely, some continents are occasionally fragmented and carried along in a conveyor belt fashion until they collide with other landmasses. Presently, Australia, which separated from Antarctica, is being rafted northward and will probably join Asia in much the same manner as India did about 45 million years ago. Thus, according to this view, fragmentation and the formation of new crustal rocks that

accompanied the reshuffling of these fragments are responsible for the present volume, structure, and configuration of continents.

A word of caution is in order. The views set forth in this section to explain the origin and evolution of the continents are still somewhat speculative. Plate tectonics appears to be the major force in crustal evolution over the last 600 million years. However, during the early history of the earth, the heat released by the decay of uranium, thorium, and potassium must have been at least twice as great as it is today. Was plate tectonics active early in the earth's history, only at a different rate, or were there much different processes in operation? Was the primitive crust composed primarily of continental rocks, or was it of the oceanic type? These are some of the questions that still require definitive answers.

REVIEW QUESTIONS

1. List three lines of evidence in support of the crustal uplift concept.

2. What happens to a floating object when weight is added? Subtracted? How does this principle apply to changes in the elevation of mountains? What term is applied to the adjustment that causes crustal uplift of this type?

3. List two lines of evidence in support of the idea that the lithosphere tries to remain in isostatic balance.

4. What conditions favor rock deformation by folding? By fracturing?

5. Contrast the movements along normal and reverse faults. What type of stress is indicated by each fault?

6. At which of the three types of plate boundaries does normal faulting predominate? Reverse faulting? Strike-slip faulting?

7. Describe a horst and a graben. Explain how a graben valley forms and name one.

8. Compare and contrast anticlines and synclines. Domes and basins. Anticlines and domes.

9. Although we classify many mountains as folded, why might this description be somewhat misleading?

10. Describe the modern-day concept of a geosyncline.

11. What is a mélange? Briefly explain its formation.

12. In what way are the islands of Japan, the Sierra Nevada, and the western Andes similar?

13. Would the discovery of a sliver of oceanic crust in a continental interior tend to support or refute the theory of plate tectonics? Why?

14. Why do geologists believe that the northward journey of India is near an end?

15. How does the plate tectonics theory help explain the existence of fossil marine life on top of the Ural Mountains?

16 In your own words, briefly enumerate the steps involved in the formation of a complex mountain system according to the plate tectonics model.

17 Based on current knowledge, describe the major difference between the evolution of the Appalachian Mountains and the North American Cordillera.

18 Contrast the opposing views on the origin of the continental crust.

KEY TERMS

anticline (p. 457)

basin (p. 460)

dip (p. 460)

dip-slip fault (p. 460)

dome (p. 460)

elastic deformation (p. 455)

eugeocline (p. 473)

eugeosyncline (p. 472)

fault-block mountain (p. 465)

folded mountain (p. 465)

geosyncline (p. 472)

graben (p. 463)

hogback (p. 470)

horst (p. 463)

isostasy (p. 453)

joint (p. 464)

mélange (p. 475)

miogeocline (p. 473)

miogeosyncline (p. 472)

monocline (p. 459)

normal fault (p. 461)

oblique-slip fault (p. 463)

orogenesis (p. 451)

plastic deformation (p. 456)

reverse fault (p. 461)

strike (p. 460)

strike-slip fault (p. 462)

syncline (p. 457)

thrust fault (p. 462)

upwarped mountain (p. 468)

19

GEOLOGIC TIME

19

I n 1869 John Wesley Powell, who was later to head the U.S. Geological Survey, led a pioneering expedition down the Colorado River and through the Grand Canyon (Figure 19.1). Writing about the strata that were exposed by the downcutting of the river, Powell said that, ". . . the canyons of this region would be a Book of Revelations in the rock-leaved Bible of geology." Powell was undoubtedly impressed with the countless millions of years of earth history exposed along the walls of the Grand Canyon (see chapter-opening photo). Interpreting earth history is a prime goal of the science of geology. Like a modern-day sleuth, the geologist must interpret the clues found preserved in the rocks. By studying rocks, especially sedimentary rocks, and the features they contain, geologists can often unravel the complexities of the past.

Events by themselves, however, have little meaning until they are put into a time perspective. Studying history, whether it be the Civil War or the Age of Dinosaurs, requires a calendar. Among the major contributions that geology has made to the knowledge of humankind is the geologic calendar and the concept that earth history is exceedingly long. Over many years geologists have devised a time scale of earth history—a calendar where geologic events can be put in their proper place. Geologists, recognizing that earth history has spanned an immense amount of time, worked at finding out just how old the earth is.

The strata exposed in the Grand Canyon contain clues to millions of years of earth history. (Photograph used by permission of Dennis Tasa)

491

A.

B.

FIGURE 19.1
A. Start of the expedition from Green River station. A drawing from Powell's 1875
book. **B.** Major John Wesley Powell, pioneering geologist and the second director of
the U.S. Geological Survey. (Courtesy of the U.S. Geological Survey)

EARLY METHODS OF DATING THE EARTH

Current methods of radiometric dating put the age of
the earth between 4.6 and 4.8 billion years. However, this great age for the earth is a relatively recent
discovery. Although James Hutton and others who
accepted the principle of uniformitarianism believed
the earth was very old, they had no way of knowing
its exact age. Solutions to this dating problem were
sought, and several methods were subsequently devised.

One method involved the rate at which sediment
is deposited. Some geologists reasoned that if they
could determine the rate that sediment accumulates,
and could further ascertain the total thickness of sedimentary rock that had been deposited during earth
history, they could accurately estimate the length of
geologic time. All that was necessary was to divide
the rate of sediment accumulation into the total thick-

ness of sedimentary rock. Unfortunately this method
was riddled with difficulties, some of which are as
follows:

1 Different sediments accumulate at different rates
   under varying conditions. Thus, determining an
   overall rate of sediment accumulation is extremely difficult. Further, if such a rate is determined, it does not necessarily mean that the same
   rate can be applied to the past.

2 Since no single locality has a complete geologic
   column, estimates of the total thickness of sedimentary rocks had to be compiled by adding together the maximum known thickness of rocks of
   each age. These estimates had to be revised each
   time a thicker section was discovered.

3 Sediment compacts when it is lithified; thus, a
   correction for compaction had to be made.

Needless to say, estimates of the earth's age varied considerably as different scientists attempted this method. The figure representing the maximum thickness of sedimentary rock ranged from 9600 meters (32,000 feet) to over 100,500 meters (330,000 feet). The amount of time for 0.3 meter (1 foot) of sediment to accumulate varied from 100 years to over 8600 years. The age of the earth as calculated by this method therefore ranged from 3 million to 1.5 billion years!

Another method for dating the earth involved the salinity of the oceans, which were assumed to originally have been fresh water. Scientists felt that if they could accurately estimate the quantity of salt being carried to the ocean each year by rivers and the total amount of salt currently in the oceans, they could determine the length of geologic time by dividing the latter figure by the former. Near the turn of the twentieth century John Joly calculated the age of the earth at about 90 million years using this method. Joly, however, had no accurate notion of the amount of salt lost from the oceans because of deposition and winds blowing salt inland. It is also probable that the rate of salt accumulation has not always been constant. Thus, Joly's estimate for the age of the earth was not accurate. However, both of the methods for dating the earth that have just been described indicated that the earth was considerably older than the 6000 years given it by Archbishop Usher.[1]

Perhaps the most influential estimates of the age of the earth were compiled by the well-known and highly respected physicist Lord Kelvin in the latter part of the nineteenth century. Since Kelvin's estimates required few assumptions and were based on precise measurements, they were widely accepted for a time. One of Kelvin's methods was founded on the widely held assumption that the earth had originally been molten and had cooled to its present condition. Although his data and calculations were limited, Kelvin still made it quite obvious that the earth could not be more than 100 million years old, and likely much less. The second of Kelvin's estimates was based on the fact that the source of the sun's tremendous output of energy was of a conventional nature (nuclear fusion and radioactivity had not yet been discov-

ered). His calculations indicated that the sun could only have illuminated the earth for a few tens of millions of years. Furthermore, he said that in the past it had been much hotter and in the future it would become much cooler. He believed the earth was inhabitable for organisms for a period of only 20–40 million years. Kelvin's apparently irrefutable estimates had a rather profound impact:

> **Evolutionists found it virtually impossible to accept these figures, but all they had were educated guesses in the face of Kelvin's potent mathematics. Darwin and others compromised their original theories in their later years in an effort to reconcile evolution and uniformitarianism with the physicists' estimates. Eventually, however, they were vindicated.[1]**

RADIOACTIVITY AND RADIOMETRIC DATING

Most atoms are stable and do not change. However, some are unstable, constantly releasing heat as their nuclei break apart or decay. This is the heat that helps maintain the high temperatures in the earth's interior and is the source of the heat which Kelvin was measuring when he thought he was measuring the "cooling" earth.

In Chapter 2 we learned that an atom is composed of electrons, protons, and neutrons. *Electrons* have a negative charge and *protons* have a positive charge. Since a *neutron* is actually a proton and electron combined, it has no charge. Protons and neutrons are found in the center, or *nucleus*, of the atom, and electrons spin around the nucleus in definite paths, or orbits. Practically all (99.9 percent) of the mass of an atom is found in the nucleus, indicating that electrons have practically no mass at all. By adding together the number of protons and neutrons in the nucleus, the *mass number* of the atom is determined. The *atomic number* (the atom's identifying number) is equal to the number of protons. Each one of the more than 100 known elements has a different number of protons in the nucleus, and thus a different atomic number. Atoms of the same element may

---

[1]See the section entitled *Catastrophism* in Chapter 1.

[1]Leigh W. Mintz. *Historical Geology: The Science of a Dynamic Earth,* 2nd ed. (Columbus, Ohio : Merrill, 1977), pp. 84–85.

have different numbers of neutrons in the nucleus. Such atoms, called *isotopes,* have different mass numbers but the same atomic number.

The forces which bind protons and neutrons together in the nucleus are very strong; however, the nature of these forces is still poorly understood. Some isotopes have unstable nuclei; that is, the forces which bind the protons and neutrons together are not sufficiently strong. As a result, the nuclei spontaneously break apart, or decay, a process called **radioactivity.** What happens when unstable nuclei break apart? Two common types of radioactive decay are illustrated in Figure 19.2 and are summarized as follows:

1   *Alpha particles* (α particles) may be emitted from the nucleus. An alpha particle is composed of 2 protons and 2 neutrons. Thus, the emission of an alpha particle means that the mass number of the isotope is reduced by 4 and the atomic number is lowered by 2.

2   When a *beta particle* (β particle), or electron, is given off from a nucleus, the mass number re-

mains unchanged, because electrons have practically no mass. However, since the electron must have come from a neutron (remember, a neutron is a combination of a proton and an electron), the nucleus contains one more proton than before. Therefore, the atomic number increases by 1.

The radioactive isotope is often referred to as the **parent,** and the isotopes resulting from the decay of the parent are termed the **daughter products.** Figure 19.3 provides an example of radioactive decay. Here it may be seen that when the radioactive parent, uranium-238 (atomic number 92, mass number 238) decays, it emits 8 alpha particles and 6 beta particles before becoming the stable daughter product lead-206 (atomic number 82, mass number 206).

Certainly among the most important results of the discovery of radioactivity is that it provided a reliable means of calculating the ages of rocks and minerals which contain radioactive isotopes, a procedure referred to as **radiometric dating.** Why is radiometric dating reliable? The answer lies in the fact that the

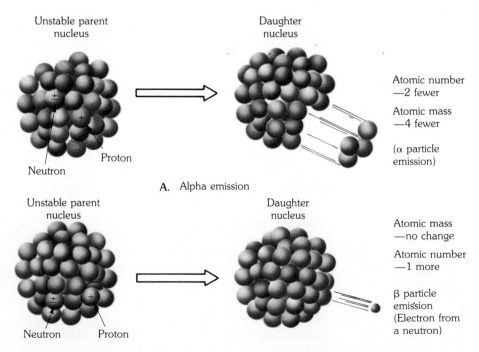

FIGURE 19.2
Common types of radioactive decay. Notice that in each case the number of protons (atomic number) in the nucleus changes, thus producing a different element.

rate at which radioactive isotopes decay is constant and unaffected by any chemical or physical agents.

The amount of time required for one-half of the nuclei in a sample to decay, called its **half-life,** is a common way of expressing the rate of radioactive disintegration. If we began with a pound of radioactive material, half a pound would decay after one half-life, half the remaining amount would break down after another half-life, and so on.

Figure 19.4 illustrates the principle of radiometric dating using a hypothetical radioactive parent that decays directly into the stable daughter product. Its half-life is 1 million years. By calculating the percentages of radioactive parent and stable daughter product, the age of the specimen can be determined. In this example, when the quantities of parent and daughter are equal (ratio 1:1), we know that one half-

life has transpired and that the specimen is 1 million years old. When the ratio of parent to daughter reaches 1:15, we know the sample is 4 million years old.

Of the many radioactive isotopes that exist in nature, five have proven significant in providing radiometric ages for ancient rocks. Table 19.1 summarizes these most frequently used isotopes. Others are either very rare or have half-lives that are too short or much too long to be useful. Rubidium-87 and the two isotopes of uranium are used only for dating rocks that are millions of years old, but potassium-40 is more versatile. Although the half-life of potassium-40 is 1.3 billion years, recent analytic techniques have made it possible to detect the tiny amounts of its stable daughter product, argon-40, in rocks as young as 50,000 years.

To date more recent events, carbon-14 (also called **radiocarbon**), the radioactive isotope of carbon, is used. Since it has a half-life of only 5730 years, it can be used for dating events from the historic past, as well as those from recent geologic history. Until the late 1970s radiocarbon was useful in dating events only as far back as 40,000–50,000 years. However, as was the case with potassium-40, the development of more sophisticated analytical techniques has increased the usefulness of this "clock." Carbon-14 can now be used to date events as far back as 75,000 years. This is a significant accomplishment because it means that geologists can

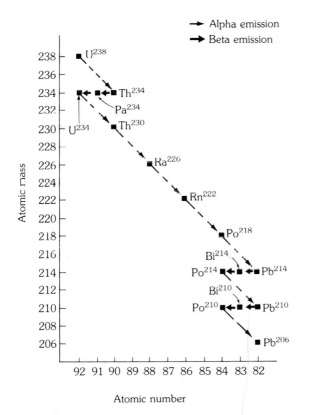

FIGURE 19.3

The most common isotope of uranium (U-238) is an example of a radioactive decay series. Before the stable end product (Pb-206) is reached, many different isotopes are produced as intermediate steps.

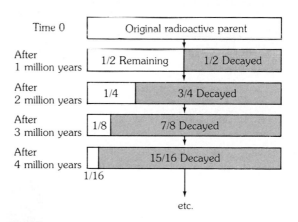

FIGURE 19.4

Decay of a hypothetical radioactive isotope with a half-life of 1 million years.

TABLE 19.1
Frequently used radioactive isotopes.

Radioactive Parent	Stable Daughter Product	Currently Accepted Half-life Values
Uranium-238	Lead-206	4.5 billion years
Uranium-235	Lead-207	713 million years
Thorium-232	Lead-208	14.1 billion years
Rubidium-87	Strontium-87	47.0 billion years
Potassium-40	Argon-40	1.3 billion years

now date many ice-age phenomena that previously could not be dated accurately.

Carbon-14 is continuously produced in the upper atmosphere as a consequence of cosmic ray bombardment, in which cosmic rays (high-energy nuclear particles) shatter the nuclei of gases to release neutrons. The neutrons are absorbed by nitrogen (atomic number 7, mass number 14), causing its nucleus to emit a proton. Thus, the atomic number drops by 1 (to 6), and a new element, carbon-14, is created (Figure 19.5A). This isotope of carbon is quickly incorporated into carbon dioxide, circulates in the atmosphere, and is absorbed by living matter. As a result, all organisms contain a small amount of carbon-14.

While an organism is alive, the decaying radiocarbon is continually replaced. As a result, the ratio of carbon-14 to carbon-12 (the most common isotope of carbon) remains constant. However, when the plant or animal dies, the amount of carbon-14 gradually decreases as it decays to nitrogen-14 by beta emission (Figure 19.5B). Therefore, by comparing the proportions of carbon-14 and carbon-12 in a sample, radiocarbon dates can be determined. Although carbon-14 is only useful in dating the last small fraction of geologic time, it has become a very valuable tool for anthropologists, archeologists, and historians, as well as for geologists who study very recent earth history. In fact, the development of radiocarbon dating was considered so important that

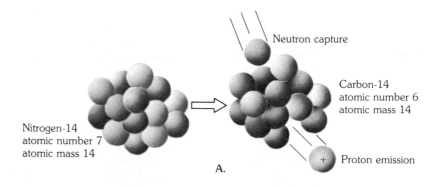

Neutron capture

Carbon-14
atomic number 6
atomic mass 14

Nitrogen-14
atomic number 7
atomic mass 14

+ Proton emission

A.

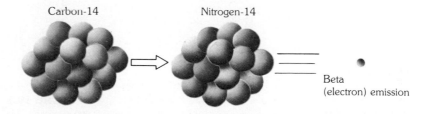

Carbon-14

Nitrogen-14

Beta
(electron) emission

B.

FIGURE 19.5
A. Production and **B.** decay of carbon-14.

the chemist who discovered this application, Willard F. Libby, received a Nobel Prize.

Bear in mind that although the basic principle of radiometric dating is relatively simple, the actual procedure is quite complex, for the chemical analysis which determines the quantities of parent and daughter that are present must be painstakingly precise. In addition, some radioactive materials do not decay directly into the stable daughter product as was the case with our hypothetical example, a fact which may further complicate the analysis. In the case of uranium-238, there are thirteen intermediate unstable daughter products formed before the fourteenth, and last daughter product, the stable isotope lead-206, is produced (Figure 19.3).

Radiometric dating methods have produced literally thousands of dates for events in earth history. Rocks from several localities have been dated at more than 3 billion years, and geologists realize that still older rocks exist. For example, a granite from South Africa which has been dated at 3.2 billion years contains inclusions of quartzite. Quartzite is a metamorphic rock which originally was the sedimentary rock sandstone. Since sandstone is the product of the lithification of sediments produced by the weathering of pre-existing rocks, we have a positive indication that older rocks existed.

Radiometric dating has vindicated the ideas of Hutton, Darwin, and others who over 150 years ago assumed that geologic time must be immense. Indeed, it has proven that there has been enough time for the slow processes we observe to have accomplished tremendous tasks.

RELATIVE DATING

Radiometric dating results in specific dates for rock units which represent various events in the earth's distant past. We can now state with some confidence that particular geologic events took place a certain number of years ago. Such dates are referred to as **absolute dates,** for they pinpoint the time in history when something took place. Prior to the discovery of radioactivity and the development of the technology of radiometric dating, geologists had no precise method of absolute dating and had to rely solely on relative dating. **Relative dating** means that rocks are placed in their proper sequence or order. Relative dating will not tell us how long ago something took place, only that it followed one event and preceded another. The relative dating techniques which were developed are still widely used. Absolute dating methods did not replace these techniques; they simply supplemented them. To establish a relative time scale, a few simple principles or rules had to be discovered and applied. Although they may seem rather obvious to us today, their discovery was a very important scientific achievement.

Nicolaus Steno, a physician in Florence, Italy, is credited with being the first to recognize a sequence of historical events in an outcrop of sedimentary rock layers. Working in the mountains of western Italy, Steno applied a very simple rule that has come to be the most basic principle of relative dating—the **law of superposition.** The law simply states that in an undeformed sequence of sedimentary rocks, each bed is older than the one above it and younger than the one below. Although it may seem obvious that a layer could not be deposited with nothing beneath it for support, it was not until 1669 that Steno clearly stated the principle. This rule also applies to other surface-deposited materials such as lava flows and beds of ash from volcanic eruptions. Applying the law of superposition to the beds shown in Figure 19.6, we can easily place the layers in their proper order. The sandstone is youngest and the shale is oldest.

FIGURE 19.6
Applying the law of superposition to this cross section, the shale bed is oldest and the sandstone is youngest.

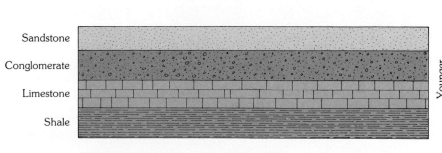

Sandstone

Conglomerate

Limestone

Shale

Younger

Steno is also credited with recognizing the importance of another basic principle, called the **principle of original horizontality.** Simply stated, it means that layers of sediment are generally deposited in a nearly horizontal position. Thus, if we observe rock layers that are inclined at a steep angle, they must have been moved into that position by crustal disturbances sometime after their deposition.

When igneous intrusions or faults cut through other rocks, they are assumed to be younger than the rocks they cut. For example, when two dikes intersect, the older one must have been opened up in order to allow the younger one to cut through it. The younger dike would be continuous, while the older dike would be interrupted at the point of their intersection. Figure 19.7 illustrates this principle of **crosscutting.**

Layers of rock are said to be **conformable** when they are found to have been deposited without interruption. However, there is no place on earth that contains a complete set of conformable strata. Even for a particular span of time, many locations do not have a complete sequence of rocks representing the entire period. All such breaks in the rock record are termed **unconformities.** Figure 19.8 illustrates some of the ways in which unconformities may develop. Perhaps the most easily recognized type of unconformity consists of tilted or folded sedimentary rocks that are overlain by other, more flat-lying strata. These are called **angular unconformities** and indicate that the period of deformation (folding or tilting) and erosion is not represented by sedimentary rocks (Figure 19.9). **Disconformities** may be more difficult to recognize because the strata on either side of these unconformities are essentially parallel. Disconformities may represent either a period of nondeposition or a period of erosion.

By applying the principles of relative dating to the hypothetical geologic cross section shown in Figure 19.10, the rocks and the events in earth history they represent may be placed in their proper sequence. The following statements summarize the logic used to interpret the cross section:

1   Applying the law of superposition, beds A, B, C, and E were deposited, in that order. Since bed D is a sill (a concordant igneous intrusion), it is younger than the rocks that were intruded. Further evidence that the sill is younger than beds C and E are the inclusions in the sill of fragments from these beds. If the igneous mass contains pieces of surrounding rock, the surrounding rock must have been there first.

2   Following the intrusion of the sill (D), the intrusion of the dike (F) occurred. Since the dike cuts through beds A through E, it must be younger than all of them.

3   Next, the rocks were tilted and then eroded. We know the tilting happened first because the upturned ends of the strata have been eroded. The tilting and erosion, followed by further deposition, produced an angular unconformity.

4   Beds G, H, I, J, and K were deposited in that order, again using the law of superposition. Although the lava flow (bed H) is not a sedimentary rock layer, it is a surface-deposited layer, and thus superposition may be applied.

5   Finally, the irregular surface and the stream valley indicate that another gap in the rock record is being produced by erosion.

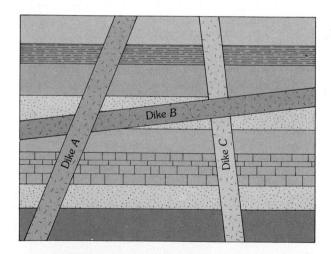

FIGURE 19.7

Cross-cutting relationships. All of the dikes are younger than the rock into which they were intruded. Since Dike B cuts through Dike C and Dike A cuts through Dike B, the order of intrusion, from oldest to youngest, is Dike C, Dike B, Dike A.

In the foregoing example our goal was to establish a relative time scale for the rocks and events in the area of the cross section. Remember, we do not have any idea how many years of earth history are represented, nor do we know how the ages of the strata in this area compare to any other area.

CORRELATION

In order to develop a geologic calendar that is applicable to the whole earth, rocks of similar age in different regions must be matched up. Such a task is referred to as **correlation**. Within a limited area there are several methods of correlating the rocks of one locality with those of another. A bed or series of beds may be traced simply by walking along the outcropping edges. However, this may not be possible when the continuity of the bed is interrupted. Correlation over short distances is often achieved by noting the place of a bed in a sequence of strata, or a bed may be identified in another location if it is composed of very distinctive minerals (Figure 19.11). By corre-

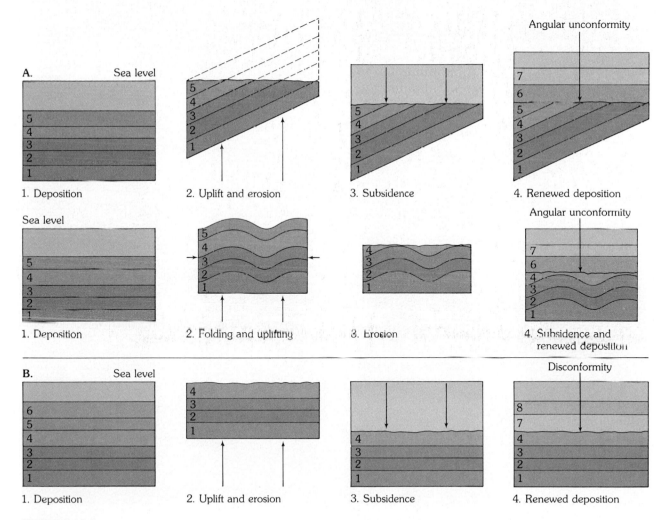

FIGURE 19.8
Development of unconformities. **A.** Two examples of angular unconformity. **B.** Erosion resulting in a disconformity. In each case a gap in the rock record has been created.

FIGURE 19.9
Angular unconformity as seen from Desert View, Grand Canyon National Park. Here tilted and eroded Precambrian rocks are overlain by younger horizontal strata of Paleozoic age. (Photo by E. J. Tarbuck)

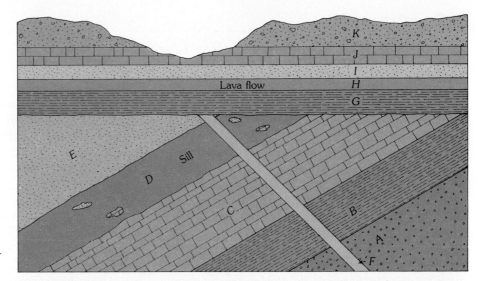

FIGURE 19.10
Geologic cross section of a hypothetical region.

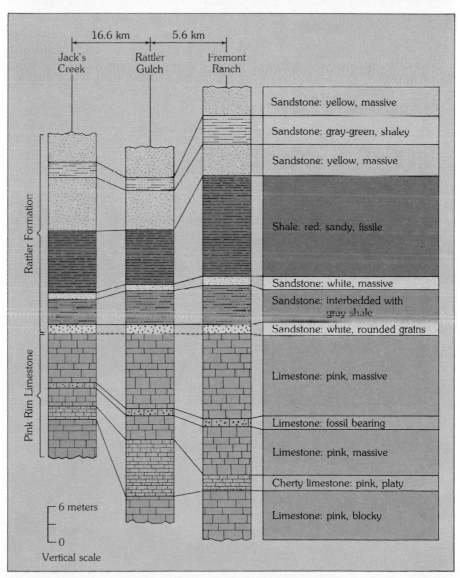

FIGURE 19.11
Correlation of strata within a small area. (After U.S. Geological Survey)

lating the rocks from one place to another, a more comprehensive view of the geologic history of a region is possible. Figure 19.12, for example, shows the correlation of strata at several places on the Colorado Plateau. No single locale exhibits a complete sequence, but correlation reveals the total extent of sedimentary rocks.

Most geologic studies involve rather small areas. Although they are usually important in their own right, their full value is realized only when they are correlated with other regions. Although the methods just described may be sufficient to trace a rock formation over relatively short distances, they are not adequate for matching up rocks at great distance. When correlation between widely separated areas or between continents is the objective, the geologist must rely upon fossils.

Although the existence of fossils had been known for centuries, it was not until the late 1700s and early 1800s that their significance as geologic tools was made evident. During this period an English engineer and canal builder, William Smith, discovered that each rock formation in the canals contained fossils unlike those in the beds either above or below. Further, he noted that sedimentary strata in widely separated areas could be identified by their distinctive fossil content. Based upon Smith's classic observations and the findings of many geologists who followed, one of the most important and basic principles in historical geology was formulated: fossil organisms succeed one another in a definite and determinable order, and therefore any time period can be recognized by its fossil content. This has come to be known as the **principle of faunal succession.** In other words, when fossils are arranged according to their age, they do not present a random or haphazard picture. To the contrary, fossils show progressive changes from simple to complex and reveal the

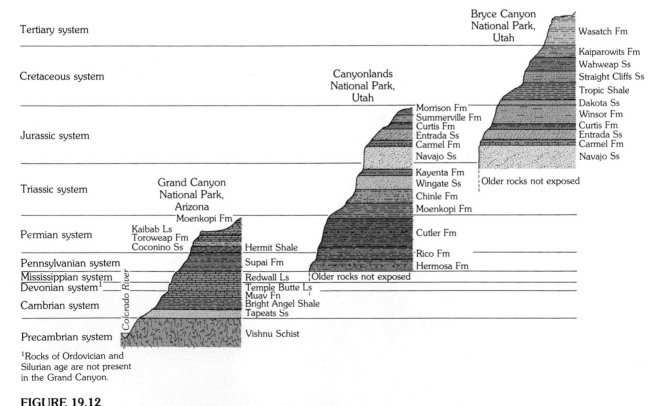

¹Rocks of Ordovician and Silurian age are not present in the Grand Canyon.

FIGURE 19.12
Correlation of strata at three locations on the Colorado Plateau reveals the total extent of sedimentary rocks in the region. (After U.S. Geological Survey)

advancement of life through time. For example, an Age of Trilobites is recognized quite early in the fossil record. Then, in succession, paleontologists recognize an Age of Fishes, an Age of Coal Swamps, an Age of Reptiles, and an Age of Mammals. These "ages" pertain to groups that were especially plentiful and characteristic during particular time periods. Within each of the "ages" there are many subdivisions based, for example, on certain species of trilobites, and certain types of fish, reptiles, and so on. This same succession of dominant organisms, never out of order, is found on every major landmass.

Since fossils were found to be time indicators, they became the most useful means of correlating rocks of similar age in different regions. Geologists pay particular attention to certain fossils called **index, or guide, fossils.** Since these fossils are widespread geographically and are limited to a short span of geologic time, their presence provides an important method of matching rocks of the same age. Rock formations, however, do not always contain a specific index fossil. In such situations, groups of fossils are used to establish the age of the bed. Figure 19.13 illustrates how a group of fossils may be used to date rocks more precisely than could be accomplished by the use of any one of the fossils.

THE GEOLOGIC CALENDAR

The whole of geologic history has been subdivided into units of varying magnitude which together comprise the calendar of earth history (Figure 19.14). The

major units of the calendar were delineated during the nineteenth century, principally by workers in western Europe and Great Britain. Since absolute dating was not a reality during this time, the entire calendar was created using methods of relative dating. It has only been recently that absolute dates have been added to the calendar.

By examining Figure 19.14, you can see that the largest of the subdivisions of the geologic calendar are called **eras.** Three eras are currently recognized: the **Paleozoic** ("ancient life"), the **Mesozoic** ("middle life"), and the **Cenozoic** ("recent life"). As the names imply, the eras are bounded by quite profound worldwide changes in life forms. Each era is subdivided into time units known as **periods.** The Paleozoic has seven, the Mesozoic three, and the Cenozoic two. Since we are currently living in the Cenozoic era, there may be more periods yet to come. Each period is characterized by a somewhat less profound change in life forms as compared with the eras. The major divisions, with brief explanations of each, are shown in Table 19.2. Finally, each of the twelve periods are further divided into still smaller units called **epochs.** Except for the seven epochs which have been named for the periods of the Cenozoic era, those of other periods are not named.

Notice that the detail of the geologic calendar does not begin until about 600 million years ago, the date for the beginning of the first period of the Paleozoic era, the Cambrian period. The more than 1 billion years prior to the Cambrian is simply referred to as

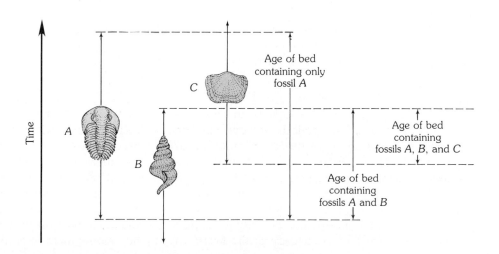

FIGURE 19.13
Overlapping ranges of fossils help date rocks more exactly than using a single fossil.

Time

A

B

C

Age of bed containing only fossil A

Age of bed containing fossils A, B, and C

Age of bed containing fossils A and B

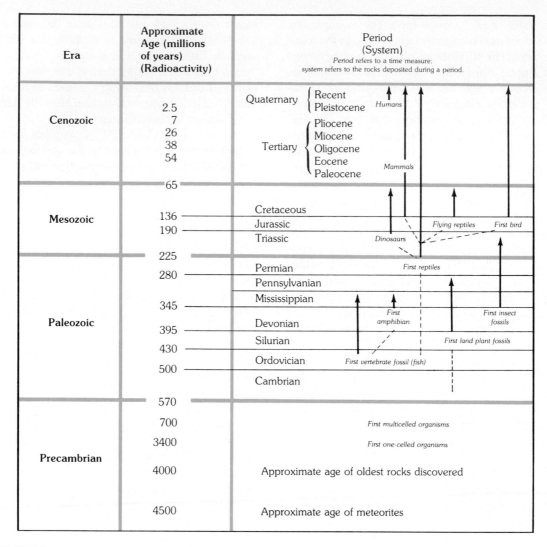

FIGURE 19.14
The geologic calendar. The absolute dates were added quite recently, long after the calendar had been established using relative dating techniques. (From Foster, *Physical Geology*, 4th ed., Columbus, Ohio: Charles E. Merrill, 1983)

the **Precambrian.** Why is the Precambrian not subdivided into numerous eras, periods, and epochs? The reason is that Precambrian history is not known in great enough detail. The quantity of information geologists have deciphered about the earth's past is somewhat analogous to the detail of human history. The farther back we go, the less that is known. Certainly more data and information exist about the past ten years than for the first decade of the twentieth century; the events of the nineteenth century have been documented much better than the events of the first century A.D.; and so on. So it is with earth history. The more recent past has the freshest, least disturbed, and most observable record. The farther back in time the geologist goes, the more fragmented the record and clues become. There are other reasons to explain our lack of a detailed time scale for this vast segment of earth history:

TABLE 19.2
Major divisions of geologic time.

CENOZOIC ERA (Age of Recent Life)	Quaternary period	The several geologic eras were originally named Primary, Secondary, Tertiary, and Quaternary. The first two names are no longer used; Tertiary and Quaternary have been retained but used as period designations.
	Tertiary period	
MESOZOIC ERA (Age of Middle Life)	Cretaceous period	Derived from Latin word for chalk (creta) and first applied to extensive deposits that form white cliffs along the English Channel.
	Jurassic period	Named for the Jura Mountains, located between France and Switzerland, where rocks of this age were first studied.
	Triassic period	Taken from word ''trias'' in recognition of the threefold character of these rocks in Europe.
PALEOZOIC ERA (Age of Ancient Life)	Permian period	Named after the province of Perm, U.S.S.R., where these rocks were first studied.
	Pennsylvanian period	Named for the state of Pennsylvania where these rocks have produced much coal.
	Mississippian period	Named for the Mississippi River valley where these rocks are well exposed.
	Devonian period	Named after Devonshire County, England, where these rocks were first studied.
	Silurian period	Named after Celtic tribes, the Silures and the Ordovices, that lived in Wales during the Roman Conquest.
	Ordovician period	
	Cambrian period	Taken from Roman name for Wales (Cambria) where rocks containing the earliest evidence of complex forms of life were first studied.
PRECAMBRIAN		The time between the birth of the planet and the appearance of complex forms of life. More than 80 percent of the earth's estimated 4.6 billion years falls into this era.

SOURCE: U.S. Geological Survey.

1 The first abundant fossil evidence does not appear in the geologic record until the beginning of the Cambrian period. Prior to the Cambrian, very simple life forms such as algae, bacteria, fungi, worms, and sponges predominated. All of these organisms lack hard parts, an important prerequisite for fossilization. Therefore, there is only a meager Precambrian fossil record. Many exposures of Precambrian rocks have been studied in some detail, but correlation is exceedingly difficult

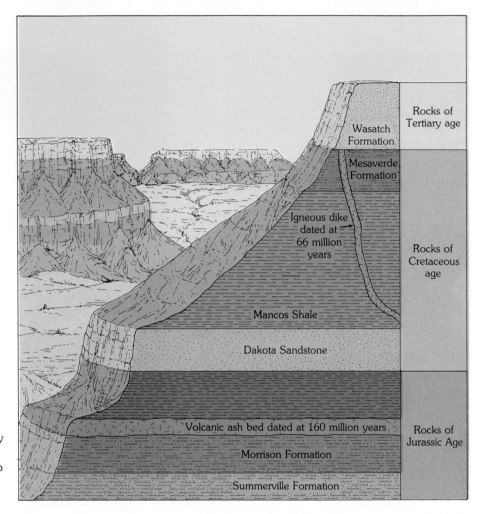

FIGURE 19.15
Absolute dates for sedimentary layers are usually determined by examining their relationship to igneous rocks. (After U.S. Geological Survey)

Rocks of Tertiary age
Wasatch Formation
Mesaverde Formation
Igneous dike dated at 66 million years
Rocks of Cretaceous age
Mancos Shale
Dakota Sandstone
Volcanic ash bed dated at 160 million years
Morrison Formation
Summerville Formation
Rocks of Jurassic Age

without fossils. Consequently, there is no general agreement on the subdivisions of the Precambrian.

2 Because Precambrian rocks are very old, most have been subjected to a great many changes. The bulk of the Precambrian rock record is composed of highly distorted metamorphic rocks. This makes the interpretation of past environments very difficult, because many of the clues found in sedimentary rocks have been destroyed.

With the development of radiometric dating methods, a solution to the troublesome task of dating and correlating Precambrian rocks is at hand. Untangling the complex Precambrian record, however, is still many years away.

DIFFICULTIES IN DATING THE GEOLOGIC CALENDAR

Although reasonably accurate absolute dates have been worked out for the periods of the geologic calendar (see Figure 19.14), the task is not without its difficulties. The primary difficulty in assigning absolute dates to units of time lies in the fact that radioactive elements are typically restricted to igneous rocks. Even if a detrital rock included sediment which contained a radioactive mineral, the rock could not be dated, for the grains in a detrital sedimentary rock are not the same age as the rock in which they occur. The sediments composing such a rock may have been weathered from rocks of diverse ages. Thus, the age of a mineral in a sedimentary rock only tells us that the rock can be no older.

On the other hand, in igneous rocks the minerals and rock form simultaneously; the age of the mineral containing a radioactive isotope is the same age as the rock. Therefore, in order to date sedimentary strata, the geologist must relate them to igneous masses, as in Figure 19.15. In this example, the age of the volcanic ash bed within the Morrison Formation and the dike cutting the Mancos Shale and Mesaverde Formation are known. The sedimentary beds below the ash are obviously older than the ash,

and all the layers above the ash are younger. The dike is younger than the Mancos Shale and the Mesaverde Formation but older than the Wasatch Formation because the dike does not intrude the Tertiary rocks. From this kind of evidence, geologists estimate that a part of the Morrison Formation was deposited about 160 million years ago as indicated by the ash bed. Further, they can conclude that the beginning of the Tertiary period began after the intrusion of the dike, 66 million years ago. This is one example of the literally thousands that illustrates how dated materials are used to bracket the various episodes in the history of the earth within specific time periods and illustrates the necessity of combining laboratory dating methods with geological information.

REVIEW QUESTIONS

1 Describe two early methods for dating the earth. How old was the earth thought to be according to these estimates? List some weaknesses of each method.

2 If a radioactive isotope of thorium (atomic number 90, mass number 232) emits 6 alpha particles and 4 beta particles during the course of radioactive decay, what is the atomic number and mass number of the stable daughter product?

3 Why is radiometric dating the most reliable method of dating the geologic past?

4 A hypothetical radioactive isotope has a half-life of 10,000 years. If the ratio of radioactive parent to stable daughter product is 1:3, how old is the rock containing the radioactive material?

5 Assume that the age of the earth is 5 billion years.
   (a) What fraction of geologic time is represented by recorded history (assume 5000 years for the length of recorded history)?
   (b) The first abundant fossil evidence does not appear until the beginning of the Cambrian period (600 million years ago). What percent of geologic time is represented by abundant fossil evidence?

6 Distinguish between absolute and relative dating.

7 What is the law of superposition? How are cross-cutting relationships used in relative dating?

8 When you observe an outcrop of steeply inclined sedimentary layers, what principle allows you to assume that the beds were tilted after they were deposited?

9 What is meant by the term *correlation?*

10 Describe William Smith's important contribution to the science of geology.

11 Why are fossils such useful tools in correlation?

12 What subdivisions make up the geologic calendar?

13 Why is Precambrian time not divided into smaller units?

14 Briefly describe the difficulties in assigning absolute dates to layers of sedimentary rock.

KEY TERMS

absolute date (p. 497)

angular unconformity (p. 498)

Cenozoic era (p. 503)

conformable (p. 498)

correlation (p. 499)

cross-cutting (p. 498)

daughter product (p. 494)

disconformity (p. 498)

epoch (p. 503)

era (p. 503)

faunal succession, principle of (p. 502)

guide (index) fossil (p. 503)

half-life (p. 495)

index (guide) fossil (p. 503)

Mesozoic era (p. 503)

original horizontality, principle of (p. 498)

Paleozoic era (p. 503)

parent (p. 494)

period (p. 503)

Precambrian (p. 504)

radioactivity (p. 494)

radiocarbon (p. 495)

radiometric dating (p. 494)

relative dating (p. 497)

superposition, law of (p. 497)

unconformity (p. 498)

20

PLANETARY GEOLOGY

20

The sun is the hub of a huge rotating system consisting of nine planets, their satellites, and numerous small but nevertheless interesting bodies, including asteroids, comets, and meteoroids. An estimated 99.85 percent of the mass of the solar system is contained within the sun, while the planets collectively make up most of the remaining 0.15 percent. The planets, in order from the sun, are Mercury, Venus, Earth, Mars, Jupiter, Saturn, Uranus, Neptune, and Pluto. Under the control of the sun's gravitational force, each planet maintains an almost circular orbit and travels in a counterclockwise direction around the sun. Further, the orbits of all the planets lie within 3 degrees of the plane of the sun's equator, except for those of Mercury and Pluto, which are inclined 7 and 17 degrees, respectively.

When man first came to recognize that the planets are "worlds" much like the earth, a great deal of interest was generated. A primary concern has always been the possibility of intelligent life existing elsewhere in the universe. This expectation has not as yet come to pass. Nevertheless, since all of the planets most probably formed from the same primordial cloud of dust and gases, they should provide valuable information concerning the earth's history. Recent space explorations have been organized with this goal in mind. To date, Mercury, Venus, Mars, Jupiter, Saturn, and the moon have been explored by space probes and in 1986 *Voyager 2* is expected to rendezvous with Uranus.

THE PLANETS: AN OVERVIEW

Careful examination of Table 20.1 will show that the planets fall quite nicely into two groups: the **terrestrial** (earthlike) **planets** of Mercury, Venus, Earth, and Mars; and the **Jovian** (Jupiterlike) **planets** of Jupiter, Saturn, Uranus, and Neptune. Pluto is not included in either category. Pluto's position at the far edge of the solar system and its small size make this planet's true nature a mystery. The most obvious difference between these groups is the size of their members. The largest terrestrial planet (Earth) has a diameter only one-quarter as great as the diameter of the smallest Jovian planet (Neptune), and its mass is only one-seventeenth as great. Hence, the Jovian planets are often called *giants*. Also, because of their relative locations, the four Jovian planets are referred to as the *outer planets,* while the terrestrial planets are called the *inner planets*. As we shall see, there appears to be a correlation between the locations of these planets and their sizes.

Other dimensions along which the two groups markedly differ include density, composition, and rate of rotation. The densities of the terrestrial planets average about 5 times the density of water, whereas the Jovian planets have densities that average only 1.5 times that of water. One of the outer planets, Saturn, has a density only 0.7 that of water, which means that if a large enough ocean existed, Saturn could float in it. The variations in the compositions of the planets is largely responsible for these differences.

The substances of which both groups of planets are composed can be divided into three groups based upon their melting points. They are called *gases, rocks,* and *ices*. The gases are those materials with melting points near absolute zero[1], $-273°C$, and consist of hydrogen and helium. The rocky materials are made principally of silicate minerals and metallic iron, which have melting points exceeding 700°C.

Saturn photographed from 18 million kilometers by Voyager 1. (Courtesy of NASA)

---

[1]Absolute zero is the lowest possible temperature. All molecular motion ceases at that point.

TABLE 20.1
Planetary data.

Planet	Symbol	Mean Distance from Sun: AU	Mean Distance from Sun: Millions of miles	Mean Distance from Sun: Millions of kilometers	Period of Revolution	Inclination to Ecliptic	Orbital velocity: mi/s	Orbital velocity: km/s
Mercury	☿	0.387	36	58	88^d	7° 00′	29.5	47.9
Venus	♀	0.723	67	108	225^d	3° 24′	21.8	35.0
Earth	⊕	1.000	93	150	365.25^d	0° 00′	18.5	29.8
Mars	♂	1.524	142	228	687^d	1° 51′	14.9	24.1
Jupiter	♃	5.203	483	778	12^{yr}	1° 19′	8.1	13.1
Saturn	♄	9.539	886	1427	29.5^{yr}	2° 30′	6.0	9.6
Uranus	⛢	19.180	1780	2869	84^{yr}	0° 46′	4.2	6.8
Neptune	♆	30.060	2790	4498	165^{yr}	1° 46′	3.3	5.4
Pluto	♇	39.440	3670	5900	248^{yr}	17° 12′	2.9	4.7

Planet	Period of Rotation	Diameter: miles	Diameter: kilometers	Relative Mass (Earth = 1)	Average Density (g/cm^3)	Polar Flattening (%)	Eccentricity	Number of Known Satellites
Mercury	59^d	3015	4878	0.056	5.1	0.0	0.206	0
Venus	243^d	7526	12,112	0.82	5.3	0.0	0.007	0
Earth	23^{h}56^{m}04^s	7920	12,742	1.00	5.52	0.3	0.017	1
Mars	24^{h}37^{m}23^s	4216	6800	0.108	3.94	0.5	0.093	2
Jupiter	~9^{h}50^m	88,700	143,000	318.000	1.34	6.5	0.048	15
Saturn	~10^{h}25^m	75,000	121,000	95.200	0.70	10.5	0.056	17
Uranus	10^{h}45^m	29,000	47,000	14.600	1.55	7.0	0.047	5
Neptune	18^h(?)	28,900	45,000	17.300	2.27	2.5	0.008	2
Pluto	6.4^d	~1500	~2400	~0.01(?)	~1.5(?)	?	0.250	1

The ices have intermediate melting points and include ammonia (NH_3), methane (CH_4), carbon dioxide (CO_2), and water (H_2O).

The terrestrial planets are composed mostly of dense rocky and metallic material with minor amounts of gases. The Jovian planets, on the other hand, contain a large percentage of hydrogen and helium, with varying amounts of ices (mostly water, ammonia, and methane), which accounts for their low densities. The outer planets are also thought to contain as much rocky and metallic material as the terrestrial planets, and this material may be concentrated in a small central core.

The Jovian planets have very thick atmospheres consisting of varying amounts of hydrogen, helium, methane, and ammonia. By comparison, the terrestrial planets have meager atmospheres at best. A planet's ability to retain an atmosphere depends on its temperature and mass. Simply stated, a gas molecule can "evaporate" from a planet if it reaches a speed known as the *escape velocity*. For the earth, this velocity is 11 kilometers (7 miles) per second. Any material, including a rocket, must reach this speed before it can leave the earth and go into space. The Jovian planets, because of their greater mass, have higher escape velocities than the terrestrial planets. Consequently, it is more difficult for gases to "evaporate" from them. Also, because the molecular motion of a gas is temperature dependent, at the low temperatures of the Jovian planets even the lightest gases are unlikely to acquire the speed needed to escape. On the other hand, a comparatively warm body with a small mass, like our moon, is unable to hold even the heaviest gas and thus lacks an atmosphere. The slightly larger terrestrial planets of Earth, Venus, and Mars retain some heavy gases (as compared to hydrogen), but even their atmospheres make up only an infinitesimally small portion of their total mass.

It is hypothesized that the primordial cloud of dust and gas from which all the planets are thought to have condensed had a composition somewhat similar to that of Jupiter. However, unlike Jupiter, the terrestrial planets are nearly void of light gases and ices. Were the terrestrial planets once much larger? Did they contain these materials but lose them because of their close proximity to the sun? In the following section we will consider the evolutionary histories of these two diverse groups of planets in an attempt to answer these questions.

ORIGIN AND EVOLUTION OF THE PLANETS

The orderly revolution of all nine planets along the sun's equatorial plane leads most astronomers to conclude that the planets formed at essentially the same time and from the same primordial material as the sun. This **nebular hypothesis** suggests that all bodies of the solar system formed from an enormous nebular cloud consisting of approximately 80 percent hydrogen, 15 percent helium, and a few percent of all the other heavier elements known to exist (Figure 20.1). The heavier substances in this frigid cloud of dust and gases consisted mostly of elements such as

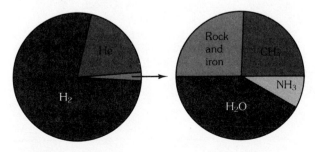

FIGURE 20.1
Composition of the primordial cloud of dust and gases from which the solar system is thought to have evolved. (Data from W. B. Hubbard)

silicon, aluminum, iron, and calcium—the substances of the common rocky materials. Also prevalent were the other familiar elements, including oxygen, carbon, and nitrogen. Astronomical studies indicate that these latter substances existed in a variety of organic compounds. Thus, the stuff out of which life has formed was present long before the solar system came to be.

About 5 billion years ago, and for reasons not yet fully understood, this huge cloud of minute rocky fragments and gases began to contract under its own gravitational influence. The contracting clump of material is assumed to have had some component of rotational motion, which rotated faster and faster as it gravitationally contracted. This rotation caused the nebular cloud to assume a disk-like shape. Within this rotating disk, relatively small eddy-like contractions formed the nuclei from which the planets would eventually develop. However, the greatest concentration of material was gravitationally pulled toward the center, forming the *protosun*.

As more and more of this gas plunged inward, the temperature of the central mass continued to increase. The nebular material located near the protosun reached temperatures of several thousand degrees and was completely vaporized. However, at distances beyond the orbit of Mars, the temperatures probably always remained very low. Here, at −200°C, the dust fragments were most likely covered with a thick layer of water ice, and ices of carbon dioxide, ammonia, and methane. The disk-shaped cloud also contained appreciable amounts of the lighter gases, namely hydrogen and helium, which had not been consumed by the protosun.

In a relatively short time after the protosun formed, the temperature in the inner portion of the nebula dropped significantly. This temperature decrease caused those substances with high melting points to condense into perhaps sand-sized particles. Materials such as iron and nickel solidified first. Next to condense were the elements of which the rock-forming minerals are composed. As these fragments collided they joined into larger asteroid-sized objects which in a few tens of millions of years accreted into the four inner planets we call Mercury, Venus, Earth, and Mars. As more and more of the nebular debris was swept up by these *protoplanets*, the inner solar system began to clear, allowing sunlight to heat the planets' surfaces. Due to their relatively high temperatures and weak gravitational fields, the inner planets were unable to accumulate an appreciable amount of the lighter components of this nebular cloud. These materials, namely hydrogen, ammonia, methane, and water, were eventually wisked from the inner solar system by the solar winds.

Shortly after the four terrestrial planets formed, the decay of radioactive isotopes within them plus the heat from the colliding particles produced at least some melting of the planets' interiors. Melting, in turn, caused the heavier elements, principally iron and nickel, to sink, while the lighter silicate minerals floated upward. During this period of chemical differentiation, gaseous materials were allowed to escape from the planets' interiors, much like what happens during a volcanic event on Earth. The hottest and second smallest planet, Mercury, was unable to retain

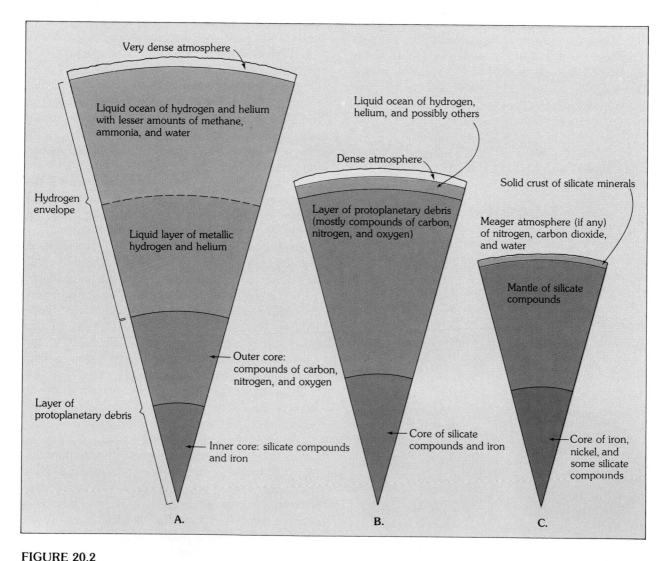

FIGURE 20.2
Idealized models for the internal structure of the Jovian and terrestrial planets. **A.** Jupiter and Saturn. **B.** Uranus and Neptune. **C.** Terrestrial planets. (Data from W. B. Hubbard et al)

even the heaviest of these gases. Mars, on the other hand, being slightly larger and cooler than Mercury, retained a thin layer of carbon dioxide and some water in the form of ice. The largest of the terrestrial planets, Venus and Earth, have surface gravitations strong enough to retain a substantial amount of the heavier gases. However, when compared to the four Jovian planets, even the atmospheres of these planets must be looked upon as meager at best.

At the same time that the terrestrial planets were forming, the larger Jovian planets, along with their extensive satellite systems, were also developing. However, because of the frigid temperatures existing far from the sun, the fragments out of which these planets formed contained a high percentage of ices— water, carbon dioxide, ammonia, and methane. Perhaps by random chance, two of the outer planets, Jupiter and Saturn, grew many times larger (by mass) than Uranus and Neptune. For comparison, Jupiter is 318 and Saturn 95 times more massive than the earth. However, Uranus and Neptune have masses about 14 and 17 times greater, respectively, than the earth. When Jupiter and Saturn reached a certain size, estimated to be about 10 earth masses, their surface gravitation was sufficient to attract and hold even the lightest materials—hydrogen and helium. It is thought that these gases gravitationally collapsed onto these large protoplanets as they swept through their region of the solar system. Thus, much of their size is attributable to the large envelope of light elements, which exists as a dense liquid below a thick hydrogen-rich atmosphere. Jupiter and Saturn therefore consist of a central core of ices and rock, and a much larger outer envelope containing mostly hydrogen and helium (Figure 20.2).

By contrast, the smaller Jovian planets, Uranus and Neptune, grew more slowly and contain proportionately much smaller amounts of hydrogen and helium. Nevertheless, hydrogen, methane, and ammonia are still the major constituents of their dense atmospheres and perhaps a thin outer ocean of hydrogen exists on these planets as well. Thus, Uranus and Neptune are proposed to have a small rocky-iron core and a large mantle of water, ammonia, and methane surrounded by a thin ocean of liquid hydrogen (Figure 20.2). Consequently, these planets structurally resemble Jupiter and Saturn without their large hydrogen-helium envelopes.

In many respects the development of the outer planets with their large satellite systems roughly parallels the events which formed the solar system as a whole. Like their parents, the satellites of the outer planets are composed primarily of icy materials with lesser amounts of rocky substances. However, because of their small size, they could not contain appreciable amounts of hydrogen and helium.

In the remainder of this chapter, we will consider each planet in more detail, as well as some minor members of the solar system. First, however, a discussion of the moon, the earth's companion in space, is appropriate.

THE MOON

Only one natural satellite, the moon, accompanies the earth on its annual flight around the sun. This planet-satellite system is unique in the solar system, because the moon is unusually large compared to its parent planet. The diameter of the moon is 3475 kilometers, and from the calculation of its mass, its density is 3.3 times that of water. This density is comparable to that of crustal rocks on earth but a fair amount less than the earth's average density. Geologists have suggested that this difference can be accounted for if the moon's iron core is rather small. The gravitational attraction at the lunar surface is one-sixth of that experienced on the earth's surface. This difference allows an astronaut to lift a "heavy" life-support system with relative ease. When not carrying such a load, an astronaut could jump six times higher than on Earth, easily clearing a one-story building.

THE LUNAR SURFACE

When Galileo first pointed his telescope toward the moon, he saw two different types of terrain (Figure 20.3). The dark areas he observed are now known to be fairly smooth lowlands, while the bright regions are densely cratered highlands. Because the dark regions resembled seas on earth, they were later named **maria** (singular, *mare:* Latin for "sea"). This name is unfortunate because the moon's surface is totally void of water.

Today we know that the moon has no atmosphere, and lacks water as well. Therefore, the

FIGURE 20.3
Telescopic view of the lunar surface. (Courtesy of Lick Observatory)

processes of weathering and erosion which continually modify the earth are virtually lacking. In addition, tectonic events such as earthquakes and volcanic eruptions do not occur on the moon. However, since the moon is unprotected by an atmosphere, tiny particles (micrometeorites) continually bombard its surface and ever so gradually smooth the landscape. Rocks, for example, can become slightly rounded on top if exposed at the lunar surface for a long enough period. Nevertheless, it is unlikely, except for the addition of a few large craters, that the moon has changed appreciably in the last 3 billion years.

The most obvious features of the lunar surface are craters. They are so profuse that craters within craters within craters are the rule. The larger ones seen in the lower portion of Figure 20.3 are about 250 kilometers (150 miles) in diameter, and these often overlap. Most craters were produced by the impact of rapidly moving debris (meteoroids), which was considerably more abundant in the early history of the solar system than it is today. By contrast, the earth has only about a dozen recognized impact craters. This difference can be attributed to the earth's atmosphere, which burns up small debris before it reaches the ground. In

516

addition, evidence for many of the craters which did form early in the earth's history has since been destroyed by erosion or tectonic processes.

The formation of an impact crater is illustrated in Figure 20.4. Upon impact, the high-speed particle compresses the material it strikes, then almost instantaneously the compressed rock rebounds, ejecting material from the crater. This process is analogous to the splash that occurs when a rock is dropped into water, and it often results in the formation of a central peak as seen in the large crater in Figure 20.5. Most of the ejected material *(ejecta)* lands near the crater, building a rim around it. The heat generated by the impact is sufficient to melt some of the impacted rock. Astronauts have brought back samples of glass beads produced in this manner, as well as rock formed when angular fragments and dust were welded together by the impact. The latter material is called **lunar breccia.**

A meteoroid only 3 meters wide can blast out a 150-meter (500-foot) crater. A few of the large craters such as Kepler and Copernicus, shown in Figure 20.3, formed from the impact of bodies a kilometer, or more, in diameter. These two large craters are thought to be relatively young because of the bright **rays** ("splash" marks) that radiate outward for hundreds of kilometers. These bright rays consist of fine debris ejected from the primary crater, including impact-generated glass beads, as well as material displaced during the formation of smaller, secondary craters.

The densely pock-marked highland areas make up most of the lunar surface. In fact, all of the "back" side of the moon is characterized by such topography. Within the highland regions are mountain ranges that have been named for mountainous terrains on earth. The highest lunar peaks reach

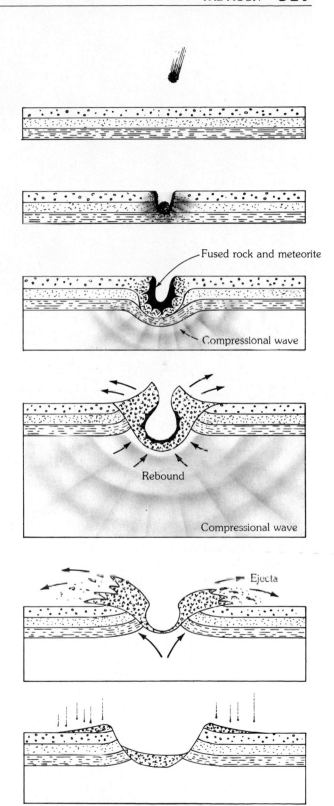

FIGURE 20.4
Formation of an impact crater. The energy of the rapidly moving meteoroid is transferred into heat energy and compressional waves. The rebound of the compressed rock causes debris to be ejected from the crater, and the heat melts some material, producing glass beads. Small secondary craters are formed by the material "splashed" from the impact crater. (After E. M. Shoemaker)

Crater ray

Secondary crater chain

Central peak

Continuous ejecta

Discontinuous ejecta

FIGURE 20.5
The 20-kilometer wide lunar crater Euler located in the southwestern Mare Imbrium. Clearly visible are the bright rays, central peak, secondary craters, and the large accumulation of ejecta near the crater rim. (Courtesy of NASA)

elevations approaching 8 kilometers, only 1 kilometer less than Mount Everest.

Although highlands predominate, the less rugged maria have attracted most of the interest. The origin of maria basins as enormous impact craters produced by the violent impact of at least a dozen asteroid-sized bodies was hypothesized before the turn of the century by the noted American geologist G. K. Gilbert (Figure 20.6A). However, it remained for the *Apollo* missions to determine what filled these depressions to produce the relatively flat topography. Apparently, the craters were flooded, layer upon layer, by very fluid basaltic lava, in which case they somewhat resemble the Columbia Plateau in the northwestern United States (Figure 20.6B). The photograph in Figure 20.7 reveals the forward edge of a lava flow that is "frozen" in place. Astronauts have also viewed and

photographed the layered nature of maria. The layers are often over 30 meters thick, and the total thickness of the material that fills the maria must approach thousands of meters.

On several occasions the lava flowed beyond the impact crater, engulfing the surrounding lowlands. If the rim of a remnant crater can be seen above the lava, an estimate of the flow's thickness can be made. Many geologists believe that the impacts that produced the maria basins were great enough to fracture the lunar crust some distance away. Examples of basins in which all of the disturbed regions were filled to overflowing include Mare Tranquillitatis (Sea of Tranquility), the site where man first put a footprint on lunar soil, and Mare Imbrium (Sea of Rains). Some maria basalts fill only the central crater; these appear as dark, smooth crater floors in Figure 20.3.

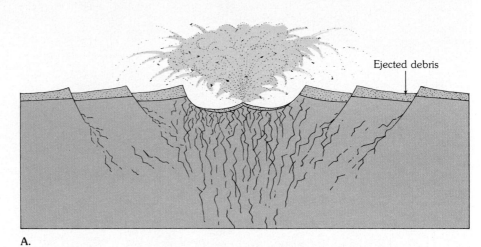

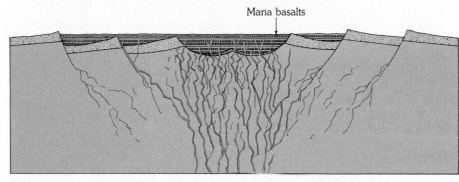

FIGURE 20.6

Formation of lunar maria.
A. Impact of an asteroid-sized mass produced a huge crater hundreds of kilometers in diameter and disturbed the lunar crust beyond the crater.
B. Filling of the impact area with fluid basalts, perhaps derived from partial melting deep within the lunar mantle.

A.

Maria basalts

Ejected debris

B.

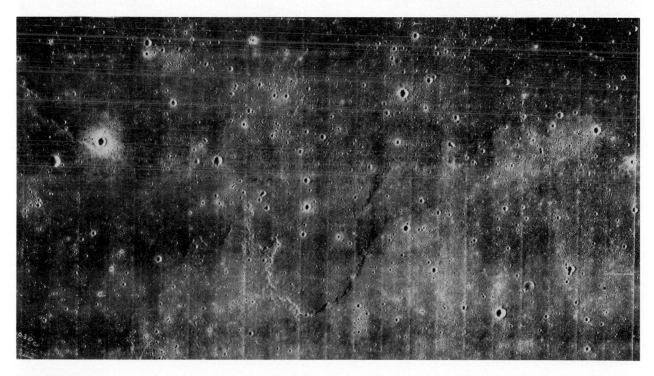

FIGURE 20.7

Margin of a lava flow on the surface of Mare Imbrium. (Courtesy of National Space Data Center)

519

FIGURE 20.8
Astronaut Harrison Schmitt sampling the lunar surface. Notice the footprints left in the lunar "soil." (Courtesy of NASA)

All lunar terrains are mantled with a layer of gray, unconsolidated debris derived from a few billion years of meteoric bombardment (Figure 20.8). This soil-like layer, properly called **lunar regolith,** is composed of igneous rocks, breccia, glass beads, and fine particles commonly called *lunar dust.* As meteoroid after meteoroid collided with the lunar surface, the thickness of the regolith increased, while the size of the bombarded debris diminished. In the maria that have been explored by *Apollo* astronauts, the lunar regolith is apparently just over 3 meters thick, but it is believed to form a thicker mantle upon the older highlands.

LUNAR HISTORY

Planetary geologists have been able to work out some of the general details of the moon's history,

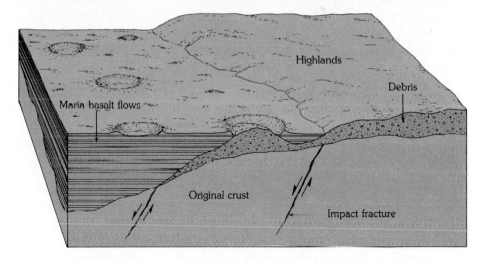

FIGURE 20.9
Mare Imbrium and associated mountainous region. A few isolated peaks are standing above the lava plain, and numerous craters dot the highlands and adjacent mare. (Photo courtesy of NASA)

using among other things variations in crater density (quantity per unit area). Simply stated, the higher the crater count, the longer that topographic feature has been in existence.

Although the early history of the moon can only be hypothesized, it most likely paralleled that of the earth and the other planets of the solar system. Originally, the moon was much smaller, but as it moved through the nebula it swept up debris and grew by accretion. Continuous bombardment and perhaps radioactive decay generated enough heat to melt the moon's outer shell, and quite possibly the rest of the moon as well. When a large percentage of the debris

had been gathered, the outer layer of the moon began to cool and form a crystalline crust. From samples obtained by *Apollo* astronauts, the rocks of the primitive lunar crust are thought to be composed of a high percentage of a calcium-rich feldspar (anorthosite). This feldspar mineral crystallized early and, because it was less dense than the remaining melt, floated to the top and formed a surface scum. While this process was taking place, iron and other heavy metals probably sank to form a small central core. Even after the crust had solidified, its surface was continually bombarded. Remnants of the original crust occupy the densely cratered highlands, which have been estimated to be as much as 4.5 billion years old.

The last period of heavy bombardment recorded in the lunar highlands occurred almost 500 million years after the crust had formed. It is not known with certainty whether this final episode of bombardment was simply a clean-up phase where the remaining large particles in the earth-moon orbit were swept up or whether it was an influx of bodies from farther out in the solar system.

The next major event in the moon's evolution was the formation of maria basins (Figure 20.9). The meteoroids that produced these huge pits ejected mountainous quantities of lunar rock into piles rising 5 kilometers or more. The Apennine mountain range, which typifies such an accumulation, was produced in conjunction with the formation of the Imbrium basin, the site of the exploration conducted by the *Apollo 15* astronauts. The crater density of the ejected material is greater than that of the surface of the associated mare, confirming that an appreciable time elapsed between the formation and filling of these basins. Radiometric dating of the maria basalts puts their age between 3.2 and 3.8 billion years, somewhat younger than the initial crust. In places, the lava flows overlap the highlands, another testimonial to the lesser age of the maria deposits.

The last prominent features to form on the lunar surface were the rayed craters, as exemplified by the crater Copernicus (Figure 20.3). Material ejected from these "young" depressions is clearly seen blanketing the surface of the maria and many older rayless craters. By contrast, the older craters have rounded rims, and their rays have been erased by the impact of small debris. However, even a relatively young crater like Copernicus must be millions of years old. Had it formed on the earth, erosional forces would have long since removed it.

Evidence carried back from the lunar landings indicates that most, if not all, of the moon's tectonic activity ceased about 3 billion years ago. The youngest maria lava flows are about equivalent in age to the oldest rocks so far discovered on the earth. If photos of the moon taken several hundreds of millions of years ago were available, they would reveal that the moon has changed little in the intervening years. By all standards of measure the moon is a dead body wandering lifelessly through space and time.

MERCURY: THE INNERMOST PLANET

Mercury, the innermost and swiftest planet, has a diameter of 4878 kilometers, not much larger than the moon. Also like the moon, it absorbs most of the sunlight that strikes it, reflecting only 6 percent into space. This is characteristic of a terrestrial body without an atmosphere. By way of comparison one of the icy moons of Saturn reflects more than 60 percent of the light which it encounters.

Mercury's close proximity to the sun makes viewing from earthbound telescopes difficult at best. The first good glimpse of this planet came in the spring of 1974, when *Mariner 10* passed within 800 kilometers of its surface (Table 20.2). Its striking resemblance to the moon was immediately evident from the high-resolution images that were radioed back. In fact, the similarity was so great that a project scientist remarked that these photos could be substituted for ones of the back side of the moon and most laymen would not know the difference.

Mercury has densely cratered highlands much like the moon and vast smooth terrains which resemble maria. However, unlike the moon, Mercury is a very dense planet, which implies that this planet contains an iron core perhaps larger than the earth's. In addition, Mercury has very long scarps that cut across the plains and craters alike. One proposal is that these scarps resulted from crustal shortening as the planet cooled and shrank.

A typical period of daylight and darkness on Mercury lasts 88 days. This causes the night temperatures to drop as low as $-173°C$ ($-280°F$) and the noontime temperatures to exceed $427°C$ ($800°F$), hot enough to melt tin and lead. Mercury has the greatest temperature extremes of any planet. The odds of life as we know it existing on Mercury are nil.

VENUS: THE VEILED PLANET

Venus, second only to the moon in brilliance in the night sky, is named for the goddess of love and beauty. It orbits the sun in a nearly perfect circle once every 225 days. Because Venus is similar to the earth in size, density, and mass, and is located in the same part of the solar system it has been referred to as "Earth's twin." Because of these similarities, it is

TABLE 20.2
Summary of significant space probes.

Mariner 2	1962	Fly-by of Venus (first to any planet)
Mariner 4, 6, 7	1965 1969 1969	Fly-by missions to Mars
Mariner 9	1971	Orbiter of Mars
Apollo 8	1968	Astronauts circled the moon and returned to the earth
Apollo 11	1969	First astronaut landed on the moon
Apollo 17	1972	Last of six Apollo missions to carry man to the moon
Mariner 10	1974	Orbited the sun, allowing it to pass Mercury several times; fly-by mission to Venus
Pioneer 10, 11	1973 1974	First close-up views of Jupiter
Venera 8, 9, 10	1972 1975	Soviet landers on Venus (operative about one hour each)
Venera 13, 14	1982	First color images of Venus
Viking 1, 2	1976	Orbiters and landers of Mars
Voyager 1	1979 1980	Fly-by of Jupiter Fly-by of Saturn
Voyager 2	1979 1981 1986 1989	Fly-by of Jupiter Fly-by of Saturn Fly-by of Uranus Fly-by of Neptune
Galileo[1]	1986	Orbiter to Jupiter
VOIR[2]	?	Radar imaging orbiter to Venus

[1] In the developmental stage
[2] Proposed

hoped a detailed study of Venus will provide geologists with a better understanding of the earth's evolutionary history.

Although Venus is shrouded in a thick cloud cover, radar mapping by the Pioneer/Venus orbiter, as well as by earthbound instruments, has revealed a rather varied topography with features somewhat between those of Earth and Mars. Simply, radar impulses are sent toward the Venusian surface and the heights of features such as plateaus and mountains are measured by timing the return of the radar echo. A high percentage of the Venusian surface consists of a rather subdued rolling topography (Figure 20.10). Within this area of rolling plains are several large valleys believed to have a tectonic origin. This region also has the highest density of circular features, which most probably are impact craters. Only 8 percent of the surface consists of highlands that may be likened to continental areas on Earth. The highlands of Venus are dominated by large plateau-like structures; however, discrete mountain ranges, including Maxwell Montes, which rises 11 kilometers above the lowlands, also exist. In addition, two topographic features rising nearly 5 kilometers above the adjacent terrain resemble large shield volcanoes like those found on Earth and Mars.

The resolution obtained from the Pioneer orbiter was rather low; consequently, only the largest features have been mapped. However, more detailed information about the surface of Venus may eventually be provided by a more sophisticated space probe (VOIR) that is currently being designed.

Before the advent of space vehicles, Venus and Mars were considered the most hospitable sites in the solar system (Earth excluded) and the logical places to expect living organisms. Unfortunately, evidence from Mariner fly-by space probes and Soviet unmanned landings on Venus indicate differently. The surface of Venus reaches temperatures of 480°C (900°F), and the Venusian atmosphere is composed mostly of carbon dioxide (97 percent). Only minor amounts of water vapor and nitrogen have been detected. Its atmosphere contains an opaque cloud deck about 25 kilometers thick which begins approximately 70 kilometers from the surface. Although the unmanned Soviet Venera 8 survived less than an hour on the Venusian surface, it determined that the atmospheric pressure on this planet is 90 times greater than that found on the earth's surface. This hostile environment makes it unlikely that life as we know it exists on Venus and makes manned space flights to Venus improbable in the foreseeable future.

Later Venera missions transmitted a few images of Venus before they too succumbed to the tremendous heat and pressure. The landing site of Venera 9 showed numerous angular rocks, a setting similar to what the Viking lander found on Mars. Venera 10, on the other hand, photographed rather flat, rounded rocks, possibly the result of longer exposure to

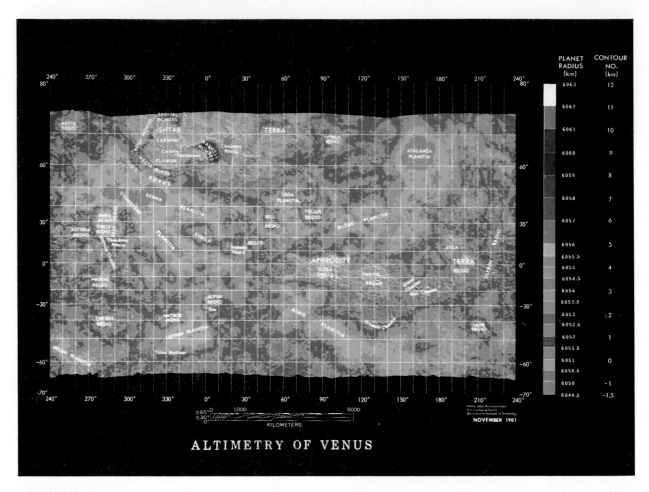

FIGURE 20.10
Map of the topography of Venus generated by computer and radar altimeter data from the *Pioneer/Venus* spacecraft. (Courtesy of U.S. Geological Survey)

erosion than those found by *Venera 9*. Soil measurements taken by the *Venera* landers indicated a composition similar to the basalts found on the earth, the moon, and Mars.

Carl Sagan, one of the foremost experts on extraterrestrial life, in a description of planetary environments, called Earth "the Heaven of the solar system" and Venus "the Hell." Why should the environments of two planets that are nearly the same size, which evolved in a similar manner, and are located in close proximity to one another, be so dramatically different as to foster Sagan's remarks? The primary reason is the blistering temperature of the Venusian atmo-

sphere, which is caused by a runaway greenhouse effect. The **greenhouse effect** is also experienced on Earth on a sunny day by anyone entering a greenhouse or a car with its windows rolled up. The dense Venusian atmosphere acts like the glass in a greenhouse, allowing visible solar energy to penetrate to the surface. Planets typically lose energy at nearly the same rate as it is received by reradiating it back to space. However, the dense Venusian atmosphere, like the glass in the greenhouse, is quite opaque to the outgoing radiation. Thus, the surface of Venus is very hot because the heat is temporarily trapped.

Carbon dioxide, which makes up 97 percent of

the Venusian atmosphere, is the main contributor to the greenhouse effect. The primary source of carbon dioxide is outgassing during volcanic eruptions. Why does the earth, which has numerous active volcanoes, have an atmosphere which consists of only a small fraction of one percent carbon dioxide? The answer can be found in the earth's abundant plant life. Plants use carbon dioxide and water in photosynthesis to generate organic matter, while oxygen is released as a by-product. Thus, over millions of years plant life has altered our atmosphere, making it carbon dioxide-poor and oxygen-rich. In addition, carbon dioxide dissolved in seawater is used by a vast number of organisms for the production of carbonate shells. These shells are eventually deposited as sediment on the ocean floor. Consequently, huge quantities of carbon dioxide from the earth's atmosphere are continually being converted into organic matter and carbonate sediments. Apparently, Venus does not possess living organisms or a chemical mechanism capable of removing atmospheric carbon dioxide. Hence, our neighboring planet has a carbon dioxide concentration that has reached extreme proportions.

We should expect that studies of the Venusian atmosphere will provide some insight into how the earth's atmosphere evolved and how it will respond to changes induced by human activity. This insight could keep our planet from becoming, as some have suggested, "hothouse earth."

MARS: THE RED PLANET

Mars has evoked greater interest than any other planet, for astronomers as well as for laymen. When we think of intelligent life on other worlds, "little green Martians" come to mind. Interest in Mars stems mainly from this planet's accessibility to observation. All other planets within telescopic range have their surfaces hidden by clouds, except for Mercury, whose nearness to the sun makes viewing difficult. Through the telescope, Mars appears as a reddish ball interrupted by some permanent dark regions that change in intensity during the Martian year. The most prominent telescopic features of Mars are its brilliant white polar caps.

The dramatic photographs of Mars radioed back from the first successful fly-by probe, *Mariner 4,* revealed a crater-pocked planet more akin to the moon than to the earth. As it turned out, these images were taken of the highly cratered southern hemisphere which exhibits a topography sharply different from that of the "younger" northern hemisphere. Follow-up missions revealed a dynamic planet which at some time in its history experienced volcanic activity, wind erosion, and crustal movements with associated "marsquakes." Further, *Viking* orbital images provided unmistakable evidence for erosion caused by flowing water.

The Martian atmosphere is only 1 percent as dense as that of the earth, much less than that found at the top of Mount Everest, and is composed primarily of carbon dioxide with only very small amounts of water vapor. Data radioed back to earth confirmed speculations that the polar caps of Mars are made of water ice covered by a relatively thin layer of frozen carbon dioxide. As the winter nears in a given hemisphere, we see the equatorward growth of that ice cap as additional carbon dioxide is deposited. This finding is compatible with the temperatures observed at the polar caps. Here temperatures reach a chilly −125°C, cold enough to solidify carbon dioxide.

Although the atmosphere of Mars is very thin, extensive dust storms do occur and may be responsible for the color changes observed from earth-based telescopes. Winds up to 270 kilometers (170 miles) per hour can persist for weeks. Images radioed back by *Viking 1* revealed a Martian landscape remarkably similar to a rocky desert on Earth (Figure 20.11). Sand dunes are abundant, and many Martian impact craters have flat bottoms because they are partially filled with dust. Thus, unlike the craters of the moon, the oldest craters on Mars are completely obscured by deposits of windblown material.

When *Mariner 9,* the first artificial satellite to orbit another planet, reached Mars in 1971, a large dust storm was raging, and only the ice caps were initially observable. It was spring in the southern hemisphere, and that polar cap was receding rapidly. About midsummer, when the rate of evaporation should have been greatest, the size of the polar cap showed litte change. A residual cap remained through the summer, providing strong evidence that the residual polar

FIGURE 20.11
This spectacular picture of the Martian landscape by the *Viking 1* lander shows a dune field with features remarkably similar to many seen in the deserts of the earth. The dune crests indicate that recent wind storms were capable of moving sand over the dunes in the direction from upper left to lower right. The large boulder at the left is about 10 meters from the spacecraft and measures 1 by 3 meters. (Courtesy of NASA)

cap is made of water ice. (Water ice has a much lower rate of evaporation than frozen carbon dioxide.)

When the dust cleared, images of the northern hemisphere revealed numerous large volcanoes. The largest, called Mons Olympus, covers an area the size of the state of Ohio and is no less than 23 kilometers (75,000 feet) in elevation, over twice as high as any mountain on the earth. This gigantic volcano and others that were observed closely, resemble the shield volcanoes found on earth, being similar to those of Hawaii (Figure 20.12). Their extreme size is thought to result from the absence of plate movements on Mars. Therefore, rather than a chain of volcanoes forming as we find in Hawaii, single, larger cones developed.

Impact craters are notably less abundant in the region where the volcanic activity appears most prevalent. This indicates that at least some of the volcanic topography formed more recently in that planet's his-

tory, somewhat after the early period of heavy bombardment. Nevertheless, age determinations based on crater densities obtained from *Viking* photographs indicate that most Martian surface features are old by earth standards. The highly cratered Martian southern hemisphere is probably similar in age to the comparable lunar highlands (3.5–4.5 billion years old). The discovery of several highly cratered and weathered volcanoes on Mars further indicates that volcanic activity began early and had a long history. However, even the relatively fresh-appearing volcanic features of the northern hemisphere may be older than 1 billion years. This fact, coupled with the absence of "marsquake" recordings by *Viking* seismographs, points to a tectonically dead planet.

Another surprising find made by *Mariner 9* was the existence of several canyons, which dwarf even the Grand Canyon of the Colorado. One of the largest, Valles Marineris, is roughly 6 kilometers deep, up to

A.

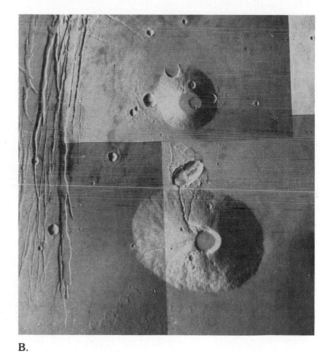

B.

FIGURE 20.12
Inactive shield volcanoes of Mars. **A.** Mons Olympus, a truly gigantic volcanic mountain. **B.** Ceranius Tholus (lower) and Uranius Tholus (upper) are volcanoes that have been flooded by somewhat younger lava flows. (Courtesy of NASA)

160 kilometers wide, and extends for almost 5000 kilometers along the Martian equator (Figure 20.13). This vast chasm is thought to have formed by slippage of crustal material along huge faults in the crustal layer. In this respect, it would be comparable to the rift valleys of Africa.

Not all valleys on Mars have a tectonic origin. Many Martian valleys have tributaries exhibiting drainage patterns similar to those of stream valleys found on Earth. In addition, *Viking* orbiter photographs have also revealed streamlined features that are unmistakably ancient islands located in what is

FIGURE 20.13
The vast canyon of Mars called Valles Marineris is 5000 kilometers long, nearly the distance across the United States. Notice the huge landslides on the canyon's far (upper) wall which have acted to enlarge the structure. On the opposite (lower) wall, smaller valleys form a branching pattern which has dissected a portion of the flat plateau. (Courtesy of NASA)

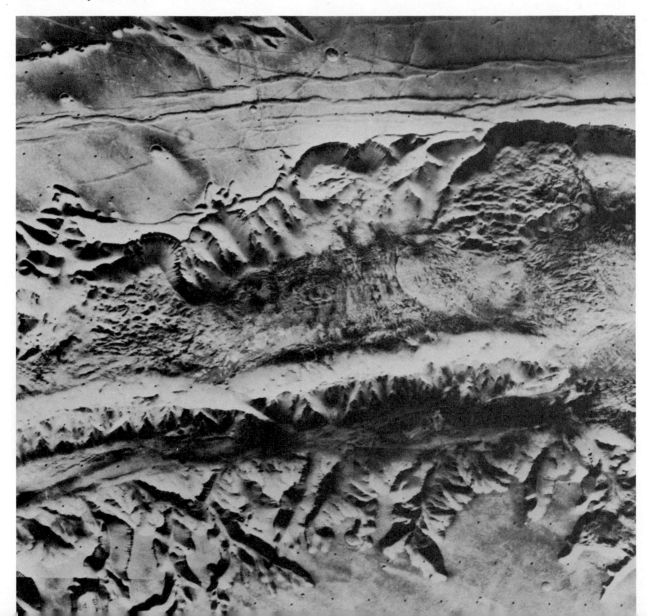

now a dry stream bed (Figure 20.14). When these streamlike channels were first discovered, some observers speculated that a thick water-laden atmosphere capable of generating torrential downpours once existed on Mars. But what happened to this water? The present Martian atmosphere contains

FIGURE 20.14
Stream-like channels discovered by *Viking* cameras. Note the teardrop-shaped features that are unmistakably ancient islands in what is now a dry stream bed. (Courtesy of NASA)

only trace amounts of water. Moreover, the environment of Mars is far too harsh to allow water to exist as a liquid. Despite these difficulties, the work of flowing water still remains the most acceptable explanation for many of the Martian channels.

Many planetary geologists do not accept the premise that Mars once had an active water cycle similar to that on Earth. Rather they believe that many of the large streamlike valleys were created by the collapse of surface material caused by the slow melting of subsurface ice (Figure 20.15). If this is the case, these large valleys would be more akin to features formed by mass wasting processes on the earth. Some of the smaller channels, they contend, have been cut by the gradual release of subsurface water, which flowed out of the ground like springs. Further, the heat from volcanic activity or meteoroid impact may have rapidly melted subsurface ice which, in turn, caused local flooding. These floods may have carved some of the channels. An occasional flood would also account for the streamlined features found in what are presently dry river beds. The question of whether rainfall, gradual seepage of groundwater, or the collapse of surface material created the branching valleys of Mars will be debated for some time. The answer might not come until we obtain images of Mars which reveal greater detail than those presently available.

Because of their small size, the two satellites of Mars, Phobos and Deimos, were not discovered until 1877. Phobos is closer to its parent than any other natural satellite. Only 5500 kilometers from the Martian surface, Phobos requires just 7 hours and 39 minutes for one revolution. Deimos, which is smaller and 20,000 kilometers away, revolves in 30 hours and 18 minutes. *Mariner 9* revealed that both satellites are irregularly shaped and have numerous impact craters, much like their parent. The maximum diameter of Phobos is 24 kilometers and the maximum diameter of Deimos is only about 15 kilometers. Undoubtedly, these moons are asteroids which were captured by Mars. One of the most interesting coincidences in astronomy is the close resemblance between Phobos and Deimos and the two fictional satellites of Mars described by Jonathan Swift in *Gulliver's Travels,* written about 150 years before they were actually discovered.

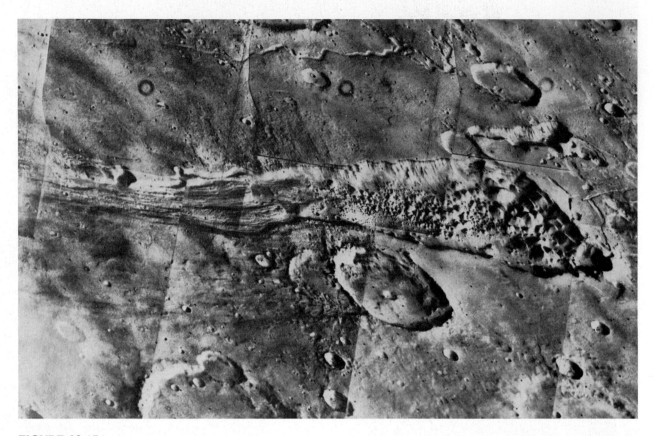

FIGURE 20.15
Extending across the middle of this photograph is a "stream" channel that most probably formed as a result of the melting of subsurface ice. As the ice melted, the unsupported land above collapsed to produce the hummocky topography to the right. The flood of meltwater carved the channel to the left. (Courtesy of NASA)

JUPITER: THE LORD OF THE HEAVENS

Jupiter, the largest planet in our solar system, has a mass 2½ times greater than the combined mass of all the remaining planets, satellites, and asteroids. It is truly a giant among planets. In fact, had Jupiter been about ten times larger, it would have evolved into a small star. Despite its great size, however, it is only $1/1000$ as massive as the sun. Jupiter also rotates more rapidly than any other planet, completing one rotation in slightly less than ten hours. The effect of Jupiter's rapid rotation causes the equatorial region to be slightly bulged and the polar dimension to be flattened.

When viewed through a telescope or binoculars, Jupiter appears to be covered with alternating bands of multicolored clouds aligned parallel to its equator. The most striking feature on the disk of Jupiter is the great red spot (Figure 20.16). Although its color varies greatly in intensity, it has been an ever-present feature since it was first discovered more than three

FIGURE 20.16
A mosaic of Jupiter and its four Galilean satellites as seen from *Voyager 1*. Ganymede is located at lower left, Callisto at lower right, Europa is adjacent to Jupiter, and Io appears to the left in the distance. (Courtesy of NASA)

centuries ago. When *Voyager 2* swept by Jupiter in July, 1979, the size of the red spot was 11,000 kilometers by 22,000 kilometers, which equals two earth-sized circles placed side by side. On occasion it has grown even larger. Although the great red spot varies in size, it does remain the same distance from the Jovian equator. The cause of the spot has been attributed to everything from volcanic activity to a large cyclonic storm. Images obtrained by *Pioneer 11* as it moved to within 42,000 kilometers of Jupiter's cloud tops in December, 1974, support the latter view. The great red spot apparently is a counterclockwise-rotating storm caught between two jetstream-like bands of atmosphere flowing in opposite directions. This huge hurricane-like storm rotates once every 12 earth days. Although several smaller storms have been observed in other regions of Jupiter's atmosphere, none has survived for more than a few days.

STRUCTURE OF JUPITER

Jupiter's atmosphere is composed primarily of hydrogen and helium, with methane, ammonia, water, and sulfur compounds as minor constituents. The wind systems generate the light and dark colored bands that encircle this giant. Unlike the winds on Earth, which are driven by solar energy, Jupiter gives off nearly twice as much heat as it receives from the sun. Thus, it is the heat emanating from Jupiter's interior that produces huge convection currents in the atmosphere (Figure 20.17). The light colored zones are regions where gases are ascending and cooling. The light color is thought to be produced by ammonia crystallizing into "snowflakes" near the cloud tops where temperatures of −120°C have been measured. The cold material from the light zones spills over onto the lower dark belts, where air is descending and heating. The reason for the rusty-brown color of the dark belts is not known for certain; however,

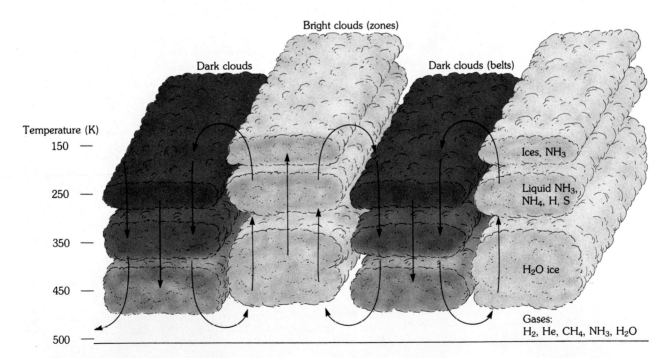

FIGURE 20.17
The structure of Jupiter's atmosphere. The bright zones are clouds composed of ice particles located higher in the atmosphere than the relatively warm clouds of the dark bands. The dark color of the belts may be caused by sulfur compounds or perhaps organic molecules.

they may be colored by sulfur compounds.

Atmospheric pressure at the top of the clouds is equal to sea-level pressure on Earth, but because of Jupiter's immense gravity, the pressure increases rapidly toward its surface. At 1000 kilometers below the clouds, the pressure is great enough to liquify hydrogen. Consequently, the surface of Jupiter is thought to be a gigantic ocean of liquid hydrogen. Less than halfway into Jupiter's interior, pressures of unimaginable magnitude cause liquid hydrogen to turn into liquid metallic hydrogen. Although scientists have never seen this unusual material, its properties are thought to be predictable. Jupiter is also believed to contain as much rocky and metallic material as is found in the terrestrial planets, such as Earth. Whether this "earthy" material is scattered as small bits or makes up a central core is not known with certainty, but the latter idea is generally accepted.

JUPITER'S MOONS

Jupiter's satellite system, consisting of fifteen moons, resembles a miniature solar system. The four largest satellites were discovered by Galileo and travel in nearly circular orbits around the parent with periods of from two to seventeen days (Figure 20.16). The two largest of the Galilean satellites, Callisto and Ganymede, surpass Mercury in size, while the two smaller ones, Europa and Io, are about the size of the earth's moon. These Galilean moons can be observed with a small telescope and are interesting in their own right. Because their orbits are along Jupiter's equatorial plane and since they all have the same orbital direction, these moons most probably formed from "leftover" debris in much the same way as the planets did. By contrast, the four outermost satellites are very small (20 kilometers in diameter), revolve in a direction that is opposite the other moons, and have orbits that are steeply inclined to the Jovian equator. These satellites appear to be asteroids that passed near enough to be captured gravitationally by Jupiter.

The images obtained by *Voyager 1* and *2* in 1979, revealed to the surprise of almost everyone that each of the four Galilean satellites has a character all its own. The entire surface of Callisto, the outermost of the Galilean satellites, is densely cratered, much like the surfaces of Mercury and the moon. However, in this instance the impacts occurred in a crust which appears to be a dirty, frozen ocean of water ice. The largest features discovered were sets of bright concentric rings surrounding impact craters. These bright rings are thought to resemble the ripples produced when a pebble is dropped into calm water. They are probably composed of blocks of ice, which were deformed and uplifted by the impact that generated the central crater.

Ganymede, the largest Jovian satellite, contains the most diverse terrain. Like our moon, it has densely cratered regions, and other very smooth areas, where a younger icy layer covers the older cratered surface. In addition, Ganymede has numerous parallel grooves. This suggests some type of tectonic activity has occurred in the distant past (Figure 20.18). In some places structural features show evidence of lateral displacement along strike-slip faults,

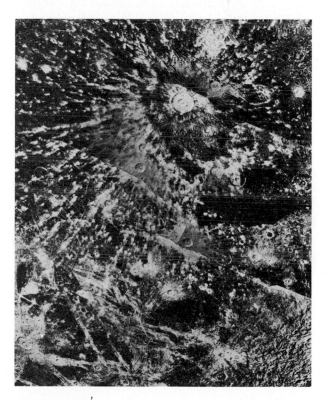

FIGURE 20.18
Ganymede, Jupiter's largest satellite. This view shows numerous impact craters, many with bright rays. Grooved terrain is visible at the bottom center.

a phenomenon previously found only on the earth.

Europa, smallest of the Galilean satellites, has an icy surface that is crisscrossed by many linear features. Although some of these linear markings are thousands of kilometers long, they have very little surface expression. Further, this satellite is notably void of large impact craters. Europa is as smooth as a billiard ball on the same scale. Therefore, the present surface of Europa must have formed sometime after the early period of bombardment, when rocky chunks were far more numerous in the solar system. The crust of Europa may be a thick, frozen ice layer which caps a slushy ocean.

The innermost of the Galilean moons, Io, is the only volcanically active body other than Earth so far discovered in our solar system. To date, eight active sulfurous volcanic centers have been discovered (Figure 20.19). Umbrella-shaped plumes have been seen rising from the surface of Io to heights approaching 200 kilometers. The surface of Io is very colorful, a result of its sulfurous "rocks" which change colors from red to yellow, and from black to white depending on temperature. The source of heat for this volcanic activity is thought to be tidal energy generated by a "tug-of-war" between Jupiter and the Galilean satellites. Since Io is gravitationally locked to Jupiter, the same side always faces this giant. The

gravitational influence of Jupiter and the other nearby satellites pulls and pushes on Io's tidal bulge as its slightly eccentric orbit takes it alternately closer and farther from Jupiter. This gravitational flexing of Io is transformed into heat energy.

One of the most interesting discoveries made by *Voyager 1* was the ring system of Jupiter. Thought to be less than 30 kilometers thick, the apparently continuous ring may extend outward from the surface of Jupiter to a distance equal to twice the diameter of the planet. This ring system is not believed to be like that found on Saturn. Rather than particles held in planetary-type orbits, the particles composing Jupiter's ring appear to be temporarily entrapped by the planet's intense magnetic field. The source of the ring material may be sulfur from the volcanoes of Io, which is ionized and pulled toward Jupiter by its magnetic field.

Years of investigation will be needed to unravel all of the mysteries uncovered by the *Voyager 1* and *2* missions. Even when this task is accomplished, our knowledge of Jupiter and its satellites will be far from complete. Unfortunately, many of the remaining secrets and questions will have to go undiscovered and unanswered for many years because only a single planetbound probe is currently being developed in the United States. The *Galileo* orbiter is expected to drop instrument packages into the Jovian atmosphere no earlier than 1986, if it ever gets operational.

FIGURE 20.19
One of Io's volcanoes, Ra Patera, covered by what appears to be "lava" flows. The central caldera is about 30 kilometers across. (Courtesy of NASA)

SATURN: THE ELEGANT PLANET

Requiring 29½ years to make one revolution, Saturn is almost twice as far from the sun as Jupiter, yet its atmosphere, composition, and internal structure are thought to be remarkably similar. The most prominent feature of Saturn is its system of rings (Figure 20.20). These rings were discovered by Galileo and appeared to him as two smaller bodies adjacent to the planet because he could not resolve them with his primitive telescope. Their ring nature was revealed 50 years later by the Dutch astronomer Christian Huygens. Until the recent discovery that Jupiter and Uranus have very faint ring systems, this spectacular phenomenon was thought to be unique to Saturn.

In 1980 and 1981, fly-by missions of the nuclear-powered *Voyagers 1* and *2* unmanned space vehicles came within one hundred thousand kilometers of the surface of Saturn. More information on Saturn and its satellite system was gained in a few days than had been acquired since Galileo first viewed this elegant planet telescopically in 1610. Some of the information acquired by these space probes is as follows:

1 Saturn's atmosphere is very dynamic with winds roaring at speeds approaching 1500 kilometers per hour.

2 Large cyclonic "storms" similar although much smaller than Jupiter's great red spot occur in Saturn's atmosphere.

3 Verification was made of the existence of the sixth and innermost ring system, and evidence was gathered for a seventh ring system.

4 The icy rings of Saturn were discovered to be more complex than expected. Each of the seven rings is made of numerous ringlets resembling the grooves on a phonograph record, while the rings in the faint *F* ring are intertwined in stable kinked

FIGURE 20.20
A view of Saturn's rings taken from a distance of 8 million kilometers by *Voyager 1*. At this distance almost 100 individual rings can be seen, including several narrow bands in the dark Cassini division. (Courtesy of NASA)

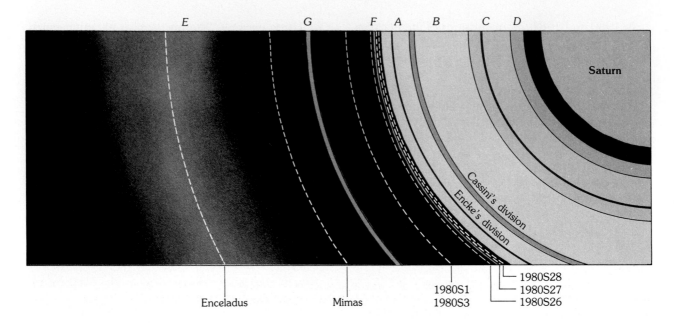

FIGURE 20.21
A polar view of the ring system of Saturn is shown diagrammatically to scale. (From "Rings in the Solar System," by James B. Pollack and Jeffrey N. Cuzzi. Copyright © 1981 by Scientific American, Inc. All rights reserved.)

and braidlike configurations. Further, the *B* ring develops perplexing outwardly radiating spokes which survive for hours at a time (Figure 20.20).

5 Three additional moons were discovered, bringing the total so far detected to *seventeen*. The larger two of these icy worlds, although similar in size, display a surprising range of geologic evolution.

Voyager 1 has completed its designated mission and will continue on an endless journey through space. In 1998, it will exit the realm of our solar system. Following *Voyager 1*, *Voyager 2* carries along a gold-plated plaque which contains a message from Earth. *Voyager 2* has also been programmed for a fly-by probe of Uranus in 1986, and may, if all goes well, encounter Neptune in 1989, before also leaving our solar system.

THE RINGS OF SATURN

When viewed from Earth, Saturn's rings appear to consist of three rather distinct, concentric bands, which have classically been called the *A, B,* and *C* rings (Figure 20.21). The *A* ring is the outermost of the bright rings and is separated from the brightest ring (*B* ring) by a large gap. This space, called the **Cassini gap,** is easily seen in a photo of Saturn (see chapter-opening photo). Having a width of 5000 kilometers, the Cassini gap would be large enough to accommodate our moon. During this century evidence has been gathered for the existence of four other very faint bands; the *D* ring is located inside and the *F, G,* and *E* rings are located outside the classically known rings. In 1980, *Voyager 1* revealed that the seven rings are actually composed of hundreds or perhaps thousands of smaller ringlets. Even the Cassini gap is not empty, but rather contains a number of very faint bands. From the earth, Saturn's rings can be viewed on edge once every 15 years and appear as an extremely fine line. Satellite images reveal the thickness of the ring system to be no more than a few hundred meters while its lateral extent exceeds 200,000 kilometers.

Although none of the images obtained so far has the resolution needed to "see" the fine structures of the rings, the rings are undoubtedly composed of relatively small particles (moonlets) which orbit the planet much like any other satellite. Radar observations indicate that most of the particles of the A, B, and C rings are no larger than 10 meters and the more abundant particles are perhaps as small as 10 centimeters. Further studies of reflected light obtained from *Voyager* data revealed that a large fraction of the particles in the F ring and certain portions of the B ring are extremely minute. Information on the rings' composition has also been obtained from studies of how these particles reflect or absorb light of varying wavelengths. It has been determined that these moonlets are good reflectors of visible light and poor reflectors of near-infrared wavelengths, which is characteristic of water ice. Hence, it is thought that at least the outer surfaces of these particles are coated with a highly reflective layer of water ice. Saturn's rings therefore are actually swarms of small ice-covered debris that orbit the planet in nearly circular paths that are aligned with the equatorial plane of the planet. Most probably they are composed of material which failed to accrete into moons at the time the Saturnian System was forming.

The origin of the rings is believed to be related to their distance from the surface of Saturn. Since they are very close, the disruptive gravitational force of the parent prevented the individual particles from accreting into a larger satellite. Stated another way, objects cannot be held together by self-gravitation when they are within the influence of a stronger gravitational field. In fact, should one of Saturn's large icy satellites approach closer than the outer edge of the bright rings, it would be destroyed and its remains distributed among the rings. The gravitational (tidal) force of Saturn pulling on the near side of such a satellite would be enough greater than the force pulling on its far side that it would tear the satellite apart. Although the rings could be the remains of a satellite destroyed

in this fashion, these moonlets are more likely part of the primordial material from which Saturn formed.

Beyond the outermost bright ring (A ring), some moonlets have accreted to form very small satellites having diameters on the order of 100 kilometers. Five of these asteroid-sized moons have been discovered orbiting within the faint outer rings, and others probably exist. The gravitational influence of these so-called shepherding satellites is believed to be responsible for keeping the moonlets confined within the rings, thereby producing the sharp edges which are observed. Planetary geologists are very interested in the gravitational interaction of the objects that comprise Saturn's ring system. It is hoped that this information will reveal how material from the primordial cloud of dust and gases condensed to produce the planets. This information would be valuable in reconstructing the earth's early history.

THE MOONS OF SATURN

The Saturnian satellite system consists of seventeen known bodies, all but two of which have nearly circular, counterclockwise orbits along Saturn's equatorial plane (Figure 20.22). The other two, Iapetus and Phoebe, have orbits inclined 15 degrees and 150 degrees, respectively, to the plane of the system, with the latter exhibiting clockwise motion. Based on estimates of their densities, which range from 1.1–1.5 gram per cubic centimeter (except for Titan) and their high reflectivity, these satellites are probably composed mostly of water ice as well as ices of ammonia and methane, with lesser amounts of rocky material. Thus, located over 1.4 billion kilometers from the sun, these moons must be frozen, lifeless worlds. Despite their frigid character and similar compositions, these bodies display surprising diversity.

The eight smallest satellites are irregularly shaped and have diameters that range from roughly 200 kilometers to less than 30 kilometers. The outer three are trapped in stable regions located either 60 degrees ahead of or behind the orbit of a larger satellite. Two others, 1980S1 and 1980S3, are locked in the same orbit (co-orbital) but travel at slightly different velocities. As the faster innermost satellite begins to overtake its companion, it gravitationally acquires orbital momentum from its companion which thrusts it into a

FIGURE 20.22
Montage of the Saturnian System. Dione foreground; Tethys, Mimas distant right; Enceladus, Rhea off ring's left; Titan upper right. (Courtesy of NASA)

larger, slower orbit. At the same time, its companion drops into a lower, faster orbit. Thus, the faster inner moon becomes the slower outer moon and vice versa. This "celestial dance" is repeated about once every four years. The remaining three unnamed moons are called ring shepherds because of their apparent role of "herding" the particles within the rings. Two are found on each side of the narrow *F* ring and a third is located on the outer edge of the *A*

ring. These satellites may be responsible for the sharpness of these ring edges.

Except for Titan, which has a diameter of 5000 kilometers, the largest satellites of Saturn are all smaller than the earth's moon. Phoebe, the smallest, is only about 200 kilometers across, while the two largest, Rhea and Iapetus, are about 1500 kilometers in diameter. Because of their small size, it seems unlikely that an internal heat engine such as that

538

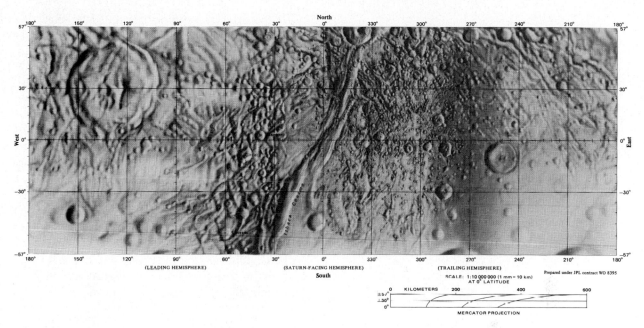

FIGURE 20.23
Map of Saturn's satellite Tethys, based on data from *Voyagers 1* and *2*. (Courtesy of NASA)

found on Earth could have operated within these bodies. However, several of these moons exhibit structural features such as fractures, rift-like valleys, and even folded terrain, indicating that some type of ancient tectonic activity occurred (Figure 20.23). It has been suggested that early in these moons' histories, ammonia and methane ices, which have a much lower melting point than water ice, could have been set in motion by a very small internal heat source.

The largest Saturnian moon is the only satellite in the solar system known to have a substantial atmosphere. Because of its dense gaseous cover, the atmospheric pressure at the surface of Titan is about one and one-half times that experienced at the earth's surface. Although Titan's atmosphere was predicted to be composed largely of methane, data from *Voyager 1* revealed that this was not the case. Rather, scientists discovered that as much as 80 percent of Titan's atmosphere is nitrogen, with methane probably accounting for less than 6 percent. The orange color of Titan's atmosphere may be the result of photochemical "smog" composed of hydrocarbon molecules, including ethylene and hydrogen cyanide.

Further, this planet-sized moon appears to have polar ice caps which show seasonal variations in size. Its surface, if unfrozen, would be an ocean of liquid nitrogen.

URANUS AND NEPTUNE: THE TWINS

If any two planets in the solar system can be considered twins, Uranus and Neptune can. Besides being similar in size, they appear a pale green color, attributable to the methane in their atmospheres. Their structure and composition are believed to be similar, as well, but because of its greater orbital distance, Neptune experiences somewhat lower temperatures.

The unique feature of Uranus is that its axis of rotation lies only 8 degrees from the plane of its orbit (Figure 20.24). Its rotational motion, therefore, has the appearance of rolling, rather than spinning like a top as the other planets. Because the axis of Uranus is inclined almost 90 degrees, the sun is nearly

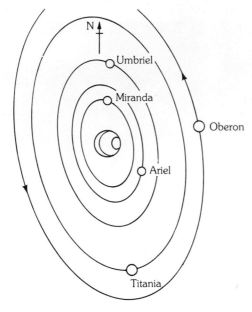

FIGURE 20.24
Orientation of Uranus and its satellites. Note the "north" arrow. (After Wyatt and Kaler, *Principles of Astronomy*, Newton, Mass.: Allyn and Bacon, 1974, p. 183)

overhead at one of the poles once each revolution, and then half a revolution later, it is nearly overhead at the other.

A surprise discovery in 1977 revealed that Uranus is surrounded by rings, much like those encircling Jupiter. This find occurred as Uranus passed in front of a distant star and blocked its view, a process called **occultation.** Observers saw the star "wink" briefly five times before the primary occultation and again five times afterward. Later studies have indicated that Uranus has at least nine distinct belts of debris orbiting its equatorial region.

PLUTO: PLANET X

Pluto lies on the fringe of the solar system almost 40 times farther from the sun than the earth. It is 10,000 times too dim to be visible with the unaided eye. Because of its great distance and slow orbital speed, it takes Pluto 248 years to orbit the sun. Since its discovery in 1930, it has completed less than one-fifth of a revolution. Pluto's orbit is noticeably elongated (highly eccentric), causing it to occasionally travel inside the orbit of Neptune. There is little likelihood that Pluto and Neptune will ever collide, because their orbits are inclined to each other and do not actually cross.

In June, 1978, a moon was discovered orbiting Pluto. Although this satellite is too small to be observed visually, it appears as an elongated bulge on photographic plates taken of Pluto. The satellite is about 20,000 kilometers from the planet, or 20 times closer than our moon. This discovery greatly altered earlier estimations of Pluto's size. Current data suggest that Pluto has a diameter of less than 3000 kilometers, making it the smallest planet in the solar system.

The average temperature of Pluto is estimated at −210°C, cold enough to solidify any gas that might be present. Pluto could not have an atmosphere. Consequently, Pluto might best be described as a large, dirty iceball made up of a mixture of frozen gases with lesser amounts of rocky substances.

A recent proposal suggests that Pluto was once a satellite of Neptune and was displaced from its original orbit when it collided with a large foreign object. The discovery of a satellite around Pluto is considered evidence that this event broke the Neptunian satellite into two pieces and sent them into an elongated orbit around the sun.

MINOR MEMBERS OF THE SOLAR SYSTEM

ASTEROIDS

Asteroids are relatively small bodies that have been likened to "flying mountains." The largest, Ceres, is 800 kilometers (500 miles) in diameter, but most of the 50,000 that have been observed are only about one kilometer across. The smallest asteroids are assumed to be no larger than grains of sand. Most asteroids lie between the orbits of Mars and Jupiter and have periods ranging from three to six years. Some asteroids have very eccentric orbits and travel very near the sun, while a few larger ones regularly pass close to the earth and moon. Many of the most recent impact craters on the moon were probably caused by collisions with asteroids.

Because many asteroids have irregular shapes, planetary geologists first speculated that they may

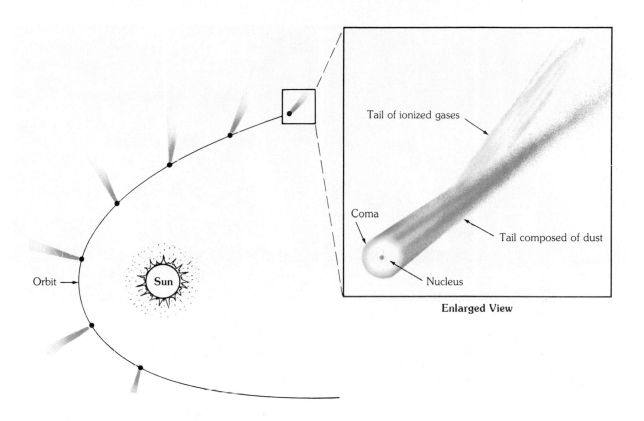

FIGURE 20.25
Orientation of a comet's tail as it orbits the sun.

have formed from the breakup of a planet that once occupied an orbit between Mars and Jupiter. However, the total mass of the asteroids is estimated to be only one one-thousandth that of the earth, which itself is not a large planet. What, then, happened to the remainder of the original planet? Others have hypothesized that several larger bodies once co-existed in close proximity and that their collisions produced numerous smaller ones. The existence of several "families" of asteroids has been used to support this latter explanation. However, no conclusive evidence has been found for either hypothesis.

COMETS

Comets are among the most spectacular and unpredictable bodies in the solar system. They have been compared to large, dirty snowballs, since they are made of frozen gases (water, ammonia, methane,

and carbon dioxide) which hold together small pieces of rocky and metallic materials. Many comets travel along very elongated orbits that carry them beyond Pluto. On their return, these comets are visible only after they are within the orbit of Saturn.

When first observed, comets appear very small, but as they approach the sun, solar energy begins to vaporize the frozen gases, producing a glowing head called the **coma** (Figure 20.25). The size of the coma varies greatly from one comet to another. Some exceed the size of the sun, but most approximate the size of Jupiter. Within the coma, a small glowing nucleus with a diameter of only a few kilometers can sometimes be detected. As they approach the sun, some comets develop a tail that extends for millions of kilometers. Despite the enormous size of their tail and coma, comets are thought to have insignificant masses.

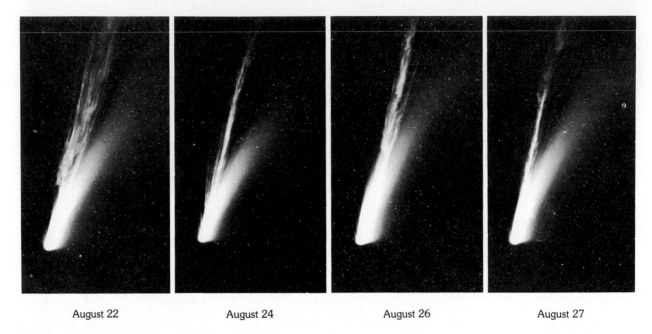

August 22	August 24	August 26	August 27

FIGURE 20.26
Comet Mrkos photographed with the 48-inch Schmidt telescope on four nights in 1957. (Courtesy of NASA)

The tail of a comet points away from the sun in a slightly curved manner (Figure 20.25). This fact led early astronomers to propose that the sun had a repulsive force that pushed the particles of the coma to form the tail. Today, two solar forces are known to contribute to the formation of the tail. One, radiation pressure, pushes dust particles away from the coma, and the second, solar wind, is responsible for moving the ionized gases, particularly carbon monoxide. Usually, a single tail composed of both types of materials is produced, but two somewhat separate tails like those of the comet Mrkos can form (Figure 20.26).

As a comet moves away from the sun, the gases begin to condense, the tail disappears, and the comet once again returns to "cold storage." The material that was blown from the coma to form the tail is lost from the comet forever. Consequently, it is believed that most comets cannot survive more than a few hundred close encounters with the sun. Once all the gases are expended, the remaining material—a swarm of disconnected metallic and stony particles—continues to orbit the sun, but without a coma or a tail.

Little is known about the origin of comets. The most widely accepted hypothesis considers them to be members of the solar system that formed at great distances from the sun. Accordingly, millions of comets are believed to form a spherical cloud located beyond the orbit of Pluto. It is proposed that the gravitational effect of stars passing nearby sends some of them into highly eccentric orbits which carry them toward the center of our solar system. Here, the gravitation of the larger planets, particularly Jupiter, alters their orbit and reduces their period of revolution. Many short-period comets of this type have been discovered. However, since they have a short life expectancy, we can be reasonably certain that they are always being replaced by other long-period comets which are gravitationally deflected toward the sun.

One of the most spectacular events of modern times has been attributed to the collision of our planet with a comet. In 1908, in a remote region of Siberia, a brilliant fireball exploded with such a violent force as to create a shock wave that rattled windows and was heard hundreds of kilometers away. Trees were scorched and flattened for a distance of 30 kilometers from the impact zone. Although numerous craters

FIGURE 20.27
Meteor Crater, about 32 kilometers (20 miles) west of Winslow, Arizona. (Courtesy of Meteor Crater Enterprises, Inc.)

resulted, some 50 meters wide, no large metallic fragments were found. The notable lack of impact fragments points strongly to the explosion of the nucleus of a comet that was vaporized as it penetrated our atmosphere.

METEOROIDS

Nearly everyone has seen a meteor, popularly called a "shooting star." This streak of light, which lasts for a few seconds at most, occurs when a small solid particle, a **meteoroid,** enters the earth's atmosphere from interplanetary space. The friction between the meteoroid and the air heats both and produces visible light. Most meteoroids are about the size of sand grains and weigh less than one-thousandth of a gram. Consequently, they vaporize before reaching the earth's surface. Some, called **micrometeorites,** are so tiny and their rate of fall so slow that they drift down like dust. Each day, the total number of meteoroids that enter the earth's atmosphere must reach into the thousands. After sunset, a half dozen or more are bright enough to be seen with the naked eye each hour from a single spot on earth.

Occasionally, the number of meteor sightings increases dramatically to 60 or more per hour. These spectacular displays, called **meteor showers,** result when the earth encounters a swarm of meteoroids traveling in the same direction and at nearly the same speed. The close association of these swarms to the orbits of some short-term comets strongly suggests that they represent material lost by the comets. Some swarms not associated with the orbits of known comets are probably the remains of the nucleus of a defunct comet. The meteor showers that occur regularly each year around August 12 are believed to be the remains of the tail of Comet 1862 III, which has a period of 110 years. Meteoroids associated with comets are small and not known to reach the ground. Most meteoroids large enough to survive the fall are thought to originate in the belt of asteroids, where a chance collision sends them toward the earth. The earth's gravitational force does the rest.

The remains of meteoroids, when found on the earth, are referred to as **meteorites.** A few very large meteorites have blasted out craters on the earth's surface that are not much different in appearance from those found on the lunar surface. Perhaps the most famous of these is Meteor Crater in Arizona (Figure 20.27). This huge crater is about 1.2 kilometers across, 170 meters deep, and has an upturned rim that rises 50 meters above the surrounding countryside. Over 30 tons of iron fragments have been found in the immediate area, but attempts to locate the main metallic body have been unsuccessful. Judging from the amount of erosion, it appears that the impact occurred within the last 20,000 years.

Prior to lunar exploration, meteorites were the only samples of extraterrestrial material that could be

543

directly examined (Figure 20.28). Depending upon their composition, meteorites can generally be put into one of three categories: (1) **irons**—mostly iron with 5–20 percent nickel; (2) **stony**—silicate minerals with inclusions of other minerals; and (3) **stony-irons**—mixtures. Although stony meteorites are probably more common, most meteorite finds are irons. This is understandable, since irons tend to withstand impact, weather more slowly, and are much easier for a lay person to distinguish from terrestrial rocks than are stony meteorites. One rare kind of meteorite, called a *carbonaceous chondrite,* was found to contain some simple amino acids, which are the basic building blocks to life. This discovery confirms similar findings in observational astronomy which indicate that numerous organic compounds exist in the frigid realm of outer space.

If the composition of meteorites is representative of the material that makes up the ''earthlike'' planets, as some planetary geologists believe, then the earth must contain a much larger percentage of iron than is indicated by the surface rock. This is one of the rea-

FIGURE 20.28
Iron meteorite found near Meteor Crater, Arizona.

sons why geologists suggest that the core of the earth may be mostly iron and nickel. In addition, the dating of meteorites has indicated that our solar system has an age which certainly exceeds 4 billion years. This ''old age'' has also been confirmed by data obtained from lunar samples.

REVIEW QUESTIONS

1 By what criteria are the planets categorized as either the Jovian or terrestrial?

2 What three types of materials are thought to make up the planets? How are they different? How does their distribution account for the density differences between the terrestrial and Jovian planetary groups?

3 Briefly describe the events which are thought to have lead to the formation of the solar system.

4 Why are large-rayed craters considered to be relatively young features on the lunar surface?

5 Outline the steps in the evolutionary history of the moon.

6 How are the maria of the moon similar to the Columbia Plateau?

7 Why are meteorite craters more common on the moon than on the earth, even though the moon is much smaller?

8 Although some Martian valleys appear to be the products of stream erosion, what fact makes it unlikely that Mars ever had a water cycle like that found on Earth?

9 What surface features does Mars have that are also common on Earth?

10 How fast does a location on the equator of Jupiter rotate? Note that Jupiter's circumference equals 416,000 kilometers (260,000 miles).

11 Briefly describe the four Galilean satellites of Jupiter.

12 Why are the four outer satellites of Jupiter thought to have been captured?

13 What evidence indicates that Saturn's rings are composed of individual moonlets rather than solid disks?

14 Examine the orbits of the satellites of Uranus in Figure 20.24. Do their orbits indicate that they were formed at the same time as Uranus or that they were captured at a later date?

15 What do you think would happen if the earth passed through the tail of a comet?

16 How do meteoroids and meteorites differ? How are they alike?

KEY TERMS

asteroid (p. 540)

Cassini gap (p. 536)

coma (p. 541)

comet (p. 541)

greenhouse effect (p. 524)

iron meteorite (p. 544)

Jovian planet (p. 511)

lunar breccia (p. 517)

lunar regolith (p. 520)

maria (p. 515)

meteorite (p. 543)

meteoroid (p. 543)

meteor shower (p. 543)

micrometeorite (p. 543)

nebular hypothesis (p. 513)

occultation (p. 540)

ray (p. 517)

stony meteorite (p. 544)

stony-iron meteorite (p. 544)

terrestrial planet (p. 511)

APPENDIX A
METRIC AND ENGLISH UNITS COMPARED

UNITS

1 kilometer (km)	= 1000 meters (m)
1 meter (m)	= 100 centimeters (cm)
1 centimeter (cm)	= 0.39 inches (in.)
1 mile (mi)	= 5280 feet (ft)
1 foot (ft)	= 12 inches (in.)
1 inch (in.)	= 2.54 centimeters (cm)
1 square mile (mi^2)	= 640 acres (a)
1 kilogram (kg)	= 1000 grams (g)
1 pound (lb)	= 16 ounces (oz)
1 fathom	= 6 feet (ft)

CONVERSIONS

When you want to convert:	Multiply by:	To find:
Length		
inches	2.54	centimeters
centimeters	0.39	inches
feet	0.30	meters
meters	3.28	feet
yards	0.91	meters
meters	1.09	yards
miles	1.61	kilometers
kilometers	0.62	miles
Area		
square inches	6.45	square centimeters
square centimeters	0.15	square inches
square feet	0.09	square meters

547

°F	°C
210	100
200	90
190	
180	80
170	
160	70
150	
140	60
130	
120	50
110	
100	40
90	30
80	
70	20
60	
50	10
40	
30	0
20	
10	−10
10	
0	−20
−10	

When you want to convert:	Multiply by:	To find:
square meters	10.76	square feet
square miles	2.59	square kilometers
square kilometers	0.39	square miles

Volume

cubic inches	16.38	cubic centimeters
cubic centimeters	0.06	cubic inches
cubic feet	0.028	cubic meters
cubic meters	35.3	cubic feet
cubic miles	4.17	cubic kilometers
cubic kilometers	0.24	cubic miles
liters	1.06	quarts
liters	0.26	gallons
gallons	3.78	liters

Masses and Weights

ounces	20.33	grams
grams	0.035	ounces
pounds	0.45	kilograms
kilograms	2.205	pounds

When you want to convert:	Multiply by:	To find:

Temperature

When you want to convert degrees Fahrenheit (°F) to degrees Celsius (°C), subtract 32 degrees and divide by 1.8.

When you want to convert degrees Celsius (°C) to degrees Fahrenheit (°F), multiply by 1.8 and add 32 degrees.

When you want to convert degrees Celsius (°C) to kelvins (K), delete the degree symbol and add 273.

When you want to convert kelvins (K) to degrees Celsius (°C), add the degree symbol and subtract 273.

APPENDIX B
PERIODIC TABLE OF
THE ELEMENTS

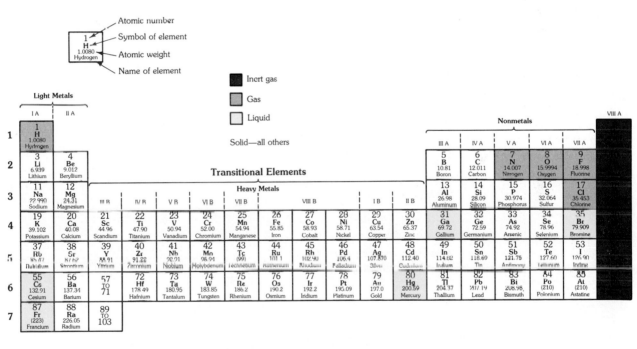

APPENDIX C
COMMON MINERALS OF THE EARTH'S CRUST

Mineral or Group Name	Composition	Cleavage/ Fracture	Color	Hardness	Other Properties/Comments
Albite	See *Plagioclase feldspar*				
Amphibole (common member: hornblende)	Complex family of hydrous, Ca, Na, Mg, Fe, Al silicates	Two at 60 and 120 degrees	Deep green to black	5–6	Forms elongated crystals. Commonly found in igneous and metamorphic rocks.
Anorthite	See *Plagioclase feldspar*				
Augite	See *Pyroxene*				
Bauxite	Mixture of weathered clay minerals	Irregular fracture	Varied, reddish-brown common	Variable	Earthy luster commonly contains small spheres. Ore of aluminum.
Biotite	$K(Mg,Fe)_3(AlSi_3O_{10})(OH)_2$	Perfect cleavage in one direction	Black to dark brown	2–2.5	Splits into thin, flexible sheets. Common mica found in igneous and metamorphic rocks.
Bornite	Cu_5FeS_4	Uneven fracture	Brownish bronze on a fresh surface	3	Tarnishes to a variegated purple blue; hence, called peacock ore. High specific gravity (5). Ore of copper.
Calcite	$CaCO_3$ Calcium carbonate	Three perfect cleavages at 75 degrees	White or colorless	2.5–3	Common in sedimentary rocks. When transparent exhibits double refraction. Reacts with weak acid.
Chalcedony	SiO_2 Silicon dioxide	Conchoidal fracture	White when pure. Often multicolored	5–6.5	Microcrystalline form of quartz. Multicolored. Called agates when banded. Opal is an amorphous variety.

Mineral or Group Name	Composition	Cleavage/Fracture	Color	Hardness	Other Properties/Comments
Chalcopyrite	$CuFeS_2$	Irregular fracture	Brass yellow	3.5–4	Usually massive. Specific gravity 4–4.5. Ore of copper.
Chlorite	$(Mg,Fe)_5(Al,Fe)_2Si_3O_{10}(OH)_8$	One direction of cleavage	Light to dark green	2–2.5	Occurs as mass of flaky scales. Common in metamorphic rocks.
Cinnabar	HgS	One direction, but not generally observed	Scarlet red	2.5	Occurs in masses mixed with other materials. Often dull earthy luster. Important ore of mercury.
Clay minerals (common member: kaolinite)	Complex group of hydrous aluminum silicates	Irregular fracture	Buff to brownish gray	1–2.5	Found in earthy masses as the main constituent of soil. Also abundant in shales and other sedimentary rocks.
Corundum	Al_2O_3	Two good cleavages with striations	Variable; red, blue, yellow and green	9	Important gemstone. Red variety called ruby; blue variety is sapphire. Also used as an abrasive.
Dolomite	$CaMg(CO_3)_2$	Three good cleavages at 75 degrees	Variable; white when pure	3.5–4	Similar to calcite, but will effervesce only when powdered. Common in sedimentary rocks.
Epidote	Complex Ca, Fe, and Al silicate	One good cleavage, one poor	Yellow-green to dark green	6–7	Commonly occurs as small elongated crystals in metamorphic rocks.
Feldspar See *Orthoclase feldspar* and *Plagioclase feldspar*					
Fluorite	CaF_2	Perfect cleavage in 4 directions	Colorless; violet, green, or yellow	4	Commonly found with ores of metals.
Galena	PbS	Three cleavages at right angles	Silver gray	2.5	Shiny metallic mineral with high specific gravity (7.6). Ore of lead.
Garnet	Complex family of silicate minerals containing Ca, Mg, Fe, Mn, Al, Ti, Cr	Uneven to conchoidal fracture	Various colors; commonly deep red to brown	6.5–7.5	Forms 12- or 24-sided crystals commonly found in metamorphic rocks.
Graphite	C	One direction of cleavage	Steel gray	1–2	Occurs in scaly, foliated masses. Used as a lubricant. Greasy feel.

Mineral or Group Name	Composition	Cleavage/ Fracture	Color	Hardness	Other Properties/Comments
Gypsum	$CaSO_4 \cdot 2H_2O$	Cleavage good in one direction, poor in two others	Colorless to white	2	Occurs as tabular crystals, or fibrous or finely crystalline masses. Common in sedimentary layers. Used for plaster.
Halite	NaCl	Three cleavages at right angles	Colorless to white	2.5	Common table salt. Occurs as granular masses. Common sedimentary mineral.
Hematite	Fe_2O_3	Uneven fracture	Reddish brown to steel gray	5.5–6.5	Occurs as earthy masses. High specific gravity (4.8–5.5). Important ore of iron.
Hornblende	See Amphibole				
Kaolinite	See Clay minerals				
Kyanite	Al_2SiO_5	One good direction of cleavage	White to light blue	5–7	Forms long, bladed or tabular crystals. Common mineral in metamorphic rocks.
Labradorite	See Plagioclase feldspar				
Limonite (goethite)	Mixture of hydrous iron oxides	Uneven fracture	Yellowish to brown	1–5.5	Earthy masses. Forms from the alteration of other iron-rich minerals. Gives rock surfaces and soils a yellow color.
Magnetite	Fe_3O_4	Uneven fracture	Black	5.5–6.5	Submetallic to metallic luster. Magnetic. High specific gravity (5). Generally occurs in granular masses. Ore of iron.
Malachite	$Cu_2CO_3(OH)_2$	Uneven fracture	Bright green	3.5–4	Effervesces in acid. Ore of copper.
Mica	See Biotite and Muscovite				
Muscovite	$KAl_3Si_3O_{10}(OH)_2$	Perfect cleavage in one direction	Colorless to light gray	2–2.5	Splits into thin elastic sheets. Transparent in thin sheets. Common in all rock types.
Olivine	$(Mg,Fe)_2SiO_4$	Conchoidal fracture	Olive to dark green	6.5–7	Occurs as granular masses or some grains in dark colored igneous rocks.
Orthoclase feldspar (K feldspar)	$KAlSi_3O_8$	Two cleavages at right angles	White to gray. Frequently salmon pink	6	Forms elongated crystals in igneous rocks. Also, commonly found in sedimentary and metamorphic rocks.

Mineral or Group Name	Composition	Cleavage/ Fracture	Color	Hardness	Other Properties/Comments
Plagioclase feldspar	$NaAlSi_3O_8$ (albite) $CaAl_2Si_2O_8$ (anorthite)	Two cleavages at nearly right angles	White to gray	6	Forms elongated crystals in igneous rocks. Also commonly found in sedimentary and metamorphic rocks. Striations on some cleavage planes.
Pyrite	FeS_2	Uneven fracture	Brass yellow	6–6.5	Occurs as granular masses or well-formed cubic crystals. High specific gravity (4.8–5.2). Often called "fool's gold."
Pyroxene (common member: augite)	Complex family of Mg, Fe, Ca, Na, and Al silicate	Good cleavage in two directions at nearly right angles	Green to black	5–6	Occurs as individual grains in igneous and metamorphic rocks.
Quartz	SiO_2	Conchoidal fracture	Colorless when pure	7	Common in all rock types. Often lightly colored, including gray, pink, yellow, and violet.
Serpentine	$Mg_3Si_2O_5(OH)_4$	Uneven fracture	Light to dark green	2.5–5	Fibrous variety is asbestos. Occurs most often in metamorphic rocks.
Sillimanite	Al_2SiO_5	One direction of cleavage	White to gray	6–7	High grade metamorphic mineral.
Sphalerite	ZnS	Six directions of cleavage	Yellow to brown	3.5–4	Moderate specific gravity (4.1–4.3). Smell of sulfur when powdered. Ore of zinc.
Staurolite	$FeAl_4(SiO_4)_2(OH)_2$	Cleavage not prominent	Brown to reddish brown	7	Elongated crystals, occasionally twinned to form a cross-shaped crystal. Commonly found in metamorphic rocks.
Sulfur	S	Irregular fracture	Yellow	1.5–2.5	Bright yellow mineral most often associated with sedimentary deposits, in coal, and near volcanoes.
Talc	$Mg_3(Si_4O_{10})(OH)_2$	Good cleavage in one direction	White to light green	1–1.5	Soapy feel. Found in foliated masses consisting of thin flakes or scales. Most often associated with metamorphic rocks.
Wollastonite	$CaSiO_3$	Two perfect cleavages	Colorless to white	4.5–5	Forms fibrous or bladed crystals. Common in contact metamorphic rocks.

APPENDIX D
TOPOGRAPHIC MAPS

A map is a representation on a flat surface of all or a part of the earth's surface drawn to a specific scale. Maps are often the most effective means for showing the locations of both natural and manmade features, their sizes, and their relationships to one another. Like photographs, maps readily display information that would be impractical to express in words.

While most maps show only the two horizontal dimensions, geologists, as well as other map users, often require that the third dimension, elevation, be shown on maps. Maps that show the shape of the land are called **topographic maps.** Although various techniques may be used to depict elevations, the most accurate method involves the use of contour lines.

Contour Lines

A **contour line** is a line on a map representing a corresponding imaginary line on the ground that has the same elevation above sea level along its entire length. While many map symbols are pictographs, resembling the objects they represent, a contour line is an abstraction that has no counterpart in nature. It is, however, an accurate and effective device for representing the third dimension on paper.

Some useful facts and rules concerning contour lines are listed as follows. This information should be studied in conjunction with Figure D.1.

1   Contour lines bend upstream or upvalley. The contours form Vs that point upstream, and in the upstream direction the successive contours represent higher elevations. For example, if you were standing on a stream bank and wished to get to the point at the same elevation directly opposite you on the other bank, without stepping up or down, you would need to walk upstream along the contour at that elevation to where it crosses the stream bed, cross the stream, and then walk back downstream along the same contour.

2   Contours near the upper parts of hills form closures. The top of a hill is higher than the highest closed contour.

3   Hollows (depressions) without outlets are shown by closed, hatched contours. Hatched contours are contours with short lines on the inside pointing downslope.

4   Contours are widely spaced on gentle slopes.

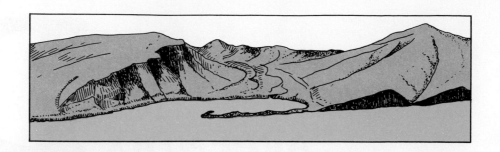

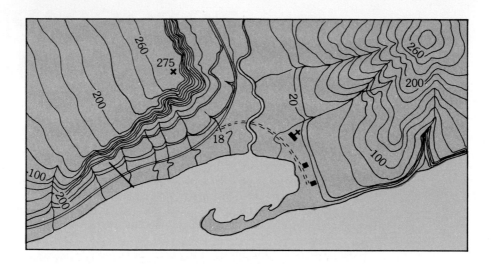

FIGURE D.1

Perspective view of an area and a contour map of the same area. These illustrations show how features are depicted on a topographic map. The upper illustration is a perspective view of a river valley and the adjoining hills. The river flows into a bay, which is partly enclosed by hooked sandbar. On either side of the valley are terraces through which streams have cut gullies. The hill on the right has a smoothly eroded form and gradual slopes, whereas the one on the left rises abruptly in a sharp precipice, from which it slopes gently, and forms an inclined plateau traversed by a few shallow gullies. A road provides access to a church and the two houses situated across the river from a highway that follows the seacoast and curves up the river valley. The lower illustration shows the same features represented by symbols on a topographic map. The contour interval (vertical distance between adjacent contours) is 20 feet. (After U.S. Geological Survey)

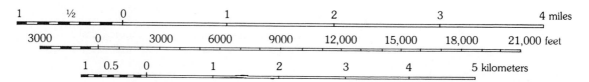

FIGURE D.2

Graphic scale.

556

5 Contours are closely spaced on steep slopes.

6 Evenly spaced contours indicate a uniform slope.

7 Contours usually do not cross or intersect each other, except in the rare case of an overhanging cliff.

8 All contours eventually close, either on a map or beyond its margins.

9 A single high contour never occurs between two lower ones, and vice versa. In other words, a change in slope direction is always determined by the repetition of the same elevation either as two different contours of the same value or as the same contour crossed twice.

10 Spot elevations between contours are given at many places, such as road intersections, hill summits, and lake surfaces. Spot elevations differ from control elevation stations, such as bench marks, in not being permanently established by permanent markers.

Relief

Relief refers to the difference in elevation between any two points. Maximum relief refers to the difference in elevation between the highest and lowest points in the area being considered. Relief determines the **contour interval,** which is the difference in elevation between succeeding contour lines that is used on topographic maps. Where relief is low, a small contour interval, such as 10 or 20 feet, may be used. In flat areas, such as wide river valleys or broad, flat uplands, a contour interval of 5 feet is often used. In rugged mountainous terrain, where relief is many hundreds of feet, contour intervals as large as 50 or 100 feet are used.

Scale

Map **scale** expresses the relationship between distance or area on the map to the true distance or area on the earth's surface. This is generally expressed as a ratio or fraction, such as 1:24,000 or 1/24,000. The numerator, usually 1, represents map distance, and the denominator, a large number, represents ground distance. Thus, 1:24,000 means that a distance of 1 unit on the map represents a distance of 24,000 such units on the surface of the earth. It does not matter what the units are.

Often, the graphic or bar scale is more useful than the fractional scale, because it is easier to use for measuring distances between points. The graphic scale (Figure D.2) consists of a bar divided into equal segments, which represent equal distances on the map. One segment on the left side of the bar is usually divided into smaller units to permit more accurate estimates of fractional units.

Topographic maps, which are also referred to as quadrangles, are generally classified according to publication scale. Each series is intended to fulfill a specific type of map need. To select a map with the proper scale for a particular use, remember that large-scale maps show more detail and small-scale maps show less detail. The sizes and scales of topographic maps published by the U.S. Geological Survey are shown in Table D.1.

TABLE D.1
National topographic maps.

Series	Scale	1 Inch Represents	Standard Quadrangle Size (latitude-longitude)	Quadrangle Area (square miles)	Paper Size E-W N-S Width Length (inches)
7½-minute	1:24,000	2000 feet	7½′ × 7½′	49–70	22 × 27*
Puerto Rico 7½-minute	1:20,000	about 1667 feet	7½′ × 7½′	71	29½ × 32½
15-minute	1:62,500	nearly 1 mile	15′ × 15′	197–282	17 × 21*
Alaska 1:63,360	1:63,360	1 mile	15′ × 20′ − 36′	207–281	18 × 21**
U.S. 1:250,000	1:250,000	nearly 4 miles	1° × 2°†	4580–8669	34 × 22‡
U.S. 1:1,000,000	1:1,000,000	nearly 16 miles	4° × 6°†	73,734–102,759	27 × 27

SOURCE: U.S. Geological Survey.

*South of latitude 31 degrees, 7½-minute sheets are 23 × 27 inches; 15-minute sheets are 18 × 21 inches.

**South of latitude 62 degrees, sheets are 17 × 21 inches.

†Maps of Alaska and Hawaii vary from these standards.

‡North of latitude 42 degrees, sheets are 29 × 22 inches; Alaska sheets are 30 × 23 inches.

Color and Symbol

Each color and symbol used on the U.S. Geological Survey topographic map has significance. Common topographic map symbols are shown in Figure D.3. The meaning of each color is as follows:

Blue—water features
Black—works of man, such as homes, schools, churches, roads, and so forth
Brown—contour lines
Green—woodlands, orchards, and so forth
Red—urban areas, important roads, public land subdivision lines

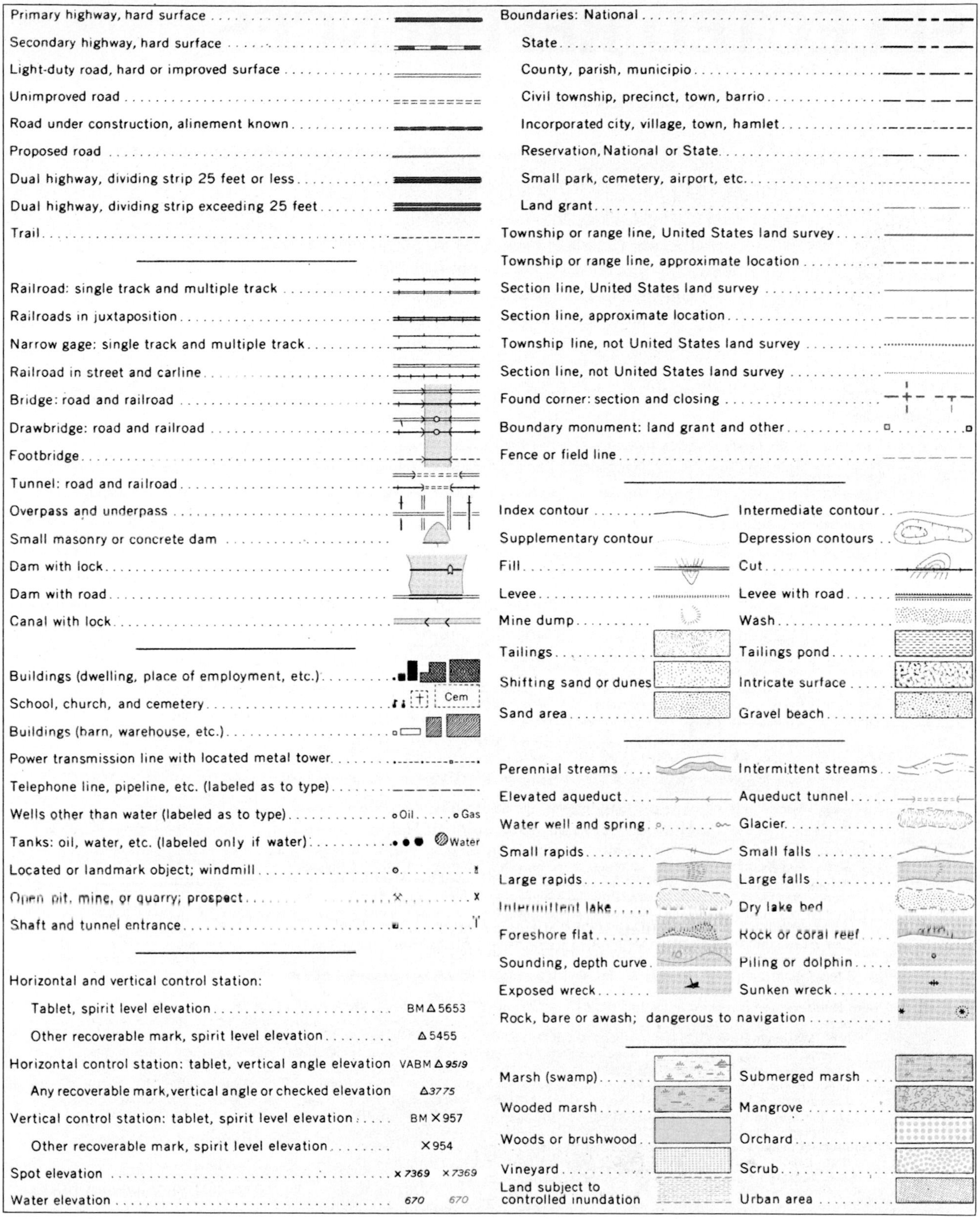

Primary highway, hard surface	
Secondary highway, hard surface	
Light-duty road, hard or improved surface	
Unimproved road	
Road under construction, alinement known	
Proposed road	
Dual highway, dividing strip 25 feet or less	
Dual highway, dividing strip exceeding 25 feet	
Trail	

Railroad: single track and multiple track	
Railroads in juxtaposition	
Narrow gage: single track and multiple track	
Railroad in street and carline	
Bridge: road and railroad	
Drawbridge: road and railroad	
Footbridge	
Tunnel: road and railroad	
Overpass and underpass	
Small masonry or concrete dam	
Dam with lock	
Dam with road	
Canal with lock	

Buildings (dwelling, place of employment, etc.)	
School, church, and cemetery	
Buildings (barn, warehouse, etc.)	
Power transmission line with located metal tower	
Telephone line, pipeline, etc. (labeled as to type)	
Wells other than water (labeled as to type)	∘Oil ∘Gas
Tanks: oil, water, etc. (labeled only if water)	∙ ● ● ◎Water
Located or landmark object; windmill	
Open pit, mine, or quarry; prospect	✕ x
Shaft and tunnel entrance	

Horizontal and vertical control station:	
Tablet, spirit level elevation	BM △ 5653
Other recoverable mark, spirit level elevation	△ 5455
Horizontal control station: tablet, vertical angle elevation	VABM △ 95/9
Any recoverable mark, vertical angle or checked elevation	△3775
Vertical control station: tablet, spirit level elevation	BM ✕957
Other recoverable mark, spirit level elevation	✕954
Spot elevation	✕ 7369 ✕ 7369
Water elevation	670 670

Boundaries: National	
State	
County, parish, municipio	
Civil township, precinct, town, barrio	
Incorporated city, village, town, hamlet	
Reservation, National or State	
Small park, cemetery, airport, etc.	
Land grant	
Township or range line, United States land survey	
Township or range line, approximate location	
Section line, United States land survey	
Section line, approximate location	
Township line, not United States land survey	
Section line, not United States land survey	
Found corner: section and closing	
Boundary monument: land grant and other	
Fence or field line	

Index contour		Intermediate contour ..
Supplementary contour		Depression contours ...
Fill		Cut ....
Levee		Levee with road .....
Mine dump		Wash ....
Tailings		Tailings pond ....
Shifting sand or dunes		Intricate surface .....
Sand area		Gravel beach ....

Perennial streams		Intermittent streams ..
Elevated aqueduct		Aqueduct tunnel .....
Water well and spring .		Glacier ....
Small rapids		Small falls ....
Large rapids		Large falls ....
Intermittent lake		Dry lake bed ....
Foreshore flat		Rock or coral reef ...
Sounding, depth curve .		Piling or dolphin .....
Exposed wreck		Sunken wreck ....
Rock, bare or awash; dangerous to navigation		

Marsh (swamp)		Submerged marsh ....
Wooded marsh		Mangrove ....
Woods or brushwood .		Orchard ....
Vineyard		Scrub ....
Land subject to controlled inundation		Urban area ....

FIGURE D.3

U.S. Geological Survey topographic map symbols. (Variations will be found on older maps)

559

SELECTED REFERENCES

CHAPTER 1

Adams, F. D. *The Birth and Development of the Geological Sciences.*
New York: Dover, 1954.

Albritton, C. C., Jr., (ed.). *The Fabric of Geology.*
Reading, MA: Addison-Wesley, 1963.

Fenton, C. L., and **Fenton, M. A.** *Giants of Geology.*
Garden City, NY: Doubleday, 1952.

Gillispie, C. C. *Genesis and Geology.*
New York: Harper & Row, 1951.

Harrington, J. W. *To See a World.*
St. Louis: C. V. Mosby, 1973.

Matthews, W. H., III. *Invitation to Geology: The Earth Through Time and Space.*
Garden City, NY: Natural History Press, 1971.

McPhee, J. *Basin and Range.*
New York: Farrar, Straus, & Giroux, 1981.

Sullivan, W. *Continents in Motion.*
New York: McGraw-Hill, 1974.

CHAPTER 2

Ahrens, L. H. *Distribution of the Elements in our Planet.*
New York: McGraw-Hill, 1965.

Dietrich, R. V., and **Skinner, B. J.** *Rocks and Rock Minerals.*
New York: Wiley, 1979.

Ernst, W. G. *Earth Materials.*
Englewood Cliffs, NJ: Prentice-Hall, 1969.

Holden, A., and **Singer, P.** *Crystals and Crystal Growing.*
Garden City, NY: Doubleday, 1960.

Hurlbut, C. S., Jr. *Minerals and Men.*
New York: Random House, 1969.

Hurlbut, C. S., Jr., and **Klein, C.** *Manual of Mineralogy,* 19th ed.
(After James Dana), New York: Wiley, 1977.

Turekian, K. K. *Chemistry of the Earth.*
New York: Holt, Rinehart, & Winston, 1972.

Vanders, I., and **Kerr, P. F.** *Mineral Recognition.*
New York: Wiley, 1967.

CHAPTER 3

Bowen, N. L. *The Evolution of the Igneous Rocks.*
Princeton, NJ: Princeton University Press, 1928. (Reprinted by Dover
Publications, New York, 1956.)

Cox, K. G., Bell, J. D., and **Pankhurst, R. J.** *The Interpretation of Igneous Rocks.*
Winchester, MA: Allen and Unwin, 1979.

Ernst, W. G. *Earth Materials.*
Englewood Cliffs, NJ: Prentice-Hall, 1969.

Hamilton, **W. B.**, and **Myers, W. B.** *The Nature of Batholiths.*
U.S. Geological Survey Professional Paper 554-C.

Hyndman, D. W. *Petrology of Igneous and Metamorphic Rocks.*
New York: McGraw-Hill, 1972.

Yoder, H. S., Jr., (ed.). *The Evolution of Igneous Rocks, Fiftieth Anniversary Perspectives.*
Princeton, NJ: Princeton University Press, 1979.

CHAPTER 4

Bolt, B. A., MacDonald, G. A., and **Scott, R. F.** *Geological Hazards.*
New York: Springer-Verlag, 1977.

Bryson, R. A., and **Goodman, B. M.** "Volcanic Activity and Climatic Changes."
Science, Vol. 207 (1980), pp. 1041–4.

Bullard, F. M. *Volcanoes of the Earth,* 2nd ed.
Austin, TX: University of Texas Press, 1976.

Decker, R., and **Decker, B.** *Volcanoes.*
San Francisco: W. H. Freeman, 1981.

Fiske, R. S., Hopson, C. A., and **Waters, A. C.** *Geology of Mount Rainier National Park.*
U.S. Geological Survey Professional Paper 444, 1963.

Green, J., and **Short, N. M.,** (eds.). *Volcanic Landforms and Surface Features: A Photographic Atlas and Glossary.*
New York: Springer-Verlag, 1977.

MacDonald, G. A. *Volcanoes.*
Englewood Cliffs, NJ: Prentice Hall, 1972.

Rosenfeld, C. L. "Observations on the Mount St. Helens Eruptions."
American Scientist, Vol. 68 (1980), pp. 494–509.

Sheets, P. D., and **Grayson, D. K.,** (eds.) *Volcanic Activity and Human Ecology.*
New York: Academic Press, 1979.

Williams, H. *The Geology of Crater Lake National Park.*
Washington, DC: Carnegie Institution, 1942.

CHAPTER 5

Birkeland, P. W. *Pedology, Weathering, and Geomorphological Research.*
New York: Oxford University Press, 1974.

Foth, H. D., and **Turk, L. M.** *Fundamentals of Soil Science.*
New York: Wiley, 1972.

Hunt, C. B. *Geology of Soils: Their Evolution, Classification, and Uses.*
San Francisco: W. H. Freeman, 1972.

Loughnan, F. C. *Chemical Weathering of the Silicate Minerals.*
New York: Elsevier, 1969.

Ollier, C. *Weathering.*
New York: Elsevier, 1969.

Ritter, D. F. *Process Geomorphology.*
Dubuque, IA: Wm. C. Brown, 1978.

CHAPTER 6

Blatt, H., Middleton, G., and **Murray, R.**   *Origin of Sedimentary Rocks,* 2nd ed.
Englewood Cliffs, NJ: Prentice-Hall, 1980.

Ernst, W. G.   *Earth Materials.*
Englewood Cliffs, NJ: Prentice-Hall, 1969.

Garrels, R. M., and **Mackenzie, F. T.**   *Evolution of Sedimentary Rocks.*
New York: W. W. Norton, 1971.

LaPorte, L. F.   *Ancient Environments,* 2nd ed.
Englewood Cliffs, NJ: Prentice-Hall, 1979.

Pettijohn, F. J.   *Sedimentary Rocks,* 3rd ed.
New York: Harper & Row, 1975.

Reineck, H. E., and **Singh, I. B.**   *Depositional Sedimentary Environments,* 2nd ed.
New York: Springer-Verlag, 1980.

Shelley, R. C.   *An Introduction to Sedimentology.*
New York: Academic Press, 1976.

Skinner, B. J.   *Earth Resources,* 2nd ed.
Englewood Cliffs, NJ: Prentice-Hall, 1976.

CHAPTER 7

Ernst, W. G.   *Earth Materials.*
Englewood Cliffs, NJ: Prentice-Hall, 1969.

————, (ed.).   *Metamorphism and Plate Tectonic Regimes.*
Benchmark Papers in Geology, Stroudsburg, PA: Dowden, Hutchinson, and Ross, 1975.

Hyndman, D. W.   *Petrology of Igneous and Metamorphic Rocks.*
New York: McGraw-Hill, 1972.

Miyashiro, A.   *Metamorphism and Metamorphic Belts.*
New York: Halsted Press, 1978.

Turner, F. J.   *Metamorphic Petrology,* 2nd ed.
New York: McGraw-Hill, 1980.

CHAPTER 8

Alden, W. C.   "Landslide and Flood at Gros Ventre, Wyoming."
Transactions, American Institute of Mining and Metallurgical Engineers, Vol. 76 (1928), pp. 347–58. (Reprinted in Tank, R. W. [ed.]. *Focus on Environmental Geology.* New York: Oxford University Press, 1973.)

Carson, M. A., and **Kirkby, M. J.**   *Hillslope Form and Process.*
New York: Cambridge University Press, 1972.

Ericksen, G. E., and **Plafker, G.**   *Preliminary Report on the Geologic Events Associated with the May 31, 1970, Peru Earthquake.*
U.S. Geological Survey Circular 639, 1970.

Fleming, R. W., and **Taylor, F. A.**   *Estimating Costs of Landslide Damage in the United States.*
U.S. Geological Survey Circular 832, 1980.

Kiersch, G. A. "Vaoint Reservoir Disaster."
 Civil Engineering, Vol. 34 (1964), pp. 32–39. (Reprinted in Tank, R. W. [ed.].
 Focus on Environmental Geology. New York: Oxford University Press, 1973.)

Péwé, T. L. *Permafrost and Its Effect on Life in the North.*
 Corvallis, OR: Oregon State University Press, 1970.

Sharpe, C. F. S. *Landslides and Related Phenomena.*
 New York: Columbia University Press, 1938.

CHAPTER 9

Bloom, A. L. *Geomorphology: A Systematic Analysis of Late Cenozoic Landforms.*
 Englewood Cliffs, NJ: Prentice-Hall, 1978.

Emerson, J. W. "Channelization: A Case Study."
 Science, Vol. 173 (1971), pp. 325–26.

Judson, S. "Erosion of the Land."
 American Scientist, Vol. 56 (1968), pp. 356–74.

Larimer, O. J. "Flood of June 9–10, 1972, at Rapid City, South Dakota."
 U.S. Geological Survey *Hydrologic Investigations Atlas* HA-511, 1973.

Leopold, L. B. *Water: A Primer.*
 San Francisco: W. H. Freeman, 1974.

Leopold, L. B., Wolman, M. G., and **Miller, J. P.** *Fluvial Processes in
 Geomorphology.*
 San Francisco: W. H. Freeman, 1964.

Morisawa, M. *Streams: Their Dynamics and Morphology.*
 New York: McGraw-Hill, 1968.

Schumm, S. A. *The Fluvial System.*
 New York: Wiley, 1979.

Thornbury, W. D. *Principles of Geomorphology.*
 New York: Wiley, 1969.

CHAPTER 10

Baldwin, H. L., and **McGuinness, C. L.** *A Primer on Groundwater.*
 U.S. Geological Survey, 1963.

Fetter, C. W., Jr. *Applied Hydrogeology.*
 Columbus, OH: Charles E. Merrill, 1980.

Freeze, R. A., and **Cherry, J. A.** *Groundwater.*
 Englewood Cliffs, NJ: Prentice-Hall, 1979.

Monroe, W. H. *The Karst Landforms of Puerto Rico.*
 U.S. Geological Survey Professional Paper 899, 1976.

Rinehart, J. S. *Geysers and Geothermal Energy.*
 New York: Springer-Verlag, 1980.

Sweeting, M. M. *Karst Landforms.*
 New York: Columbia University Press, 1973.

White, D. F., and **Williams, D. L.** *Assessment of Geothermal Resources of the
 United States.*
 U.S. Geological Survey Circular 726, 1975.

CHAPTER 11

Embleton, C., and **King, C. A. M.** *Glacial and Periglacial Geomorphology,* 2nd ed.
New York: Wiley, 1975.

Flint, R. F. *Glacial and Quaternary Geology.*
New York: Wiley, 1971.

Hays, J. D., Imbrie, J., and **Shackleton, N. J.** "Variations in the Earth's Orbit:
Pacemaker of the Ice Ages."
Science, Vol. 194 (1976), pp. 1121–32.

Imbrie, J., and **Imbrie, K. P.** *Ice Ages: Solving the Mystery.*
Hillside, NJ: Enslow Publishers, 1979.

Matthes, F. E. *Geologic History of the Yosemite Valley.*
U.S. Geological Survey Professional Paper 137, 1930.

Post, A. S., and **LaChapelle, E. R.** *Glacier Ice.*
Seattle, WA: University of Washington Press, 1971.

Sharp, R. P. *Glaciers.*
Eugene, OR: University of Oregon Press, 1960.

Wright, H. E., and **Frye, D. G.,** (eds.). *The Quaternary of the United States.*
Princeton, NJ: Princeton University Press, 1965.

CHAPTER 12

Bagnold, R. A. *Physics of Blown Sand and Desert Dunes.*
London: Methuen, 1941. (Reprinted by Halsted Press, New York, 1965.)

Denny, C. S. "Fans and Pediments."
American Journal of Science, Vol. 265 (1967), pp. 81–105.

Glennie, K. W. *Desert Sedimentary Environments.*
New York: Elsevier, 1970.

Mabbutt, J. A. *Desert Landforms.*
Cambridge, MA: MIT Press, 1977.

McGinnies, W. G., et al (eds.). *Deserts of the World.*
Tucson, AZ: University of Arizona Press, 1968.

McKee, E. D. (ed.). *A Study of Global Sand Seas.*
U.S. Geological Survey Professional Paper 1052, 1979.

Norris, R. M. "Barchan Dunes of the Imperial Valley, California."
Journal of Geology, Vol. 74 (1966), pp. 292–306.

Schultz, C. B., and **Frye, J. C.** *Loess and Related Eolian Deposits of the World.*
Lincoln, NE: University of Nebraska Press, 1968.

CHAPTER 13

Bascom, W. *Waves and Beaches.*
Garden City, NY: Doubleday, 1964.

Bird, E. C. *Coasts.*
Cambridge, MA: MIT Press, 1969.

King, C. A. M. *Beaches and Coasts,* 2nd ed.
London: Edward Arnold, 1972.

Komar, P. D.   *Beach Processes and Sedimentation.*
    Englewood Cliffs, NJ: Prentice-Hall, 1976.

Shepard, F. P., and **Wanless, H. R.**   *Our Changing Coastlines.*
    New York: McGraw-Hill, 1971.

Strahler, A. N.   *A Geologist's View of Cape Cod.*
    Garden City, NY: Natural History Press, 1966.

Turekian, K. K.   *Oceans,* 2nd ed.
    Englewood Cliffs, NJ: Prentice-Hall, 1976.

CHAPTER 14

Bolt, B. A.   *Earthquakes: A Primer.*
    San Francisco: W. H. Freeman, 1978.

Evans, D. M.   "Man-made Earthquakes in Denver."
    Geotimes, (May–June, 1966), pp. 11–18.

Hodgson, J. H.   *Earthquakes and Earth Structure.*
    Englewood Cliffs, NJ: Prentice-Hall, 1964.

Iacopi, R., (ed.).   *Earthquake Country: California.*
    Menlo Park, CA: Lane, 1964.

Nichols, D. R., and **Buchanan-Banks, J. M.**   *Seismic Hazards and Land-Use Planning.*
    U.S. Geological Survey Circular 690.

Press, F.   "Earthquake Prediction."
    Scientific American, Vol. 232 (May, 1975), pp. 14–23.

Raleigh, C. B., et al.   "An Experiment in Earthquake Control at Rangely, Colorado."
    Science, Vol. 191 (1976), pp. 1230–36.

U.S. Geological Survey.   *The Alaska Earthquake.*
    Professional Papers 541 (1965)–546 (1970).

U.S. Geological Survey and National Oceanic and Atmospheric Administration.
    The San Fernando, California, Earthquake of February 9, 1971.
    Professional Paper 733 (1971).

CHAPTER 15

Bolt, B. A.   "The Fine Structure of the Earth's Interior."
    Scientific American, Vol. 228 (March, 1973), pp. 24–33.

———.   *Inside the Earth.*
    San Francisco: W. H. Freeman, 1982.

Clark, S. P.   *Structure of the Earth.*
    Englewood Cliffs, NJ: Prentice-Hall, 1971.

Garland, G. D.   *Introduction to Geophysics,* 2nd ed.
    Toronto: W. B. Saunders, 1979.

Gutenburg, B., and **Richter, C. F.**   *Seismicity of the Earth and Associated Phenomena,* 2nd ed.
    Princeton, NJ: Princeton University Press, 1954.

Jacobs, J. A., Russell, R. D., and **Wilson, J. T.**   *Physics and Geology,* 2nd ed.
    New York: McGraw-Hill, 1974.

Stacey, F. D.   *Physics of the Earth,* 2nd ed.
    New York: Wiley, 1977.

CHAPTER 16

Clague, D. A., and **Jarrard, R. D.** "Tertiary Pacific Plate Motion Deduced from the Hawaiian-Emperor Chain."
Bulletin of the Geological Society of America, Vol. 84 (1973), pp. 1135–54.

Condie, K. C. *Plate Tectonics and Crustal Evolution.*
Elmsford, NY: Pergamon, 1976.

Cox, A., (ed.). *Plate Tectonics and Geomagnetic Reversals.*
San Francisco: W. H. Freeman, 1973.

Mathews, S. W. "This Changing Earth."
National Geographic, Vol. 143 (January, 1973), pp. 1–37.

Sullivan, W. *Continents in Motion: The New Earth Debate.*
New York: McGraw-Hill, 1974.

Uyeda, S. *The New View of the Earth: Moving Continents and Moving Oceans.*
San Francisco: W. H. Freeman, 1978.

Wegener, A. *The Origin of Continents and Oceans.*
New York: Dover, 1929. (Reprinted 1966)

Wyllie, P. J. *The Way the Earth Works.*
New York: Wiley, 1976.

CHAPTER 17

Corliss, J. B., et al. "Submarine Thermal Springs on the Galapagos Rift."
Science, Vol. 203 (1979), pp. 1073–83.

Heezen, B. C., and **Ewing, M.** "Turbidity Currents and Submarine Slumps and the Grand Banks Earthquake."
American Journal of Science, Vol. 250 (1952), pp. 849–73.

Heezen, B. C., and **Hollister, C. D.** *The Face of the Deep.*
New York: Oxford University Press, 1971.

Hess, H. H. "Mid-Oceanic Ridges and Tectonics of the Sea-floor."
In Whittard, W. F., and Bradshaw, R., (eds.). *Submarine Geology and Geophysics.* Proceedings of the 17th Symposium Colston Research Society.
London: Butterworths, 1965.

Kennett, J. P. *Marine Geology.*
Englewood Cliffs, NJ: Prentice-Hall, 1982.

MacDonald, K. C., and **Luyendyk, B. P.** "The Crest of the East Pacific Rise."
Scientific American, Vol 244 (May, 1981), pp. 100–116.

Menard, H. W. *Marine Geology of the Pacific.*
New York: McGraw-Hill, 1964.

Shepard, F. P. *Submarine Geology,* 3rd ed.
New York: Harper & Row, 1973.

Thurman, H. V. *Introductory Oceanography,* 3rd ed.
Columbus, OH: Charles E. Merrill, 1981.

CHAPTER 18

Cook, F. A., Brown, L. D., and **Oliver, J. E.** "The Southern Appalachians and the Growth of Continents."
Scientific American, (October, 1980), pp. 156–68.

Dennis, J. G. *Structural Geology.*
New York: Wiley, 1972.

Dewey, J. F., and **Bird, J. M.** "Mountain Belts and the New Global Tectonics."
Journal of Geophysical Research, Vol. 75 (1980), pp. 2625–47.

Hobbs, B. E., Means, W. D., and **Williams, P. F.** *An Outline of Structural Geology.*
New York: Wiley, 1976.

King, P. B. *The Evolution of North America,* 2nd ed.
Princeton, NJ: Princeton University Press, 1977.

Shelton, J. *Geology Illustrated.*
San Francisco: W. H. Freeman, 1966.

Spencer, E. W. *Introduction to the Structure of the Earth,* 2nd ed.
New York: McGraw-Hill, 1977.

CHAPTER 19

Berry, W. B. N. *Growth of the Prehistoric Time Scale.*
San Francisco: W. H. Freeman, 1968.

Breed, W. J., and **Roat, E. C.,** (eds.). *Geology of the Grand Canyon.*
Flagstaff, AZ: Museum of Northern Arizona and Grand Canyon Natural History Association, 1974.

Burchfield, J. D. *Lord Kelvin and the Age of the Earth.*
New York: Science History Pub., 1975.

Eicher, D. L. *Geologic Time,* 2nd ed.
Englewood Cliffs, NJ: Prentice-Hall, 1976.

Harbaugh, J. W. *Stratigraphy and Geologic Time.*
Dubuque, IA: Wm. C. Brown, 1968.

Hurley, P. M. *How Old is the Earth?*
Garden City, NY: Doubleday, 1959.

CHAPTER 20

Beatty, J. K. *The New Solar System.*
London: Cambridge University Press, 1981.

Cameron, A. G. W. "The Outer Solar System."
Science, Vol. 180 (1973), pp. 701–8.

Carr, M. H. *The Surface of Mars.*
New Haven, CT: Yale University Press, 1981.

Guest, P., Butterworth, P., Murray, J., and **O'Donnell, W.** *Planetary Geology.*
New York: Halsted Press, 1980.

Murray, B., Malin, M. C., and **Greeley, R.** *Earthlike Planets: Surfaces of Mercury, Venus, Earth, Moon, Mars.*
San Francisco: W. H. Freeman, 1981.

NASA. *Mars as Viewed by* Mariner 9.
Special Publication 329, 1976.

Ringwood, A. E. *Origin of the Earth and Moon.*
New York: Springer-Verlag, 1979.

Short, N. M. *Planetary Geology.*
Englewood Cliffs, NJ: Prentice-Hall, 1975.

GLOSSARY

Aa   A type of lava flow that has a jagged, blocky surface.

Ablation   A general term for the loss of ice and snow from a glacier.

Abrasion   The grinding and scraping of a rock surface by the friction and impact of rock particles carried by water, wind, or ice.

Absolute dating   Determination of the number of years since the occurrence of a given geologic event.

Abyssal plain   Very level area of the deep-ocean floor, usually lying at the foot of the continental rise.

Active layer   The zone above the permafrost that thaws in summer and refreezes in winter.

Aftershock   A smaller earthquake that follows the main earthquake.

Alluvial fan   A fan-shaped deposit of sediment formed when a stream's slope is abruptly reduced.

Alluvium   Unconsolidated sediment deposited by a stream.

Alpine glacier   A glacier confined to a mountain valley, which in most instances had previously been a stream valley.

Angle of repose   The steepest angle at which loose material remains stationary without sliding downslope.

Angular unconformity   An unconformity in which the older strata dip at an angle different from that of the younger beds.

Antecedent stream   A stream that continued to downcut and maintain its original course as an area along its course was uplifted by faulting or folding.

Anthracite   A hard, metamorphic form of coal that burns clean and hot.

Anticline   A fold in sedimentary strata that resembles an arch.

Aphanitic   A texture of igneous rocks in which the crystals are too small for individual minerals to be distinguished with the unaided eye.

Aquiclude   An impermeable bed that hinders or prevents groundwater movement.

Aquifer   Rock or sediment through which groundwater moves easily.

Arête   A narrow, knifelike ridge separating two adjacent glaciated valleys.

Arkose   A feldspar-rich sandstone.

Artesian well   A well in which the water rises above the level where it was initially encountered.

Asteroid   One of thousands of small planetlike bodies, ranging in size from a few hundred kilometers to less than one kilometer across. Most asteroids' orbits lie between those of Mars and Jupiter.

Asthenosphere   A subdivision of the mantle situated below the lithosphere. This zone of weak material exists below a depth of about 100 kilometers and in some regions extends as deep as 700 kilometers. The rock within this zone is easily deformed.

Astronomical theory   A theory of climatic change first developed by the Yugoslavian astronomer Milankovitch. It is based upon changes in the shape of the earth's orbit, variations in the obliquity of the earth's axis, and the wobbling of the earth's axis.

Atmosphere   The gaseous portion of a planet; the planet's envelope of air. One of the traditional subdivisions of the earth's physical environment.

Atoll   A continuous or broken ring of coral reef surrounding a central lagoon.

Atom   The smallest particle that exists as an element.

Atomic number   The number of protons in the nucleus of an atom.

Atomic weight   The average of the atomic masses of isotopes for a given element.

Aureole   A zone or halo of contact metamorphism found in the country rock surrounding an igneous intrusion.

Azoic zone   A well-known but incorrect theory formulated around 1850 by Edward Forbes stating that no life existed in the ocean below a depth of about 550 meters.

Back swamp   A poorly drained area on a floodplain resulting when natural levees are present.

Bajada   An apron of sediment along a mountain front created by the coalescence of alluvial fans.

Barchan dune   A solitary sand dune shaped like a crescent with its tips pointing downwind.

Barrier island   A low, elongate ridge of sand that parallels the coast.

Basal slip   A mechanism of glacial movement in which the ice mass slides over the surface below.

Basalt   A fine-grained igneous rock of mafic composition.

Base level   The level below which a stream cannot erode.

Basin   A circular downfolded structure.

Batholith   A large mass of igneous rock that formed when magma was emplaced at depth, crystallized, and was subsequently exposed by erosion.

Baymouth bar   A sandbar that completely crosses a bay, sealing it off from the main body of water.

Beach drift   The transport of sediment in a zigzag pattern along a beach caused by the uprush of water from obliquely breaking waves.

Bedding plane   A nearly flat surface separating two beds of sedimentary rock. Each bedding plane marks the end of one deposit and the beginning of another having different characteristics.

Bed load   Sediment rolled along the bottom of a stream by moving water, or particles rolled along the ground surface by wind.

Belt of soil moisture   A zone in which water is held as a film on the surface of soil particles and may be used by plants or withdrawn by evaporation. The uppermost subdivision of the zone of aeration.

Benioff zone   The zone of inclined seismic activity that extends from a trench downward into the asthenosphere.

Biogenous sediment   Sea-floor sediments consisting of material of marine-organic origin.

Bituminous coal   The most common form of coal, often called soft, black coal.

Blowout (deflation hollow)   A depression excavated by wind in easily eroded materials.

Body wave   A seismic wave that travels through the earth's interior.

Bottomset bed   A layer of fine sediment deposited beyond the advancing edge of a delta and then buried by continued delta growth.

Braided stream   A stream consisting of numerous intertwining channels.

Breakwater A structure protecting a nearshore area from breaking waves.

Breccia A sedimentary rock composed of angular fragments that were lithified.

Cactolith A quasi-horizontal chonolith composed of anastomosing ductoliths, whose distal ends curl like a harpolith, thin like a sphenolith, or bulge discordantly like an akmolith or ethmolith.

Caldera A large depression typically caused by collapse or ejection of the summit area of a volcano.

Caliche A hard layer, rich in calcium carbonate, that forms beneath the *B* horizon in soils of arid regions.

Calving Wastage of a glacier that occurs when large pieces of ice break off into water.

Capacity The total amount of sediment a stream is able to transport.

Capillary fringe A relatively narrow zone at the base of the zone of aeration. Here water rises from the water table in tiny threadlike openings between grains of soil or sediment.

Cap rock A necessary part of an oil trap. The cap rock is impermeable and hence keeps upwardly mobile oil and gas from escaping at the surface.

Cassini gap A wide gap in the ring system of Saturn between the *A* ring and the *B* ring.

Catastrophism The concept that the earth was shaped by catastrophic events of a short-term nature.

Cavern A naturally formed underground chamber or series of chambers most commonly produced by solution activity in limestone.

Cenozoic era A time span on the geologic calendar beginning about 65 million years ago following the Mesozoic era.

Chemical weathering The processes by which the internal structure of a mineral is altered by the removal and/or addition of elements.

Cinder cone A rather small volcano built primarily of pyroclastics ejected from a single vent.

Cirque An amphitheater-shaped basin at the head of a glaciated valley produced by frost wedging and plucking.

Clastic A sedimentary rock texture consisting of broken fragments of pre-existing rock.

Cleavage The tendency of a mineral to break along planes of weak bonding.

Col A pass between mountain valleys where the headwalls of two cirques intersect.

Column A feature found in caves that is formed when a stalactite and stalagmite join.

Columnar joints A pattern of cracks that forms during cooling of molten rock to generate columns.

Coma The fuzzy, gaseous component of a comet's head.

Comet A small body which generally revolves about the sun in an elongated orbit.

Competence A measure of the largest particle a stream can transport; a factor dependent on velocity.

Composite cone A volcano composed of both lava flows and pyroclastic material.

Compound A substance formed by the chemical combination of two or more elements in definite proportions and usually having properties different from those of its constituent elements.

Concordant A term used to describe intrusive igneous masses that form parallel to the bedding of the surrounding rock.

Cone of depression A cone-shaped depression immediately surrounding a well.

Conformable layers Rock layers that were deposited without interruption.

Conglomerate A sedimentary rock composed of rounded gravel-sized particles.

Contact metamorphism Changes in rock caused by the heat from a nearby magma body.

Continental drift A hypothesis, credited largely to Alfred Wegener, that suggested all present continents once existed as a single supercontinent. Further, beginning about 200 million years ago, the supercontinent began breaking into smaller continents which then "drifted" to their present positions.

Continental glacier A massive accumulation of ice that covers extensive land areas and whose flow is not usually controlled by the underlying topography.

Continental margin That portion of the sea floor adjacent to the continents. It may include the continental shelf, continental slope, and continental rise.

Continental rise The gently sloping surface at the base of the continental slope.

Continental shelf The gently sloping submerged portion of the continental margin extending from the shoreline to the continental slope.

Continental slope The steep gradient that leads to the deep-ocean floor and marks the seaward edge of the continental shelf.

Convergent boundary A boundary in which two plates move together, causing one of the slabs of lithosphere to be consumed into the mantle as it descends beneath an overriding plate.

Correlation Establishing the equivalence of rocks of similar age in different areas.

Covalent bond A chemical bond produced by the sharing of electrons.

Crater The depression at the summit of a volcano, or that which is produced by a meteorite impact.

Creep The slow downhill movement of soil and regolith.

Crevasse A deep crack in the brittle surface of a glacier.

Cross-cutting A principle of relative dating. A rock or fault is younger than any rock (or fault) through which it cuts.

Crust The very thin outermost layer of the earth.

Crystal An orderly arrangement of atoms.

Crystal form The external appearance of a mineral as determined by its internal arrangement of atoms.

Crystallization The formation and growth of a crystalline solid from a liquid or gas.

Curie point The temperature above which a material loses its magnetization.

Cut bank The area of active erosion on the outside of a meander.

Cutoff A short channel segment created when a river erodes through the narrow neck of land between meanders.

Darcy's law When permeability is uniform, the velocity of groundwater increases as the slope of the water table increases. It is expressed by the formula: $V = K\frac{h}{l}$, where V is velocity, h the head, l the length of flow, and K the coefficient of permeability.

Daughter product An isotope resulting from radioactive decay.

Deep-focus earthquake   An earthquake focus at a depth of more than 300 kilometers.

Deep-ocean basin   The portion of sea floor that lies between the continental margin and the oceanic ridge system. This region comprises almost 30 percent of the earth's surface.

Deflation   The lifting and removal of loose material by wind.

Delta   An accumulation of sediment formed where a stream enters a lake or ocean.

Dendritic pattern   A stream system that resembles the pattern of a branching tree.

Density   The weight per unit volume of a particular material.

Desalination   The removal of salts and other chemicals from seawater.

Desert pavement   A layer of coarse pebbles and gravel created when wind removed the finer material.

Detrital sedimentary rocks   Rocks that form from the accumulation of materials that originate and are transported as solid particles derived from both mechanical and chemical weathering.

Dike   A tabular-shaped intrusive igneous feature that cuts through the surrounding rock.

Dip   The angle at which a rock layer is inclined from the horizontal. The direction of dip is at a right angle to the strike.

Dip-slip fault   A fault in which the movement is parallel to the dip of the fault.

Discharge   The quantity of water in a stream that passes a given point in a period of time.

Disconformity   A type of unconformity in which the beds above and below are parallel.

Discontinuity   A sudden change with depth in one or more of the physical properties of the material making up the earth's interior. The boundary between two dissimilar materials in the earth's interior as determined by the behavior of seismic waves.

Discordant   A term used to describe plutons that cut across existing rock structures, such as bedding planes.

Dissolved load   That portion of a stream's load carried in solution.

Distributary   A section of a stream that leaves the main flow.

Diurnal tide   A tide characterized by a single high and low water height each tidal day.

Divergent boundary   A boundary in which two plates move apart, resulting in upwelling of material from the mantle to create new sea floor.

Divide   An imaginary line that separates the drainage of two streams; often found along a ridge.

Dome   A roughly circular upfolded structure.

Drainage basin   The land area that contributes water to a stream.

Drawdown   The difference in height between the bottom of a cone of depression and the original height of the water table.

Drift   The general term for any glacial deposit.

Drumlin   A streamlined asymmetrical hill composed of glacial till. The steep side of the hill faces the direction from which the ice advanced.

Dry climate   A climate in which yearly precipitation is less than the potential loss of water by evaporation.

Dune   A hill or ridge of wind-deposited sand.

Earthflow   The downslope movement of water-saturated, clay-rich sediment. Most characteristic of humid regions.

Earthquake   Vibration of the earth produced by the rapid release of energy.

Ebb tide   The movement of tidal current away from the shore.

Echo sounder   An instrument used to determine the depth of water by measuring the time interval between emission of a sound signal and the return of its echo from the bottom.

Effluent stream   A stream channel that intersects the water table. Consequently, groundwater feeds into the stream.

Elastic deformation   Nonpermanent deformation in which rock returns to its original shape when the stress is released.

Elastic rebound   The sudden release of stored strain in rocks that results in movement along a fault.

Electron   A negatively charged subatomic particle that has a negligible mass and is found outside an atom's nucleus.

Element   A substance that cannot be decomposed into simpler substances by ordinary chemical or physical means.

Emergent coast   A coast where land formerly below sea level has been exposed either by crustal uplift or a drop in sea level or both.

End moraine   A ridge of till marking a former position of the front of a glacier.

Entrenched meander   A meander cut into bedrock when uplifting rejuvenated a meandering stream.

Epicenter   The location on the earth's surface that lies directly above the focus of an earthquake.

Epoch   A unit of the geologic calendar that is a subdivision of a period.

Era   A major division on the geologic calendar; eras are divided into shorter units called periods.

Erosion   The incorporation and transportation of material by a mobile agent, such as water, wind, or ice.

Esker   Sinuous ridge composed largely of sand and gravel deposited by a stream flowing in a tunnel beneath a glacier near its terminus.

Estuary   A funnel-shaped inlet of the sea that formed when a rise in sea level or subsidence of land caused the mouth of a river to be flooded.

Eugeosyncline   The portion of a geosyncline seaward of the miogeosyncline in which predominately deep-water deposits accumulate, including graywackes, lava flows, volcanic debris, and shales.

Evaporite   A sedimentary rock formed of material deposited from solution by evaporation of the water.

Evapotranspiration   The combined effect of evaporation and transpiration.

Exotic stream   A permanent stream that traverses a desert and has its source in well-watered areas outside the desert.

Extrusive   Igneous activity that occurs at the earth's surface.

Fall   A type of movement common to mass wasting processes that refers to the free falling of detached individual pieces of any size.

Fault   A break in a rock mass along which movement has occurred.

Fault-block mountain A mountain formed by the displacement of rock along a fault.

Faunal succession Fossil organisms succeed one another in a definite and determinable order, and any time period can be recognized by its fossil content.

Fetch The distance that the wind has traveled across the open water.

Fiord A steep-sided inlet of the sea formed when a glacial trough was partially submerged.

Firn Granular recrystallized snow. A transitional stage between snow and glacial ice.

Fissure eruption An eruption in which lava is extruded from narrow fractures or cracks in the crust.

Flood basalts Flows of basaltic lava that issue from numerous cracks or fissures and commonly cover extensive areas to thicknesses of hundreds of meters.

Floodplain The flat, low-lying portion of a stream valley subject to periodic inundation.

Flood tide The tidal current associated with the increase in the height of the tide.

Flow A type of movement common to mass wasting processes in which water-saturated material moves downslope as a viscous fluid.

Fluorescence The absorption of ultraviolet light, which is re-emitted as visible light.

Focus (earthquake) The zone within the earth where rock displacement produces an earthquake.

Foliated A texture of metamorphic rocks that gives the rock a layered appearance.

Foreset bed An inclined bed deposited along the front of a delta.

Foreshocks Small earthquakes that often precede a major earthquake.

Fossil The remains or traces of organisms preserved from the geologic past.

Fractional crystallization The process that separates magma into components having varied compositions and melting points.

Frost wedging The mechanical breakup of rock caused by the expansion of freezing water in cracks and crevices.

Fumarole A vent in a volcanic area from which fumes or gases escape.

Geology The science that examines the earth, its form and composition, and the changes which it has undergone and is undergoing.

Geosyncline A large linear downwarp in the earth's crust in which thousands of meters of sediment have accumulated.

Geothermal energy Natural steam used for power generation.

Geothermal gradient The gradual increase in temperature with depth in the crust. The average is 30°C per kilometer in the upper crust.

Geyser A fountain of hot water ejected periodically from the ground.

Glacial erratic An ice-transported boulder that was not derived from the bedrock near its present site.

Glacial striations Scratches and grooves on bedrock caused by glacial abrasion.

Glacial trough A mountain valley that has been widened, deepened, and straightened by a glacier.

Glacier A thick mass of ice originating on land from the compaction and recrystallization of snow that shows evidence of past or present flow.

Glass (volcanic) Natural glass produced when molten lava cools too rapidly to permit crystallization. Volcanic glass is a solid composed of unordered atoms.

Glassy A term used to describe the texture of certain igneous rocks, such as obsidian, that contain no crystals.

Gondwanaland The southern portion of Pangaea consisting of South America, Africa, Australia, India, and Antarctica.

Graben A valley formed by the downward displacement of a fault-bounded block.

Graded bed A sediment layer characterized by a decrease in sediment size from bottom to top.

Graded stream A stream that has the correct channel characteristics to maintain exactly the velocity required to transport the material supplied to it.

Gradient The slope of a stream; generally measured in feet per mile.

Granitization The process of converting country rock into granite. The process is thought to occur when hot, ion-rich fluids migrate through a rock and chemically alter its composition.

Greenhouse effect Carbon dioxide and water vapor in a planet's atmosphere absorb and re-radiate infrared wavelengths, effectively trapping solar energy and raising the temperature.

Groin A short wall built at a right angle to the seashore to trap moving sand.

Groundmass The matrix of smaller crystals within an igneous rock that has porphyritic texture.

Ground moraine An undulating layer of till deposited as the ice front retreats.

Groundwater Water in the zone of saturation.

Guyot A submerged flat-topped seamount.

Half-life The time required for one-half of the atoms of a radioactive substance to decay.

Hanging valley A tributary valley that enters a glacial trough at a considerable height above the floor of the trough.

Hardness A mineral's resistance to scratching and abrasion.

Head The vertical distance between the recharge and discharge points of a water table.

Headward erosion The extension upslope of the head of a valley due to erosion.

Historical geology A major division of geology that deals with the origin of the earth and its development through time. Usually involves the study of fossils and their sequence in rock beds.

Hogback A narrow, sharp-crested ridge formed by the upturned edge of a steeply dipping bed of resistant rock.

Horn A pyramid-like peak formed by glacial action in three or more cirques surrounding a mountain summit.

Horst An elongate, uplifted block of crust bounded by faults.

Hot spot A proposed concentration of heat in the mantle capable of producing magma which, in turn, extrudes onto the earth's surface. The intraplate volcanism that produced the Hawaiian Islands is one example.

Hot spring A spring in which the water is 6–9°C (10–15°F) warmer than the mean annual air temperature of its locality.

Humus Organic matter in soil produced by the decomposition of plants and animals.

Hydraulic gradient The slope of the water table. Expressed as h/l, where h is the head and l the length of flow.

Hydrogenous sediment Sea-floor sediments consisting of minerals that crystallize from seawater. The principal example is manganese nodules.

Hydrologic cycle The unending circulation of the earth's water supply. The cycle is powered by energy from the sun and is characterized by continuous exchanges of water among the oceans, the atmosphere, and the continents.

Hydrolysis A chemical weathering process in which minerals are altered by chemically reacting with water and acids.

Hydrosphere The water portion of our planet; one of the traditional subdivisions of the earth's physical environment.

Hydrothermal solution The hot, watery solution that escapes from a mass of magma during the latter stages of crystallization. Such solutions may alter the surrounding country rock and are frequently the source of significant ore deposits.

Ice-contact deposit An accumulation of stratified drift deposited in contact with a supporting mass of ice.

Igneous rock A rock formed by the crystallization of molten magma.

Immature soil A soil lacking horizons.

Index fossil A fossil that is associated with a particular span of geologic time.

Index mineral A mineral that is a good indicator of the metamorphic environment in which it formed. Used to distinguish different zones of regional metamorphism.

Inertia Objects at rest tend to remain at rest and objects in motion tend to stay in motion unless either is acted upon by an outside force.

Infiltration The movement of surface water into rock or soil through cracks and pore spaces.

Infiltration capacity The maximum rate at which soil can absorb water.

Influent stream A stream channel that is above the water table level. Water seeps downward from the channel to the zone of saturation to produce an upward bulge in the water table.

Inner core The solid innermost layer of the earth, about 1216 kilometers (754 miles) in radius.

Inselberg An isolated mountain remnant characteristic of the late stage of erosion in a mountainous arid region.

Interior drainage A discontinuous pattern of intermittent streams that do not flow to the ocean.

Intermediate focus An earthquake focus at a depth of between 60 and 300 kilometers.

Intrusive rock Igneous rock that formed below the earth's surface.

Ion An atom or molecule that possesses an electrical charge.

Ionic bond A chemical bond between two oppositely charged ions formed by the transfer of valence electrons from one atom to the other.

Irons One of the three main categories of meteorites. This group is composed largely of iron with varying amounts of nickel (5–20 percent). Up to 60 percent of all meteorites are of this type.

Island arc A chain of volcanic islands generally located a few hundred kilometers from a trench where active subduction of one oceanic slab beneath another is occurring.

Isostasy   The concept that the earth's crust is "floating" in gravitational balance upon the material of the mantle.

Isotopes   Varieties of the same element that have different mass numbers; their nuclei contain the same number of protons but different numbers of neutrons.

Jetties   A pair of structures extending into the ocean at the entrance to a harbor or river that are built for the purpose of protecting against storm waves and sediment deposition.

Joint   A fracture in rock along which there has been no movement.

Jovian planet   The Jupiter-like planets Jupiter, Saturn, Uranus, and Neptune. These planets have relatively low densities.

Kame   A steep-sided hill composed of sand and gravel originating when sediment collected in openings in stagnant glacial ice.

Kame terrace   A narrow, terrace-like mass of stratified drift deposited between a glacier and an adjacent valley wall.

Karst   A topography consisting of numerous depressions called sinkholes.

Kettle holes   Depressions created when blocks of ice become lodged in glacial deposits and subsequently melt.

Laccolith   A massive igneous body intruded between pre-existing strata.

Laminar flow   The movement of water particles in straightline paths that are parallel to the channel. The water particles move downstream without mixing.

Lateral moraine   A ridge of till along the sides of an alpine glacier composed primarily of debris that fell to the glacier from the valley walls.

Laterite   A red, highly leached soil type found in the tropics that is rich in oxides of iron and aluminum.

Laurasia   The northern portion of Pangaea consisting of North America and Eurasia.

Lava   Magma that reaches the earth's surface.

Lava dome   A bulbous mass associated with an old-age volcano, produced when thick lava is slowly squeezed from the vent. Lava domes may act as plugs to deflect subsequent gaseous eruptions.

Law of superposition   In any undeformed sequence of sedimentary rocks, each bed is older than the one above it and younger than the one below.

Lithification   The process, generally cementation and/or compaction, of converting sediments to solid rock.

Lithosphere   The rigid outer layer of the earth, including the crust and upper mantle.

Loess   Deposits of windblown silt, lacking visible layers, generally buff colored, and capable of maintaining a nearly vertical cliff.

Longitudinal (seif) dunes   Long ridges of sand oriented parallel to the prevailing wind; these dunes form where sand supplies are limited.

Longitudinal profile   A cross section of a stream channel along its descending course from the head to the mouth.

Longshore current   A nearshore current that flows parallel to the shore.

Long (L) waves   These earthquake-generated waves travel along the outer layer of the earth and are responsible for most of the surface damage. L waves have longer periods than other seismic waves.

Low-velocity zone   A subdivision of the mantle located between 100 and 250 kilometers and discernible by a marked decrease in the velocity of seismic waves. This zone does not encircle the earth.

Lunar breccia   A lunar rock formed when angular fragments and dust are welded together by the heat generated by the impact of a meteoroid.

Luster   The appearance or quality of light reflected from the surface of a mineral.

Magma   A body of molten rock found at depth, including any dissolved gases and crystals.

Magnetometer   A sensitive instrument used to measure the intensity of the earth's magnetic field at various points.

Magnitude (earthquake)   The total amount of energy released during an earthquake.

Manganese nodules   A type of hydrogenous sediment scattered on the ocean floor, consisting mainly of manganese and iron, and usually containing small amounts of copper, nickel, and cobalt.

Mantle   The 2885-kilometer (1789-mile) thick layer of the earth located below the crust.

Maria   The smooth areas on our moon's surface that were incorrectly thought to be seas.

Mass number   The sum of the number of neutrons and protons in the nucleus of an atom.

Mass wasting   The downslope movement of rock, regolith, and soil under the direct influence of gravity.

Meander   A looplike bend in the course of a stream.

Meander scar   A floodplain feature created when an oxbow lake becomes filled with sediment.

Mechanical weathering   The physical disintegration of rock, resulting in smaller fragments.

Medial moraine   A ridge of till formed when lateral moraines from two coalescing alpine glaciers join.

Mélange   A highly deformed mixture of rock material formed in areas of plate convergence.

Melt   The liquid portion of magma excluding the solid crystals.

Mercalli intensity scale   A 12-point scale originally developed to evaluate earthquake intensity based upon the amount of damage to various types of structures.

Mesozoic era   A time span on the geologic calendar between the Paleozoic and Cenozoic eras—from about 225 to 65 million years ago.

Metallic bond   A chemical bond present in all metals that may be characterized as an extreme type of electron sharing in which the electrons move freely from atom to atom.

Metamorphic rock   Rock formed by the alteration of pre-existing rock deep within the earth (but still in the solid state) by heat, pressure, and/or chemically active fluids.

Metamorphism   The changes in mineral composition and texture of a rock subjected to high temperature and pressure within the earth.

Meteorite   Any portion of a meteoroid that survives its traverse through the earth's atmosphere and strikes the surface.

Meteoroid   Any small solid particle that has an orbit in the solar system.

Meteor shower   Numerous meteoroids traveling in the same direction and at nearly the same speed. They are thought to be material lost by comets.

Micrometeorite   A very small meteorite that does not create sufficient friction to burn up in the atmosphere, but slowly drifts down to the earth.

Mid-ocean ridge   A continuous mountainous ridge on the floor of all the major ocean basins and varying in width from 500–5000 kilometers (300–3000 miles). The rifts at the crests of these ridges represent divergent plate boundaries.

Migmatite   A rock exhibiting both igneous and metamorphic rock characteristics. Such rocks may form when light-colored silicate minerals melt and then crystallize, while the dark silicate minerals remain solid.

Mineral   A naturally occurring, inorganic crystalline material with a unique chemical structure.

Miogeosyncline   The portion of a geosyncline landward of the eugeosyncline and characterized by deposits of clean sandstones, limestones, and shales.

Mohorovičić discontinuity (Moho)   The boundary separating the crust and the mantle, discernible by an increase in seismic velocity.

Mohs scale   A series of ten minerals used as a standard in determining hardness.

Monocline   A one-limbed flexure in strata. The strata are usually flat lying or very gently dipping on both sides of the monocline.

Mud crack   A feature in some sedimentary rocks that forms when wet mud dries out, shrinks, and cracks.

Mudflow   The flowage of debris containing a large amount of water; most characteristic of canyons and gullies in dry, mountainous regions.

Natural levees   The elevated landforms composed of alluvium that parallel some streams and act to confine their waters, except during floodstage.

Neap tide   The lowest tidal range, occurring near the times of the first and third quarters of the moon.

Neutron   A subatomic particle found in the nucleus of an atom. The neutron is electrically neutral with a mass approximately equal to that of a proton.

Nonclastic   A term for the texture of sedimentary rocks in which the minerals form a pattern of interlocking crystals.

Nonfoliated   Metamorphic rocks that do not exhibit foliation.

Normal fault   A fault in which the rock above the fault plane has moved down relative to the rock below.

Normal polarity   A magnetic field the same as that which presently exists.

Nucleus   The small heavy core of an atom that contains all of its positive charge and most of its mass.

Nuée ardente   Incandescent volcanic debris buoyed up by hot gases that moves downslope in an avalanche fashion.

Oblique-slip fault   A fault having both vertical and horizontal movement.

Occultation   The disappearance of light resulting when one object passes behind an apparently larger one. For example, the passage of Uranus in front of a distant star.

Octet rule   Atoms combine in order that each may have the electron arrangement of a noble gas; that is, the outer energy level contains eight electrons.

Oil trap   A geologic structure that allows for significant amounts of oil and gas to accumulate.

Ophiolite complex   The sequence of rocks that make up the oceanic crust. The three-layer sequence includes an upper layer of pillow basalts, a middle zone of sheeted dikes, and a lower layer of gabbro.

Original horizontality   Layers of sediment are generally deposited in a horizontal or nearly horizontal position.

Orogenesis   The processes that collectively result in the formation of mountains.

Outer core   A layer beneath the mantle about 2270 kilometers (1410 miles) thick which has the properties of a liquid.

Outwash plain   A relatively flat, gently sloping plain consisting of materials deposited by meltwater streams in front of the margin of an ice sheet.

Oxbow lake   A curved lake produced when a stream cuts off a meander.

Oxidation   The removal of one or more electrons from an atom or ion. So named because elements commonly combine with oxygen.

Pahoehoe   A lava flow with a smooth-to-ropy surface.

Paleomagnetism   The natural remnant magnetism in rock bodies. The permanent magnetization acquired by rock which can be used to determine the location of the magnetic poles and the latitude of the rock at the time it became magnetized.

Paleontology   The systematic study of fossils and the history of life on earth.

Paleozoic era   A time span on the geologic calendar between the Precambrian and Mesozoic eras—from about 600 million to 225 million years ago.

Pangaea   The proposed supercontinent which 200 million years ago began to break apart and form the present landmasses.

Parabolic dune   A sand dune similar in shape to a barchan dune except that its tip points into the wind. These dunes often form along coasts that have strong onshore winds, abundant sand, and vegetation that partly covers the sand.

Parasitic cone   A volcanic cone which forms on the flank of a larger volcano.

Parent material   The material upon which a soil develops.

Partial melting   The process by which most igneous rocks melt. Since individual minerals have different melting points, most igneous rocks melt over a temperature range of a few hundred degrees. If the liquid is squeezed out after some melting has occurred, a melt with a higher silica content results.

Pater noster lakes   A chain of small lakes in a glacial trough that occupy basins created by glacial erosion.

Pedalfer   Soil of humid regions characterized by the accumulation of iron oxides and aluminum-rich clays in the *B* horizon.

Pediment   A sloping bedrock surface fringing a mountain base in an arid region, formed when erosion causes the mountain front to retreat.

Pedocal   Soil associated with drier regions and characterized by an accumulation of calcium carbonate in the upper horizons.

Pegmatite   A very coarse-grained igneous rock (typically granite) commonly found as a dike associated with a large mass of plutonic rock that has smaller crystals. Crystallization in a water-rich environment is believed to be responsible for the very large crystals.

Peneplain   In the idealized cycle of landscape evolution in a humid region, an undulating plain near base level associated with old age.

Perched water table   A localized zone of saturation above the main water table created by an impermeable layer (aquiclude).

Peridotite   An igneous rock of ultramafic composition thought to be abundant in the upper mantle.

Period   A basic unit of the geologic calendar that is a subdivision of an era. Periods may be divided into smaller units called epochs.

Permafrost   Any permanently frozen subsoil. Usually found in the subarctic and arctic regions.

Permeability   A measure of a material's ability to transmit water.

Phaneritic   An igneous rock texture in which the crystals are roughly equal in size and large enough so that individual minerals can be identified with the unaided eye.

Phenocryst   Conspicuously large crystals imbedded in a matrix of finer-grained crystals.

Physical geology   A major division of geology that examines the materials of the earth and seeks to understand the processes and forces acting beneath and upon the earth's surface.

Pillow lava   Basaltic lava that solidifies in an underwater environment and develops a structure that resembles a pile of pillows.

Plastic deformation   Permanent deformation that results in a change in size and shape through folding or flowing.

Plastic flow   A type of glacial movement that occurs within the glacier, below a depth of approximately 50 meters, in which the ice is not fractured.

Plate   One of numerous rigid sections of the lithosphere that moves as a unit over the material of the asthenosphere.

Plate tectonics   The theory which proposes that the earth's outer shell consists of individual plates which interact in various ways and thereby produce earthquakes, volcanoes, mountains, and the crust itself.

Playa   The flat central area of an undrained desert basin.

Playa lake   A temporary lake in a playa.

Playfair's law   A well-known and oft-quoted statement by John Playfair that states a valley is the result of the work of the stream that flows in it.

Pleistocene epoch   An epoch of the Quaternary period beginning about 2.5 million years ago and ending about 10,000 years ago. Best known as a time of extensive continental glaciation.

Plucking (quarrying)   The process by which pieces of bedrock are lifted out of place by a glacier.

Pluton   A structure that results from the emplacement and crystallization of magma beneath the surface of the earth.

Pluvial lake   A lake formed during a period of increased rainfall. For example, this occurred in many nonglaciated areas during periods of ice advance elsewhere.

Point bar   A crescent-shaped accumulation of sand and gravel deposited on the inside of a meander.

Polar wandering hypothesis   As the result of paleomagnetic studies in the 1950s, researchers proposed that either the magnetic poles migrated greatly through time or the continents had gradually shifted their positions.

Polymorphs   Two or more minerals having the same chemical composition but different crystalline structures. Exemplified by the diamond and graphite forms of carbon.

Porosity   The volume of open spaces in rock or soil.

Porphyritic   An igneous rock texture characterized by two distinctively different crystal sizes. The larger crystals are called phenocrysts while the matrix of smaller crystals is termed the groundmass.

Porphyry   An igneous rock with a porphyritic texture.

Pothole   A depression formed in a stream channel by the abrasive action of the water's sediment load.

Precambrian   All geologic time prior to the Paleozoic era.

Principle of faunal succession   Fossil organisms succeed one another in a definite and determinable order, and any time period can be recognized by its fossil content.

Principle of original horizontality   Layers of sediment are generally deposited in a horizontal or nearly horizontal position.

Proton   A positively charged subatomic particle found in the nucleus of an atom.

P wave   The fastest earthquake wave, which travels by compression and expansion of the medium.

Pyroclastic material   The volcanic rock ejected during an eruption. Pyroclastics include ash, bombs, and blocks.

Pyroclastic   An igneous rock texture resulting from the consolidation of individual rock fragments that are ejected during a violent eruption.

Radial drainage   A system of streams running in all directions away from a central elevated structure, such as a volcano.

Radioactivity   The spontaneous decay of certain unstable atomic nuclei.

Radiocarbon (carbon-14)   The radioactive isotope of carbon, which is produced continuously in the atmosphere and used in dating events as far back as 75,000 years.

Radiometric dating   The procedure of calculating the absolute ages of rocks and minerals that contain certain radioactive isotopes.

Rainshadow   A dry area on the lee side of a mountain range.

Rays   Bright streaks that appear to radiate from certain craters on the lunar surface. The rays consist of fine debris ejected from the primary crater.

Recessional moraine   An end moraine formed as the ice front stagnated during glacial retreat.

Rectangular pattern   A drainage pattern characterized by numerous right angle bends that develops on jointed or fractured bedrock.

Refraction   A change in direction of waves as they enter shallow water. The portion of the wave in shallow water is slowed, which causes the wave to bend and align with the underwater contours.

Regional metamorphism   Metamorphism associated with large-scale mountain building.

Regolith   The layer of rock and mineral fragments that nearly everywhere covers the earth's land surface.

Rejuvenation   A change in relation to base level, often caused by regional uplift, that causes the forces of erosion to intensify.

Relative dating   Rocks are placed in their proper sequence or order. Only the chronological order of events is determined.

Reservoir rock   The porous, permeable portion of an oil trap that yields oil and gas.

Residual soil   Soil developed directly from the weathering of the bedrock below.

Reverse fault   A fault in which the material above the fault plane moves up in relation to the material below.

Reverse polarity   A magnetic field opposite to that which presently exists.

Richter scale   A scale of earthquake magnitude based on the motion of a seismograph.

Rift   A region of the earth's crust along which divergence is taking place.

Ripple marks   Small waves of sand that develop on the surface of a sediment layer by the action of moving water or air.

Roche moutonnée   An asymmetrical knob of bedrock formed when glacial abrasion smoothes the gentle slope facing the advancing ice sheet and plucking steepens the opposite side as the ice overrides the knob.

Rock   A consolidated mixture of minerals.

Rock avalanche   The very rapid downslope movement of rock and debris. These rapid movements may be aided by a layer of air trapped beneath the debris, and they have been known to reach speeds in excess of 200 kilometers per hour.

Rock cleavage   The tendency of rock to split along parallel, closely spaced surfaces. These surfaces are often highly inclined to the bedding planes in the rock.

Rock flour   Ground-up rock produced by the grinding effect of a glacier.

Rockslide   The rapid slide of a mass of rock downslope along planes of weakness.

Runoff   Water that flows over the land rather than infiltrating into the ground.

Salinity   The proportion of dissolved salts to pure water, usually expressed in parts per thousand ($^0/_{00}$).

Saltation   Transportation of sediment through a series of leaps or bounces.

Salt flat   A white crust on the ground produced when water evaporates and leaves its dissolved materials behind.

Schistosity   A type of foliation characteristic of coarser-grained metamorphic rocks. Such rocks have a parallel arrangement of platy minerals such as the micas.

Scoria   Hardened lava which has retained the vesicles produced by the escaping gases.

Sea arch   An arch formed by wave erosion when caves on opposite sides of a headland unite.

Sea-flooring spreading   The hypothesis first proposed in the 1960s by Harry Hess which suggested that new oceanic crust is produced at the crests of mid-ocean ridges, which are the sites of divergence.

Seamount   An isolated volcanic peak that rises at least 1000 meters (3300 feet) above the deep-ocean floor.

Sea stack   An isolated mass of rock standing just offshore, produced by wave erosion of a headland.

Sediment   Unconsolidated particles created by the weathering and erosion of rock, by chemical precipitation from solution in water, or from the secretions of organisms, and transported by water, wind, or glaciers.

Sedimentary rock   Rock formed from the weathered products of pre-existing rocks that have been transported, deposited, and lithified.

Seiche   The rhythmic sloshing of water in lakes, reservoirs, and other smaller enclosed basins. Some seiches are initiated by earthquake activity.

Seismic sea wave   A rapid moving ocean wave generated by earthquake activity which is capable of inflicting heavy damage in coastal regions.

Seismogram   The record made by a seismograph.

Seismograph   An instrument that records earthquake waves.

Seismology   The study of earthquakes and seismic waves.

Settling velocity   The speed at which a particle falls through a still fluid. The size, shape, and specific gravity of particles influence settling velocity.

Shadow zone   The zone between 105 and 140 degrees distance from an earthquake epicenter which direct waves do not penetrate because of refraction by the earth's core.

Shallow-focus earthquake   An earthquake focus at a depth of less than 60 kilometers.

Shear   Stress that causes two adjacent parts of a body to slide past one another.

Sheeting   A mechanical weathering process characterized by the splitting off of slablike sheets of rock.

Shelf break   The point at which a rapid steepening of the gradient occurs, marking the outer edge of the continental shelf and the beginning of the continental slope.

Shield   A large, relatively flat expanse of ancient metamorphic rock within the stable continental interior.

Shield volcano   A broad, gently sloping volcano built from fluid basaltic lavas.

Silicate   Any one of numerous minerals that have the silicon-oxygen tetrahedron as their basic structure.

Silicon-oxygen tetrahedron   A structure composed of four oxygen atoms surrounding a silicon atom that constitutes the basic building block of silicate minerals.

Sill   A tabular igneous body that was intruded parallel to the layering of pre-existing rock.

Sinkhole   A depression produced in a region where soluble rock has been removed by groundwater.

Slide   A movement common to mass wasting processes in which the material moving downslope remains fairly coherent and moves along a well-defined surface.

Slip face   The steep, leeward surface of a sand dune which maintains a slope of about 34 degrees.

Slump   The downward slipping of a mass of rock or unconsolidated material moving as a unit along a curved surface.

Snowfield   An area where snow persists throughout the year.

Snowline   Lower limit of perennial snow.

Soil   A combination of mineral and organic matter, water, and air; that portion of the regolith that supports plant growth.

Soil horizon   A layer of soil that has identifiable characteristics produced by chemical weathering and other soil-forming processes.

Soil profile A vertical section through a soil showing its succession of horizons and the underlying parent material.

Solifluction Slow, downslope flow of water-saturated materials common to permafrost areas.

Solum The *A* and *B* horizons in a soil profile. Living roots and other plant and animal life are largely confined to this zone.

Solution The change of matter from the solid or gaseous state into the liquid state by its combination with a liquid.

Specific gravity The ratio of a substance's weight to the weight of an equal volume of water.

Speleothem A collective term for the dripstone features found in caverns.

Spheroidal weathering Any weathering process that tends to produce a spherical shape from an initially blocky shape.

Spit An elongate ridge of sand that projects from the land into the mouth of an adjacent bay.

Spring A flow of groundwater that emerges naturally at the ground surface.

Spring tide The highest tidal range. Occurs near the times of the new and full moons.

Stalactite The iciclelike structure that hangs from the ceiling of a cavern.

Stalagmite The columnlike form that grows upward from the floor of a cavern.

Stock A pluton similar to but smaller than a batholith.

Stony-irons One of the three main categories of meteorites. This group represents between 1 and 2 percent of all meteorites and, as the name implies, is a mixture of iron and silicate minerals.

Stony meteorite One of the three main categories of meteorites. This group constitutes about 92 percent of all meteorites and is composed largely of silicate minerals with inclusions of other minerals.

Strata Parallel layers of sedimentary rock.

Stratified drift Sediments deposited by glacial meltwater.

Stratovolcano See *Composite cone.*

Streak The color of a mineral in powdered form.

Stream A general term to denote the flow of water within any natural channel. Thus, a small creek and a large river are both streams.

Stream piracy The diversion of the drainage of one stream resulting from the headward erosion of another stream.

Stress The force per unit area acting on any surface within a solid. Also known as *directed pressure.*

Striations (glacial) Scratches or grooves in a bedrock surface caused by the grinding action of a glacier and its load of sediment.

Strike The compass direction of the line of intersection created by a dipping bed or fault and a horizontal surface. Strike is always perpendicular to the direction of dip.

Strike-slip fault A fault along which the movement is horizontal.

Subduction The process of thrusting oceanic lithosphere into the mantle along a convergent zone.

Submarine canyon A seaward extension of a valley that was cut on the continental shelf during a time when sea level was lower, or a canyon carved into the outer continental shelf, slope, and rise by turbidity currents.

Submergent coast A coast whose form is largely the result of the partial drowning of a former land surface either due to a rise of sea level or subsidence of the crust, or both.

Subsoil A term applied to the *B* horizon of a soil profile.

Surf A collective term for breakers; also the wave activity in the area between the shoreline and the outer limit of breakers.

Surface soil The uppermost layer in a soil profile: the *A* horizon.

Surface waves Seismic waves that travel along the outer layer of the earth.

Surge A period of rapid glacial advance. Surges are typically sporadic and short lived.

Suspended load The fine sediment carried within the body of flowing water or air.

S wave An earthquake wave, slower than a P wave, that travels only in solids.

Swells Wind-generated waves that have moved into an area of weaker winds or calm.

Syncline A linear downfold in sedimentary strata; the opposite of anticline.

Talus An accumulation of rock debris at the base of a cliff.

Tarn A small lake in a cirque.

Tectonics The study of the large-scale processes that collectively deform the earth's crust.

Temporary (local) base level The level of a lake, resistant rock layer, or any other base level that stands above sea level.

Terminal moraine The end moraine marking the farthest advance of a glacier.

Terrace A flat, benchlike structure produced by a stream, which was left elevated as the stream cut downward.

Terrestrial planet The Earth-like planets Mercury, Venus, Earth, and Mars. These planets have similar densities.

Terrigenous sediment Sea-floor sediments derived from terrestrial weathering and erosion.

Texture The size, shape, and distribution of the particles that collectively constitute a rock.

Thrust fault A low-angle reverse fault.

Tidal current The alternating horizontal movement of water associated with the rise and fall of the tide.

Tidal flat A marshy or muddy area that is alternately covered and uncovered by the rise and fall of the tide.

Tide Periodic change in the elevation of the ocean surface.

Till Unsorted sediment deposited directly by a glacier.

Tillite A rock formed when glacial till is lithified.

Tombolo A ridge of sand that connects an island to the mainland or to another island.

Topset bed An essentially horizontal sedimentary layer deposited on top of a delta during floodstage.

Transform fault boundary   A boundary in which two plates slide past one another without creating or destroying lithosphere.

Transpiration   The release of water vapor to the atmosphere by plants.

Transported soil   Soils that form on unconsolidated deposits.

Transverse dunes   A series of long ridges oriented at right angles to the prevailing wind; these dunes form where vegetation is sparse and sand is very plentiful.

Travertine   A form of limestone ($CaCO_3$) that is deposited by hot springs or as a cave deposit.

Trellis drainage   A system of streams in which nearly parallel tributaries occupy valleys cut in folded strata.

Trench   An elongate depression in the sea floor produced by bending of oceanic crust during subduction.

Truncated spurs   Triangular-shaped cliffs produced when spurs of land that extend into a valley are removed by the great erosional force of an alpine glacier.

Tsunami   The Japanese word for a seismic sea wave.

Turbidite   Turbidity current deposit characterized by graded bedding.

Turbidity current   A downslope movement of dense, sediment-laden water created when sand and mud on the continental shelf and slope are dislodged and thrown into suspension.

Turbulent flow   The movement of water in an erratic fashion often characterized by swirling, whirlpool-like eddies. Most streamflow is of this type.

Ultimate base level   Sea level; the lowest level to which stream erosion could lower the land.

Unconformity   A surface that represents a break in the rock record, caused by erosion or nondeposition.

Uniformitarianism   The concept that the processes that have shaped the earth in the geologic past are essentially the same as those operating today.

Valence electron   The electrons involved in the bonding process; the electrons occupying the highest principal energy level of an atom.

Valley train   A relatively narrow body of stratified drift deposited on a valley floor by meltwater streams that issue from the terminus of an alpine glacier.

Ventifact   A cobble or pebble polished and shaped by the sandblasting effect of wind.

Vesicles   Spherical or elongated openings on the outer portion of a lava flow that were created by escaping gases.

Vesicular   A term applied to igneous rocks that contain small cavities called vesicles, which are formed when gases escape from lava.

Viscosity   A measure of a fluid's resistance to flow.

Volcanic arc   Mountains formed in part by igneous activity associated with the subduction of oceanic lithosphere beneath a continent. Examples include the Andes and the Cascades.

Volcanic bomb   A streamlined pyroclastic fragment ejected from a volcano while molten.

Volcanic neck   An isolated, steep-sided, erosional remnant consisting of lava that once occupied the vent of a volcano.

Volcano   A mountain formed from lava and/or pyroclastics.

Wash   A desert stream course that is typically dry except for brief periods immediately following rainfall.

Water gap   A pass through a ridge or mountain in which a stream flows.

Water table   The upper level of the saturated zone of groundwater.

Wave-cut cliff   A seaward-facing cliff along a steep shoreline formed by wave erosion at its base and mass wasting.

Wave-cut platform   A bench or shelf along a shore at sea level, cut by wave erosion.

Wave height   The vertical distance between the trough and crest of a wave.

Wave length   The horizontal distance separating successive crests or troughs.

Wave period   The time interval between the passage of successive crests at a stationary point.

Weathering   The disintegration and decomposition of rock at or near the surface of the earth.

Welded tuff   A pyroclastic deposit composed of particles fused together by the combination of heat still contained in the deposit after it has come to rest and the weight of overlying material.

Well   An opening bored into the zone of saturation.

Wilson cycle   The complex cycle of ocean basin openings and closings named in honor of J. Tuzo Wilson, the Canadian geologist who proposed their existence.

Wind gap   An abandoned water gap. These gorges typically result from stream piracy.

Xenolith   An inclusion of unmelted country rock in an igneous pluton.

Xerophyte   A plant highly tolerant of drought.

Yazoo tributary   A tributary that flows parallel to the main stream because a natural levee is present.

Zone of accumulation   The part of a glacier characterized by snow accumulation and ice formation. The outer limit of this zone is the snowline.

Zone of aeration   Area above the water table where openings in soil, sediment, and rock are not saturated but filled mainly with air.

Zone of fracture   The upper portion of a glacier consisting of brittle ice.

Zone of saturation   Zone where all open spaces in sediment and rock are completely filled with water.

INDEX